Handbook of
SPECIALTY ELASTOMERS

Handbook of
SPECIALTY ELASTOMERS

Edited by
Robert C. Klingender

CRC Press
Taylor & Francis Group
Boca Raton   London   New York

CRC Press is an imprint of the
Taylor & Francis Group, an **informa** business

CRC Press
Taylor & Francis Group
6000 Broken Sound Parkway NW, Suite 300
Boca Raton, FL 33487-2742

First issued in paperback 2019

ISBN-13: 978-1-57444-676-0 (hbk)
ISBN-13: 978-0-367-38780-8 (pbk)

This book contains information obtained from authentic and highly regarded sources. Reasonable efforts have been made to publish reliable data and information, but the author and publisher cannot assume responsibility for the validity of all materials or the consequences of their use. The authors and publishers have attempted to trace the copyright holders of all material reproduced in this publication and apologize to copyright holders if permission to publish in this form has not been obtained. If any copyright material has not been acknowledged please write and let us know so we may rectify in any future reprint.

Library of Congress Cataloging-in-Publication Data

Klingender, Robert C.
    Handbook of specialty elastomers / Robert C. Klingender.
        p. cm.
    Includes bibliographical references and index.
    ISBN 978-1-57444-676-0 (alk. paper)
        1. Elastomers--Handbooks, manuals, etc. I. Title.

TS1925.K46 2007
620.1'94--dc22   2007020182

Visit the Taylor & Francis Web site at
http://www.taylorandfrancis.com

and the CRC Press Web site at
http://www.crcpress.com

Contents

Preface

The *Handbook of Specialty Elastomers* was conceived as a single reference source for the rubber compounder with some experience in designing parts in the rubber industry. The definition of specialty elastomers referenced in this publication is heat, oil, fuel, and solvent-resistant polymers that include polychloroprene (CR), nitrile rubber (NBR), hydrogenated nitrile rubber (HNBR), fluoroelastomer (FKM), polyacrylate (ACM), ethylene acrylic elastomer (AEM), polyepichlorohydrin (CO, ECO), chlorinated polyethylene (CPE), chlorosulfonated polyethylene (CSM), ethylene vinyl acetate (EAM), and thiokol (T).

In addition to the information on the specialty elastomers, chapters on the more important ingredients used with them are included. These are plasticizers, vulcanization agents, antioxidants and antiozonants, and process aids.

The final chapter, in three sections, provides one example of industry requirements for rubber parts, considerations to be made concerning the life expectancy of elastomer compounds and processing factors to be taken into account in the molding operation of a rubber factory.

It is the desire of the editor and contributing authors that this book provide a comprehensive insight into the process of designing rubber formulations based on specialty elastomers.

Editor

Robert C. Klingender, a graduate of the University of Toronto with a BASc degree in chemical engineering, is retired after serving over 54 years in the rubber industry. During that time he worked at Gutta Percha & Rubber Ltd., a mechanical rubber goods manufacturer, as assistant chief chemist; Polysar Ltd., a synthetic rubber producer, as technical service manager, technical service and sales district manager, technical director of custom mixing; Goldsmith & Eggleton, a distributor for Nippon Zeon, as vice president, technical products; and Zeon Chemicals, LLC, a synthetic rubber producer in various technical sales and marketing functions. Bob's career focused on specialty elastomer applications in the mechanical and automotive products industries.

Service to the rubber industry has been Klingender's passion over the years, having served in many capacities in the Rubber Division, ACS as well as the Chicago, Wisconsin, Twin Cities and Northeast Ohio rubber groups.

In his various capacities, Klingender authored or coauthored over 15 technical papers for the Rubber Division, ACS and various local rubber groups. In addition he wrote some 25 technical bulletins and contributed a chapter on "Miscellaneous Elastomers" to *Rubber Technology*, third edition, edited by Maurice Morton.

After retirement Robert has concentrated more on golf (with not too much success), playing bridge, and gourmet cooking (a skilled rubber compounder can also work well with food recipes).

Contributors

Pascal Ferrandez
DuPont Performance Elastomers, LLC
Wilmington, Delaware, U.S.A.

Stephen K. Flanders (Deceased)
Morton International, Inc.
Woodstock, Illinois, U.S.A.

Klaus Kammerer
DuPont Performance Elastomers
    International S.A.
Geneva, Switzerland

Robert W. Keller
Consultant
Lexington, Kentucky, U.S.A.

Robert C. Klingender
Specialty Elastomer Consulting
Arlington Heights, Illinois, U.S.A.

Ray Laakso
The Dow Chemical Company
Plaquemine, Lousiana, U.S.A.

Hans Magg
Bayer Corporation
Leverkusen, Germany

Russell A. Mazzeo
Mazzeo Enterprises
Waterbury, Connecticut, U.S.A.

Edward McBride
DuPont Packaging and Industrial
    Polymers
Wilmington, Delaware, U.S.A.

Hermann Meisenheimer (Retired)
Bayer Corporation
Leverkusen, Germany

Rudiger Musch (Retired)
Bayer Corporation
Leverkusen, Germany

Lawrence C. Muschiatti
DuPont Performance Elastomers LLC
Wilmington, Delaware, U.S.A.

Robert F. Ohm
Lion Copolymer, LLC
Baton Rouge, Louisiana, U.S.A.

Peter C. Rand
Merrand International Corporation
Portsmouth, New Hampshire, U.S.A.

Jerry M. Sherritt (Retired)
Struktol Company
Barberton, Ohio, U.S.A.

John Vicic
Weatherford International, Inc.
Houston, Texas, U.S.A.

Yun-Tai Wu
DuPont Packaging and Industrial
    Polymers
Wilmington, Delaware, U.S.A.

Andrea Zens
Bayer Corporation
Leverkusen, Germany

1 Polychloroprene Rubber

Rudiger Musch and Hans Magg

CONTENTS

1.1   INTRODUCTION

Polychloroprene was one of the first synthetic rubbers and has played an important role in the development of the rubber industry as a whole, a fact that can be attributed to its broad range of excellent characteristics.

In terms of consumption, polychloroprene has become a most important specialty rubber for non-tire applications.

1.2   HISTORY, POLYMERIZATION, STRUCTURE, AND PROPERTIES

1.2.1   HISTORY

The polychloroprene story started in 1925, with the synthesis of the monomer by Father Nieuwland [1]. The first successful polymerization under economically feasible conditions was discovered in 1932 by Carothers, Collins, and coworkers using emulsion polymerization techniques [2]. In the same year DuPont began marketing the polymer first under the trade name Duprene and since 1938 as Neoprene. A wide range of polychloroprene grades has since been developed to meet changing market demands

1940	A breakthrough in 1939 due to the development of a copolymer with sulfur (Neoprene GN) featuring more desirable viscosity and processing behavior
1950	Soluble, sulfur-free homo- and copolymers using mercaptans as chain transfer agents (M-grades) offering improved heat resistance were invented and, in the case of copolymers, these had reduced tendency to crystallization (DuPont)
1960	Precrosslinked grades for improved processability, in particular where reduced nerve and die swell is of prime concern (DuPont)
1970	Precrosslinked and soluble grades with improved physical and mechanical properties (DuPont) sulfur-modified grades with higher dynamic load-bearing capacity and better heat stability (DuPont)
1980	Commercially successful soluble homo- and copolymers using special Xanthogen-disulfides as chain modifiers (XD-grades) with improved processability and vulcanizate properties (Bayer AG/Distugil); soluble copolymers with excellent performance under adverse climatic conditions (extremely slow crystallization with a higher service temperature) (Bayer AG/Denki)
1990	Newly developed M- and XD-grades combining low-temperature flexibility, improved heat resistance, and dynamic properties as well as low mold fouling (Bayer AG)

TABLE 1.1

Production Facilities for Chloroprene Rubber (IISRP Worldwide Rubber Statistic 2001)

Company	Location	Country	Capacity[a] (in Metric Tons)
DuPont	Louisville	USA	64,000
	Pontchartrain	USA	36,000
Bayer AG	Dormagen	Germany	65,000
Denki Kagaku	Omi	Japan	48,000
Enichem	Champagnier	France	40,000
Showa	Kawasaki	Japan	20,000
Tosoh	Shinnayo	Japan	30,000
Razinoimport	Eravan	Armenia	5,000
People's Republic of China[b]	Chang Zhou	People's Republic of China	5,000[b]
	Daiton		10,000[b]
	Qindau		

[a] Latex and adhesive grades included.
[b] Estimated by editor.

Since 1933, when DuPont started up their first production plant, several other companies have also joined the list of producers.

The current list of polychloroprene producers is shown in Table 1.1. Name plate capacity for all plants worldwide, former Soviet Union included, is estimated to be 348,000 metric tons (2001). DuPont announced the closure of the Louisville, KY plant by 2005, reducing worldwide capacity by 64,000 metric tons.

1.2.2 CHLOROPRENE MONOMER PRODUCTION

From the very beginning up to the 1960s, chloroprene was produced by the older energy-intensive "acetylene process" using acetylene, derived from calcium carbide [3]. The acetylene process had the additional disadvantage of high investment costs because of the difficulty of controlling the conversion of acetylene into chloroprene. The modern butadiene process, which is now used by nearly all chloroprene producers, is based on the readily available butadiene [3].

Butadiene is converted into monomeric 2-chlorobutadiene-1,3(chloroprene) via 3,4-dichlorobutene-1 involving reactions that are safe and easy to control.

The essential steps in both processes are listed in Figure 1.1.

1.2.3 POLYMERIZATION AND COPOLYMERIZATION

In principle, it is possible to polymerize chloroprene by anionic-, cationic-, and Ziegler Natta catalysis techniques [4] but because of the lack of useful properties, production safety, and economical considerations, free radical emulsion polymerization is

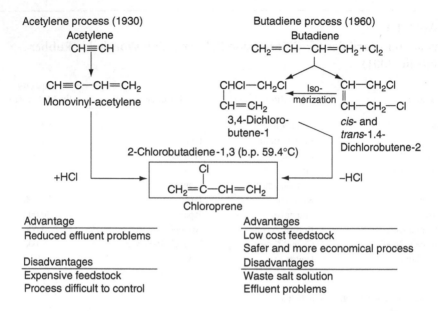

FIGURE 1.1 Acetylene and butadiene route to chloroprene.

exclusively used today. It is carried out on a commercial scale using both batch and continuous processes.

A typical production flow diagram is shown in Figure 1.2.

Chloroprene in the form of an aqueous emulsion is converted with the aid of radical initiators into homopolymers or, in the presence of comonomers, into copolymers [5].

Comonomers, which have been used with success, are those with chemical structures similar to that of chloroprene, in particular

1. 2,3-Dichloro-butadiene to reduce the crystallization tendency, that is, the stiffness of the chain.
2. Acrylic or methacrylic acid esters of oligo functional alcohols to produce the desired precrosslinked gel polymers.
3. Unsaturated acids, for example, methacrylic acid, to produce carboxylated polymers.
4. Elemental sulfur to produce polymer chains with sulfur segments in the backbone, facilitating peptization.

1.2.4 STRUCTURE AND STRUCTURAL VARIABLES

Polychloroprene is highly regular in structure and consists primarily of *trans*-units; however, there are sufficient *cis*-units to disturb the backbone symmetry and maintain a rubbery state [6].

Therefore, the physical, chemical, and rheological properties of polychloroprene are, to a large extent, dependent on the ability to change the molecular structure,

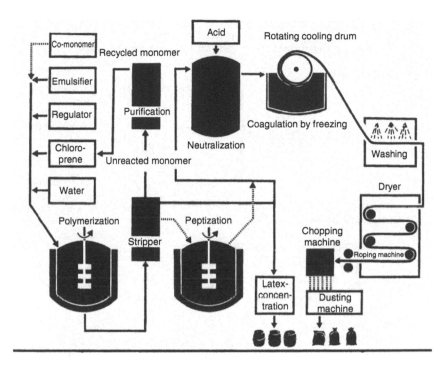

FIGURE 1.2 Flow diagram for the polymerization and isolation of polychloroprene.

for example, the *cis/trans* ratio, long chain branching, and the amount of cross-linking. Key roles in changing the molecular structure are played by

1. Polymerization conditions: polymerization temperature, monomer conversion, polymerization process [7]
2. Polymerization aids: concentration and type of chain modifier, comonomers, and emulsifier [8]
3. Conditions during finishing

Figure 1.3 compares the structural units of commercially available polychloroprene. In this polymer the 1,4 addition [1], in particular the 1,4-*trans*-addition (lb), is dominant. In addition, small proportions of the 1,2-(II) and 3,4-(III) structures are also present. These polymer structures are combined in sequential isomers derived from head to tail (IV), head to head (VI), and tail to tail (V) addition [9].

In addition, the preparation of stereoregular polychloroprene by unusual polymerization conditions has demonstrated that the glass transition temperature and the melting temperature of the polymer are inverse functions of polymerization temperature [10,11] as seen in Figure 1.4).

Using standard polymerization conditions, crystallization is an inherent property of all polychloroprene rubbers [12]. A homopolymer manufactured at 40°C has a *trans*-1,4-content of ca. 90%, a degree of crystallization of ca. 12%, and a crystalline melting temperature of ca. 45°C. A reduction in the rate of crystallization is possible

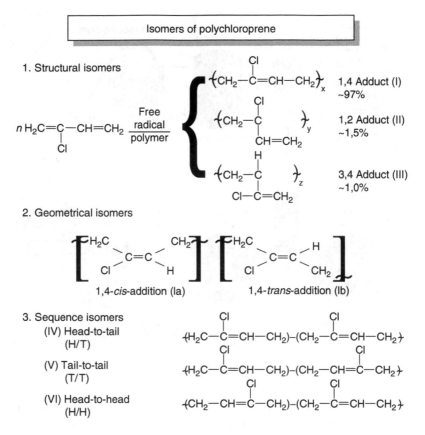

FIGURE 1.3 Structural units in the polychloroprene chain (typical commercial rubber grade) crosslinking site.

by either decreasing *trans*-1,4-content or increasing non-1,4-content or by introducing comonomers. In practice, the latter is the easiest. The crystallinity in polychloroprene makes processing difficult and the vulcanizate increases in hardness with age. Therefore, polychloroprene polymers are normally produced at high polymerization temperatures (30°C–60°C) or using additional comonomers interfering with crystallization. Through such measures, the crystallizing tendency of polychloroprene in both the raw and vulcanized states is reduced.

The crystallization process is temperature dependent and has its maximum rate at −5°C to −10°C. This effect is responsible for the hardening and the reduction in elasticity of chloroprene rubber (CR) polymer compounds and vulcanizates during storage at low temperatures. Crystallization is completely reversible by heat or dynamic stress. In general, the raw polymers crystallize 10 times faster than vulcanized, plasticizer-free compounds (ISO 2475, ASTM D 3190-90).

Variations in microstructure are responsible for significant changes in polymer properties. Figure 1.5 shows the main modifications of the polychloroprene chain

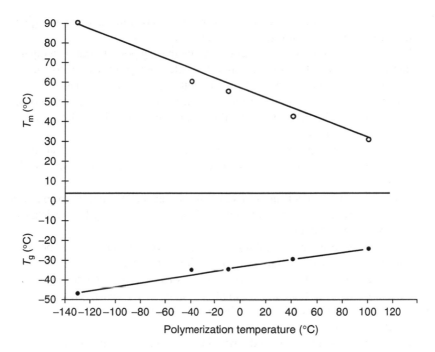

FIGURE 1.4 Glass transition temperature (T_g) and melt temperature (T_m) of polychloroprene prepared at various temperatures.

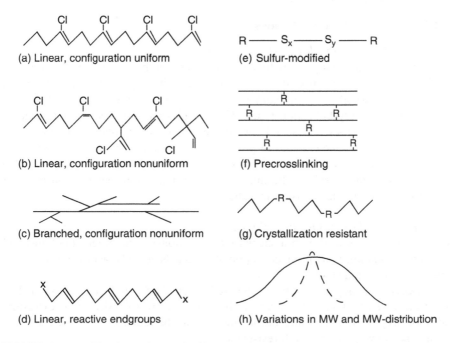

(a) Linear, configuration uniform

(b) Linear, configuration nonuniform

(c) Branched, configuration nonuniform

(d) Linear, reactive endgroups

(e) Sulfur-modified

(f) Precrosslinking

(g) Crystallization resistant

(h) Variations in MW and MW-distribution

FIGURE 1.5 Modifications of the polychloroprene chain.

1. Increasing *trans*-content with decreasing polymerization temperature gives increasing crystallization tendency (adhesive grades).
2. Increasing 1,2- and *cis*-1,4-additions with increasing polymerization temperature reduces crystallization and provides faster curing, necessary for rubber grades (1,2-structures important for crosslinking with metal oxides).
3. Chain branching with high polymerization temperature and high monomer conversion results in reduced stability in polymer viscosity and processing properties deteriorate.
4. Reactive end groups using XD-chain modifier provide reduced branching, easy processing, and elastomers with more homogeneous networks, for example, high tensile strength.
5. Polymer chains with sulfur atoms in multiple sequences ranging from 2 to 8 show improved breakdown during mastication, outstanding tear resistance, and dynamic behavior.
6. Specially induced precrosslinking yields sol/gel type blends yielding processing and extrusion advantages with increasing gel content (10%–50%).
7. Reduced stereoregularity using comonomers leads to reduced crystallization tendency and level, thus the so called "crystallization resistant grades."
8. Increasing molecular weight results in increasing the polymer viscosity and tensile strength of vulcanizates.
9. Increasing molecular weight distribution gives improved processability and reduced tensile strength.

1.2.5 STRUCTURE AND PROPERTIES

1.2.5.1 General Purpose Grades

1.2.5.1.1 Mercaptan-Modified (M-Grades)
This group contains non-precrosslinked, sulfur-free, soluble, homo- and copolymers and is the most important in terms of the quantity used. It comprises the standard grades with polymer viscosities of approximately 30–140 Mooney units (ML4 at 100°C) and slight to medium crystallization types. These grades are also known as mercaptan grades. Their property profiles tend to be influenced mainly by polymer viscosity.

Table 1.2 shows the changes in properties as a function of Mooney viscosity.

Grades with slight to very slight crystallization should be used in parts intended for low-temperature service. The influence of crystallization tendency on polymer and elastomer properties is listed in Table 1.3.

1.2.5.1.2 Xanthogen-Disulfide Grades
XD-grades are produced with a special modifier. Some of them are copolymerized with other monomers to produce copolymers that have only a medium or slight tendency to crystallization.

Processing behavior: They are generally less elastic (reduced "nerve") than M-grades and are, therefore, more easily processed by calendering or extrusion. Additionally, the ram pressure during mixing can be reduced and as a result the compounds have greater scorch resistance.

TABLE 1.2
Influence of Mooney Viscosity

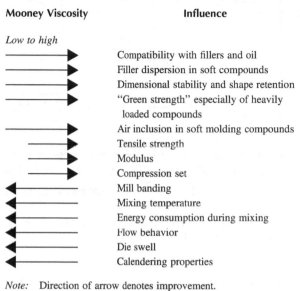

Mooney Viscosity	Influence
Low to high	
→	Compatibility with fillers and oil
→	Filler dispersion in soft compounds
→	Dimensional stability and shape retention
→	"Green strength" especially of heavily loaded compounds
→	Air inclusion in soft molding compounds
→	Tensile strength
→	Modulus
→	Compression set
←	Mill banding
←	Mixing temperature
←	Energy consumption during mixing
←	Flow behavior
←	Die swell
←	Calendering properties

Note: Direction of arrow denotes improvement.

Vulcanizate properties: If M-grades are substituted with XD-grades in a given recipe, vulcanizates with improved mechanical properties will result, that is, higher tensile strength and tear.

Strength, rebound resilience, and resistance to dynamic stress are obtained. The importance of these differences is emphasized in Figure 1.6.

In contrast to M-grades, the tensile strength of vulcanizates based on XD-grades is essentially independent of the viscosity of the starting material within a broad

TABLE 1.3
Influence of Crystallization on Properties

Crystallization	Influence
Slight to strong	
→	"Green strength"
→	Cohesive strength
→	Setting rate (adhesives)
→	Tensile strength
→	Modulus
←	Tack and building tack
←	Retention of rubberlike properties at low temperatures over long periods of time

Note: Direction of arrow denotes improvement.

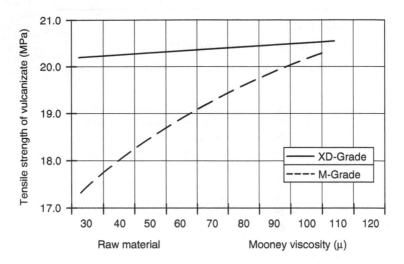

FIGURE 1.6 Tensile strength—Mooney viscosity relationship of M- and XD-modified general purpose grades (Recipe ISO 2475).

viscosity range. This improved performance permits heavier filler and plasticizer loadings, thereby reducing compound cost.

More recently developed M- and XD-grades show reduced nerve, significant reduction in mold fouling, higher tensile strength, better aging characteristics, significantly improved dynamic properties, and better low-temperature behavior.

1.2.5.2 Precrosslinked Grades

Precrosslinked grades have proven particularly suitable for extruded and calendered goods and, in special cases, for injection molding. The precrosslinking that occurs during the production of the polymer improves processability, because it reduces the elasticity or "nerve" of the raw rubber and its compounds.

Typical characteristics are improved mill banding, low die swell, smooth surfaces, excellent dimensional stability, and in the case of XD-precrosslinked grades, no decrease in tensile strength.

As the degree of precrosslinking rises, several properties of the compounds and vulcanizates change as shown in Table 1.4.

1.2.5.3 Sulfur-Modified Grades (S-Grades)

Sulfur-modified grades are obtained by copolymerization of chloroprene with small amounts of sulfur, followed by peptization of the resulting copolymer in the presence of tetra alkyl thiuram disulfide. Sulfur is built into the polymer chains in short sequences.

Sulfur modification improves the breakdown of the rubber during mastication, permitting the production of low-viscosity compounds with good building tack. Only zinc oxide and magnesium oxide are needed for vulcanization. In many cases

TABLE 1.4
Relationship between Precrosslinking and Properties

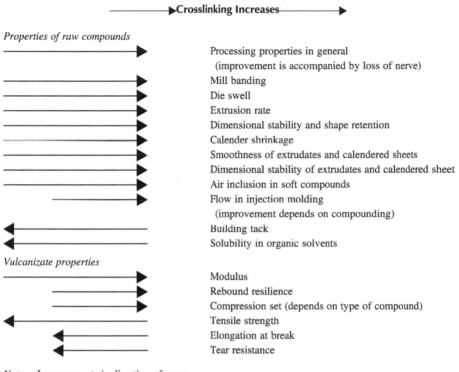

Note: Improvements in direction of arrow.

the vulcanizates have better tear resistance and adhesion to fabrics than those based on general purpose grades.

Disadvantages: Polymers are less stable during storage and vulcanizates are less resistant to aging.

Differences in the property profile of commercially available S-grades are caused by different combinations of sulfur level, comonomers, soap system, polymerization and peptization reactions, and staining or nonstaining stabilizers.

More recently developed grades give elastomers with a higher tear propagation resistance, greater resistance to dynamic stress, and better aging behavior. They also cause less mold fouling.

1.2.6 COMMERCIALLY AVAILABLE CR RUBBERS

Table 1.5 lists the most commonly used grades marketed in 2002 by the main suppliers in the western hemisphere [13]. The available grades are divided into three groups: general purpose grades (non-precrosslinked), precrosslinked grades, and sulfur-modified grades.

TABLE 1.5
Cross Reference of Polychloroprene Grades

DuPont Neoprene		Bayer Baypren		Denki Kagaku Kogyo Denka Chloroprene		Enichem Butaclor		Showa DDE Mfg. Neoprene		TOSOH Corporation Skyprene	
Grade	ML 4	Grade	ML 4	Grade	ML 4	Grade	ML 4	Grade	ML 4	Grade	ML 4
						General Purpose Grades					
Slow Crystallization Types											
WRT-M1	38	110	45	S-40V	48	MC-10	45	WRT	46	B-5	49
WRT-M2	46	111	42							B-5A	45
		112	42	S-40/41	48	MC-20	46	WX-J	46	B-10	51
		116[a]	46			MC-122[a]	43			B-11	49
		126[a]	70							B-10H	75
WD	110	130	105								
Medium Crystallization Types											
W	49	210	45	M-40/41	48	MC-30	46	W	46	B-30	49
WM-1	38	211	39	M-30/31	38	MC-31	38	WM-1	37	B-31	40
		216[a]	46			MC-322	43			P-90	45
		226[a]	75	M-70	70	MC-323	59				
WHV-100	97	230	100	M-100	100	MH-31	94	WHV-100	100	Y-31	100
WHV	115			M-120	120	MH-30	114	WHV	120	Y-30S	123

Precrosslinked Grades

Slow Crystallization Types											
WB	47	114	62	ES-70	75			WXKT	110		
TRT	47	214	55	ES-40	43			WXK	80	E-20H	64
		215	50			DE-102	48	TRT	46	E-20	48
Medium Crystallization Types											
TW	48			MT-40	48	DE-302	48	TW	46	E-33	48
						ME-20	52			Y-20E	48
TW-100	93	235	95	MT-100	95	DE-305	92	TW-100	95		
Sulfur-Modified Grades											
Slow Crystallization Types											
GW	45	510	45	DCR-45	45	SC-102	45	GW	43		
GRT	45	611	45	PS-40	50	SC-202	45	GRT	47	R-10	45
				PS-40A	42	SC-10	43				
						SC-132	43				
						SC-22	43				
Medium Crystallization Types											
GNA	50	712	43	PM-40	50			GS	47	R-22	45
		711	45	PM-40NS	50						

[a] XD, Xanthogen disulfide modified grades.

TABLE 1.6

Selection of Compound Properties versus CR Grades

Desired Property	Grades
Optimum processing	Grades of low viscosity, precrosslinked grades
Best mastication	S-grades
Best tackiness	S-grades; grades of low crystallization tendency
Best green strength	Medium fast crystallizing grades, high viscous grades
Highly extended compounds	Grades of high viscosity; XD-grades
Best extrudability	Precrosslinked grades

1.2.7 COMPOUNDING AND PROCESSING

Chloroprene rubber (CR) vulcanizates can be made using fillers, plasticizers, antioxidants, and processing aids commonly used in diene rubber compounding.

Principles related to compounding and processing are discussed in subsequent sections.

1.2.7.1 Selection of Chloroprene Rubber Grades

To achieve the best compromises in compounds and vulcanizate properties, a proper selection of grades is essential. Table 1.6 shows the best selection of elastomer to achieve desired processing properties. Table 1.7 illustrates a number of properties and the corresponding best choice of grade of elastomer for various vulcanizate properties.

1.2.7.2 Blends with Other Elastomers

Blends of CR and other elastomers are desirable in order to achieve special properties either of a CR-based compound or of a compound mainly based on the second

TABLE 1.7

Selection of Vulcanizate Properties versus CR Grades

Desired Property	Grades
Best tensile and tear resistance	M-grades; XD-grades
Best compression set	M-grades
Optimum heat resistance	M-grades; XO-grades
Best low-temperature properties	M-grades; XD-grades. Both of slow crystallization
Lowest dynamic loss factor, highest elasticity	S-grades
Best dynamic behavior	S-grades; XD-grades
Best adhesion to textile and metal	S-grades

component. In many cases general purpose diene rubbers, such as SBR, BR, or NR, are also used to reduce compound costs.

It is advantageous to select compatible polymers as blending components to form alloys during the mixing process. With respect to CR crosslinking systems that are of dissimilar reactivity to that used in the blending elastomers, it is unsuitable in most cases, thus resulting in an inhomogeneous network. Accelerator systems based on thiurams and amines are best for an effective co-cure.

In the assessment of polymer blends, a somewhat lower level of physical properties than a similar formulation based on pure polymers has to be taken into account. In any case blending requires a well-adjusted mixing procedure.

A number of blends are used in the rubber industry, the most important of which are summarized as follows:

1. Natural Rubber (NR) improves building tack, low-temperature flexibility, elasticity, and reduces cost.
2. Butadiene Rubber (BR) added at levels of up to 10% to improve processing of S-grades (reduced mill sticking); however, a reduction in flex-fatigue life may be observed. BR also improves low-temperature brittleness.
3. Styrene-Butadiene-Rubber (SBR) has a predominant benefit of reducing cost. It reduces crystallization hardening as well.
4. Acrylonitrile-Butadiene Rubber (NBR) is used for improved oil resistance and (less importantly) for better energy-uptake in a microwave cure.
5. Ethylene-Propylene-Rubber (EPDM) with CR can be used in EPDM-vulcanizates to achieve a certain degree of oil resistance. It improves adhesion of EPDM to reinforcing substrates. In blends where CR is dominant, price reduction and better ozone-resistance are obtained by EPDM.

Further details are given elsewhere in the literature.

1.2.7.3 Accelerators

CR can be crosslinked by metal oxides alone. Thus, there is a major difference between general purpose diene rubbers and CR. Suitable accelerators help to achieve a sufficient state of crosslinking under the desired conditions.

Zinc oxide (ZnO) and magnesium oxide (MgO) are the most frequently used metal oxides; lead oxides are used instead for optimal water/acid/alkaline resistance.

Figure 1.7 refers to some curing characteristics and physical properties attainable by varying the amounts of ZnO and MgO.

In the absence of zinc oxides the rheometer curve is rather flat. Although the state of cure is increased, the crosslinking density remains low if zinc oxide is used alone. Best results are obtained with a combination of zinc oxide and magnesium oxide.

There is a tendency to "marching modulus" characteristics if high levels of both metal oxides are used.

The combination of 5 pphr ZnO and 4 pphr MgO is particularly favorable. In principle, the conditions described for M-grades are also valid for XD-grades.

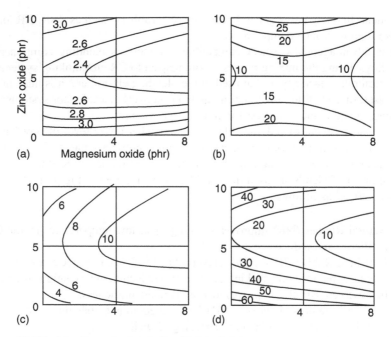

(a) Effect on rheometer incubation time t_{10} (min)
(b) Effect on rheometer cure time t_{80} (min)
(c) Effect of tensile (MPa) of the vulcanizates
(d) Effect of on compression set (70 h/100 D.C:) (%)

FIGURE 1.7 Effect of zinc oxide and magnesium oxide levels on compound and vulcanizate properties of CR. (*Notes:* (a) All vulcanizates were cured 30 min at 150°C. (b) Formulation: CR (medium fast crystallization grade) 100, stearic acid 0.5, PBNA 2, SRF N762 30, ETU 0.5. (c) Excerpt from technical information bulletin Baypren 2.2.1, Bayer AG.)

S-grades are highly reactive with metal oxides so that no further accelerators are necessary to obtain a sufficient state of cure (although they are often used to adjust curing characteristics or to enhance the level of physical properties).

Various types of lead oxides are used in large amounts especially if resistance against water, acids, and alkaline solutions is required. With lead oxides, scorch times can be reduced; therefore, particular caution is required in formulation, mixing, and processing. Lead oxides, on the other hand, enable "self curing" CR compounds. A dispersed form should be used for health reasons.

With respect to the crosslinking mechanism reference must be made to the work of R. Pariser [14], who recommends sequences of chemical reactions, which are basically influenced by

1. Amount (approximately 1.5 mol%) and statistical distribution of allylic chlorine atoms in the main chain
2. Presence of ZnO/MgO
3. Certain organic accelerators to form monosulfidic bridges

S-grades and, to a lesser extent, XD-grades contain inherent structures, which are able to play the role of the organic accelerator in Pariser's mechanism [15] and lead to measurable crosslinking density without further components.

As an organic accelerator, ethylene thiourea (ETU), which is preferably used in non-dusty forms, is widely used. Different derivatives of thiourea, such as diethyl thiourea (DETU) and diphenyl thiourea (DPTU), are typical ultrafast accelerators, especially suitable for continuous cure.

In 1969, it was disclosed that under certain conditions ETU can cause cancer and birth defects in some laboratory animals. As a result, a number of substitutes have been developed of which N-methyl-thiazolidine-2-thione (MTT: Vulkacit CRV/LG) [16] has gained technical importance.

Systems free of thioureas, or their substitutes, exhibit slower cure and give vulcanizates with higher set properties and lower heat resistance.

Best tear resistance is achieved by a combination of sulfur, thiurams, guanidine-based accelerators, and methyl mercapto benzimidazole (so-called MMBI system).

Levels of 0.5–1.0 pphr methyl mercapto benzimidazole have been shown to improve resistance to flex cracking of CR vulcanizates, but tend to be scorchy. The zinc salt of MMBI (ZMMBI, Vulkanox ZMB-2) is more effective in this respect.

A summary of important accelerator systems for CR M or XD-grades is compiled in Table 1.8. A wide variety of other accelerators have been used with CR, but most have not achieved widespread acceptance.

For the sake of completeness it should be noted that peroxide crosslinking instead of metal oxide crosslinking is also possible. However, the properties of the vulcanizates are inferior (e.g., heat resistance) to those achieved with a metal oxide/ETU system. Therefore, application remains limited.

1.2.7.4 Antioxidants, Antiozonants

Vulcanizates of CR need to be protected by antioxidants against thermal aging and by antiozonants to improve ozone resistance. Some of these ingredients also improve flex-fatigue resistance.

Slightly staining antioxidants, which are derivatives of diphenylamine, such as octylated diphenylamine (ODPA), styrenated diphenylamine (SDPA), or 4,4-bis (dimethylbenzyl)-diphenylamine are especially effective in CR compounds.

Trimethyl dihydroquinoline (TMQ) is not recommended because of its pronounced accelerator effect, which causes scorchiness.

MMBI is used to improve flex cracking resistance, but it tends to reduce the scorch time of compounds. Pronounced synergistic effects with ODPA or similar chemicals in order to optimize hot air aging have not been observed, so that it is not advisable to use this chemical where optimal heat resistance is required.

A strong dependence of antioxidant on dosage of diphenylamine antioxidants was found, revealing that a level of 2–4 pphr is sufficient for most applications (Figure 1.8). Similar relationships have been described by Brown and Thompson [17].

In accordance with general experience, nonstaining antioxidants from the class of stearically hindered phenols or bisphenols are less effective.

TABLE 1.8
Typical Accelerator Systems for CR M and XD-Grades

System		Dosage (phr)	Characteristics
I	ZnO/MgO	5–4	
	ETU+[a]	0.5–1.5	Good heat resistance
	MBTS or TMTD	0–1	Good compression set
II	MTT	0.5–1.5	Similar to I[b]
III	S+	0.5–1	Slow curing, inferior heat
	TMTM or TMTD	0.5–1.5	resistance to I and II
	DPG or DOTG	0.5–1.5	
IV	S+	0.5–1	Medium fast curing,
	TMTM or TMTD	0.5–1.5	optimum tear resistance
	DPG or DOTG	0.5–1.5	
	MMBI[c]	0.5–1.5	
V	ETU+	1.5–2.5	Ultrafast curing system,
	DETU or DPTU	0.5–1.5	suitable for continuous vulcanization
	ZDEC	0.5–1.5	
VI	S	0.1–1.5	For food contact[d]
	TMTM or TMTD	0.5–1.5	
	OTBG[d]	0.5–1.5	
VII	Lead oxide (instead of ZnO/MgO)	20	"self curing"-compounds
	S	0–1	
	Aldehyd-Amine[e]	1.5–2.5	
	DPTU	1.5–2.5	

[a] For industrial hygiene reasons, polymer-bound ETU is recommended.
[b] Nontoxic alternative to ETU (N-methyl thiazolidine thione-2)Vulkacit CRV.
[c] or similarly MBI.
[d] BGA only. The actual status of legislation in different countries must be considered.
[e] For example, butyraldehyde-amine reaction product (Vulkacit 576, suppl. Bayer AG).

p-Phenylenediamines are used as staining antioxidants/antiozonants and also improve fatigue resistance, for which the DPTD type gives the most favorable results. Other than DPTD, p-phenylenediamines tend to impair storage stability and processing safety.

A comparison of the different types of antioxidants is presented in Figures 1.9 and 1.10.

Figure 1.9 representing the class of nonstaining antiozonants, which is described as "cyclic enole" derivatives, are compared with the p-phenylenediamines to their influence on storage stability. Together with a second grade, described as "phenol ether," this class of rubber chemicals serves to give sufficient ozone protection under static and, to a limited extent, dynamic conditions.

A study of the influence of various p-phenylenediamines on flex cracking resistance is shown in Figure 1.10.

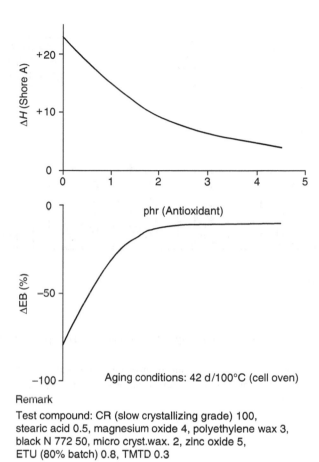

FIGURE 1.8 Effect of antioxidant levels on aged properties of a CR compound.

It must be added that ozone protection is improved if antiozonants are used together with microcrystalline waxes. Optimized CR vulcanizates have been shown to resist outdoor conditions for several years. Long-term tests with several antiozonant/wax combinations in an outdoor test yielded the results presented in Figure 1.11.

1.2.7.5 Fillers

Generally, CR can be treated as a diene rubber as far as fillers are concerned. Carbon black and mineral fillers, of either synthetic or natural origin, can be employed.

Active fillers serve to improve physical properties, whereas less active or predominantly inactive fillers are used to reduce compound cost.

Figure 1.12 illustrates the typical influence of carbon black types and levels in CR. Carbon black is easily incorporated in CR compounds. In most cases N 550, (FEF)-blacks or even less active types are sufficient to meet most requirements.

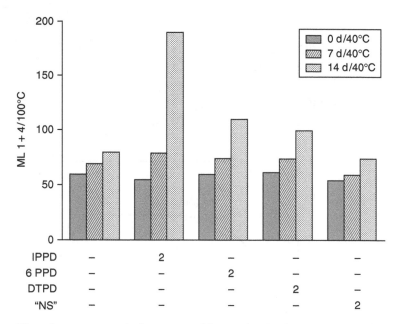

Viscosity measurement after storage of the unvulcanized compound at 40°C

FIGURE 1.9 The influence of antioxidant/antiozonants on storage stability of CR compounds. (*Note:* NS refers to nonstaining antiozonant; described as "cyclic acetal," Vulkazon AFS/LG, Bayer AG.)

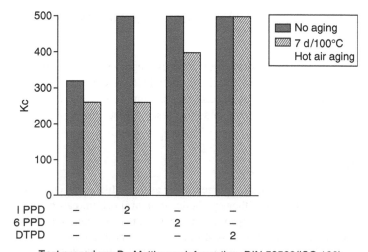

Test procedure: De Mattia crack formation, DIN 53522(ISO 132)

FIGURE 1.10 PPDA Type antioxidant influence of flex fatigue resistance of CR vulcanizates.

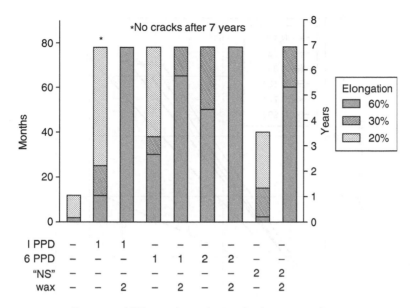

Exposure of CR-samples under tension in an open-air
test (Engerfeld, Germany)
The figures indicate cracks observed after time of exposure

FIGURE 1.11 Influence of antiozonants on weather resistance of CR. (*Note:* NS here refers to
Enol-ether type nonstaining antiozonant, Vulkazon AFD, Bayer AG.)

If active carbon blacks are necessary, dispersion problems may arise, but often these
can be rectified by proper mixing techniques.

Active silica (BET-surface of approximately 170 m^2/g) improves tear resistance
and also gives rise to better fatigue resistance. Microtalc may be used if optimum
heat resistance or resistance against mineral acids or water is required. Silane
coupling agents are often used in conjunction with silica, silicate, and clay fillers
to improve these properties. In this case mercapto silanes or chloro silanes are
preferred.

Clays, talcs, and whitings are often used for cost reduction, either alone or in
combination with reinforcing fillers.

Although CR is inherently flame retardant, for certain applications it is necessary to further improve this property. This can be achieved with aluminum trihydrate,
zinc borate, and antimony trioxide. A chlorinated paraffin instead of a mineral oil
plasticizer is also beneficial.

1.2.7.6 Plasticizers

Mineral oils, organic plasticizers, and special synthetic plasticizers can be used in
typical CR compounds in varying amounts between 5 to approximately 50 pphr.

These plasticizers can have the following effects:

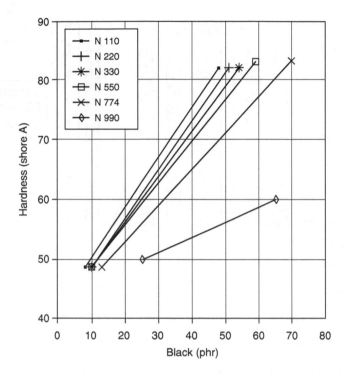

FIGURE 1.12 Relationship of carbon black loading on hardness of CR compounds. (*Note:* Medium fast crystallizing grade CR with 10 pphr aromatic oil (Technical Information Bulletin Baypren 2.3).)

1. Lowering of the glass transition temperature
2. Reduction in tendency to crystallize
3. Lowering of compound cost

A summary of typical plasticizers and their effects is given in Table 1.9.

Special plasticizers are very effective in lowering the glass transition temperature and improving rebound resilience. These products are needed for articles in which resilience is required down to approximately −45°C.

Unfortunately, such plasticizers also promote the crystallization rate, so that polymers with low crystallization tendency have to be chosen.

Highly aromatic mineral oils can be recommended for compounds where a reduction in crystallization rate is required. This class of plasticizers is also compatible, so that 50 pphr or even more can be used without exudation effects. Among other mineral oil plasticizers, napthenic oils have gained importance where staining due to leaching and migration must be avoided. Their compatibility is somewhat limited depending on the compound formulation. Check with local health regulations on the use of highly aromatic oils as some are suspected carcinogens. Paraffinic oils are of very limited compatibility, so that they find only restricted application. These are the most economical plasticizers to use.

TABLE 1.9

Influence of Plasticizers on Low-Temperature Behavior of CR

	Max Dosage (phr)	Effect on Low Temperature			
		Glass Transition Temperature	Brittleness Temperature	Crystallization Reduction	Compound Price
Mineral oils					
Aromatic	<90			++	++
Naphthenic	<30		+		++
Paraffinic	<10				+
Carboxylic-acid ester-plasticizers					
DOP (Dioctyl phthaiate)	<30	+	+	−	+
DOS (Dioctyl sebazate)	<30	++	++	− −	−
DOA (Dioctyl adipate)	<30	++	++	− −	
Special synthetic plasticizers					
Thioethers, thioesters or related compounds	<30	++	++	− −	−
Phenol alkyl-sulfonic acid ester	<20	+	+	−	+
Chlorinated-paraffin plasticizer	<15	−	−	+	+

No significant effect.
+ improvement.
++ highly effective.
− significantly disadvantageous.
− − highly disadvantageous.

At higher cost, synthetic plasticizers such as dioctyl phthalate (DOP), butyl oleate (Plasthall 503), or phenol alkyl-sulfonic acid esters can be used if aromatic mineral oils are not possible. These offer improved low-temperature flexibility and are nondiscoloring and nonstaining.

If higher heat resistance is needed, polymeric, chlorinated paraffins, polyesters, and low volatility mineral oils are used.

Good flame resistance is obtainable with liquid CR, chlorinated paraffins, and phosphate esters.

1.2.7.7 Miscellaneous Compounding Ingredients

This section gives a brief description of other common compounding ingredients such as stearic acid and derivatives, resins, processing aids, and blowing agents.

Stearic acid at levels of 0.5–1.0 pphr is recommended in CR compounds to improve processing and to reduce mill sticking. Zinc stearate acts as an accelerator;

therefore, one must take care to avoid higher temperatures during mixing and processing where it may be formed through the reaction of zinc oxide and stearic acid.

Resins such as cournarone resins are able to act as dispersants and tackifiers. Sometimes, reactive reinforcing phenolic resins are also used, in which case if the crosslinking component (e.g., hexamethylene tetramine or other formaldehyde donors) is used, it must be added in the second stage together with the accelerators.

For some applications CR compounds must be adhered to textiles or metals. Bonding resins of the resorcinol type are normally used as internal bonding agents. Because of the scorching effect of resorcinol, modified grades, such as resorcinol diacetate, are recommended to preserve processing safety.

There are no objections to the use of certain processing aids, of which there are many on the market. For applications and handling, the suppliers' recommendations must be followed. In addition to stearic acid and commercial process aids, low-molecular weight polyethylene, waxes, and wax-like materials, and blends with other elastomers (e.g., BR) are commonly employed.

Vulcanized vegetable oils are used for soft compounds since they permit the use of high plasticizer levels while maintaining good green strength with calendering and extrusion properties. There are specially developed products on the market, such as Faktogel Asolvan, which do not cause a drop in swelling resistance.

Blowing agents commonly used in other diene rubbers are also suitable for CR, for example, azodicarbonamide and sulfohydrazide types.

1.2.8 PROCESSING

Chloroprene rubber is typically supplied in chip form and is normally coated with talc to prevent blocking during shipping and storage. These chips can be processed on open mills or internal mixers using conventional or upside-down techniques. Crystallized chips cause no problems in processing because the crystallites melt at temperatures above 40°C–60°C.

It is recommended that magnesium oxide be added in the early stage of the mixing cycle and not to exceed dump temperatures of 130°C to prevent undesirable side reactions (cyclization, scorch).

Processing safety requires the incorporation of all ingredients with crosslinking activity, for example, zinc oxide, lead oxide, accelerators, and others, in the later stages of mixing, if the compound temperature is not too high, or in the second stage (productive mix).

For high-quality, lightly loaded compounds, a two-stage mixing is recommended with a 1 day rest period between the stages.

CR compounds, especially of low viscosity, mineral filled, or those based on S-types, show a tendency to mill sticking. To overcome this effect, low friction ratios and low-temperature processing are recommended. Process aids may also assist in providing better mill release. If the compound temperature exceeds 70°C, CR compounds become somewhat grainy in appearance, lose cohesive strength, and stick to metal surfaces.

S-grades breakdown in viscosity under shear, which is beneficial for tackiness, and is important for articles such as belts and some hoses. Another consequence of this phenomenon is that S-grades are the preferred basic material for low-viscosity friction and skim-compounds.

CR may also be used in "dough processes" employed in coated fabrics for various applications. It is important to ensure that regulations related to solvent vapors are followed for those processes where combinations of solvents such as naphtha/methyl ethyl ketone (MEK) or naphtha/toluene are employed. The solutions may contain special bonding agents, for example, polyisocyanates. Pot lifetimes of such compounds are fairly short (1–4 h), because of the crosslinking activity of such materials.

CR compounds can be used in all vulcanization processes, such as compression and injection molding, hot air, steam autoclaves, and continuous vulcanization (salt baths, microwave-hot air cure, CV-cure).

Reversion is not a problem for CR, so curing temperatures of up to 240°C are possible.

1.2.9 Properties and Applications

1.2.9.1 General

The attraction of CR lies in its combination of technical properties, which are difficult to match with other types of rubber for a comparable price. With the correct compound formulation, CR vulcanizates are capable of yielding a broad range of excellent properties as shown below:

1. Good mechanical properties, independent of the use of reinforcing agents
2. Good ozone, sunlight, and weather resistance
3. Good resistance to chemicals
4. High dynamic load-bearing capacity
5. Good aging resistance
6. Favorable flame resistance
7. Good resistance to fungi and bacteria
8. Good low-temperature resistance
9. Low gas permeability
10. Medium oil and fuel resistance
11. Adequate electrical properties for a number of applications
12. Vulcanizable over a wide temperature range with different accelerator systems
13. Good adhesion to reinforcing and rigid substrates, such as textiles and metals

1.2.9.2 Physical Properties

Polychloroprene vulcanizates possess good physical strength, and with optimum formulations, the level is comparable to that of NR, SBR, or NBR.

Tear resistance of CR vulcanizates is better than that of SBR. Tear propagation resistance of CR vulcanizates containing active silica may be greater than that of those with natural rubber. CR vulcanizates show good elasticity, although they do

TABLE 1.10

Comparison of Typical Vulcanizate Properties of CR, NBR, NR, and SBR

Basic Properties	Chloroprene Rubber	Nitrile Rubber	Natural Rubber	Butadiene-Styrene Rubber
Tensile strength (MPa)	Up to 25	Up to 25	Up to 28	Up to 25
Hardness (Shore A)	30–90	From 20 to ebonite hardness	From 20 to ebonite hardness	From 20 to ebonite hardness
Abrasion resistance	A	A	B–C	B
Tear propagation resistance	B	C	A	C
Fatigue resistance	A	A	A	A
Rebound	B	B–D	A	B
Hot air resistance, temperature limit for continuous stressing (°C) (VDE 0304, Part 2)	+80	+80	+60	+70
Low-temperature flexibility	B–C	B–D	B	B–C
Weathering and ozone resistance	B	D	D	C
Flame retardance	A–B	E–C	E–C	E–C
Compression set (22 h at 70°C)	B	B	C	B
Oil resistance	B	A	D	D

Note: A = Excellent; B = Very Good; C = Good; D = Fair; E = Unsatisfactory. The ratings are compound composition dependent, hence all optimum values may not be obtained simultaneously.

not reach the level of NR. Table 1.10 provides a comparison of CR with NBR, NR, and SBR.

The compression set of CR is low over a wide range of temperatures from −10°C to +145°C, as given in Figure 1.13. The low-temperature compression set is one of the key values employed for the assessment of vulcanizates used in seals. For CR, testing is commonly run at −10°C, the temperature at which optimum crystallization occurs. It is possible to improve the low-temperature compression set to less than 50% at −30°C by using the most crystallization resistant CR and low-temperature plasticizers.

At higher temperatures, where aging also plays a role, the compression set curves are at a lower level than for a large number of other elastomers. It is important to use M-types of CR, heat-resistant antioxidants, and nonvolatile plasticizers.

The abrasion resistance of CR is comparable to that of NBR.

The gas permeability is roughly equivalent to NBR of medium ACN content. Thermal conductivity and thermal expansion are comparable to other elastomers.

1.2.9.3   Aging and Heat Resistance

CR M-type vulcanizates, especially those that contain optimized antioxidants and crosslinking systems and low volatility plasticizers, display good heat resistance.

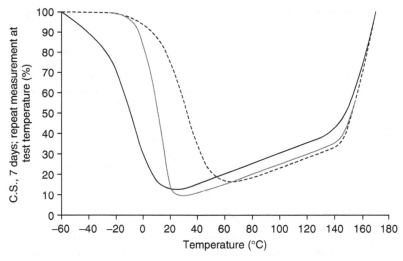

FIGURE 1.13 Relationship of temperature on compression set of CR vulcanizates.

They neither soften nor harden over a long period of stress; remaining serviceable and elastic.

In the ASTM D 2000 and SAE J 200 systems, CR is positioned with respect to thermal aging between NR and CSM.

A more relevant description of heat resistance is shown in Figure 1.14, where the Arrhenius equation is applied to the thermal aging of optimized CR vulcanizates.

The continuous service temperature in accordance with VDE 0304, Part 2 (25,000 h), is 80°C. Optimized vulcanizates for automotive application can perform for 1000 h at 100°C–110°C and will survive short- or medium-term exposure up to 120°C.

1.2.9.4 Low-Temperature Flexibility

Apart from crystallization effects, the differential scanning calorimeter reveals a glass transition temperature for polychloroprene at around −40°C, which is practically independent of the type of polymer tested. Compounding ingredients can shift the glass transition temperature further to lower temperatures. Typical data are summarized in Table 1.11. Low crystallization grades of CR need to be used.

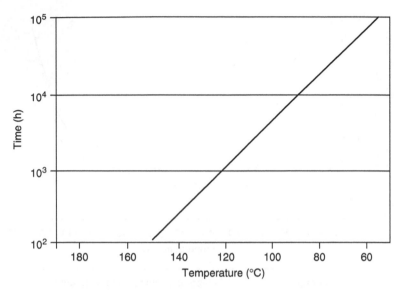

Parameter: Aging time to 100% elongation
CR- M-grade, cable sheathing compound

FIGURE 1.14 Arrhenius plot of air aged CR vulcanizates (VDE 0304/Part 2/7.59).

Synthetic low-temperature plasticizers allows CR vulcanizates to exhibit elastic behavior down to temperatures of around −45°C to −50°C (depending on the formulation).

1.2.9.5 Flame Retardance

CR vulcanizates are inherently flame resistant. The good flame resistance properties of the polymer itself mean that stringent end-user specifications can be fulfilled by the use of appropriate compounding materials such as chlorinated paraffins instead of mineral oils, mineral fillers plus aluminum trihydrate, zinc borate, and antimony trioxide.

TABLE 1.11
Typical Low-Temperature Properties of CR

	Polychloroprene	
Glass transition temperature (DSC Test, 2nd heating cycle)	−40°C	
Vulcanizate properties 65–75 Shore A	*Without plasticizer*	*With plasticizer*
Torsion pendulum test DIN 53445 (°C)	−30°	−43°
Brittleness point ASTM D 736 (°C)	−36°	−52°

Limited oxygen index values of 50% can be attained with CR and construction materials can be manufactured to meet, for example, DIN 4102, Part I, Class B1.

However, like all organic substances, CR vulcanizates will decompose at high temperatures such as those encountered in open fires. In addition to decomposition products such as carbon dioxide and water, corrosive hydrogen chloride gas is also formed.

1.2.9.6 Resistance to Various Fluids

CR possesses medium oil resistance making the polymer suitable for articles resistant to intermittent oil exposure or exposure to less aggressive oils such as paraffinic and napthenic oils or corresponding hydraulic oils. The resistance to regular fuels is limited, and insufficient in fuels with high aromatic content.

Additives in oils may cause hardening of vulcanizates. CR, however, proves to be more stable than a typical NBR vulcanizate.

Unless a lead oxide cure is used, CR compounds can fail to meet severe water resistance requirements. The use of a lead oxide cure allows limited swell in water, so that volume changes of only a few percent can be obtained. Properly compounded CR also exhibits good resistance to dilute acids and alkaline solutions at moderate temperatures.

A list of the swell resistance in various chemicals is given in Table 1.12.

1.2.9.7 Resistance to Fungi and Bacteria

Rubber articles in contact with soil for longer periods of time are liable to attack by soil bacteria and fungi. This can lead to underground cables being destroyed. In contrast to the majority of other rubber types, CR shows a surprisingly higher level of resistance to these microorganisms. This resistance can be further enhanced by the use of fungicides such as Vancide 51Z and a fungus-resistant plasticizer polyether-[di (butoxy-ethoxy-ethyl) formal].

1.2.10 APPLICATIONS

CR is one of the dominant specialty elastomers and is the basis for a wide variety of technical rubber goods. The estimated consumption of CR solid rubber, CR adhesive raw materials, and CR latex is approximately 300,000 tons per year (excluding former Soviet Union and PR of China). Approximately two-thirds of this consumption is for typical rubber applications. Thus, the CR Market can be analyzed in terms of the market sector as shown in Figure 1.15.

Alternatively, the CR market can be analyzed as shown in Figure 1.16 in terms of the article type.

Examples for some applications are summarized subsequently.

1.2.10.1 Hoses

CR is the classical elastomer for hose covers. Industrial hydraulic hoses, either medium- or high-pressure types, currently contain CR covers. For cost reasons,

TABLE 1.12
Comparing the Swell of CR, NBR, SBR, and NR in Various Fluids

Chemicals	Swelling Conditions	CR	NBR[a]	SBR	NR
Acetic acid, 10%	50°C, 12 weeks	E	E	E	E
Acetone	20°C, 20 d	C–D	E	C–D	D
Acetone	30°C, 20 d	D	E	D	D
ASTM fuel No. 3	50°C, 28 d	E	E	E	E
ASTM oil No. 1	70°C, 28 d	B–C	A	D–E	E
ASTM oil No. 1	100°C, 28 d	B–C	A	D–C	E
ASTM oil No. 2	70°C, 28 d	C–D	A	E	E
ASTM oil No. 2	100°C, 28 d	C–D	A	E	E
ASTM oil No. 3	70°C, 28 d	E	C	E	E
ASTM oil No. 3	100°C, 28 d	E	C	E	E
Ethylene glycol	100°C, 20 d	A	A	A	A–B
Fatty acid[b]	—	B–C	A	D	D
R 11	20°C, 28 d	E	E	E	E
R 12	20°C, 21 d	C–D	C–D	D	E
Glycerol	50°C, 20 d	A	A	A	A–B
Glycerol	100°C, 20 d	B	B	B	E
Methanol	50°C, 20 d	C	C–D	B	B
Methyl ethyl ketone	20°C, 28 d	E	E	D–E	D–E
Paraffin[b]	—	B	A	D–E	D–E
Sulfuric acid, 25%	50°C, 12 weeks	B–C	B	C	C
Sulfuric acid, 50%	100°C, 12 weeks	E	E	E	E
Toluene	20°C, 28 d	E	E	E	
City gas[b]	—	A–B	A	C–D	C–D
Water, distilled	20°C, 2/4 years	B/C	B/B	B/B	C/C

[a] Perbunan N 3310 Bayer AG.
[b] Data from "Dichtelemente", Vol. II (1965), Asbest-u. Gummiwerk, Martin Merkel KG, Hamburg.
Notes: Ratings: A, extremely resistant; B, highly resistant; C, resistant; D, partially resistant; E, not resistant.

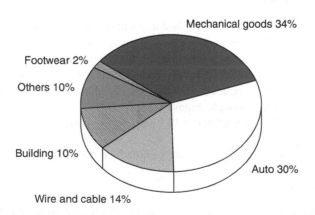

FIGURE 1.15 Polychloroprene end use application survey.

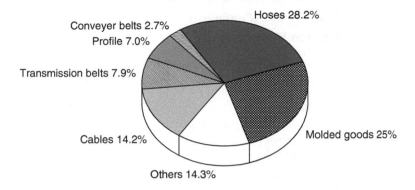

FIGURE 1.16 Polychloroprene application survey.

blending with SBR is practiced. CR covers are also used where hoses resistant to oil and ozone are required. A typical hose cover formulation is given in Table 1.13. Baypren 210 is a medium crystallization general purpose grade of CR and if better low-temperature resistance is needed, Baypren 110 is suggested.

TABLE 1.13
Baypren Hose Cover

Baypren 210	100
Maglite D	4
Stearic acid	1
Vulkanox OCD	3
Vulkanox 3100	1.5
Antilux 111	2
N-660 Black	100
Aromatic oil	20
Mesamoll	20
Aflux 42	3
Zinc oxide	5
Vulkacit thiuram MS (TMTM)	0.8
Vulkacit D (DOTG)	0.8
Sulfur	1
Total	262.1
Compound properties	
ML 1 + 4 at 100°C	37
t_5 at 120°C (min)	>45
Vulkameter data at 160°C	
$F_{max} - F_{min}$ (N·m)	38.1
t_{10} (minutes)	3.2
t_{80} (minutes)	16.6
t_{90} (minutes)	21.5

(*continued*)

TABLE 1.13   (Continued)
Baypren Hose Cover

Physical properties, cured 30 min at 160°C
Hardness (Shore A)	74
Tensile (MPa)	14.0
Elongation (%)	215

Compression set B (ASTM D395B)
70 h at 100°C (%)	31
70 h at 125°C (%)	53

Aged in air 7 days at 100°C-change
Hardness, pts.	+10
Tensile (%)	−10
Elongation (%)	−28
Brittle point	
BP (°C)	−36

Other applications of importance are suction and discharge hose covers and tubes and general tubing for the automotive industry.

1.2.10.2   Molded Goods

Bellows and seals for various applications, of which axle boots are a typical example, are made from CR. A suggested compound is illustrated in Table 1.14. The compound is designed to have excellent low temperature, flex, ozone, and weather resistance.

TABLE 1.14
Baypren Automotive Axle Boot

Baypren 126	100.0
Maglite DE	4.0
Stearic acid	0.5
Vulkanox OCD	2.0
Vulkanox 4020	2.0
Antilux 110	2.5
FEF (N550) black	35.0
SRF-HM (N774) black	40.0
DOS	25.0
ZnO	5.0
Rhenogran ETU 80	0.8
Vulkacit Thiuram/C	0.8
Total	217.6
Density (g/cm^3)	1.35

TABLE 1.14   (Continued)
Baypren Automotive Axle Boot

Compound properties

ML 1 + 4 at 100°C	50
ML min at 145°C	38
t_5 at 120°C (minutes)	24
t_5 at 145°C (minutes)	5

Vulkameter data at 170°C

M_{min} (N · m)	1.9
$M_{max} - F_{min}$ (N · m)	38
T_s 10 (minutes)	1.8
T_c 80 (minutes)	8.5
T_c 90 (minutes)	13

Vulcanizate properties cured 13 min at 170°C

Hardness, Shore A	66
100% Modulus (MPa)	4.6
200% Modulus (MPa)	11.8
Tensile strength (Mpa)	17.2
Elongation (%)	285
Tear die "C" (kN/mm)	35

Aged in AIR 7 days at 100°C-change[a]

Hardness (pts)	+7
100% Modulus (%)	45
Tensile (%)	−1
Elongation (%)	−11
Tear die "C" (%)	−7

DeMattia Flex (DIN 53522)[b] cured t_c 90 + 2 min

Unaged

100% Modulus (MPa)	4.9
Kilocycles	500

Aged in air 7 days at 100°C

100% Modulus (MPa)	6.9
Kilocycles	1000

[a]   Average of samples cured t_{90} and t_{90} + 2 min.
[b]   Kc to crack rating 6 (crack size >3 mm).

CR has been used for many years as the elastomer of choice for bearings in machinery and bridges.

No problems occur with CR rubber-to-metal bonding using conventional techniques of metal preparation and a commercial adhesive such as Chemlok 855 or 8560 single coat system or a two coat combination of Chemlok 205/220 or 805/8200.

TABLE 1.15

Core Compound for Cut Edge V-Belts (ID 0044.CR)

Baypren 711	100.00
TAKTENE 1203	5.00
Magnesium oxide	4.00
Stearic acid	3.00
Vulkanox OCD	2.00
Vulkanox 3100	1.50
Silica	10.00
N-660 Black	30.00
Sundex 8125	5.00
Polyester fiber (1 mm)	15.00
Zinc oxide	5.00
Rhenogran ETU-75	0.25

Sulfur grade is used in this application for good dynamic properties.

1.2.10.3   Belting

CR is the dominant elastomer for power transmission and timing belting. Other uses include various industrial belts. A starting compound is given in Table 1.15. The compound is designed for excellent flex resistance. The polyester fiber is used for good dynamic compression resistance.

Mining conveyor belts are based on CR where stringent flame retardance requirements must be fulfilled. Flame retardant mineral fillers used in combination with a chlorinated wax are recommended along with a silica filler for abrasion and tear resistance.

1.2.10.4   Extruded Profiles

Automotive and building profiles in the hardness range of 50 Shore A to 90 Shore A and sponge profiles have been in use for many years. In the construction industry, some CR has been replaced by EPDM for cost reasons; however, CR is still the preferred polymer if flame retardance and some oil resistance are required. An extruded road seal formulation is provided in Table 1.16. Baypren 115 may be used in place of Baypren 111 if improved extrusion properties are wanted. Desical 85 is used to absorb moisture in the compound to prevent porosity in the extrudate.

1.2.10.5   Wire and Cable

CR is the polymer of choice for cable jackets in heavy duty applications (transport, mining, welding, and others). Typical general purpose cable jackets are shown in Table 1.17.

TABLE 1.16
Baypren Road Seal Formulation

Baypren 111	100.0
Maglite D	4.0
Vulkanox OCD/SG	1.5
Vulkanox 3100	1.5
N-660 Black	35.0
TP-90B	7.5
Sundex 790	7.5
Sunolite 666	2.0
TMTM	0.5
DOTG	0.5
Sulfur	0.5
Zinc oxide	5.0
Rhenogran ETU-75	1.5
Desical 85	3.0
Total	170.0

Physical properties—cured 20 min at 153°C

Hardness (Shore A)	60
Tensile (MPa)	14.7
Elongation (%)	315

Aged in air oven 70 h at 100°C-change

Hardness (pts)	+9
Tensile (%)	+1.9
Elongation (%)	−1.6

Aged in ASTM #3 Oil 70 h at 100°C-change

Weight (%)	+37.2

TABLE 1.17
Baypren Cable Jacket Formulations

Baypren 211	100.0	—
Baypren 226	—	100.0
Maglite D	4.0	4.0
Stearic acid	1.0	1.0
Vulkanox DDA	2.0	2.0
Suprex clay	120.0	150.0
N-774 Black	2.0	2.0
Aromatic oil	20.0	30.0
Paraffin	5.0	5.0
Zinc oxide	5.0	5.0
Rhenogran ETU-80	1.5	1.5
Vulkacit DM (MBTS)	0.5	0.5
Total	261.0	301.0

(continued)

TABLE 1.17   (Continued)
Baypren Cable Jacket Formulations

Compound properties		
ML 1 + 4 at 100°C	28	39
t_5 at 120°C (minutes)	18	17
Vulkameter data at 200°C		
M_{min} (N · m)	0.7	0.8
M_{max} (N · m)	26.0	30.0
t_s10 (minutes)	0.9	0.7
t_x80 (minutes)	3.4	3.8
Physical properties cured 90 s at 200°C in steam		
Hardness (Shore A)	56	60
100% Modulus (MPa)	2.5	2.6
300% Modulus (MPa)	4.0	4.2
Tensile (MPa)	12.6	12.7
Elongation (%)	760	710

1.2.10.6   Miscellaneous

Some other applications are rollers for the printing and textile industry, coated fabrics, membranes, air bags, tank linings, closed cell sponge surf, and diving suits.

REFERENCES

1. J.A. Nieuwland, "Acetylene Reactions, Mostly Catalytic," paper presented at First National Symposium on Organic Chemistry, ACS Dec. 29–31, 1925. Rochester, NY.
2. A.M. Collins, "The Discovery of Polychloroprene," *Rubber Chem. & Technol.*, 46(2): 48, June–July, 1973.
3. P.S. Bauchwitz, J.B. Finlay, C.A. Stewart, Jr., *Vinyl and Diene Monomers,* Part II E.C. Leonhard, ed., John Wiley and Sons, NY, pp. 1149–1184, 1971.
4. P.R. Johnson, "Polychloroprene Rubber," *Rubber Chem. & Technol.*, 49(3): 650, Jul–Aug, 1976.
5. W. Obrecht, Makromolekulare Stoffe Bd E 20 H. Bartl, J. Falke. (Methoden der organischen Chemie, Thieme Verlag Stuttgart, NY, S. 843–856, 1987 (in German).
6. W. Gobel, E. Rohde, E. Schwinum, Kautschuk + Gummi, *Kunststoffe*, 25: 11, 1982 (in German).
7. R. Musch, Polychloroprene Grades with Improved Processing Behavior and Vulcanizate Properties (140). Rubber Division, ACS Meeting, Oct., 1991, Detroit.
8. R. Musch, U. Eisele, New Polychloroprene Grades with Optimized Structure Property Relationship. (136). Rubber Division, ACS Meeting, Oct., 1989, Detroit (Kautschuk + Gummi in press).
9. R. Petioud, Q. Tho Pham, I. Pol, *Sci. Pol. Chem. Ed.*, 22: 1333–1342, 1985.
10. R.R. Garett, C.A. Hargreaves, D.N. Robinson, *I. Macromob. Sci. Chem.*, A 4.8: 1679, 1970.
11. C.A. Aufdermarsh, R. Pariser, *I. Pol. Sci.* A. 2: 4727, 1964.

12. E. Rohde, H. Bechen, M. Mezger, Kautschuk + Gummi, *Kunststoffe*, 42: 1121–1129, 1989 (in German).
13. *The Synthetic Rubber Manual*, 15th Edition 2002 IISRP, Houston, Texas.
14. R. Pariser, *Kunststoffe*, 50, Nr. 11: 623, 1960 (in German).
15. R. Musch, U. Eisele, International Rubber Conference, 90, June 12–14, 1990, Paris, (Kautschuk + Gummi, *Kunststoffe* in press).
16. U. Eholzer, Th. Kempermann, W. Warrach, *Rubber & Plastics News Technical Note Book*, Nr. 48, May, 1985.
17. D.C.H. Brown, J. Thompson, *Rubber World*, 32, Nov., 1981.

2 Acrylonitrile Butadiene Rubber

Robert C. Klingender

CONTENTS

2.1 INTRODUCTION

Acrylonitrile butadiene rubber (NBR), also known as nitrile rubber or NBR, was first developed by Konrad, Tschunkur, and Kleiner at I.G. Farbenindustrie, Ludwigshafen, then with Oppau and Hoechst as a joint development in 1930, and commercialized in

1934. The original name was Buna N and later changed to Perbunan. The Second World War prevented export to Great Britain and the United States; hence Standard Oil Company and other companies, licensees of I.G. Farbenindustrie, began production in 1941 by Goodyear, Firestone, U.S. Rubber, and B.F. Goodrich as part of the war effort in the United States. In addition, Polymer Corporation in Sarnia, Ontario, Canada, began nitrile production in 1948.

A joint company of Distillers Company and B.F. Goodrich began production of NBR in Barry, S. Wales in 1959. The production of nitrile rubber spread between that time and 1962 to other countries of the world including France, Italy, and Japan as well as Russia. The total capacity worldwide in 1962 was estimated at 167,000 metric tons, which has grown to approximately 480,000 metric tons in 2001 [1]. A listing of the various suppliers of NBR worldwide is given in Table 2.1 as of 2005 [2]. There has been considerable consolidation of producers and production facilities in recent years so this information may become outdated with time [3,4].

TABLE 2.1
Worldwide NBR Suppliers

Company	Symbol	Trade Name	Country
Dwory SA	Dwory	KER	Poland, Oswiecim
Carom SA	CO	CAROM	Romania
Eliokem Chemicals	Eliokem	Chemigum	France, Le Havre USA, Akron
Hyundai Petrochemical Co. Ltd.	Hyundai	SEETEC	Korea, Daesan
Industrias Negromex SA de CV	N	EMULPRENE	Mexico
JSR Corporation	JSR	JSR	Japan, Yokkaichi
Korea Kumho Petrochemical Co.	KKPC	KOSYN	Korea, Seoul
Lanxess Elastomers	Lanxess LE	Perbunan NT Krynac	France, La Wantzenau
Lanxess Inc.	Lanxess LINC	Krynac Perbunan NT	Canada, Sarnia
Lanzhou Chemical Industry	LZCC	NBR	China
Nantes Industry Co. Ltd.	Nantex	NANTEX	Taiwan, Kaohsiung
Negromex Industrias, SA de CV	Negromex	N-xxxx	Mexico
Nitriflex Industria e Comercio SA	Nitriflex	NITRIFLEX N	Brazil, Rio de Janeiro
ParaTec Elastomers LLC	Paratec	Paracril Paraclean Paracril OZO	Mexico, Altamira
Petrobras Energia SA	Petrobras	ARNIPOL	Argentina, Buenos Aires
PetroChina Jilin Petrochemical Co.	JIL	NBR	China, Jilin
Petroflex Industria e Comercio SA	Petroflex	PETROFLEX	Brazil

TABLE 2.1 (Continued)
Worldwide NBR Suppliers

Company	Symbol	Trade Name	Country
Polimeri Europa S. p. A.	Polimeri	Europrene N	Italy, Milan
Sibur Krasnoyarsky zavod SK	Sibur	SKN	Russia, Krasnoyarsk
Sibur Voronezhsyntezkauchuk	Sibur	Nitrilast	Russia, Voronezh
Zeon Chemicals Europe Ltd.	ZCE	Breon Nipol Polyblack	Wales, Sully South Glamorgan
Zeon Chemicals LP	ZCLP	Nipol	USA, Louisville
Zeon Corporation	ZECO	Nipol	Japan, Kawasaki

The oil, fuel, and heat resistance of nitrile rubber, or NBR, have made this elastomer very important to the automotive non-tire and industrial rubber business. NBR is considered to be the major oil, fuel, and heat resistant elastomer in the world.

Since the original hot polymerized NBRs there have been many improvements and expansion of the types and ranges of properties available to the rubber industry. Cold polymerization is the predominant process for the emulsion polymerization of NBR with Acrylonitrile (ACN) contents now ranging from 18%–50% and Mooney viscosity going from 25 to 120. In addition, modifications now include carboxylated, precrosslinked, ACN/isoprene/butadiene, liquid, carbon black masterbatches, plasticizer extended, and nitrile/pvc blends (with and without plasticizer). Nitrile latices, powdered and crumb will not be covered in this book. Hydrogenated NBR is dealt with in a separate chapter.

The structure of standard NBR is given in Figure 2.1. The chemical configuration of carboxylated NBR, XNBR, may be seen in Figure 2.2 and that of the butadiene/isoprene/ACN NIBR terpolymer is shown in Figure 2.3. The molecular weight of the various grades of NBR varies to provide a very broad range of Mooney viscosities. The molecular weight distribution also ranges from narrow to broad to provide processing characteristics required for given process conditions. In addition the polymerization chemistry has been modified in recent years for special "clean" grades of NBR to eliminate residual materials that cause mold fouling, especially for injection molding at high temperatures.

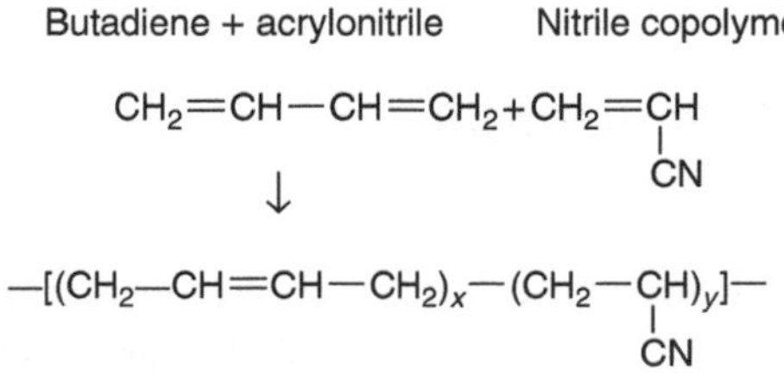

Butadiene + acrylonitrile   Nitrile copolymer unit

$$CH_2{=}CH{-}CH{=}CH_2 + CH_2{=}\underset{\underset{CN}{|}}{CH}$$

$$\downarrow$$

$$-[(CH_2{-}CH{=}CH{-}CH_2)_x{-}(CH_2{-}\underset{\underset{CN}{|}}{CH})_y]-$$

FIGURE 2.1 Conventional NBR polymers.

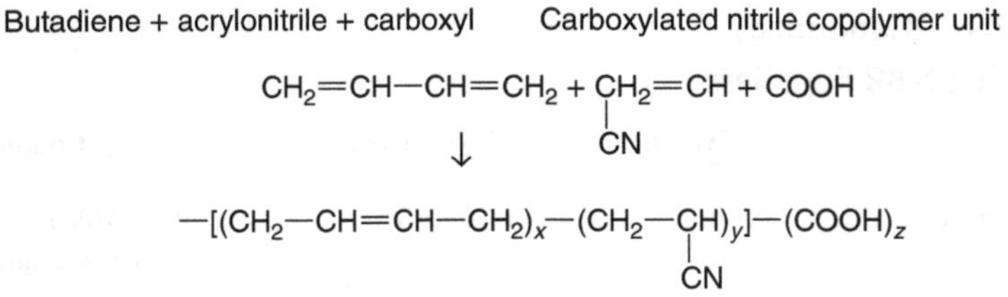

Butadiene + acrylonitrile + carboxyl   Carboxylated nitrile copolymer unit

$$CH_2{=}CH{-}CH{=}CH_2 + \underset{\underset{CN}{|}}{CH_2}{=}CH + COOH$$

$$\downarrow$$

$$-[(CH_2{-}CH{=}CH{-}CH_2)_x{-}(CH_2{-}\underset{\underset{CN}{|}}{CH})_y]{-}(COOH)_z$$

FIGURE 2.2 Carboxylated XNBR polymer.

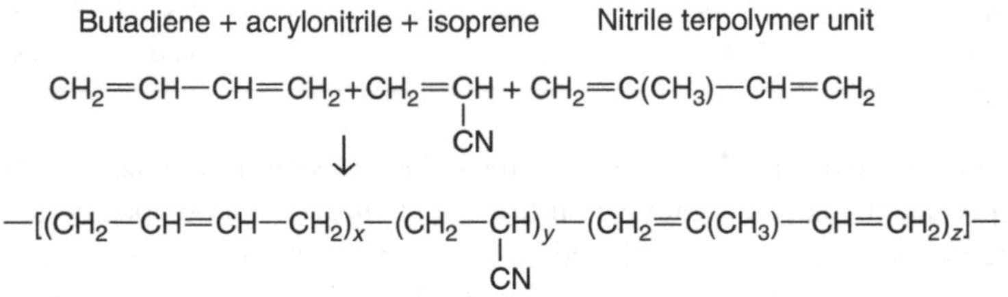

Butadiene + acrylonitrile + isoprene   Nitrile terpolymer unit

$$CH_2{=}CH{-}CH{=}CH_2 + \underset{\underset{CN}{|}}{CH_2}{=}CH + CH_2{=}C(CH_3){-}CH{=}CH_2$$

$$\downarrow$$

$$-[(CH_2{-}CH{=}CH{-}CH_2)_x{-}(CH_2{-}\underset{\underset{CN}{|}}{CH})_y{-}(CH_2{=}C(CH_3){-}CH{=}CH_2)_z]{-}$$

FIGURE 2.3 Butadiene isoprene ACN terpolymer NBIR.

The almost infinite number of variations in NBR, including batch polymerization, continuous polymerization, hot polymerized, cold polymerized, narrow molecular weight distribution, wide molecular weight distribution, various coagulation and emulsion systems, clean polymers, and so on has led to the proliferation of grades, which challenges the rubber chemist to select the correct grade for a given application and process. As an example, the most common ACN content of NBR is 31%–35% and in this range 188 grades are available from the various manufacturers worldwide. This is further complicated by the fact that certain grades from a given manufacturer may be produced at various production facilities. The subtle, and sometimes not so subtle, differences in grades with the same ACN content and Mooney viscosity often make the substitution of one for another grade of NBR difficult without some adjustment in the formulation to compensate for these differences.

Despite these challenges to the rubber compounder the ease of processing, economics, tremendous versatility make NBR one of the favorite elastomers for moderately high heat, oil, and fuel resistance applications.

2.2 GRADES AND TYPES OF NBR

2.2.1 STANDARD NBR

The various grades of NBR from the two major North American suppliers are given in Tables 2.2 through 2.4. Of course there are many other sources of equivalent grades available, but for simplicity this chapter is confined to these suppliers as the complete gambit of variations is available from these sources [5,6].

TABLE 2.2
Standard NBR Polymers from Zeon Chemicals LLC, Trade Name Nipol

Product Grade	% ACN	Mooney Viscosity	Specific Gravity	Hot/Cold	AO Type	Special Properties/Applications
DN401LL	19	32–44	0.94	C	NS	Very good low-temperature flexibility and low viscosity for superior processing properties
DN401L	19	59–71	0.94	C	NS	Very good low-temperature flexibility and good processing properties. Higher viscosity version of Nipol DN401LL
1034-60	21	55–70	0.95	C	SS	Provides low temperature and very good water resistance
1014	21	75–90	0.95	H	SS	Good low-temperature properties. Blends with natural rubber
1094-80	22	65–80	0.95	C	NS	Low-temperature resistance with outstanding physical properties
N917	23	55–70	0.95	C	SS	Medium-high viscosity for compression/transfer molding. Special grade for molded goods, hose and belts requiring oil resistance and operating service of +125°C to −50°C. High resilience. Low water swell
1053	29	45–60	0.97	C	NS	Used where low temperature and good mold flow are required. Easier processing than Nipol 1043
1053HM	29	60–75	0.97	C	NS	Higher Mooney version of 1053
1043	29	75–90	0.97	C	SS	Provides better physical properties than the 1050 series and better processing than the 1000 series
30-5	30	42–52	0.97	C	NS	General-purpose, intermediate-viscosity nitrile rubber for industrial and automotive hose and seals, printing rolls and applications requiring easy processing
30-8	30	60–74	0.97	C	NS	General-purpose, high-viscosity nitrile rubber recommended for applications requiring improved physical properties, such as footwear, hose jackets, belt covers, and so on
1092-80	32	70–85	0.98	C	NS	Provides a good blend of tack, physical properties, and water resistance
1052-30	33	25–40	0.98	C	NS	Low Mooney version of Nipol 1052
N624B	33	38–50	0.98	C	NS	Medium viscosity/medium resistance to fuels and oils. Specialty grade for extruded/calendered flat goods, sponge, hose, and belting. Well-suited for mill mixing. Low water swell
1022	33	40–55	0.98	H	NS	Low viscosity, directly soluble with wide FDA acceptance
1032-45	33	40–55	0.98	C	NS	Slightly lower Mooney grade of Nipol 1032
1052	33	45–60	0.98	C	NS	Excellent general purpose nitrile. Provides exceptional processing and blending with other polymers

(continued)

TABLE 2.2 (Continued)

Standard NBR Polymers from Zeon Chemicals LLC, Trade Name Nipol

Product Grade	% ACN	Mooney Viscosity	Specific Gravity	Hot/ Cold	AO Type	Special Properties/Applications
1032	33	55–70	0.98	C	NS	Excellent water resistance with very low metal corrosion. Very good building tack. FDA applications
1002	33	75–100	0.98	H	NS	The original medium-high ACN. Provides good long-term water resistance
1042	33	75–90	0.98	C	NS	Well suited for graphic arts, rolls, and other rubber products requiring high durability
35–5	35	43–53	0.98	C	NS	General-purpose, intermediate-viscosity nitrile rubber. Combines superior physical properties and solvent resistance. Recommended for industrial and automotive applications, transfer molded goods, footwear, and so on
35–8	35	66–80	0.98	C	NS	General-purpose, high Mooney viscosity nitrile rubber. Combines exceptional physical properties and outstanding oil resistance. Recommended for critical industrial and automotive extruded goods
40–5	41	43–53	1.00	C	NS	Combines excellent physical properties and processability with very high oil resistance. Recommended for oil fields parts and other applications requiring high oil resistance
1031	41	55–70	1.00	C	NS	Excellent fuel and water resistance. Very low corrosion to metals
1051	41	60–75	1.00	C	SS	Easy processing version of Nipol 1041 with similar properties. Widely used in the petroleum industry
1001LG	41	70–90	1.00	H	NS	Similar to 1001CG, but can be dissolved in solvents without milling. Contains no fatty acids or soaps
1001CG	41	70–95	1.00	H	SS	Excellent oil and fuel resistance. Has controlled cement viscosity. Useful in adhesives
1041	41	75–90	1.00	C	SS	Polymerized at low temperatures to give better processing. Provides good tack for rolls and belting
DN4265	42	58–72	1.00	C	NS	High ACN for balance of low temperature, fuel resistance, and low fuel permeability
1000x88	43	70–90	1.00	H	SS	Excellent in adhesives when blended with phenolic resins
DN4555	45	48–63	1.00	C	NS	High ACN for balance of low temperature, fuel resistance, and low fuel permeability
DN4580	45	73–87	1.00	C	NS	High ACN for balance of low temperature, fuel resistance, and low fuel permeability
DN003	50	70–85	1.02	C	SS	Very high ACN level for excellent resistance to oils and fuels. Low fuel permeability
1000 × 132	51	45–65	1.02	C	SS	Ultra-high ACN level for maximum oil and fuel resistance and low gas permeability

TABLE 2.3
Standard NBR Polymers from Lanxess, Trade Name Krynac

Product Grade	% ACN	Mooney Viscosity	Specific Gravity	Hot/ Cold	AO Type	Special Properties/ Applications
2255 C	22	57	0.96	C	NS	Good low-temperature flexibility
2455 C	24	54	0.96	C	NS	
2645 F	26	45	0.96	C	NS	
2750 C	26.7	48	0.97	C	NS	
2840 F	28	40	0.97	C	NS	
2865 C/F	28	65	0.97	C	NS	Low water swell
3035 C	30	35	0.97	C	NS	Low water swell
3330 F	33	30	0.97	C	NS	
3335 C	33	33	0.97	C	NS	
3345 C/F	33	46	0.97	C	NS	
3355 F	33	55	0.97	C	NS	
3370 C/F	33	69	0.97	C	NS	
33110 F	33	110	0.97	C	NS	
3950 F	39	50	0.99	C	NS	
4060 C	40	61	0.99	C	NS	
4450 F	43.5	50	1.00	C	NS	Best fuel resistance
4560 C	45	64	0.99	C	SS	Best fuel resistance
4970 C	48.9	71	1.00	C	NS	Best fuel resistance
4975 F	48.5	75	1.01	C	NS	Best fuel resistance

Note: C designates grades made in Canada; F denotes grades made in France.

TABLE 2.4
Standard NBR Grades from Lanxess, Trade Name Perbunan

Product Grade	% ACN	Mooney Viscosity	Specific Gravity	Hot/ Cold	AO Type	Special Properties/Applications
2845 C/F	28	45	0.96	C	NS	Good resilience and low-temperature flexibility
2870 F	28	70	0.96	C	NS	Good resilience and low-temperature flexibility
2895 F	28	95	0.96	C	NS	Good resilience, compression set, low-temperature flexibility, and high viscosity
28120 F	28	120	0.96	C	NS	Good resilience, compression set, low-temperature flexibility, and high viscosity
3430 F	34	32	0.97	C	NS	Good balance of oil and fuel resistance and low-temperature flexibility

(*continued*)

TABLE 2.4   (Continued)
Standard NBR Grades from Lanxess, Trade Name Perbunan

Product Grade	% ACN	Mooney Viscosity	Specific Gravity	Hot/ Cold	AO Type	Special Properties/Applications
3435 C	34	35	0.98	C	NS	Good balance of oil and fuel resistance and low-temperature flexibility
3445 C/F	34	45	0.97	C	NS	Good balance of oil and fuel resistance and low-temperature flexibility
3470 F	34	70	0.99	C	NS	Good balance of oil and fuel resistance and low-temperature flexibility
3945 F	39	45	0.99	C	NS	Optimum oil and fuel resistance
3965 F	39	65	0.99	C	NS	Optimum oil and fuel resistance

Note: C denotes grades made in Canada; F designates grades made in France.

2.2.1.1   ACN Content

It is well known, as the ACN content varies from 18% to 50%, the oil and fuel resistance improves and conversely the low-temperature properties deteriorate. This influence is represented graphically in Figures 2.4 and 2.5. These graphs provide a general indication of ACN levels influence on these properties; however, these properties may vary depending on the polymerization conditions and the resulting structure of the NBR. A random distribution of ACN and butadiene will give a better balance of low-temperature properties and oil resistance than elastomers that are produced with connected blocks of ACN and butadiene. These differences in structure of NBR are more apparent at the extreme low or high levels of ACN.

The design of the formulation also plays a major role in the balance of oil and fuel resistance versus the low-temperature flexibility. High inert filler loadings will

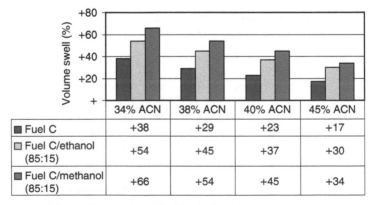

	34% ACN	38% ACN	40% ACN	45% ACN
■ Fuel C	+38	+29	+23	+17
☐ Fuel C/ethanol (85:15)	+54	+45	+37	+30
■ Fuel C/methanol (85:15)	+66	+54	+45	+34

Acrylonitrile content

FIGURE 2.4 Influence of ACN on fuel swell (48 h at 23°C).

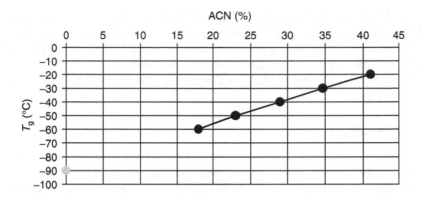

FIGURE 2.5 Influence on the low-temperature properties of NBR.

reduce volume swell in NBR. Active fillers will also reduce swell to a lesser extent, as these will tend to absorb oils and fuels and may not be used at as high a level as the inert fillers. Another major compounding factor is the plasticizer type and content. Plasticizers will give a significant improvement in low-temperature properties, depending on the type and amount used. Unfortunately good low-temperature plasticizers are often more easily extracted by oils and fuels and are more volatile, thus providing poorer high-temperature resistance. It is highly recommended that the chapter on plasticizers be consulted.

The influence of ACN on oil, fuel resistance, and low-temperature flexibility in other classes of NBR will be the same as with "Standard NBR."

2.2.1.2 Mooney Viscosity

Standard nitrile elastomers range in Mooney viscosity from 26 to 120 thus providing the chemist with a broad choice of rheological properties to meet any process conditions and compound design. The higher the loading of both filler and plasticizer the higher the viscosity needed for satisfactory uncured strength. This approach is commonly used in hose and tubing, which requires good "green strength" to maintain shape, size, and to resist imbedding of reinforcing filaments in the tubing. If a compound design is based on a high filler loading but a low plasticizer content, a low-viscosity polymer is needed. In addition, if injection or transfer molding is employed to produce the part then again a low viscosity is needed.

It should be noted that higher molecular weight NBR will provide better physical properties and lower compression set.

2.2.1.3 Polymerization Temperature

The vast majority of NBR is produced by "cold polymerization," which gives a more linear polymer with better processing properties. Hot polymerized elastomers are tougher due to more branching, which results in compounds with higher green strength, and these may require more care in handling. They will build up more heat in processing, hence are more scorchy. Branching may also lead to polymers with an undesired high amount chain entanglement, but high loadings of filler and plasticizer will help to alleviate the problem. Some grades of hot polymers by design have crosslinked domains known as gel. These are most often blended with other nitrile grades for improved green strength and dimensional stability.

Hot polymerized NBR is often used in solvent adhesives because of the better green strength.

2.2.1.4 Stabilizer

The stabilizer or antioxidant added during polymerization functions as a control of oxidation during drying of the polymer and in storage. A nonstaining type is often used to allow the NBR to be used in light-colored articles and in contact with various surfaces without causing staining or certain types that prevent the crazing of plastic. Staining stabilizers generally exhibit better protection in storage. A stabilizer that inhibits free radical formation in processing and storage of the NBR and uncured compound helps in preventing an increase in viscosity of the polymer and compound.

The stabilizer is not intended to provide protection of the vulcanized part, hence antioxidants or antiozonants are needed in the compound design to achieve satisfactory service life.

2.2.1.5 Specific Gravity

The specific gravity of NBR ranges from 0.94 for polymers in the 17%–20% ACN content range up to 1.02 for a 50% ACN type. This factor needs to be taken into account when establishing batch sizes for an internal mixer as well as preform weights when molding.

The characteristics listed in Section 2.2.1 also apply to the following specialty types of nitrile rubber.

2.2.2 Low Mold-Fouling NBR Grades

The demand for high quality at lower cost has expanded the use of injection molding at high temperatures, especially for automotive applications. This has led to the development of "low mold-fouling NBR grades" by many manufacturers of nitrile rubber. Through the proper selection of emulsifiers and coagulation systems and proprietary processes, families of clean polymers are now marketed by Zeon as seen in Table 2.5, Lanxess shown in Table 2.6, as well as other producers such as Polimeri Europa S. p. A. green types. The selection of grades in this category is not as broad as the standard grades with ACN levels ranging from 18% to 44% and Mooney viscosities from 35% to 80%. These grades are also faster curing than most standard NBR polymers.

2.2.3 Precrosslinked NBR Grades

Precrosslinked grades of NBR contain a small amount of a third monomer, which joins the butadiene/ACN chains during polymerization to create a terpolymer. The crosslinking greatly reduces the elastic memory of the polymer providing a compound with reduced shrinkage during calendering and swell while extruding the

TABLE 2.5
Low Mold-Fouling NBR Polymers from Zeon Chemicals LLC,
Trade Name Nipol

Product Grade	% ACN	Mooney Viscosity	Specific Gravity	Hot/ Cold	AO Type	Special Properties/Applications
DN2835	28	30–40	0.97	C	NS	Low mold fouling, fast curing, and easy processing. Good balance of low-temperature flex and solvent resistance
DN2850	28	45–55	0.97	C	NS	Low mold fouling, fast curing, and easy processing. Good balance of low-temperature flex and solvent resistance
DN2880	28	75–85	0.97	C	NS	Low mold fouling, fast curing, and easy processing. Good balance of low-temperature flex and solvent resistance
33-3	33	25–34	0.98	C	NS	General purpose, low-viscosity nitrile rubber. Designed for injection molding and resistance to mold fouling
DN3335	33	30–40	0.98	C	NS	Low mold fouling, fast curing, and easy processing. Good balance of low-temperature flex and solvent resistance
DN3350	33	45–55	0.98	C	NS	Low mold fouling, fast curing, and easy processing. Good balance of low-temperature flex and solvent resistance
33-5SR	33	41–51	0.98	C	NS	Special-purpose, intermediate-viscosity nitrile rubber designed to provide maximum processing safety without fouling molds. Recommended for high-temperature injection molding of parts
33-5HM	33	47–57	0.98	C	NS	Low mold fouling, high-modulus, 55 Mooney nitrile rubbers. Recommended for high-temperature injection molding
DN3380	33	75–85	0.98	C	NS	Low mold fouling, fast curing, and easy processing. Good balance of low-temperature flex and solvent resistance
DN3635	36	30–40	0.98	C	NS	Low mold fouling, fast curing, and easy processing. Good balance of low-temperature flex and solvent resistance

(*continued*)

TABLE 2.5   (Continued)
Low Mold-Fouling NBR Polymers from Zeon Chemicals LLC,
Trade Name Nipol

Product Grade	% ACN	Mooney Viscosity	Specific Gravity	Hot/ Cold	AO Type	Special Properties/Applications
DN3650	36	45–55	0.98	C	NS	Low mold fouling, fast curing, and easy processing. Good balance of low-temperature flex and solvent resistance
DN4050	40	45–60	1.00	C	NS	Low mold fouling, fast curing, and easy processing. High oil and fuel resistance
DN4080	40	73–88	1.00	C	NS	Low mold fouling, fast curing, and easy processing. High oil and fuel resistance

product. These elastomers are normally blended at the 5%–15% level with standard NBR polymers. Improved green strength and compression set are also benefits of incorporating a percentage of these polymers. Tables 2.7 and 2.8 contain the grades of precrosslinked NBR from Zeon Chemicals and Lanxess, respectively. Other producers also have this class of NBR available.

Another use for precrosslinked NBR is in conjunction with plastics such as PVC (polyvinyl chloride) and ABS (ACN/butadiene/styrene) resins to reduce elastic memory for better impression and retention of a surface pattern.

2.2.4 CARBOXYLATED NBR POLYMERS

The addition of a carboxyl group to NBR during polymerization, as seen in Figure 2.2, provides an elastomer with greatly improved abrasion and tear resistance.

TABLE 2.6
Low Mold-Fouling NBR Polymers from Lanxess, Trade Name Perbunan

Product Grade	% ACN	Mooney Viscosity	Specific Gravity	Hot/ Cold	AO Type	Special Properties/Applications
1846 F	18	45	0.93	C	NS	Optimal low-temperature flexibility and compression set
2831 F	28.6	30	0.96	C	NS	Good resilience and low-temperature flexibility
2846 F	28.6	42	0.96	C	NS	Good resilience and low-temperature flexibility
3446 F	34.7	42	0.97	C	NS	Favorable balance of oil and fuel resistance and low-temperature flexibility
3481 F	34.7	78	0.97	C	NS	
4456 F	44	55	1.00	C	NS	Optimum oil and fuel resistance

TABLE 2.7
Precrosslinked NBR Grades from Zeon Chemicals LLC, Trade Name Nipol

Product Grade	% ACN	Mooney Viscosity	Specific Gravity	Hot/ Cold	AO Type	Special Properties/Applications
1022x59	33	53–68	0.98	H	NS	Precrosslinked to provide low nerve and minimum die swell in extruded goods. Excellent as a compounding ingredient in PVC and ABS
1042x82	33	75–95	0.98	C	NS	Precrosslinked to give low die swell and nerve to extruded goods. May be blended with other polymers to improve extrusion

The greater the number of carboxyl groups, the better the tear, toughness, and abrasion properties. Tables 2.9 and 2.10 list the various grades manufactured by Zeon and Lanxess, respectively.

The presence of the carboxyl groups also may cause scorching during factory processing. Hence, care must be taken to use very low activity zinc oxides or zinc peroxide dispersion with most grades of this family typically adding these during the second pass. The exception is NX775, which will tolerate normal grades of zinc oxide without causing scorch.

2.2.5 TERPOLYMER OF ACRYLONITRILE-BUTADIENE-ISOPRENE ELASTOMER

The terpolymer acrylonitrile-butadiene-isoprene (ACN/BR/IR) supplied by Zeon Chemicals, as shown in Table 2.11, exhibits very good tensile strength without the use of reinforcing fillers, which makes it ideal for rubber thread that is resistant to dry cleaning solvents. It also finds use in covered rolls due to the better physical properties and resistance to hardness increase during service. It is theorized that the isoprene portion tends to soften slightly upon aging to counterbalance the hardening effect of the butadiene groups.

TABLE 2.8
Precrosslinked NBR Grades from Lanxess, Trade Name Krynac

Product Grade	% ACN	Mooney Viscosity	Specific Gravity	Hot/ Cold	AO Type	Special Properties/Applications
XL 3025	29.5	70	0.98	C	NS	For extrudates with excellent surface appearance, dimensional stability, and die swell
XL 3410	34	78	0.99	C	NS	For extrudates with excellent surface appearance, dimensional stability, and die swell
TRP 0005	34	70	0.99			

TABLE 2.9

Carboxylated NBR Polymers from Zeon Chemicals LLC, Trade Name Nipol

Product Grade	% ACN	Mooney Viscosity	Specific Gravity	Hot/ Cold	AO Type	Special Properties/Applications
NX775	26	38–52	0.98	C	NS	Excellent processing with conventional coated zinc oxides. High performance injection molding, rolls, calendered belting, and extruded hose. Fast cure rate. Carboxyl content is 0.083 EPHR
1072	27	40–55	0.98	C	NS	Oil resistant mechanical goods with outstanding abrasion resistance Carboxyl content is 0.075 EPHR
1072CGX	27	22–35	0.98	C	NS	Cement grade version of Nipol 1072
1072x28	27	35–55	0.98	C	NS	Precrosslinked version of Nipol 1072 Exhibits very low nerve for extrusion and calendered goods
DN631	33	48–63	0.99	C	NS	Exhibits good oil resistance and high tensile strength, and is used in belt, seal, and roll applications

2.2.6 LIQUID NBR

Zeon Chemicals and JSR produce liquid NBR in a couple of viscosity ranges. Zeon also makes a carboxylated liquid NBR called Nipol DN601, which is used mainly in resin modification for coatings and adhesives. Table 2.12 lists the liquids available from Zeon Chemicals. The major advantage of these products is to act as a nonvolatile

TABLE 2.10

Carboxylated NBR Polymers from Lanxess, Trade Name Krynac

Product Grade	% ACN	Mooney Viscosity	Specific Gravity	Hot/ Cold	AO Type	Carboxylic Acid (%)	Special Properties/ Applications
X146	32.5	45	0.97	C	NS	1	Very good abrasion resistance, high modulus, and good adhesion
X160	32.5	58	0.97	C	NS	1	Very good resistance, high modulus, and good adhesion
X740	26.5	38	0.99	C	NS	7	Best abrasion resistance, high modulus, and good adhesion
X750	27	47	0.99	C	NS	7	Best abrasion resistance, high modulus, and good adhesion

TABLE 2.11

Terpolymer NBR from Zeon Chemicals LLC, Trade Name Nipol

Product Grade	% ACN	Mooney Viscosity	Specific Gravity	Hot/ Cold	AO Type	Special Properties/Applications
DN1201	35	72.5–82.5	0.98	C	NS	ACN/BR/IR terpolymer. Has better physicals and processing than conventional nitrile rubber. Used for rolls, diaphragms, and rubber thread
DN1201L	35	40–52	0.98	C	NS	Low Mooney version of DN-1201

and nonextractable plasticizer in NBR. Many factories do not like to handle the viscous liquid NBR and will use it in the form of a dry liquid concentrate. Another use is as a nonvolatile, nonextractable plasticizer in PVC plastisols.

2.2.7 NBR CARBON BLACK MASTERBATCHES

NBR carbon black masterbatches are available from Zeon Chemicals as may be seen in Table 2.13. These are produced through the addition of the carbon black to the NBR latex and then dried, forming a crumb. These masterbatches may be compounded and processed in the same manner as SBR carbon black masterbatches. These may be used as is or blended with clear polymer or other masterbatches to adjust carbon black amounts, blends, or even ACN levels. The advantages of masterbatches are improved dispersion of the carbon black, cleaner working conditions, and very good physical properties. These are often used for pigmenting compounds using light-colored fillers, which prevent contaminating subsequent mixes. Note that care must be taken to sweep early in the mix, whether open mill or internal mixer to prevent the crumbs from dropping into the batch late and not getting dispersed.

TABLE 2.12

Liquid NBR from Zeon Chemicals LLC, Trade Name Nipol

Product Grade	% ACN	Specific Gravity	Hot/ Cold	AO Type	Viscosity (cps)	Special Properties/ Applications
1312	28	0.96	H	NS	20,000–30,000	Plasticizer used for nitrile, neoprene, and PVC compounds. Improves knitting and flow. May also be used in plastisols and phenolic resins
1312LV	26	0.96	H	NS	9,000–16,000	Same as 1312 with lower viscosity
DN601	20	0.98	C	NS	5,300–6,300	Carboxylated liquid nitrile suggested for resin modification in adhesives and coatings

TABLE 2.13
NBR Carbon Black Masterbatches from Zeon Chemicals LLC, Trade Name Nipol

Product Grade	% ACN	Black Type	Black (phr)	Specific Gravity	Nipol Base Polymer	Hot/ Cold	AO Type	Special Properties/Applications
9040	40	N330	50	1.2	40–5	C	NS	Recommended for use in molded goods requiring high tensile, solvent, and oil resistance
9010	35	N550	50	1.19	35–8	C	NS	Recommended for use in mechanical goods requiring medium–high solvent and oil resistance
9025	35	N660	75	1.25	35–5	C	NS	Recommended for use in extruded goods requiring medium–high solvent and oil resistance
DN120	33	N234	50	1.19	33–3	C	NS	Recommended for mechanical goods requiring medium–high solvent and oil resistance
DN127	33	N660	70	1.24	33–3	C	NS	Recommended for mechanical goods requiring medium–high solvent and oil resistance

TABLE 2.14
Plasticizer-Extender NBR from Zeon Chemicals LLC, Trade Name Nipol

Product Grade	% ACN	Mooney Viscosity	Specific Gravity	Hot/ Cold	AO Type	Special Properties/Applications
1082V	N/A	34	30–45	0.98	PP	Used for soft rolls and other low Durometer goods. Contains 50 phr DIDP plasticizer

2.2.8 PLASTICIZER-EXTENDED NBR

Both Zeon Chemicals and Lanxess make plasticizer-extended NBR with approximately 50 phr of an ester plasticizer added to a high viscosity NBR latex, which is then dried and compressed into bales. The primary application for this family of extended NBR is in soft roll compounds. The inclusion of the very large amounts of plasticizer greatly reduces mixing time in the factory and the high viscosity base elastomer provides better green strength to the compound. The various grades of extended NBR are given in Tables 2.14 and 2.15.

2.2.9 NITRILE PVC BLENDS

NBR to which PVC resin is added results in a compound with much improved ozone and abrasion resistance as well as green strength during processing. A minimum of 25% PVC resin provides ozone resistance, hence most simple NBR/PVC blends are 70/30 blends. The presence of PVC resin detracts from both the low- and high-temperature resistance.

The 70/30 blend grades are often used in hose jackets or cable covers. An added benefit is flame resistance when compounded with flame-retardant fillers and plasticizers. These 70/30 NBR/PVC blends also offer improved abrasion resistance and lower coefficient of friction.

Blends are available containing large amounts of plasticizer as well as higher levels of PVC, which greatly reduces mixing time and offers improved abrasion resistance in very low hardness compounds. Zeon Chemicals grades and the Sivic family they sell are shown in Table 2.16. In addition to other NBR producers some

TABLE 2.15
Plasticizer-Extended NBR from Lanxess, Trade Name Krynac

Product Grade	% ACN	Mooney Viscosity	Specific Gravity	Hot/ Cold	AO Type	Special Properties/Applications
E 3340 C	34	39	0.97	C	NS	Contains 50 phr DOP plasticizer
E 3338 F	34	30	0.98	C	NS	Contains 52 phr DOP plasticizer

Note: C, Canada; F, France.

TABLE 2.16

NBR/PVC Blends from Zeon Chemicals LLC

Product Grade	NBR/PVC Ratio	% ACN	Mooney Viscosity	Specific Gravity	Type	Special Properties/Applications
Nipol Types						
DN508SCR	70/30	38	40–50	1.07	PB	Improved ozone resistance and physical property version of P-70
P70	70/30	33	55–70	1.07	PB	A 70/30 fully fluxed nitrile-PVC blend. Provides outstanding ozone resistance. Used in cable jackets, hose, and shoe soles
DN171	70/30	30	66–81	1.06	PB	A fully fluxed NBR/PVC blend with improved ozone and abrasion resistance. Used in hose covers, shoe soles, and cable jackets
Sivic Types						
Z760	70/30	45	49–61	1.07	PB	Highest ACN content for maximum fuel resistance
Z740	70/30	41	45–57	1.06	PB	High ACN content for improved fuel resistance in automotive applications
Z730	70/30	33	66–78	1.06	PB	Medium ACN, high viscosity base polymer for improved physical properties and good extrudability. Used for industrial cable and hose
Z730M60	70/30	33	56–68	1.06	PB	A lower viscosity version of Z730 for faster processing

Z711	70/30	28	62–74	1.06	PB	Low ACN base polymer, good extrudability, and cold flex properties. Used for industrial and automotive cable and hose applications
Z711LV	70/30	28	57–69	1.06	PB	A low-viscosity version of Z711 for faster processing
Z700PX	70/30	26	59–71	1.06	PB	Carboxylated NBR base polymer. Improved abrasion resistance for printing rolls, conveyor belts, and shoe soles
Z702	70/30	23	61–73	1.06	PB	Lowest ACN content for improved cold flexibility. Automotive and Industrial cable, hose, and belt applications
Z620	60/40	33	48–58	1.09	PB	Fuel, ozone, and fire resistant blend for automotive and general purpose cable, belting, and hose applications
Z530	50/50	33	55–65	1.13	PB	Improved fuel, ozone, and fire resistance for hose covers, cable jackets, conveyor belts, and cellular goods
Z8401	80/20	33	28–38	1.04	PB/PP	A preplasticized blend with 40 phr DINP and 10 phr silica. Designed for low to medium Durometer applications
Z8901	80/20	33	8–16	1.02	PB/PP	Additional plasticizer at 90 phr DINP with 10 phr silica for very low Durometer applications
Z2710	60/40	33	16–26	1.04	PB/PP	Preplasticized with 70 phr DINP. Excellent for low Durometer applications such as soft printing rolls

custom mixing companies such as Excel Polymers manufacture various NBR/PVC blends, with and without added plasticizer.

This is particularly beneficial for roll cover compounds in which large amounts of plasticizer are often used to obtain very low Shore A hardness and the compound still needs a degree of green strength for calendering and wrapping onto a steel shaft, or for extruding to build a roll. The presence of PVC resin in the NBR also results in a compound that gives better back pressure when tape is applied to rolls or hose on a mandrel during vulcanization in an autoclave.

2.2.10   COMPOUND DESIGN OF STANDARD NBR

The first step in designing an NBR compound is to obtain COMPLETE information on the application, specification listings, including dynamic requirements if any, specified physical properties, fluids to be encountered, temperature ranges expected, environmental conditions, service life needed, color, factory process parameters, and any special needs. This is the most difficult part in designing a formulation as too often the customer assumes that you know as much about the application as they do and information provided to the sales/design engineer may be incomplete, misleading, or misinterpreted.

The major considerations in designing an NBR compound, in the order of importance, are the proper nitrile rubber; the plasticizer; the cure system; the antioxidant/antiozonants; the filler; and the economics, once the service conditions are defined and the factory process is established.

2.2.10.1   Selection of the Correct NBR

A comparison is made of NBR with ACN contents ranging from 51% to 21% in an unplasticized test formulation as seen in Table 2.17 [7]. This exaggerates the differences since most NBR compounds contain plasticizers. Depending on the solubility of plasticizers in the test fluids, aged physical properties will vary accordingly. The data are based on "work horse" NBR grades, so consideration should be made for the more modern NBR elastomers with lower and less soluble residual emulsifiers, which provide lower water swell and less mold fouling and fast cure rates. If the inventory of NBR grades on hand does not contain one with the required properties, then selection of the appropriate grades needs to be made with the help of your suppliers. There should be a rough screening step to determine the ACN content needed for the application.

The next consideration is the viscosity of the polymer that will provide good factory handling properties as well as the desired physical properties. High viscosity NBR accepts larger amounts of plasticizers and fillers for better economics and processability. The high molecular weight polymers give higher tensile strength and tougher compounds, but also this makes for more difficult handling as low-plasticizer levels are needed. Hot polymerized NBR grades are usually tougher and somewhat more difficult to handle, but the service characteristics may require their use. The low- or very low-viscosity NBR grades are used primarily for injection and transfer molding, especially if minimal amounts of plasticizers are used. These are slightly less tough than standard cold polymerized NBR, but when only small amounts of plasticizers are used, the performance is good in the end application.

TABLE 2.17

Influence of ACN Content on Physical and Chemical Properties of Unplasticized NBR Compounds Test Formulation; NBR 100; Zinc Oxide 5; Stearic Acid 1; AgeRite Stalite S 1.5; N550 40; TMTD 3.0; DTDM 3.0

Grade	1000 × 132	1001	1031	1041	1051	1091–50	1032	1042	1052	1092–30	1043	1053	1093 C50	1094–80
ACN%	51	41	41	41	41	41	33	33	33	33	29	29	29	21
ML1+4	55	95	60	80	75	50	55	80	50	30	80	60	50	80
Original properties														
Shore A, pts	79	68	68	66	67	70	66	64	63	64	62	62	58	67
Modulus at 100%, MPa	6.2	2.0	2.4	2.0	2.0	3.0	2.0	2.3	1.7	2.5	2.3	1.9	1.5	2.4
Tensile, MPa	15.3	22.1	23.6	22	19.6	22.8	20	20.9	20.4	18.8	20.8	18.8	19.1	21
Elongation, %	280	400	580	580	560	510	550	500	600	500	490	520	520	390
Compression set														
70 h at 125°C, %	47	46	38	38	37	52	41	34	44	62	39	38	34	38
Low temperature														
Gehman T_2, °C	8.5	−6	−3	−1	−1.5	0	−10.5	−12	−5	−11	−12	−10	−16	−27.1
Gehman T_{100}, °C	−9	−16	−14.9	−18	−16.5	−14.3	−25	−25	−23.5	−26	−28.5	−29	−34	−39.8
NBS Abrasion														
% of Standard	118	288	171	164	144	148	186	208	139	158	188	161	229	264
Bashore resilience														
Rebound, %	7	11	13	12	13	10	25	27	27	27	34	35	39	44
Aged 70 h at 125°C in air test tube-change														
Shore A, pts	5	4	5	5	5	5	4	4	6	7	6	6	4	4
Tensile, %	20	7	−6	−6	11	−1	−14	−3	3	10	−19	−3	3	−5
Elongation, %	−29	−25	−29	−40	−21	−38	−40	−30	−25	−36	−43	−27	−25	−28

(continued)

TABLE 2.17 (Continued)

Influence of ACN Content on Physical and Chemical Properties of Unplasticized NBR Compounds Test Formulation; NBR 100; Zinc Oxide 5; Stearic Acid 1; AgeRite Stalite S 1.5; N550 40; TMTD 3.0; DTDM 3.0

Grade	1000 × 132	1001	1031	1041	1051	1091–50	1032	1042	1052	1092–30	1043	1053	1093 C50	1094–80
Aged 70 h at 125°C in ASTM #1 Oil-change														
Shore A, pts	6	0	2	2	1	4	−1	−1	−1	3	−1	−4	−3	−4
Tensile, %	20	3	−2	4	14	−6	5	−5	−3	6	−2	2	2	−14
Elongation, %	−29	−25	−24	−29	−21	−35	−25	−30	−27	−30	−20	−17	−19	−20
Volume, %	−4	−4	−4	−3	−3	−4	−2	−2	−1	−2	4	−1	−1	2
Aged 70 h at 125°C in ASTM #3 Oil-change														
Shore A, pts	1	−4	−7	−6	−9	−5	−14	−12	−14	−9	−14	−20	−16	−17
Tensile, %	11	6	−2	−1	1	−1	−7	−7	−11	−5	−21	−21	−23	−37
Elongation, %	−29	−25	−24	−29	−21	−35	−25	−30	−27	−30	−20	−17	−19	−20
Volume, %	4	6	8	8.5	10	7	18	17	20	19	29	33	29	39
Aged 70 h at 125°C in ASTM Fuel B-change														
Shore A, pts	−17	−14	−23	−20	−23	−24	−22	−20	−24	−25	−20	−24	−21	−29
Tensile, %	−40	−41	−41	−43	−42	−46	−56	−57	−58	−54	−64	−64	−67	−74
Elongation, %	−36	−33	−26	−36	−29	−37	−51	−52	−50	−50	−55	−56	−56	−62
Volume, %	16	26	26	28	27	22	39	38	39	40	53	51	46	61
Aged 70 h at 125°C in ASTM Fuel C-change														
Shore A, pts	−27	−17	−28	−26	−27	−22	−26	−24	−28	−23	−25	−26	−27	−21
Tensile, %	−64	−62	−61	−61	−56	−58	−70	−68	−72	−66	−75	−74	−78	−77
Elongation, %	−57	−55	−48	−53	−46	−49	−65	−64	−63	−64	−69	−67	−67	−67
Volume, %	36	43	46	48	48	44	66	66	68	69	88	88	98	91
Aged 70 h at 125°C in distilled water-change														
Shore A, pts	−4	−7	−1	−5	−5	−2	−2	−3	−3	1	−6	−5	−1	−1
Tensile, %	−3	−3	−9	−14	−1	−7	−10	−10	−10	−1	−11	−7	−4	−7
Elongation, %	−29	−25	−29	−40	−21	−38	−40	−30	−25	−36	−43	−27	−25	−28
Volume, %	10	5	2	10	8	2	3	7	6	1	9	7	4	1

Processing properties in the mixer, mill, extruder, or calender are very important and need to be considered in the selection of the grade of nitrile rubber. This is usually based on personal experience in the factory. The supplier can provide information regarding molecular weight distribution or the best grade to use for a given factory process. Generally, a broad molecular weight distribution provides fast filler acceptance but may lead to bagging on the mill and a tendency to go to the back roll. The opposite is true for narrow molecular weight grades, but these do exhibit better green strength and carry higher loadings of filler and plasticizer. A broad molecular weight NBR is usually better for extrusion and calendering processes, whereas narrow molecular weight polymers are generally better for molding applications.

High temperature, short curing cycles in injection, transfer, or even compression molding operations require so-called "clean elastomers" with low mold-fouling properties and fast cure rates. There is a selection of grades available illustrated in Tables 2.5 and 2.6. Other suppliers such as Polimeri Europa, Nitriflex, and JSR also have developed these types of NBR.

Applications that require low water swell need nitrile elastomers with water insoluble residuals. Some of the newer low mold-fouling grades as well as several long standby grades such as Nipol 1031, 1091-50, 1032, 1092-30, 1092-80, 1094-80 and Perbunan NT grades exhibit this characteristic. Other NBR producers have grades that exhibit low swell in water as well.

Elastomers in contact with food or potable water need to meet FDA and BGA regulations. The base NBR elastomer is physiologically harmless but FDA and BGA approved stabilizers must be in the polymers and these may be found in all Perbunan NT, Perbunan N NS, Krynac standard grades except 19.65, and most Nipol polymers. Specific applications covered by FDA or BGA regulations may need a specific nitrile so it is advisable to contact your supplier for their latest information.

High green strength and resistance to sag or flow during vulcanization in open steam, hot air, or a salt bath may be a requirement and in these instances it is advisable to replace part of the base NBR with a crosslinked nitrile such as those given in Tables 2.7 and 2.8 as well as from other suppliers. Braided or knit hose that needs good green strength to reduce penetration of the fiber or wire may also contain some crosslinked NBR. Normally 5–15 phr is all that is needed. It should be noted that these precrosslinked grades reduce tensile strength, elongation, and compression set and increase modulus of the vulcanizate, hence minimum amounts are recommended.

Very high abrasion resistance and toughness dictate the use of a carboxylated NBR. Tables 2.9 and 2.10 contain grades available from Zeon and Lanxess and are available from other suppliers such as Nitriflex. In addition to sulfur, sulfur donor and peroxide crosslinking, carboxylated NBR is cured by zinc oxide, or more advisably zinc peroxide dispersions. The cure rates are fast and care must be taken to maintain low-process temperatures, since these elastomers generate more internal heat than standard NBR. The zinc peroxide or low-activity zinc oxide should be added during the second pass. Close attention to factory scheduling is advisable as bin life can be short with carboxylated nitrile compounds. The higher the carboxyl content the greater the toughness and abrasion resistance of the compound. The NBR with low levels of carboxyl groups is used for some specialty spinning cot and apron applications. One grade of carboxylated NBR, Nipol NX775 can be used with conventional coated zinc oxide without fear of premature cure or scorch. More detail on carboxylated NBR is given later in this chapter.

Covered rolls, especially those in the lower hardness range, often are designed with specialty types of nitrile. These would include:

1. ACN/BR/IR elastomers, which give better physical properties and resistance to surface hardening because of the inclusion of isoprene in the polymer chain. It is speculated that the isoprene portion tends to slightly revert to balance the hardening of the butadiene units. This polymer is also used in rubber thread that needs high gum strength and resistance to dry cleaning solvents.
2. Plasticizer-extended NBR, as given in Tables 2.14 and 2.15 allow for large amounts of plasticizer to be included in the compound without taking a very long time to incorporate it. This family of elastomers uses a high molecular weight base elastomer into which is added about 50 phr of plasticizer in the latex stage after which it is coagulated, dried, and finished in the normal manner.
3. A 100-60-120 nitrile-PVC-plasticizer blend is made by Excel Polymers LLC as NV866/20. This blend is used for soft rolls with the PVC providing better green strength, abrasion, and chemical resistance for roll applications. Zeon does provide Sivic Z2710, which is a blend of NBR-PVC-DINP in a ratio of 100-66.67-116.67. Other Sivic grades with lower PVC contents and two different amounts of DINP are Z8401 and Z8901 with 40 and 90 phr of plasticizer, respectively.
4. Carboxylated NBR is employed for rolls that encounter extreme abrasion conditions. Special rolls, such as spinning cots, usually contain nitrile with the lower level of carboxyl units. Applications where solvent coatings are applied to metal sheets would be another possible use.
5. Many roll-cover mixing operations use primarily light-colored fillers, hence are reluctant to use loose carbon black. The use of NBR black masterbatches to provide all or part of the reinforcing pigment or color overcomes the difficulty of having black contamination in a light-colored shop.
6. Hardening of highly plasticized rolls in service is a common problem, especially in the printing industry. The use of a liquid NBR, such as those in Table 2.12, to replace part of the plasticizer provides a nonvolatile plasticizer that reduces the tendency for rolls to harden in service. Polymeric plasticizers are another alternate.

Liquid nitrile rubber provided by Zeon Chemicals is illustrated in Table 2.12. JSR also produces a liquid NBR. The Nipol 1312 series comes in two viscosity levels. Due to the tackiness in this form, many plants elect to use these as dry liquid concentrates. Other uses are in PVC plastisols, where lower volatility is an asset. The liquid, carboxylated Nipol DN601 is reactive, hence finds its primary application modifying thermosetting resins.

Nitrile carbon black masterbatches are produced by incorporating the black into the NBR latex then coagulating, drying, and finishing in the conventional way.

The grades available from Zeon Chemicals in Table 2.13 contain several types and amounts of carbon black in four different base polymers. This provides a range of degree of reinforcement and one grade with a high ACN for better fuel and solvent resistance. These materials are provided as a crumb, hence it is important to mass the masterbatch first and sweep up well during the initial mixing stage before other ingredients are added. If this is not done the pellets can fall into the batch at a later stage in mixing and will not disperse, especially after incorporation of the plasticizer.

Nitrile PVC blends have been available for many years produced by either mixing the NBR latex with a PVC emulsion or latex and then coagulating, drying, and finishing the mixture, or by mechanically blending these materials in an internal mixer. The latex blends may or may not be fluxed, whereas the mechanical blends are fluxed during the mixing operation. Table 2.16 provides a listing of the various grades available from Zeon Chemicals. The special property description provides information to assist in selecting the appropriate grade to use. In addition to Zeon Chemicals, JSR, Nantex, Nitriflex, Paratec, Petroflex, and Polimeri make various NBR–PVC blends. Excel Polymers supply a range of mechanically mixed NBR–PVC blends, some based on obsolete Lanxess grades. The plasticizer-extended grades have been described previously in the brief roller section. The inclusion of PVC resin in a nitrile elastomer provides it with improved color capabilities, ozone and weathering resistance, lower coefficient of friction, better abrasion resistance, good resistance to fuel and fuel–alcohol blends, flame resistance, and better electrical resistance. The main applications for NBR–PVC blends are hose and cable jackets, shoe soles, conveyor belts, cellular insulation tubing, and rolls.

There is no other family of elastomers that provides such a myriad of types and grades as that of the nitrile group. This is a great challenge for the rubber chemist to select the ideal NBR elastomer for a given application and process, but there are so many choices that the perfect one may always be found.

2.2.10.2 Fillers

2.2.10.2.1 Carbon Black

Carbon black is the major filler for NBR compounds because of the important properties that it imparts to the compound. These properties include tensile and tear strength, abrasion resistance, chemical resistance, resilience and low compression set with larger particle size blacks such as MT N990, and good processing properties. The fine particle size carbon blacks, such as N231 and N330, provide higher modulus, hardness, tear and abrasion resistance, and tensile strength. The very fine blacks do increase viscosity and are best used with low-viscosity NBR and efficient plasticizers. The most popular grades are N660, N550, N774, N762, and N990, which result in good physical properties, easy processing compounds, and reasonable cost.

The base hardness of nitrile rubber with an ACN between 28% and 38% is 50 Shore A and a good rule of thumb is that to increase Shore A hardness by one point, add 2.5 phr of N330, 2.2 phr N231, 2.7 phr N550, 2.75 phr N650 and N326, 3.25 phr N660, 3.65 N774, 3.8 phr N762, and 5 phr N990.

Consideration should be given to the family of Spheron four digit series of carbon blacks from Cabot Corporation developed primarily for the mechanical goods industry. Grades are available that are designed for specific applications.

Special types of carbon black find a use in certain applications. Anthracite coal dust, Keystone and Austin Black, exhibits good chemical resistance, resilience, and low specific gravity for dynamic seals. Synthetic graphite is also used for this type of application for the same reason and it also reduces the coefficient of friction for better wear resistance.

Shawinigan Acetylene Black has an extremely small particle size, which imparts excellent electrical and thermal conductivity, but is fluffy and care must be taken to obtain good dispersion. The specific gravity is 1.95 compared to 1.80 for furnace types of carbon black. Another more commonly used electrical conductive black is N472 (XCF), Vulcan XC72 from Cabot Corporation, which is easier to incorporate and process.

2.2.10.2.2   Non-Black Fillers

Light-colored fillers do play an important role in NBR compounding, especially in the roller business. The main types of non-black fillers used with nitrile rubber are silica, silicate, clay, talc, and calcium carbonate. In addition there are many specialty light-colored fillers such as diatomaceous earth, barium sulfate, titanium dioxide, aluminum trioxide, antimony trioxide, magnesium hydroxide, and zinc oxide to name a few.

2.2.10.2.2.1   Silica

As with carbon black, the fine particle size fillers such as fumed and precipitated silica do provide high tear, tensile, hardness, and abrasion resistance, but these also increase compound viscosity, making them more difficult to incorporate and process. The hardness imparted is about the same as a fine particle carbon black such as N330. Low-viscosity elastomers and the inclusion of plasticizers and process aids help in processing. The high surface area and reactivity of these fillers also make it very advisable to include a silane coagent, tri-ethanolamine, or polyethylene glycol; added with the fillers to prevent absorption of the curatives or reaction with zinc oxide. For that reason the zinc oxide should be added after the filler is incorporated and the accelerators at the end of the mix or in a second pass. Fumed silica such as Cab-O-Sil from Cabot Corporation is very light and fluffy, making it the most difficult one to incorporate, but it does provide the highest reinforcement and abrasion resistance. There are hydrophobic grades, Cab-O-Sil TS-720 and Tullanox 500, which are treated with an organosilicon that also lessens the absorption of zinc oxide and accelerators, but these still should be added AFTER the silica is incorporated.

Precipitated silica is generally preferred to the fumed grades due to the somewhat better processing properties. There are various grades with a range of particle sizes, bulk densities as well as some with various surface treatments. The usual grade, HiSil 233 has a particle size of 0.022 μm and surface area of 143 m^2/g as are the HiSil 210 pelletized version and HiSil 243LD compacted form. The compacted and larger particle size versions, such as HiSil 532EP and Silene 732D, are faster to incorporate and process at some sacrifice in physical properties. Organosilicon surface treated hydrophobic versions, such as Tullanox, provide better water resistance and lessen the absorption of accelerators and zinc oxide (these still need to be added after the silica is incorporated).

Ground natural occurring silica is a semi-reinforcing filler that may be added in fairly large amounts without increasing viscosity and hardness excessively. Min-U-Sil and Tamsil are available in various particle sizes ranging from 8 to 25 μm.

2.2.10.2.2.2 Silicate

The medium particle non-black silicates, such as Zeolex 23, a synthetic sodium aluminum silicate, Hydrex N sodium magnesium aluminum silicate, and Pyrax hydrated aluminum silicate, are widely used in NBR compounds as they process well, are lower cost, and can be colored. The chemical and water resistance of these materials is poorer and needs to be a consideration in designing a formulation. These fillers impart hardness like N762 carbon black, about 3.5 phr for every Shore A point increase. The inclusion of silane, tri-ethanolamine, or polyethylene glycol is advisable with these fillers as well. If low water swell or chemical resistance is a factor a silane needs to be added or a pretreated version used.

Wollastonite, a naturally occurring calcium meta-silicate, may be used in large amounts as it is only semi-reinforcing and does not increase compound viscosity or hardness to the same extent as synthetic silicates.

2.2.10.2.2.3 Clay

There are many versions of clays (hydrated aluminum silicate) available including soft clay, hard clay, air floated, water washed, calcined, and surface-treated versions. Surface treatments include stearates, mercapto-silane, vinyl silane, and amino silane. Select the silane treatment that is compatible with the cure system to be used. Clay may be used in large quantities, is inexpensive, easy to incorporate, calender or extrude and provides some reinforcement to NBR compounds.

The water-fractionated versions have a more neutral pH, thus do not retard cure rates as much as air floated grades and are available in particle sizes varying from .02 to 4.8 μm.

The calcined clays such as Whitetex are neutral and have lower water absorption and better electrical properties, but slightly lower physical properties. Vinyl silane surface treated calcined clays such as Translink 37, amino-silane treated calcined Translink 445 have even better water resistance and electrical properties.

2.2.10.2.2.4 Whiting (Calcium Carbonate)

Ground calcium carbonate does not really reinforce but is the least expensive filler and may be used in large quantities and is invariably combined with a semi- or reinforcing filler. The ground limestone is available in various particle sizes and some grades have a stearate surface treatment for easier incorporation.

Precipitated calcium carbonate such as Multifex MM or Hakeunka CC do semi-reinforce NBR compounds, do not affect cure rates, and give good compression set properties. These are combined with other more reinforcing fillers to provide good physical properties.

Another special form of calcium carbonate is ground oyster shells that have a unique platy structure which assists in giving smooth calender sheets and extrusions.

2.2.10.2.2.5 Talc

Talc is magnesium calcium silicate and may be in a rough spherical form, such as Nytal, or a platy configuration known as Mistron Vapor. The platy talc assists in calendering and extrusion operations and provides good reinforcement. Talc is inert, hence finds use in electrical resistant applications and where good water resistance is needed. Various particle size talcs are available and the platy talc may be untreated or contain a surface treatment, Mistron ZSC has zinc stearates, Mistron CB contains a mercapto silane for sulfur cures, and Mistron PXL has a vinyl silane for peroxide cures.

2.2.10.2.2.6 Flame-Retardant Fillers

There are several fillers that impart flame-retardant properties to rubber compounds. These are magnesium hydroxide, Versamag or Zerogen 50; magnesium carbonate, Elastocarb or Magcarb L; zinc borate, Firebrake ZB; alumina trihydrate, Micral; and antimony oxide, Fireshield H, L, UF or Synpro-Ware R321. These are used primarily with NBR/PVC blends with chlorinated paraffins and/or phosphate ester plasticizers and mineral fillers for flame-resistant applications.

2.2.10.2.2.7 Diatomaceous Earth

Celite 350 is an inert filler with the ability to absorb large quantities of plasticizer. It is often used in rolls with large amounts of plasticizer to make for faster incorporation of the liquid and provide a drier surface. It is also used to produce DLC versions of plasticizers.

2.2.10.2.2.8 Miscellaneous Non-Black Fillers

1. Barium Sulfate, either the standard form, Barytes, or the precipitated version, Blanc Fixe, has specific gravity of >4.0 and is very inert, providing low water swell to compounds.
2. Titanium dioxide is used as a white pigment.
3. Zinc oxide, although an activator and crosslinking agent in chlorinated elastomers, also serves as an inert filler providing very good thermal conductivity.
4. Mica, alumina potassium silicate, is a platy, naturally occurring filler that enhances the calendering and extrusion characteristics of a compound. It may be used as a surface treatment to give some lubricity to the part.

2.2.10.3 Plasticizers

The polar nature of nitrile rubber dictates that polar plasticizers need to be used with this elastomer. Highly aromatic mineral oils can be used in limited quantities with NBR having an ACN content under 28% as a means of reducing cost. The higher the ACN level of the nitrile rubber, the less compatible the plasticizer, hence only smaller amounts can be incorporated without "bleeding" or exuding to the surface of the vulcanizate. It is advisable to combine two or even three types of ester plasticizers in an NBR compound to ensure their compatibility. Use more polar plasticizers with the higher ACN elastomers. It is recommended that Chapter 14 on plasticizers be reviewed.

When selecting a plasticizer to use with nitrile elastomers it is important to consider the environment that will be encountered and the expected service life. The most widely used plasticizer because of the relatively low price is DOP, dioctyl phthalate, and the similar DIOP, diisooctyl phthalate, and DIDP, diisodecyl phthalate; however, these are suspected carcinogens so it is recommended that the current EPA regulations be investigated.

Low-temperature applications will require a good low-temperature plasticizer, but a blend may be necessary to balance high- and low-temperature properties. Moreover, check the solubility in any liquid media that will be encountered as extraction could be a problem and again a blend of plasticizers is often necessary. A low volatility plasticizer should be selected for improved heat aging properties.

Unfortunately, it is not possible to have a single plasticizer that is good for all conditions and compromises must be made. See Table 2.18 for information on the volatility of the plasticizer, which will directly affect the high-temperature properties [8]. Moreover, examine the solubility properties in various oils, fuels, alcohol, water, or other fluids to be encountered as these can and will extract the plasticizer and lead to poor aging properties.

Table 2.19 has a comparison of several commonly used plasticizers in medium ACN rubbers which will illustrate the differences in low-temperature flexibility, aging properties in air, ASTM #1 Oil, ASTM #3 Oil, and ASTM Fuel C [9–11]. The weight and hardness change after air aging is an indication of influence of the volatility of the plasticizer on this property. An examination of the volume change after aging in ASTM #3 Oil illustrates the effect of the solubility of the plasticizer in the aromatic oil swell. This factor needs to be taken into account when selecting a plasticizer to meet specified volume swell properties for the application considered, but note that if the part is subjected to noncontinuous exposure to oil the part may shrink when oil is not contacted. This is even more evident upon exposure to ASTM Fuel C when the original volume swell is compared with the change after dry out [8].

2.2.10.3.1 Low-Temperature Flexibility
If low-temperature flexibility is the major concern and resistance to aging is not too important, it is suggested that plasticizers such as TEG-CC, TEG-2EH, and TEG-C9 are good, but these have limited compatibility. Other common ones to consider are DOA, DBS, DBEEA, DBEEF, and TBEP. These are all very soluble in fuels, alcohol, and acetone. If resistance to oil is a requirement as well as low-temperature properties, then consider TEG-2EH, glycol ether glutarate (TP-759 or Merrol 4425), DBEEA, DBEEF, and DBEES. The better ones to consider providing good air aging resistance would be DOZ, DOS, TEG-C9, DIDA, DBEES, and glycol ether glutarate.

2.2.10.3.2 High-Temperature Resistance
Polymeric plasticizers and those with high molecular weight work best under high-temperature conditions, but often there is limited compatibility with these. Polyester sebacate (Paraplex G-25), adipate (Plasthall P-650, P-643 and HA7A and Paraplex G-40, G-50, and G-54), glutarate (Plasthall P-550, P-7046, and P-7092) are suggested, depending on the other properties that must be met as well. The polyester adipate and glutarate families are insoluble or partially soluble in aliphatic fuel and partially soluble in acetone. Polyethylene glycol based ones are soluble in most fuels and acetone. Liquid NBR, Nipol 1312, is nonvolatile and resistant to extraction by fuels and oils.

TABLE 2.18
Plasticizer Properties

Generic Name	Abbreviation	Trade Name	Typical Values			Solubility					
			Avg. Mol. Weight	S. G.	Volatility 24 h at 155°C % Loss	Hexane	Toluene	Ethanol	Acetone	Water	ASTM Oil#1
Dibutoxyethoxyethyl adipate	DBEEA	Plasthall 226	434	1.012	33.3	S	S	S	S	I	PS
Diisodecyl adipate	DIDA	Plasthall DIDA	426	0.917	6.0	S	S	S	S	I	S
Dioctyl adipate	DOA	Plasthall DOA	371	0.926	24.1	S	S	S	S	I	S
Polyester adipate	—	Plasthall HA7A	3000	1.15	1.4	I	S	PS	PS	I	I
Polyester adipate	—	Plasthall P-643	2000	1.074	2.6	I	S	PS	PS	I	PS
Polyester adipate	—	Plasthall P-650	1500	1.058	2.5	I	S	PS	PS	I	PS
Polyester adipate	—	Plasthall P-670	1300	1.08	2.4	I	S	PS	PS	I	PS
Dioctyl azelate	DOZ	Plasthall DOZ	412	0.918	8.8	S	S	S	S	I	S
Epoxidized soybean oil	ESO	Paraplex G-62	1000	0.993	1.0	PS	S	S	S	I	PS
Butyl acetyl ricinoleate	—	Flexricin P-6	388	0.927	12.6	S	S	S	S	I	S
Dialkyl diether glutarate	—	Plasthall 7050	450	1.061	27.0	I	S	S	S	PS	S
Polyester glutarate	—	Plasthall P-550	1750	1.071	3.4	I	S	PS	PS	I	PS
Polyester glutarate	—	Plasthall P-7046	4200	1.107	3.0	I	S	PS	PS	I	PS
Polyester glutarate	—	Plasthall P-7092	5000	1.109	2.5	I	S	PS	PS	I	PS
Glycol ether glutarate	—	Merrox 4425	525	1.04		S	S	S	PS	I	PS

Plasticizer	Abbrev.	Trade name		Sp. gr.										
Butoxyethyl oleate	BEO	Plasthall 325	379	0.89	11.7	S	S	S	S	S	S	S	I	S
Polyester sebacate	—	Paraplex G-25	10.000	1.057	0.6	S	S	PS	PS	S	S	S	I	PS
Polyester adipate	—	Paraplex G-40	6000	1.15	0.7	I	S	PS	PS	PS	I	PS	I	PS
Polyester adipate	—	Paraplex G-50	1500	1.102	2.7	I	S	PS	PS	PS	I	PS	I	PS
Polyester adipate	—	Paraplex G-54	2100	1.094	1.8	I	S	PS	PS	PS	I	PS	I	PS
Dibutyl sebacate	DBS	DBS	316	0.934	60.4	S	S	S	S	S	S	S	I	S
Dioctyl sebacate	DOS	DOS	430	0.913	6.9	S	S	S	S	S	S	S	I	S
Triethylene glycol caprate-caprylate	TEGCC	Plasthall 4141	449	0.968	18.8	S	S	S	S	S	S	S	I	S
Triethylene glycol 2-ethylhexanoate	TEGEH	Merrox 3800	430	0.968	21.6	S	S	S	S	S	S	S	I	PS
Triethylene glycol dipelargonate	TEG-C9	Merrox 9404	458	0.963	18.0	S	S	S	S	S	S	S	I	S
Di (butoxy-ethoxy-ethyl) formal	DBEEF	TP-90B		0.968										
Trioctyl trimellitate	TOTM	Plasthall TOTM	550	0.99	1.5	S	S	I	S	S	S	S	I	S
Di 2-ethylhexyl phthalate	DOP	DOP	390	0.983	9.2[a]	S	S	S	S	—	S	S	I	S
Di n-butyl phthalate	DBP	DBP	278	1.045	24.0	S	S	S	S	S	S	—	I	S
Diisodecyl phthalate	DIDP	DIDP	446	0.965	3.8[a]	S	S	PS	PS	S	S	—	I	—
Tricresyl phosphate	TCP	TCP	368	1.16	—	S	S	S	S	S	S	S	I	S
Tributoxyethyl phosphate	TBEP	KP-140	392	1.02	—	S	S	S	S	S	S	S	I	S
Triaryl phosphate	TAP	Reofos 65	368	1.165	7.5[a]	S	S	S	S	S	S	S	I	S
Ether thioether	—	Vulkanol OT		0.965										
Dibutyl methylene bis-thioglycolate	—	Vulkanol 88		1.14	15[a]	S	S	S	S	S	S	S	I	S
Coumarone indene resin	—	Cumar P 25		1.03		S	S	S	S	S	S	—	—	S

[a] 70 h at 125°C.

TABLE 2.19

Plasticizer Comparison in Medium ACN NBR

	DOP	DOA	83SS	TOTM	TegMer 803	4141	DBEEA	DBEA	TP90B	G31	P 550	Reofos 65
Nipol 1042	100	100	100	100	100	100	—	100	—	100	100	100
Krynac 3450	—	—	—	—	—	—	100	—	100	—	—	—
Zinc oxide	5.0	5.0	5.0	5.0	5.0	5.0	5.0	5.0	5.0	5.0	5.0	5.0
AgeRite Resin D	1.0	1.0	1.0	1.0	1.0	1.0	1.0	1.0	1.0	1.0	1.0	1.0
N 660 carbon black	65	65	65	65	65	65	65	65	65	65	65	65
Stearic acid	1.0	1.0	1.0	1.0	1.0	1.0	1.0	1.0	1.0	1.0	1.0	1.0
DOP	20	—	—	—	—	—	—	—	—	—	—	—
DOA	—	20	—	—	—	—	—	—	—	—	—	—
Plasthall 83SS	—	—	20	—	—	—	—	—	—	—	—	—
TOTM	—	—	—	20	—	—	—	—	—	—	—	—
TegMer 803	—	—	—	—	20	—	—	—	—	—	—	—
Plasthall 4141	—	—	—	—	—	20	—	—	—	—	—	—
Plasthall 226	—	—	—	—	—	—	20	—	—	—	—	—
Plasthall 203	—	—	—	—	—	—	—	20	—	—	—	—
Thiokol TP90B	—	—	—	—	—	—	—	—	20	—	—	—
Paraplex G-31	—	—	—	—	—	—	—	—	—	20	—	—
Plasthall P-550	—	—	—	—	—	—	—	—	—	—	20	—
Reofos 65	—	—	—	—	—	—	—	—	—	—	—	20
Spider Sulfur	0.4	0.4	0.4	0.4	0.4	0.4	0.4	0.4	0.4	0.4	0.4	0.4
MBTS	2.0	2.0	2.0	2.0	2.0	2.0	2.0	2.0	2.0	2.0	2.0	2.0
ZDMDC	1.5	1.5	1.5	1.5	1.5	1.5	1.5	1.5	1.5	1.5	1.5	1.5
TOTAL	195.9	195.9	195.9	195.9	195.9	195.9	195.9	195.9	195.9	195.9	195.9	195.9
Mooney scorch at 121°C												
Minimum viscosity	28	26	26	30	26	26	29	44	26	48.5	44	33
T_5, minutes	13.1	13.3	8.5	12.3	8.0	9.3	8.8	14	6.5	14.3	14	11
Rheometer at 170°C												
ML, dN · m	2.26	2.03	2.26	2.26	.81	1.58	2.26	2.26	2.37	3.28	3.16	2.71

MH, dN·m	5.99	30	6.44	5.76	5.65	5.42	5.88	6.33	6.55	7.35	6.78	6.33
Ts2, minutes	2.3	2.4	2.3	2.3	1.8	1.8	1.8	2.5	1.9	1.9	2.2	2.1
T'c90, minutes	4.3	4.5	3.8	4.0	3.3	3.5	3.7	4.3	4.1	3.8	4.3	4.1
Original properties												
Shore A, pts	59	57	60	58	59	59	58	62	59	65	63	60
300% Modulus, MPa	5.5	4.9	7.4	5.4	6.3	6.0	5.4	10.7	7.4	11.0	11.2	6.0
Tensile, MPa	13.1	12.8	13.4	13.4	12.8	12.4	11.7	14.7	13.0	15.7	14.8	13.8
Elongation, %	600	580	550	600	570	590	570	420	550	440	410	560
Low temperature												
Low-temperature impact, °C	-33	-42	-39	-30	-37	-42	-40	-42	-44	-30	-33	-23
Aged in air 70 h at 125°C-change												
Shore A, pts	+19	+25	+9	+8	+19	+15	+15	+25	+20	+9	+8	+16
Tensile, %	+23	+32	+8	+10	+32	+28	+12	-14	+35	+2	+3	+21
Elongation, %	-46	-48	-44	-38	-49	-46	-43	-86	-51	-39	-37	-41
Weight, %	-8.9	-11	-2.2	-1.0	-10	-6.1	-4.1	-11	-11	-2.2	-1.8	-5.0
Aged 70 h at 125°C in ASTM # 1 Oil-change												
Shore A, pts	+13	+19	+13	+13	+14	+14	+10	+12	+14	+6	+6	+10
Tensile, %	+22	+26	+15	+10	+30	+25	+32	+19	+27	+5	+7	+18
Elongation, %	-41	-38	-38	-40	-37	-41	-33	-40	-42	-36	-32	-32
Volume, %	-13	-14	-10	-13	-13	-13	-12	-12	-13	-5.1	-4.2	-9.4
Aged 70 h at 125°C in ASTM # 3 Oil-change												
Shore A, pts	+1	+7	+1	+2	+2	+3	+2	+2	+1	-4	-6	0
Tensile, %	+15	+3	+8	+8	+19	+22	+12	+11	+16	-2	-6	0
Elongation, %	-23	-28	-18	-27	-23	-22	-19	-14	-22	-20	-24	-29
Volume, %	+92	+78	+1.3	+1.4	+90	+90	+2.6	+4.1	+1.2	+9.3	+10	+3.7
Aged 70 h at 23°C in ASTM Fuel C-change												
Shore A, pts							-21	-22	-18	-25	-23	
Tensile, %							-59	-55	-54	-58	-60	
Elongation, %							-51	-55	-56	-57	-59	
Volume, %							+40	+43	+38	+52	+52	
Volume 22 h dry out at 70°C							-13	-13	-13	-8.1	-10	

2.2.10.3.3   Fuel Resistance

Glyceryl mono- or diacetate is insoluble in fuels and oils but it may affect cure rate and other properties. The polyester adipate and glutarate are good for aliphatic and kerosene but not for aromatic fuels and solvents. Again the liquid NBR is good in all fuels and oils. It should be noted that in meeting some specification requirements it may be necessary to use a more extractable plasticizer for low-swell test results but a combination of low- and high- extractability types is recommended to attain proper performance in the application.

2.2.10.3.4   Water and Glycol Resistance

Almost all plasticizers are resistant to extraction by water or glycol. Exceptions to this generality are DBEEA and glycol ether glutarate; oleate, laurate, and polyethylene glycol based plasticizers.

2.2.10.3.5   Flame Resistance

See NBR/PVC blends.

2.2.10.3.6   Other Factors

Always attempt to obtain as much information as possible about temperatures, fluids, pressures, bonding, abrasion, coefficient of friction, and a myriad of other factors in order to develop the best formulation. Be sure that the plasticizers are compatible with the NBR at the levels needed, using two or even three different types if necessary to prevent bleeding or exuding that may interfere with bonding or affect the coefficient of friction. Additionally, consider the possible affect of the plasticizer on any plastic (crazing of acrylic or migration into PVC or ABS) or metal or wood that will be in contact with the part.

2.2.10.4   Curing of Standard NBR

There are basically six types of crosslinking mechanisms for NBR elastomers, very high sulfur for a special ebonite, normal sulfur, semi-EV (semi-efficient vulcanization or low sulfur systems), EV (efficient vulcanization with sulfur donors alone), peroxide and zinc oxide/peroxide for carboxylated nitrile. The innumerable combinations that can and are used are as numerous as there are rubber chemists in the world, but with care they all work in any given application.

2.2.10.4.1   High Sulfur Cures

The use of up to 35 phr sulfur with a butraldehyde-monobutylamine such as (Vanax 833) type accelerator may be used to produce a higher temperature resistant ebonite, primarily for submarine battery cases. This approach has virtually disappeared having been replaced with either plastics or nitrile with higher amounts of an acrylic monomer and a peroxide cure.

2.2.10.4.2   Sulfur Cures in NBR

A sulfur cure is used in many applications where heat resistance and low compression set are not major factors. Sulfur cures are also best for dynamic applications at moderate temperatures. An important thing to remember with sulfur in NBR is that it is more difficult to disperse so it should be added at the beginning of the mixing cycle and be either as a sulfur dispersion or as an oil-treated grade such as Spider Sulfur. If sulfur is not properly dispersed the surface of the cured part may have dimples or an "orange peel" effect. In colored compounds the surface may have brown spots and a higher microhardness at points where agglomerations of sulfur occur. Table 2.20 provides a number of sulfur-curing systems, but it is by no means complete [12]. The selection of the accelerator will determine the scorch safety, cure rate, and the modulus and elongation of the vulcanizate; hence, the processing and curing conditions as well as the service conditions need to be considered. The tan delta data properties are an indication of the dynamic characteristics of the compound.

2.2.10.4.3   Semi-EV Cures in NBR
Sulfur Donor or Semi-EV cures for nitrile rubber are given in Table 2.21 [13]. When selecting a cure system to obtain good heat and compression set resistance, this type of cure should be selected. Process safety and cure rate need to be considered in order to satisfy factory conditions and economics. Although dynamic properties are not as good as with normal sulfur cures, some combinations such as sulfur 0.5,

TABLE 2.20
Sulfur Cure Systems in NBR

N051-005		1	2	3	5
Nipol DN3380		100.00	100.00	100.00	100.00
Kadox 920C		5.00	5.00	5.00	5.00
Stearic acid		1.00	1.00	1.00	1.00
N774 Black		50.00	50.00	50.00	50.00
DOP		5.00	5.00	5.00	5.00
AgeRite Resin D		2.00	2.00	2.00	2.00
Spider Sulfur		1.50	1.50	1.50	1.50
CBTS		0.75	—	—	—
MBTS		—	1.50	1.50	1.50
Ethyl Zimate		—	—	0.20	—
Methyl Zimate		—	—	—	0.20
Total		165.25	166.00	166.20	166.20
Processing properties					
Mooney scorch: ML (1 + 30) at 125°C					
Viscosity	Minimum	42.1	43.2	46.4	47.1
t5	Minutes	19.0	15.2	6.6	5.5
MDR: Microdie, 100 cpm, 0.5° arc at 170°C					
ML	dN · m	1.07	1.05	1.11	1.13
MH	dN · m	10.64	10.48	11.97	12.45
Ts2	Minutes	1.57	1.26	0.74	0.66
T'90	Minutes	2.99	3.22	1.24	1.04
Tan Delta at T'90		0.071	0.075	0.062	0.058

(*continued*)

TABLE 2.20   (Continued)
Sulfur Cure Systems in NBR

N051-005		1	2	3	5
		Vulcanized properties			
Originals: Cured at 170°C					
Hardness (A)	pts	64	63	65	66
Stress 100%	MPa	2.1	2.0	2.4	2.6
Stress 200%	MPa	4.6	4.3	5.7	6.2
Tensile	MPa	24.9	20.2	20.3	21.8
Elongation	%	645	585	497	506
TEAR: Die C, W/G					
23°C	kN/m	52.9	53.6	56.0	46.1
Compression set: Method B, buttons					
22 h at 100°C	% Set	55.5	45.6	38.5	37.8
70 h at 100°C	% Set	67.5	63.9	52.2	54.1
Aged in air 70 h at 100°C-change					
Hard	pts	3	6	6	5
Tensile	%	−12	14	0	−7
Elongation	%	−33	−28	−32	−34
Aged in air 70 h at 125°C-change					
Hardness	pts	10	12	9	9
Tensile	%	−21	14	−6	−9
Elongation	%	−62	−52	−50	−53

MBTS 2.0, TMTD 0.75, and TETD 0.75 will match resilience and provide much improved compression set resistance.

2.2.10.4.4   EV Efficient Vulcanization Cures (Sulfur-Less)

Examples of sulfur-less or efficient vulcanization cure systems are shown in Table 2.22 [14]. This is not a complete list of combinations that serve as EV cures, but it does provide several suggested systems. Please note that if the amount of TMTD or TETD exceeds 1.25 phr that blooming may occur, hence a combination of sulfur donors should be considered. Again selection of a suitable combination will depend on the required process safety, cure rate, physical and aged properties, and economics. Again resilience may be sacrificed but certain combinations will give good compression set and resilience properties, such as TMTD 2.0, DTDM 2.0 (note that blooming may occur), or MBT 0.5 or TMTD 0.7, TETD 0.7, DTDM 2.0, and DPTT 0.7.

2.2.10.4.5   Peroxide Cure Systems

In nitrile rubber, peroxide cure systems provide the best in heat and compression set resistance compared with semi-EV and Efficient Vulcanization combinations. There are quite a few peroxides available as given in Table 2.23, but the most commonly used ones in NBR are Di-Cup 40KE (Trigonox BC-40K), Peroximon F40 (Varox 802-40KE), and Varox 130-XL (Trigonox 145-45B) [15]. Peroxides function best

TABLE 2.21
Semi-EV Sulfur Donor Cures in NBR

N051-015	1	2	3	5	11	1A	2A	3A	4A	5A
Nipol DN3380	100.00	100.00	100.00	100.00	100.00	100.00	100.00	100.00	100.00	100.00
Kadox 920C	5.00	5.00	5.00	5.00	5.00	5.00	5.00	5.00	5.00	5.00
Stearic acid	1.00	1.00	1.00	1.00	1.00	1.00	1.00	1.00	1.00	1.00
N774 Black	50.00	50.00	50.00	50.00	50.00	50.00	50.00	50.00	50.00	50.00
DOP	5.00	5.00	5.00	5.00	5.00	5.00	5.00	5.00	5.00	5.00
AgeRite Resin D	2.00	2.00	2.00	2.00	2.00	2.00	2.00	2.00	2.00	2.00
Spider Sulfur	0.50	0.50	0.50	0.50	0.50	0.50	0.50	0.50	0.50	0.50
Methyl Tuads	2.00	2.00	2.00	—	—	3.00	—	—	—	—
Ethyl Tuads	—	—	—	2.50	—	—	3.00	—	—	—
Isobutyl Tuads	—	—	—	—	3.40	—	—	3.00	—	—
Butyl Tuads	—	—	—	—	—	—	—	—	3.00	—
Benzyl Tuads	—	—	—	—	—	—	—	—	3.00	3.00
CBTS	1.00	—	—	—	—	—	—	—	—	—
OBTS	—	0.95	—	0.95	0.95	—	—	—	—	—
BBTS	—	—	0.90	—	—	—	—	—	—	—
Total	166.50	166.45	166.40	166.95	167.85	166.50	166.50	166.50	169.50	166.50
Processing properties										
Mooney scorch: ML (1 + 30) at 125°C										
Viscosity Minimum	46.1	45.8	44.3	42.1	40.7	45.2	43.3	42.5	42.0	43.1
t5 Minutes	7.7	7.7	7.9	12.2	13.8	6.6	11.6	14.1	11.0	14.0
MDR: Microdie, 100 cpm, 0.5° arc at 170°C										
ML dN·m	1.04	1.05	1.03	1.01	0.96	1.05	1.07	1.04	1.02	1.07
MH dN·m	13.88	14.34	14.05	13.91	13.20	14.86	12.28	10.22	10.47	8.38

(continued)

TABLE 2.21 (Continued)
Semi-EV Sulfur Donor Cures in NBR

N051-015		1	2	3	5	11	1A	2A	3A	4A	5A
MDR: Microdie, 100 cpm, 0.5° arc at 170°C											
ts2	Minutes	0.88	0.89	0.94	1.07	1.29	0.83	1.09	1.33	1.15	1.33
t'90	Minutes	2.05	2.51	2.30	2.68	3.45	2.31	2.47	2.72	2.30	2.73
Tan Delta at T'90		0.035	0.032	0.031	0.033	0.035	0.031	0.044	0.069	0.071	0.123
Vulcanized properties—cured											
Hardness (A)	pts	65	65	65	64	63	66	64	61	62	61
Stress 100%	MPa	2.8	2.9	2.7	2.7	2.4	3.1	2.5	2.0	2.0	1.8
Stress 200%	MPa	7.2	7.7	7.1	6.9	6.1	8.3	6.5	4.4	4.7	3.6
Tensile	MPa	17.3	17.7	19.2	20.7	21.4	20.1	22.0	24.2	24.5	16.0
Elongation	%	375	381	410	457	500	396	490	653	660	567
TEAR: Die C, W/G											
23°C	kN/m	50.6	45.7	45.0	43.6	48.7	43.1	47.1	51.5	54.6	50.4
Compression set: Method B											
22 h at 100°C	% Set	8.4	7.5	7.2	7.4	11.6	5.2	8.4	13.8	17.7	27.1
70 h at 100°C	% Set	12.8	10.4	11.0	11.3	14.5	7.9	13.3	20.7	24.0	36.7
Aged in air 70 h at 100°C-change											
Hardness	pts	5	5	5	5	4	4	4	5	4	3
Tensile	%	15	11	−15	4	−3	−3	−10	−4	−6	2
Elongation	%	−13	−17	−35	−15	−26	−19	−29	−21	−27	−25
Aged in air 70 h at 125°C-change											
Hardness	pts	8	8	7	7	8	7	8	9	9	8
Tensile	%	27	12	11	11	6	7	9	0	−3	9
Elongation	%	−21	−33	−34	−26	−34	−27	−30	−39	−40	−39

TABLE 2.22
EV Efficient Vulcanization Cures

N051-003		4	7	8	9	10	11	12
Nipol DN3380		100.00	100.00	100.00	100.00	100.00	100.00	100.00
Kadox 920C		5.00	5.00	5.00	5.00	5.00	5.00	5.00
Stearic acid		1.00	1.00	1.00	1.00	1.00	1.00	1.00
N774 Black		50.00	50.00	50.00	50.00	50.00	50.00	50.00
DOP		5.00	5.00	5.00	5.00	5.00	5.00	5.00
AgeRite Resin D		2.00	2.00	2.00	2.00	2.00	2.00	2.00
TMTM		—	—	0.30	1.50	—	—	—
MBTS		—	1.00	2.00	1.50	—	—	4.00
TMTD		3.50	3.50	1.00	0.25	1.50	—	3.00
CBTS		—	—	—	—	—	1.00	—
Sulfasan R		—	—	—	—	1.50	—	—
TETD		—	—	—	—	—	3.50	—
Total		166.50	167.50	166.30	166.25	166.00	167.50	170.00
				Processing properties				
Mooney scorch ML at 125°C								
Viscosity	Minimum	45.5	43.9	41.7	39.8	44.5	42.2	39.3
t5	Minutes	6.6	9.3	>30.0	>30.0	13.1	14.5	24.8
Rheometer MDR at 170°C, 0.5° arc, 100 cpm								
ML	dN·m	1.04	1.03	1.04	1.02	1.07	1.01	0.95
MH	dN·m	9.60	8.60	4.85	3.02	11.62	8.44	8.00
ts2	Minutes	0.98	1.73	3.78	5.51	2.58	1.86	2.81
t'90	Minutes	5.44	4.87	5.58	5.66	6.73	3.97	4.63
Tan Delta at T'90		0.076	0.091	0.217	0.354	0.049	0.089	0.096

(continued)

TABLE 2.22 (Continued)
EV Efficient Vulcanization Cures

N051-003		4	7	8	9	10	11	12
				Vulcanized properties				
Hardness (A)	pts	61	60	57	54	63	59	59
Stress 100%	MPa	1.9	1.8	1.3	1.0	2.3	1.7	1.7
Stress 200%	MPa	4.2	3.8	1.7	1.1	5.9	3.5	3.1
Tensile	MPa	25.0	25.3	14.7	3.6	22.5	24.8	26.6
Elongation	%	688	735	986	1139	541	749	810
Tear Die C	kN/m	46.1	53.6	51.7	29.6	53.1	51.0	58.0
Compression set B								
22 h at 100°C	% Set	12.2	19.7	54.6	74.0	10.9	19.3	22.8
70 h at 100°C	% Set	17.4	26.1	63.9	88.6	16.4	24.8	29.2
Aged 70 h at 100°C-change								
Hardness	pts	3	3	3	4	4	4	3
Tensile	%	7	4	25	137	−20	2	−30
Elongation	%	−14	−10	−8	−11	−37	−15	−52
Aged 70 h at 125°C-change								
Hardness	pts	5	8	9	7	8	10	8
Tensile	%	−1	0	46	382	−13	1	−2
Elongation	%1	−28	−34	−28	−32	−44	−31	−34

TABLE 2.23
Peroxide Curatives

Chemical Name	Trade Names	Minimum Curing Temperature (°C)	Suggested phr Peroxide 40% Active	Half Life Time versus Temperature °C			
				1 s	1 min	6 min	60 min
1,1-di(t-butylperoxy)-3,3,5- trimethyl-cyclohexane	Varox 231XL Trigonox 29-40B	150	2.6-4.5	198.8	152.8	135.3	114.9
n-butyl 4,4-di (t-butylperoxy) valerate	Varox 230XL Trigonox 17-40B	160	2.9-5.0	220.1	169.8	150.9	128.8
Dicumyl peroxide	Varox DCP-40KE Di-Cup 40KE Trigonox BC-40K	170	2.4-4.1	227.9	177.9	159.0	137.0
a,a'-bis (t-butylperoxy)- diisopropylbenzene	Varox 802-40KE Vul-Cup 40KE Peroximon F40 Perkadox 14-40K	180	1.5-2.5	231.3	180.7	161.6	139.4
2,5-dimethyl-2,5-di (t-butylperoxy)-3- hexane	Varox DBPH-50 Triganox 101-45B	180	2.3-3.9	231.6	181.4	162.4	140.3
2,5-dimethyl-2,5-di (t-butylperoxy)-3- hexyne	Varox 130XL Trigonox 145-45B	180	2.3-4.0	245.9	174.1	175.6	151.8

TABLE 2.24

Performance of Peroxide versus Sulfur Cures in Nitrile Rubber

Formulation	Sulfur Cure	Vul-Cup Cure	Di-Cup Cure
Nipol 1032	100	100	100
HAF N 330	50	50	50
Zinc oxide	5	5	5
Antioxidant	1	1	1
Stearic acid	1	—	—
MBTS	1.5	—	—
Sulfur	1.5	—	—
Vul-Cup 40KE	—	1.6	—
Di-Cup 40KE	—	—	2.5
Cured	30 min at 154°C	25 min at 171°C	25 min at 166°C
Shore A, pts	65	64	65
Modulus at 100%, MPa	3.1	3.0	3.2
Tensile, MPa	19.3	21.4	21.2
Elongation, %	380	395	345
Compression set B			
70 h at 100°C, %	66	20	20
70 h at 125°C, %	82	30	30
Aged 70 h at 125°C in air-change			
Shore A, pts	+5	+13	+10
Modulus at 100%, %	—	+186.4	+164.1
Tensile, %	−29.3	−1.6	−2.6
Elongation, %	−78.9	−45.6	−43.5
Aged 70 h. at 125°C in ASTM #3 Oil-change			
Shore A, pts	−5	−11	−9
Modulus at 100%, %	+52.2	+18.2	+5.4
Tensile, %	−68.9	−22.6	−14.6
Elongation, %	−84.2	−39.2	−17.4
Volume swell, %	10.8	16.6	16.6

when cured at temperatures at or higher than those indicated in Table 2.23 [16]. Some peroxides have a distinctive odor, such as Di-Cup 40KE; however, Peroximon F40, Varox 802-40KE, or Varox 130-XL have a minimal odor. A comparison between a sulfur cure and two peroxide cures is given in Table 2.24.

In addition, 1.5–2.5 phr of activators or coagents such as HVA 2, TAC, TAIC are commonly used with peroxides to achieve faster cures and higher modulus. In addition, TMPT, SR365, and other di- and tri-acrylate resins may be used at the 5 phr level. The acrylate resins may be used up to 40 phr in which case the hardness will increase approximately 8 points for every 10 phr of resin. The resin also acts as a plasticizer, which is very beneficial in keeping the compound viscosity low while achieving a high hardness vulcanizate. Table 2.25 illustrates the difference an EV, semi-EV, peroxide, and peroxide with SR-297 cure system.

TABLE 2.25
Comparison of EV, Semi-EV, and Peroxide with and without SR-297

Nipol 1042	100	100	100	100
Spider Sulfur	—	0.5	—	—
Zinc oxide	5	5	5	5
Stearic acid	1	1	1	1
N550 carbon black	40	40	40	40
TMTD	3.5	3.0	—	—
Di-Cup 40C	—	—	4.0	4.0
SR-297	—	—	—	20
Total	149.5	149.5	150.0	170.0
Mooney at 121.1°C				
Viscosity	78	NT	61	25
Original properties—cured 45 min at 154.4°C				
Shore A, pts	67	72	68	84
Modulus at 100%, MPa	1.9	3.8	3.4	13.9
Tensile, MPa	19.4	19.9	20.9	24.3
Elongation, %	580	300	340	170
Compression set B, 70 h at 100°C				
Set, %	29	11	17	14
Aged in air test tube 70 h at 121.1°C-change				
Shore A, pts	5	0	4	6
Tensile, %	7	−8	−23	−14
Elongation, %	−32	−19	−38	−28
Aged in ASTM #3 Oil 701 h at 148.9°C-change				
Shore A, pts	−13	−10	−9	−3
Tensile, %	−47	−44	−30	−19
Elongation, %	−61	−37	−40	−33
Volume, %	29	23	23	18

2.2.10.4.6   Antioxidants and Antiozonants

Protection of nitrile rubber depends upon the application and the environment encountered in service. If ozone is encountered it is suggested that a nitrile/PVC blend with at least 30 phr of PVC be considered as color, nonstaining, and protection will be very good. If the properties of NBR/PVC are not appropriate, then an antiozonant such as IPPD is recommended, realizing that the compound will be staining and discolor. NBC also provides good antiozonant properties but will have a green discoloration. The addition of 1.5–3.0 phr (keep to a minimum due to incompatibility) microcrystalline wax will also be effective in static conditions. A nonstaining antiozonant, Vulkazon AFS/LG is also effective, but the mixing temperature must reach 90°C for proper dispersion.

If the application is such that the rubber will contact food or potable water a stabilizer that is on the list of FDA approved materials such as BHT must be used, so the supplier should be contacted to obtain their recommendation, depending upon the contact time (conveying during the process or sealing during storage), material encountered (aqueous or oil based), temperature and solubility of the protective agent in fluids encountered.

General purpose antioxidants commonly used are octylated diphenylamine types, alkylated-arylated bisphenolic phosphate (also inhibits gel formation in the polymer), polymerized 1,2-dihydro-2,2,4 trimethylquinoline (AgeRite Resin D or Flectol TMQ). This is by no means a complete list as many other types are equally as effective.

The best heat resistance may be obtained with a combination of 1–2 phr zinc 2-mercapto-toluimidiazole (Vanox ZMTI) and 1-2 phr 4,4′-bis(α,-dimethylbenzyl) diphenylamine (Naugard 445) or diphenylamine-acetone reaction product (AgeRite Superflex).

2.2.10.5 Carboxylated Nitrile Rubbers

Carboxylated Nitrile Rubber (XNBR), which contains carboxyl groups ranging from approximately 1% to 8%, offers much higher abrasion, hardness, modulus, tensile, and tear resistance. In order to achieve the best properties the compound must contain zinc oxide or preferably zinc peroxide as well as sulfur or peroxide cure systems. The various grades available in North America are given in Tables 2.9 and 2.10.

The low carboxyl grades, Nipol DN 631 and Krynac X146 and X160 may be cured with a low-activity zinc oxide without scorch problems. The ACN content is 33% for very good oil resistance. These grades are used primarily in spinning cots and aprons and other applications needing better abrasion resistance and physical properties than standard NBR but not the supreme characteristics of the higher carboxyl content nitrile rubbers [17].

The standard grades of carboxylated NBR, Nipol 1072 and 1072CGX; and Krynac X740 and X750 are best cured with zinc peroxide in order to have good processing safety and shelf life. The Nipol 1072CGX is a gel free cement grade, but may be used in high performance mechanical goods. The ACN level is 26%–27% for good oil resistance.

Nipol 1072 × 28 is a crosslinked XNBR, which provides compounds with smoother extrusion and calendering characteristics, and may be used alone or in blends with standard XNBR.

Nipol NX775 is an inhibited grade of XNBR that allows for the use of normal zinc oxide without scorch or shelf life problems. There is a small sacrifice in abrasion and tear resistance and the oil resistance is good.

2.2.10.5.1 Compounding XNBR Rubber

Since these elastomers are used for applications requiring high abrasion resistance and excellent physical properties they are normally compounded with lower levels of plasticizer. The base polymer is tougher than standard NBR and higher hardness and tensile strength is obtained with the base rubber alone, hence lower levels of filler are usually used. In addition a two pass mix is essential for better processing safety with the zinc peroxide and accelerators added in the second pass.

2.2.10.5.1.1 Process Safety

The main concern chemists have when working with XNBR is process safety and shelf life. There are some simple rules to follow, which will help to allay any fears in this regard.

1. Always use a two pass mix holding the accelerators and zinc peroxide until the second pass at a maximum of 115°C.
2. Use zinc peroxide such as SynPro RD-80 or Struktol ZP-1014 as the zinc ion source.
3. If zinc oxide is used, a low surface area one such as Cerox 506 with a surface area of $0.5 \text{ m}^2/\text{g}$ is advisable [18].
4. Include at least 2 phr of stearic acid in the formulation as this acts as a retarder.
5. Select less reinforcing fillers to reduce heat build up during process as the unfilled polymer has a tensile strength of 15–16 MPa. In this regard, keep the stock temperature low during processing.
6. Use a blend of medium and low-viscosity XNBR polymers to achieve low compound viscosity. A blend of the crosslinked XNBR, Nipol 1078×28 will also assist in keeping process temperatures low.
7. Talc is the best selection for process safety in light-colored compounds, but if a silica or clay is used include a silane such as Silane A-189.
8. Selection of Nipol NX-775 will allow for the use of a standard zinc oxide and still maintain good processing properties.

2.2.10.5.1.2 Safe Zinc Oxide Curable XNBR

Nipol NX-775 may be cured with normal zinc oxide and yet exhibit good process safety, shelf life, and processing characteristics. This elastomer provides excellent abrasion resistance and physical properties comparable to other XNBR polymers. A comparison of three different XNBR rubbers is given in Table 2.27 [19].

TABLE 2.26
Zinc Oxide and Zinc Peroxide Study in XNBR

	930	911C	A(RZNO)D-60P	RD-50	ZP-1014
Nipol 1072	100	100	100	100	100
N660 carbon black	40	40	40	40	40
Naugard 445	1	1	1	1	1
Stearic acid	2	2	2	2	2
DOP	5	5	5	5	5
Spider Sulfur	1.5	1.5	1.5	1.5	1.5
TNTM	1.3	1.3	1.3	1.3	1.3
TMTD	0.2	0.2	0.2	0.2	0.2
Kadox 930	5				
Kadox 911C		5			
A(RZNO)D-60P			10		

(continued)

TABLE 2.26   (Continued)
Zinc Oxide and Zinc Peroxide Study in XNBR

	930	911C	A(RZNO)D-60P	RD-50	ZP-1014
SynPro RD-50				10	
ZP-1014					10
Total	156	156	161	161	161
Mooney scorch MS at 125°C—2 weeks after mixing					
Minimum viscosity	41.5	69.3	35.8	11.5	11.1
T_5, minutes	6.8	8.4	6.0	20.1	20.0
Rheometer at 170°C, 3° arc, 100 cpm					
ML, dN·m	8.9	9.8	6.9	5.9	6.2
MH, dN·m	109.6	111.4	101.8	90.4	90
ts2, minutes	2.4	2.4	3.3	4.0	4.0
$T'90$, minutes	5.3	5.3	5.5	7.0	7.8
Physical properties, cured 15 min at 170°C					
Shore A, pts	74	72	78	71	75
Modulus at 300%, MPa	20.7	22.5	18.2	14.1	16.9
Tensile, MPa	26.9	25	24.3	20.4	20.8
Elongation, %	400	340	390	420	370
Tear Die B, kN/m	66.5	68.3	57.8	47.3	54.3
Picco abrasion resistance, 1 Hz, 44 N load, cured 20 min at 170°C					
Picco Index, %	190	200	180	120	150
Aged in air test tube 70 h at 100°C-change					
Shore A, pts	8	14	4	6	8
Tensile, %	0	11	10	1	7
Elongation, %	−29	−18	−20	−30	−22
Aged in ASTM #3 Oil 70 h at 100°C-change					
Shore A, pts	5	9	−6	1	3
Tensile, %	11	4	−7	19	0
Elongation, %	−31	−20	−14	−25	−23
Volume, %	10	10	11	9	8

2.2.10.5.1.3   Curing Systems

The selection of zinc oxide or zinc peroxide dispersion is critical for good scorch safety and bin stability. A comparison of zinc oxide and zinc peroxide dispersions is given in Table 2.26 [17]. As may be seen SynPro RD-50 and Struktol ZP-1014 are the best for scorch resistance at some sacrifice in tensile and tear strength; and abrasion and compression set resistance.

In addition to sulfur cures, as used in Table 2.26 [17], peroxide cures in combination with zinc peroxide may be employed. A comparison of several peroxide crosslinking agents may be seen in Table 2.28. The appropriate one to provide the desired balance of scorch safety and cure rate may be selected. The addition of Saret SR-206 coagent reduces scorch and induction time but does increase modulus and lower compression set.

TABLE 2.27
Comparing Zinc Oxide Cure Compatible and Zinc Peroxide Compatible XNBR

Nipol NX-775	100	—	—
Nipol 1072	—	95	—
Krynac 221	—	—	95
N660 carbon black	40	40	40
Stearic acid	2	2	2
DBP	5	5	5
Wingstay 29	1	1	1
Spider Sulfur	1.5	1.5	1.5
TMTD	0.2	0.2	0.2
TMTM	1.3	1.3	1.3
Protox 169	5	—	—
50% zinc peroxide dispersion	—	10	10
Total	156	156	156
Mooney scorch, MS at 121.1°C			
Minimum viscosity	17	20	19
T_5, minutes	36.2	32.1	22.4
Rheometer at 162.8°C, 3° arc, 100 cpm			
ML, dN · m	0.8	1.1	1.1
MH, dN · m	11.0	12.8	15.1
ts2, minutes	4.5	3.5	3.0
$T'90$, minutes	13.8	13.0	9.7
Physical properties, cured 20 min at 162.8°C			
Shore A, pts	84	78	80
Modulus at 300%, MPa	17.4	18.3	18.1
Tensile, MPa	25.2	23.3	25.2
Elongation, %	422	372	410
Tear die C, kN/m	54.3	56.6	48.9
Picco abrasion 1 Hz, 44 N			
Picco rating, %	455	350	344
Compression set B 70 h at 121.1°C			
Set, %	36.7	35.9	40.6

2.2.10.5.1.4 Fillers
A comparison of carbon blacks, the preferred fillers for XNBR, is given in Table 2.29. An unfilled compound is shown to illustrate the properties of the base elastomer. The unfilled polymer has a Shore A hardness of 62 compared with a standard NBR with a hardness of 45, hence this factor needs consideration when designing an XNBR formulation. Other carbon blacks may be considered, but the enhancement of properties by reinforcing fillers is less than in standard NBR, hence the selection should be made based upon abrasion resistance, tear, and process characteristics.

Some non-black fillers are shown in Table 2.30 in which it is evident that these have a greater influence on scorch, induction time, and cure rate than carbon black.

TABLE 2.28
Comparison of Peroxides in XNBR

Nipol 1072	100	100	100	100	100
N660 carbon black	40	40	40	40	40
Naugard 445	1	1	1	1	1
Stearic acid	1	1	1	1	1
DOP	5	5	5	5	5
A(RZNO)D-60P	5	5	5	5	5
Di-Cup 40C	3.5	—	—	—	—
Vul-Cup 40KE	—	2.2	—	—	—
Luperco 101 LX	—	—	1.7	—	—
Luperco 130XL	—	—	—	1.7	1.7
SR-206 coagent	—	—	—	—	5
Total	155.5	154.2	153.7	153.7	158.7
Mooney scorch, MS at 125°C					
Minimum viscosity	34.9	36	35.8	38	30.4
T_5, minutes	11.8	13.8	18.9	12.2	8
Rheometer at 170°C, 3° arc, 100 cpm					
ML, dN·m	9.4	9.4	8.8	9.1	7.2
MH, dN·m	51.7	43.1	34.0	31.8	48.9
Ts2, minutes	1.6	2.2	2.5	2.8	2.0
T'90, minutes	12.0	14.7	15.1	15.8	15.5
Physical properties—cured 15 min at 170°C					
Shore A, pts	67	68	67	67	68
Modulus at 200%, MPa	8.5	8.1	6.8	6.8	9.4
Tensile, MPa	23.9	24.2	21.8	21.4	21.5
Elongation, %	470	490	530	500	370
Tear Die B, kN/m	71.8	75.3	71.8	70.1	68.3
Compression set B, 70 h at 125°C					
Set, %	25	30	40	40	27

The addition of a silane does change the scorch and cure rate of the silica and clay in XNBR. The modulus, tear, and tensile strength are improved in both of these fillers and the abrasion resistance is also better with a silane with HiSil 233.

2.2.10.5.1.5   General Compounding XNBR
Plasticizers, process aids, antioxidants, antiozonants used in standard NBR are also used in XNBR. A word of caution is that zinc salts often used in process aids should be avoided and zinc stearate slab dips should not be used.

2.2.11   NBR/PVC WITH AND WITHOUT PLASTICIZER AND PLASTICIZER-EXTENDED NBR

The inclusion of PVC resin in NBR elastomers provides the rubber chemist with valuable tools with which to achieve properties not available with nitrile rubber by itself. Some of the advantages of using NBR/PVC blends are

TABLE 2.29
Carbon Black in XNBR

Nipol 1072EP	100	100	100	100	100	100	100
N326 carbon black	40	—	—	—	—	—	—
N550 carbon black	—	40	—	—	—	—	—
N774 carbon black	—	—	40	—	—	—	—
N660 carbon black	—	—	—	20	40	60	—
Stearic acid	2	2	2	2	2	2	2
Naugard 445	1	1	1	1	1	1	1
DOP	5	5	5	5	5	5	—
Spider Sulfur	1.5	1.5	1.5	1.5	1.5	1.5	1.5
TMTM	1.3	1.3	1.3	1.3	1.3	1.3	1.3
TMTD	0.2	0.2	0.2	0.2	0.2	0.2	0.2
A(RZNO)D-60P	10	10	10	10	10	10	10
Total	161	161	161	141	161	181	116
Mooney scorch, MS at 125°C after 2 weeks at RT							
Minimum viscosity, dN · m	26.5	27.1	23.6	17.3	25.0	36.2	13.4
$t5$, minutes	18.8	21.7	20.0	27.4	24.0	18.3	>30
Rheometer at 170°C, 3° arc, 100 cpm							
ML, dN · m	8.8	9.1	9.4	7.4	9.1	10.7	5.0
MH, dN · m	86.2	91.5	86.8	72.1	86.6	100.4	53.0
$ts2$, minutes	2.4	2.8	2.5	3.1	2.8	2.5	3.8
$t'90$, minutes	6.8	7.4	7.3	7.8	7.2	7.0	7.0
Physical properties—cured 12 min at 170°C							
Shore A, pts	77	74	73	67	75	83	63
Modulus at 300%, MPa	18.9	18.6	17.6	10.0	17.1	21.7	3.4
Tensile, MPa	26.8	22.4	23.5	18.2	21.8	21.7	15.7
Elongation, %	410	370	420	440	390	300	520
Tear Die B, kN/m	83.4	85.3	73.9	57.9	89.5	96.7	—
Picco abrasion, 1 Hz, 44 N load							
Picco index, %	230	340	200	200	250	490	89
Compression set 70 h at 125°C							
Set, %	39	34	33	35	34	36	43
Aged in air test tube 70 h at 100°C-change							
Shore A, pts	4	7	5	5	4	2	NT
Tensile, %	−3	8	4	28	7	12	NT
Elongation, %	−22	−16	−14	−2	−13	−17	NT

1. Much improved ozone resistance in both black and light-colored compounds
2. Very good UV resistance
3. Very good colorability
4. Lower coefficient of friction and good abrasion resistance
5. Improved resistance to fuel/alcohol blends with the higher ACN versions
6. Improved solvent, fuel, and oil resistance

TABLE 2.30
Non-Black Fillers in XNBR

Nipol 1072EP	100	100	100	100	100	100	100
HiSil 532 EP	40	—	—	—	—	40	—
HiSil 243 LD	—	40	—	—	—	—	—
Cyprubond	—	—	40	—	—	—	—
Burgess KE	—	—	—	40	—	—	40
Cab-O-Sil M5	—	—	—	—	20	—	—
Stearic acid	2	2	2	2	2	2	2
Naugard 445	1	1	1	1	1	1	1
DOP	5	5	5	5	5	5	5
Spider Sulfur	1.5	1.5	1.5	1.5	1.5	1.5	1.5
Silane A-189	—	—	—	—	—	1	1
TMTM	1.3	1.3	1.3	1.3	1.3	1.3	1.3
TMTD	0.2	0.2	0.2	0.2	0.2	0.2	0.2
A(RZNO)D-60P	10	10	10	10	10	10	10
Total	161	161	161	161	141	162	162
Mooney scorch MS at 125°C after 2 weeks at RT							
Minimum viscosity, dN · m	25.4	51	16.9	19.1	23.3	26	32.7
t5, minutes	9.7	5.5	24.1	12	16.5	19.7	16.1
Rheometer at 170°C, 3° arc, 100 cpm							
ML, dN · m	7.7	12.7	6.3	6.9	6.9	8.5	8.8
MH, dN · m	68.9	84.6	61.3	84.6	68.5	74.6	75.1
ts2, minutes	2.8	1.8	3.2	2.4	3.8	4	2.7
t'90, minutes	7.3	5.4	8.7	5.5	9.7	10.7	7.8
Original properties cured 20 min at 170°C							
Shore A, pts	69	79	70	68	69	70	68
Modulus at 300%, MPa	5.2	8.7	5.6	5.1	5.7	12.5	7.8
Tensile, MPa	12.1	16.5	18.8	10.8	22.3	20.9	15.3
Elongation, %	490	450	530	420	530	480	500
Tear Die B, kN/m	49.0	76.7	49.7	37.6	54.8	81.1	51.3
Picco abrasion, 1 Hz, 44 N load							
Picco Index, %	80	130	80	80	105	130	75
Compression set B 70 h at 125°C							
Set, %	46	67	52	30	55	44	38
Aged in air test tube 70 h at 100°C							
Shore A, pts	6	5	5	6	4	5	5
Tensile, %	33	28	18	−7	−21	3	10
Elongation, %	−4	−4	−4	−17	−13	−10	−14

7. Gives good basic flame resistance
8. Provides good green strength in highly plasticized blends
9. Mixing time is greatly reduced and dispersion much improved with highly plasticized blends

Some disadvantages of NBR/PVC blends are

1. Poorer low-temperature flexibility
2. Poorer compression set resistance
3. Poorer heat resistance

The various NBR/PVC blends available from Zeon Chemicals are given in Table 2.16. This table has a description of the different grades available and will direct the chemist to select the appropriate material. It should be noted that several grades are not fluxed during manufacture so need to achieve approximately 150°C during mixing in order to obtain good physical properties and chemical and environmental resistance.

The plasticizer-extended blends are designed primarily for medium low and very low Durometer roll applications as these greatly reduce mixing time and improve processing.

The plasticizer-extended NBR grades, as illustrated in Tables 2.14 and 2.15, use a very high viscosity base nitrile rubber into which is added 50 phr plasticizer during manufacture. These elastomers allow the chemist to use very high loadings of plasticizer for soft compounds and yet have a good processing compound with much reduced mixing time.

Sulfur is difficult to disperse in NBR and particularly in soft compounds resulting in "orange peel" or dimpled surfaces or sulfur spots; hence sulfur dispersions added at the beginning of the mixing cycle are recommended. It is even better, as shown in a 50 phr DIDP extended NBR in Table 2.31, to use sulfurless cures [20]. Some of the EV cure systems match the physical properties of the sulfur and semi-EV cures and exhibit improved scorch safety.

Since NBR/PVC has basic good flame resistance it may be used in hose and belt applications where this is a requirement. A 70/30 or 50/50 NBR/PVC base elastomer should be selected and use non-black fillers, including alumina trihydrate (e.g., Hydrated Alumina 983), magnesium hydroxide, and zinc borate (e.g., Firebrake ZB) as flame retardant fillers. Calcium carbonate will assist in reducing smoke emission. In addition a phosphate plasticizer such as Kronitex 100 or chlorinated paraffin like Chlorowax 40 should be used as the only plasticizer types.

2.2.12 PROCESSING

Nitrile rubber should be mixed with full cooling on the mixer and with a fill factor of 60%, or up to 65% with highly plasticized compounds, in internal mixers. A tight, cool mill is advisable to avoid the batch going to the back roll. Always add the treated sulfur or sulfur dispersion, if present, at the beginning of the mix to ensure good dispersion, thus avoiding sulfur spots, orange peel, or surface dimpling, followed by the vulcanized vegetable oils, stabilizers, process aids, zinc oxide (except for XNBR), activators, and fillers. The plasticizers should be held back until most of the fillers are incorporated, and then in increments if larger amounts are employed. If resins are included they need to be added early and the batch

TABLE 2.31

EV Cure Systems versus Sulfur and Semi-EV in Plasticizer-Extended NBR

Nipol 1082V	150	150	150	150	150	150	150	150
Spider Sulfur	1.75	0.5	—	—	—	—	—	—
HiSil 532 EP	45	45	45	45	45	45	45	45
Zinc oxide	5	5	5	5	5	5	5	5
Stearic acid	1	1	1	1	1	1	1	1
AgeRite Resin D	2	2	2	2	2	2	2	2
DIDP plasticizer	15	15	15	15	15	15	15	15
MBTS	1.5	2.0	—	—	—	—	—	—
TMTM	0.25	—	—	—	—	—	—	—
TMTD	—	0.75	1.25	1.5	1.25	1.5	2.0	0.7
TETD	—	0.75	—	—	—	—	—	0.7
DTDM	—	—	1.25	1.0	—	—	2.0	2.0
Cure Rite 18	—	—	—	—	—	3.0	—	—
DPTT	—	—	—	—	1.25	—	—	0.7
MBT	—	—	—	—	—	—	0.5	—
Total	221.5	222.0	220.5	220.5	220.5	222.5	222.5	222.1
Mooney scorch MS at 125°C								
Minimum viscosity	16.9	15.6	18.3	19.1	18.5	12.6	11.9	12.3
T_5, minutes	29.6	>30.0	20.3	16.8	9.3	29.3	21.5	19.1
Rheometer at 160°C, 3° arc, 100 cpm								
ML, dN·m	3.5	3.2	3.5	3.7	3.9	2.7	3.0	3.0
MH, dN·n	18.8	17.9	10.6	11.0	15.8	8.8	19.5	21.0
T_5, minutes	5.4	5.1	6.7	5.8	2.7	7.2	5.4	4.6
$T'90$, minutes	7.6	8.4	16.3	16.8	7.5	17.8	15.5	15.4
Physical properties—cured at 160°C								
Cure time, minutes	10	10	20	20	10	40	30	30
Shore A, pts	27	29	22	22	27	23	27	28
Modulus at 300%, MPa	1.4	3.5	1.7	1.8	3.3	1.2	2.0	2.3
Tensile, MPa	7.2	9.9	6.0	7.0	9.9	5.3	8.0	7.7
Elongation, %	760	520	560	590	720	663	466	430
Tear Die C, kN/m	15.2	17.0	18.0	17.9	19.1	14.7	14.5	13.7
Compression set B 70 h at 70°C								
Set, %	29.9	16	34.3	33	20.5	30	16.9	15.5
Resilience ASTM D 2632								
Resilience, %	34	35	34	35	37	36	35	36

temperature needs to attain the melt temperature of the resin to ensure good incorporation. The accelerators, peroxide and zinc oxide or zinc peroxide in the case of XNBR, are added at the end if mixing temperatures are below 115°C, otherwise in a second pass.

The roll temperatures of the warm up mills and calenders should be at approximately 30°C for highly loaded compounds and up to 80°C for polymer rich stocks. A tight nip is advisable to prevent the compound from going to the fast roll.

Extrusion temperatures need to be 30°C–45°C on the barrel and up to 85°C for the head and die. The inclusion of up to 15% precrosslinked NBR in the compound will provide smoother surfaces and better green strength for subsequent processing procedures.

Building tack in nitrile compounds may be achieved by the incorporation of tackifying resins such as coumarone indene (e.g., Cumar MH); or nonactivated phenolic types such as Schenectady SP 1066; or hydrocarbon types such as Wingtack 95. It is important to use a minimum amount or no stearic acid, or replace it with lauric acid, and nonblooming, very compatible plasticizers such as liquid NBR, phosphate esters, and glycol ether esters.

Molding of NBR compounds may be accomplished by compression, transfer, or injection presses with no real difficulty. When curing at elevated temperatures it is important to use low mold-fouling types of NBR, avoid volatile plasticizers, use little or no stearic acid or replace with lauric acid, reduce the zinc oxide to 3 phr, and match the cure rate and scorch safety to the process.

REFERENCES

1. International Institute of Synthetic Rubber Producers, World Wide Rubber Statistics, 2001.
2. International Institute of Synthetic Producers, *The Synthetic Rubber Manual*, 16th Edition, 2005.
3. *2006 Product Guide*, Zeon Chemicals, LLC, PC41935.
4. Lanxess Corporation, Product Range, Technical Rubber Products, Edition 2005 12.
5. W. Hoffmann, "A Rubber Review for 1963 Nitrile Rubber," *Rubber Chemistry and Technology*, Volumes XXXVI and XXXVII, December 1963 and April 1964.
6. D. Seil and F.R. Wolf, "Nitrile and Polyacrylic Rubbers," in M. Morton (ed.) *Rubber Technology*, Chapter 11, 3rd edition, Van Nostrand Reinhold Company, New York.
7. *Hycar Nitrile Polymers Comparative Evaluation*, HM-1 BFGoodrich Chemicals Group.
8. Typical Values of Ester Plasticizers, October 2000, The HallStar Company Brochure.
9. Monomeric Plasticizers in Nitrile Rubber, Rev-2, October 30, 2003, The HallStar Company Technical Publication.
10. Evaluation of Polymeric Plasticizers in Nitrile Compounds, The HallStar Company Technical Publication.
11. Adipate Diesters Made with Propylene and Dipropylene Glycol n-Butyl Ether Compared to Plasthall's 203 and 226 in Nitrile and Neoprene Compounds, The HallStar Company Technical Publication.
12. Sulfur Cure Systems in NBR, Zeon Chemicals LLC Internal Report.
13. Semi EV Sulfur Donor Cures in NBR, Zeon Chemicals LLC Internal Report.
14. Efficient Vulcanization Cures, Zeon Chamicals LLC Internal Report.
15. Cross_Linking Peroxide and Co-Agents, Akzo Nobel Chemicals Inc. Publication 94–115.
16. Di-Cup and Vul-Cup Peroxides Vulcanizing Nitrile Rubber, GEO Specialty Chemicals Technical Information Bulletin ORC-105H.
17. Nipol Carboxylated Nitrile Elastomers, Zeon Chemicals LLC Internal Report NBR0904.1.

18. G.R. Hamed and K.-C. Hua, "Effect of ZnO Particle Size on the Curing of Carboxylated NBR and Carboxylated SBR," *Rubber Chemistry and Technology*, 77(2), 2004.
19. Chemigum NX-775, The Scorch Resistant XNBR, 10/81, Goodyear Chemicals Brochure.
20. An Evaluation of "Sulfur Free" Cure Systems in Low Durometer Roll Compounds, Zeon Chemicals LLC Internal Report TN94-104.

3 Hydrogenated Nitrile Rubber

Robert W. Keller

CONTENTS

3.1 BASIC TECHNOLOGY OF HYDROGENATED NITRILE, HNBR

As the name might suggest, Hydrogenated Nitrile (HNBR), is produced by the catalytic hydrogenation of Acrylonitrile Butadiene copolymer elastomers, NBR. The chemical structure of NBR polymers is shown in Figure 3.1. Note in Figure 3.1 that there are generally two types of Butadiene addition to the free radical polymerization process: addition in the 1,4 position which is the bulk of the Butadiene fraction as well as addition in the 1,2 position which forms a pendant carbon–carbon double bond. To understand the resultant properties of HNBR, it is necessary to recognize both types of Butadiene addition in the formation of the NBR polymer. The 1,2-Butadiene incorporation produces a vinyl type structure, which is particularly reactive in subsequent vulcanization reactions.

The subsequent formation of Hydrogenated Nitrile, HNBR, is shown in Figure 3.2. Notice that the resultant polymer is really a tetrapolymer of Acrylonitrile, ethylene formed by the hydrogenation of the 1,4-Butadiene segments, propylene formed by the hydrogenation of the 1,2-Butadiene segments, and Butadiene. Originally, HNBRs were known as HSN for Highly Saturated Nitrile, but the HNBR designation was later adopted as standard. Generally, some residual unsaturation remains from the catalytic hydrogenation reaction to provide sites for sulfur or peroxide vulcanization. The catalytic hydrogenation reaction does not affect the Carbon–Nitrogen triple bond of the Acrylonitrile group. However, the remaining unsaturation in the HNBR material is usually quite low compared to the starting NBR material.

Schematically, the polymerization of NBR and the subsequent formation of HNBR by catalytic hydrogenation of NBR are shown in Figures 3.3 and 3.4. It is important to understand Figures 3.3 and 3.4 to understand the cost differential between NBR and HNBR. For example, a medium Acrylonitrile, ACN, medium Mooney viscosity NBR material, typically 33 mole% ACN in the polymer and with a Mooney viscosity of 50 (ML1 + 4 at 100°C), may cost roughly $1.00 per pound. Because of the additional steps involved in re-dissolving, catalytic hydrogenation, solvent stripping, and drying, the resulting HNBR polymer can cost in the range of $12.00 to $16.00 per pound. This differential in cost is more easily understood using

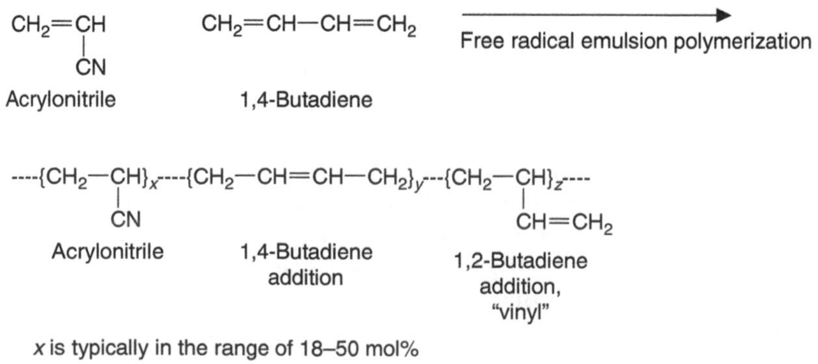

$$CH_2{=}CH \qquad CH_2{=}CH{-}CH{=}CH_2 \qquad \longrightarrow$$
$$\underset{CN}{\overset{|}{}}$$

Acrylonitrile   1,4-Butadiene   Free radical emulsion polymerization

$$----\{CH_2{-}\underset{CN}{\overset{|}{CH}}\}_x----\{CH_2{-}CH{=}CH{-}CH_2\}_y---\{CH_2{-}\underset{CH{=}CH_2}{\overset{|}{CH}}\}_z----$$

Acrylonitrile   1,4-Butadiene   1,2-Butadiene
                       addition   addition,
                                              "vinyl"

x is typically in the range of 18–50 mol%

FIGURE 3.1 NBR Polymer structure.

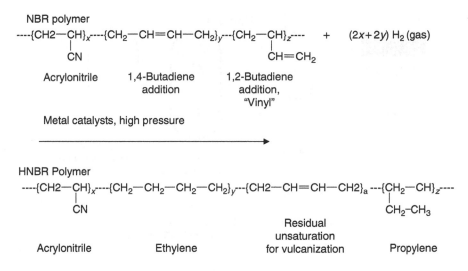

NBR polymer

----{CH2—CH}$_x$----{CH$_2$—CH=CH—CH$_2$}$_y$---{CH$_2$—CH}$_z$----   + $(2x+2y)$ H$_2$ (gas)
          | |
          CN   CH=CH$_2$

Acrylonitrile   1,4-Butadiene   1,2-Butadiene
                  addition   addition,
                                    "Vinyl"

Metal catalysts, high pressure

⟶

HNBR Polymer

----{CH2—CH}$_x$----{CH$_2$—CH$_2$—CH$_2$—CH$_2$}$_y$---{CH2—CH=CH—CH2}$_a$ ---{CH$_2$—CH}$_z$----
          | |
          CN   CH$_2$–CH$_3$

                                                     Residual
                                                     unsaturation
Acrylonitrile   Ethylene   for vulcanization   Propylene

FIGURE 3.2 HNBR Polymer structure.

Figures 3.3 and 3.4. The benefits in heat resistance and physical properties for the HNBR material make it a viable polymer in the market as will be discussed in the section on physical properties.

Initially, several of the major manufacturers of NBR developed technology to produce HNBR materials. Goodyear, Bayer, Nippon Zeon, and Polysar all developed technology for the production of HNBR. Nippon Zeon, Polysar and Bayer all

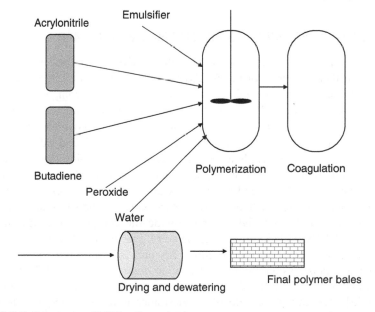

FIGURE 3.3 Schematic of NBR polymerization process.

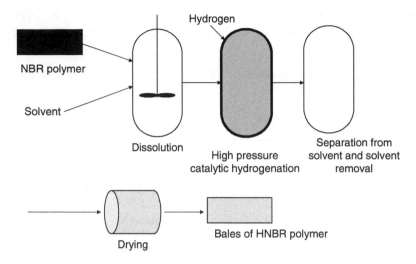

FIGURE 3.4 Schematic of HNBR manufacturing process.

developed technology for solid, bale form elastomers, while Goodyear developed an HNBR latex form for incorporation into composite packings. This technology started to become commercially available in the early to mid 1980s. Due to the relatively high cost and specialty nature of HNBRs, consolidation of this industry has occurred over the past 20 years and now two companies produce the commercial HNBR polymers: Bayer under the trademark Therban and Zeon Chemicals under the trademark Zetpol.

3.2 PHYSICAL PROPERTIES OF THE HNBR POLYMER

One way to graphically understand the niche that HNBRs fill in the elastomer market is to position them in terms of the ASTM D2000 classification system [1]. This is shown in Figure 3.5 with several other elastomers shown for comparison. Not only does the HNBR family fill a "gap" in the overall selection of properties in Figure 3.5, but HNBRs also have tear resistance, abrasion resistance, and overall toughness that polymers in similar types and classes, such as Acrylic, ACM, and Ethylene-acrylic, AEM, materials cannot match. Nitriles, NBR, as some of the oldest and most widely used oil-resistant polymers have an excellent balance of properties and durability, but they are subject to oxidation and sulfur attack due to the high level of unsaturation. The HNBR materials are much more resistant to oxidation and sulfur attack at higher temperatures compared with NBRs and provide the compounding flexibility and toughness of NBRs with improved temperature and chemical resistance. A good illustration of this is shown in Figures 3.6 and 3.7 for both engine oil and for automatic transmission fluid resistance. In Figures 3.6 and 3.7, the loss of elongation after 1008 h at 150°C is shown for two HNBRs, NBR, ACM, and CSM materials [2]. An estimate of life of an elastomer is the time required to reduce elongation to 50% of its original value. From Figures 3.6 and 3.7, the HNBR

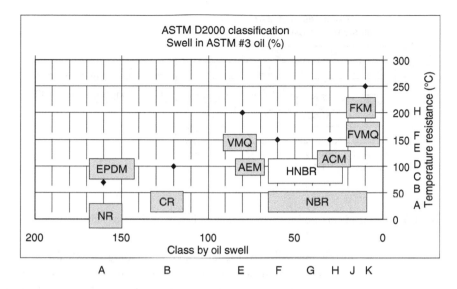

FIGURE 3.5 ASTM D2000 Comparison of elastomers.

materials have useful life remaining after the long-term aging while the NBR and the CSM materials are beyond their useful life in these fluids after the aging test. The ACM material shows excellent long-term aging resistance in these fluids, but generally does not have the tear and abrasion resistance of the HNBR materials.

The relative abrasion resistance of HNBR materials compared to other oil-resistant elastomers is shown in Figure 3.8 [3]. In Figure 3.8, traditional NBR materials are given a relative rating of 100. Abrasion, wear, and friction are not well understood, but this type of testing does give relative ratings of materials. The combination of increased high-temperature performance and improved wear resistance helps explain the niche that HNBR fills in the oil-resistant elastomer market.

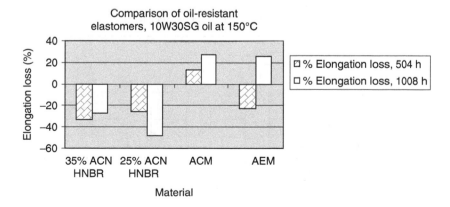

FIGURE 3.6 Comparison of oil-resistant elastomers in engine oil.

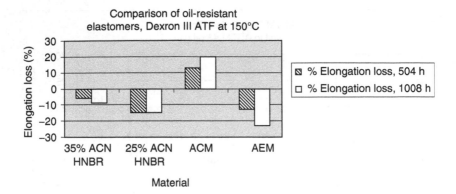

FIGURE 3.7 Comparison of oil-resistant elastomers in automotive transmission fluid.

The remainder of this chapter will cover details of compounding HNBR for various properties. This will serve as a good starting point for formulation scientists to better understand the effects of polymer selection and compounding ingredients on the final properties of the HNBR compound. It is not meant to be an exhaustive study of all possible compounding variables. Throughout the discussion of compounding, the use of statistical experimental design techniques will be emphasized and illustrated as a cost effective way to study compounding variables.

Finally, it is important to compare relative measures of cost of the various oil-resistant elastomers. Typically, the pound-volume cost method is used which is simply the cost of the compound multiplied by the specific gravity. This gives the true cost of the amount of compound needed to fill a given volume such as a mold cavity. This is shown in Table 3.1 along with the upper high-temperature limits of the various materials. While Acrylic elastomers, ACMs and AEMs, are cost effective for use up to 150°C, the HNBRs are much tougher, more abrasion-resistant materials. The Fluorocarbon, FKM, and Fluorosilicone, FVMQ, materials have excellent upper temperature limits, but the pound-volume costs are heavily influenced by the relatively

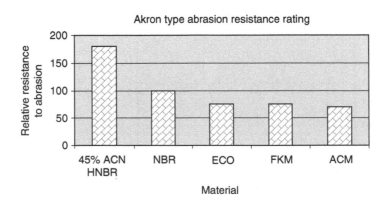

FIGURE 3.8 Relative abrasion resistance rating.

TABLE 3.1

Pound-Volume Cost Comparisons of Oil-Resistant Elastomers

Polymer	Specific Gravity of Compound	Cost of Compound, US$	Lb-Vol Cost, US$	Operating Temperature Upper Limits
NBR	1.22	1.00	1.22	100
ACM	1.32	2.30	3.04	150
AEM	1.32	2.40	3.17	150
HNBR	1.22	10.40	12.69	150
FVMQ	1.53	23.00	35.19	200
FKM	1.86	16.00	29.76	250

Note: 75 Shore A, Assumes carbon black filler, except for FVMQ.

high specific gravity of these compounds. A combination of Table 3.1 and Figures 3.6 through 3.8 gives a good picture of the niche filled by HNBR materials—HNBRs combine toughness, abrasion resistance, fluid resistance, and good cost effectiveness for applications with upper temperature limits up to 150°C requiring all these characteristics.

3.3 COMPOUNDING—POLYMER SELECTION

While the initial technology of HNBR was limited to what is considered the medium Acrylonitrile range of 33–36 mol% Acrylonitrile, advances in polymerization technology have taken HNBR products all the way from 17 mol% Acrylonitrile content to 50 mol% Acrylonitrile. Table 3.2 lists the commercially available grades of HNBR polymer for the two polymer manufacturers. The reader will see the variety of polymers available in terms of unsaturation level, Acrylonitrile content, and bulk (Mooney) viscosity. The wide variety of polymers closely mirrors the commercial polymers available in the conventional Nitrile, NBR, market.

Obviously, the start of any final compound is based on the rubber polymer selected. As with conventional NBR polymers, HNBR polymers show roughly the same effects with increasing Acrylonitrile content as shown in Table 3.3. The interesting effect in Table 3.3 is the reverse of compression set behavior. Generally, low Acrylonitrile content NBRs have superior compression set compared with high Acrylonitrile content NBRs. This effect is reversed in HNBRs and the results can be seen in Figures 3.9 and 3.10 [4,5]. The formulas used for the studies in Figures 3.9 and 3.10 are shown in Table 3.4. As can be seen in Figures 3.9 and 3.10, increasing the Acrylonitrile content of an NBR polymer gave increasing compression set while the opposite effect was observed with HNBR polymers. These studies serve as a good example of the effectiveness of statistical experimental design and response surface methods in building knowledge systems for compounding. By using these techniques, predictive equations can be developed so that changes in materials, such as changes in Acrylonitrile content of an HNBR polymer, can be evaluated before the expense of a "try it and see" approach.

TABLE 3.2
HNBR Commercial Polymers Properties

Trademark	Grade	Mole% Acrylonitrile (Approximate)	Residual Unsaturation Mole% (Approximate)	Mooney Viscosity ML 1 + 4 at 100°C (Nominal)	Comments
Zetpol	4310	17	5	80	Very low Acrylonitrile for good low-temperature flexibility, sulfur or peroxide vulcanization
Therban	LT VP KA 8882	20.5	0.9	72	Low Acrylonitrile for good low-temperature flexibility, lower viscosity for transfer and injection molding, low mold fouling, peroxide vulcanization
Therban	LT 2157	21	5.5	70	Low Acrylonitrile for good low-temperature flexibility, lower viscosity for transfer and injection molding, sulfur or peroxide vulcanization
Therban	LT VP KA 8886	21	5.5	70	Low Acrylonitrile for good low-temperature flexibility, lower viscosity for transfer and injection molding, sulfur or peroxide vulcanization, low mold fouling
Zetpol	3310	25	5	80	Medium low Acrylonitrile for good low-temperature flexibility, sulfur or peroxide vulcanization
Therban	XT VP KA 8889	33	3.5	77	Carboxylated technology for abrasion resistance, sulfur or peroxide vulcanization
Therban	HT VP KA 8805	34	0.9	45	Higher heat resistance, Medium ACN, lower viscosity for transfer and injection molding, peroxide vulcanization
Therban	A3406	34	0.9	63	General purpose, medium ACN, peroxide vulcanization
Therban	A3407	34	0.9	70	General purpose, medium ACN, higher viscosity for compression molding, peroxide vulcanization
Therban	C3446	34	3.95	56	Medium ACN, lower viscosity for transfer and injection molding, sulfur or peroxide vulcanization
Therban	VP KA 8796	34	5.5	20	Acrylate modified for toughness and abrasion resistance, peroxide or sulfur vulcanization

Therban	XQ 536	34	5.5	25	Acrylate modified for toughness and abrasion resistance, peroxide or sulfur vulcanization
Therban	C3467	34	5.5	68	Medium ACN, lower viscosity for transfer and injection molding, sulfur or peroxide vulcanization
Therban	VP KA 8837	34	18	55	Medium ACN, lower viscosity for transfer and injection molding, high level of unsaturation for sulfur or peroxide vulcanization
Zetpol	2000L	36	0.5	65	Medium high ACN, very low unsaturation level for good heat resistance, lower viscosity for transfer and injection molding, peroxide vulcanization
Zetpol	2000	36	0.5	85	Medium high ACN, very low unsaturation level for good heat resistance, peroxide vulcanization
Therban	VP KA 8918	36	0.9	66	Medium high ACN for a balance of properties and lower viscosity for transfer and injection molding, peroxide vulcanization
Therban	VP KA 8848	36	2	66	Medium high ACN, lower viscosity for transfer and injection molding, sulfur or peroxide vulcanization
Therban	VP KA 8829	36	2	87	Medium high ACN for a balance of properties and higher viscosity for compression molding, sulfur or peroxide vulcanization
Zetpol	2010L	36	4	57	Medium high ACN, good balance of heat resistance and low compression set, sulfur or peroxide vulcanization, lower viscosity for transfer and injection molding
Zetpol	2010	36	4	85	Medium high ACN, good balance of heat resistance and low compression set, sulfur or peroxide vulcanization
Zetpol	2010H	36	4	130	Medium high ACN, low compression set due to high molecular weight, sulfur or peroxide vulcanization
Zetpol	2020L	36	9	57	Medium high ACN, higher unsaturation for lower dynamic hysteresis, sulfur or peroxide vulcanization, lower viscosity for transfer and injection molding

(continued)

TABLE 3.2 (Continued)
HNBR Commercial Polymers Properties

Trademark	Grade	Mole% Acrylonitrile (Approximate)	Residual Unsaturation Mole% (Approximate)	Mooney Viscosity ML 1 + 4 at 100°C (Nominal)	Comments
Zetpol	2020	36	9	78	Medium high ACN, higher unsaturation for lower dynamic hysteresis, sulfur or peroxide vulcanization
Zetpol	ZSC 2295R	36	10	85	Zinc methacrylate modified for toughness and abrasion resistance, sulfur or peroxide vulcanization, contains silicone oil
Zetpol	ZSC 2295L	36	10	85	Zinc methacrylate modified for toughness and abrasion resistance, sulfur or peroxide vulcanization
Zetpol	ZSC 2295	36	10	95	Zinc methacrylate modified for toughness and abrasion resistance, sulfur or peroxide vulcanization
Zetpol	2030L	36	15	57	Medium high ACN, high level of unsaturation for low dynamic hysteresis in belts and dynamic applications, sulfur or peroxide vulcanization, lower viscosity for transfer and injection molding
Therban	A3907	39	0.9	70	Higher ACN content for better hydrocarbon resistance, peroxide vulcanization
Therban	A4307	42.5	0.9	63	High ACN for fuel applications, peroxide vulcanization

Therban	VP KA 8832	43	0.9	100	High ACN for fuel applications, higher viscosity for compression molding, peroxide vulcanization
Therban	C4367	43	5.5	61	High ACN, lower viscosity for transfer and injection molding, sulfur or peroxide vulcanization
Therban	VP KA 8833	43	5.5	95	High ACN and higher viscosity for compression molding, sulfur or peroxide vulcanization
Zetpol	1000L	44	2	65	High ACN for fuel resistance, low viscosity for transfer and injection molding, sulfur or peroxide vulcanization
Zetpol	1010	44	4	85	High Acrylonitrile for fuel applications and refrigerant applications, sulfur or peroxide vulcanization
Zetpol	1020L	44	9	57	High ACN for fuel applications and refrigerant applications, sulfur or peroxide vulcanization, lower viscosity for transfer and injection molding
Zetpol	1020	44	9	77	High ACN for fuel applications and refrigerant applications, sulfur or peroxide vulcanization
Zetpol	PBZ-123	44	9	125	Zetpol 1020 blended with PVC for flex fuel hoses and diaphragms, sulfur or peroxide vulcanization
Zetpol	0020	50	9	65	Very high ACN for fuel resistance, sulfur or peroxide vulcanization

Note: Zetpol supplied by Zeon Chemicals; Therban supplied by Bayer.

TABLE 3.3
Influence of Acrylonitrile Content and Unsaturation
on HNBR Properties

Property	Increasing Acrylonitrile Content	Increasing Unsaturation
Hardness	Slight increase	Slight increase
Ultimate elongation	Variable	Decrease
Elastic modulus	Negligible	Increase
Compression set	Better (decreases)	Better (decreases)
% Swell in Petroleum fluids	Decrease	Slight decrease
Low-temperature flex	Worse	Slight improvement

The amount of unsaturation in the polymer also has a major impact on the properties of the final HNBR compound. As one might imagine, the greater the level of residual unsaturation remaining, the more efficient the vulcanization. However, as the unsaturation level increases, the resulting vulcanizate can also be more susceptible to oxygen, ozone, or active sulfur attack in the end-use environment. These effects are compiled in Table 3.4. As may be found in Refs. 4 and 5, a disciplined statistical experimental design approach with response surface methods would be useful in sorting out the predictive effects of varying, for example, Acrylonitrile content and residual unsaturation.

The development of the low Acrylonitrile content HNBR polymers presented a very difficult challenge. The crystallinity of the resultant ethylene blocks after hydrogenation actually gave worse low-temperature flexibility. An additional co-monomer is incorporated into the initial NBR polymer to disrupt the crystallinity of the ethylene blocks formed during hydrogenation.

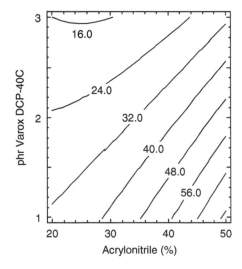

FIGURE 3.9 Influence of variable acrylonitrile content and cross-linking on compression set of NBR.

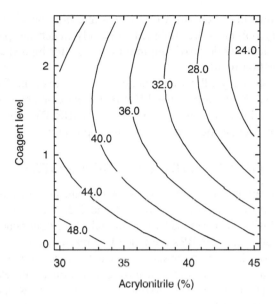

FIGURE 3.10 Influence of variable acrylonitrile content and cross-linking on compression set of HNBR.

The Zinc Methacrylate modification of HNBR gives materials of very high toughness and abrasion resistance suitable for durable power transmission belts and conveyor belts as well as any other dynamic application requiring abrasion resistance. Both Zeon, Zetpol ZSC 2295, and Bayer, VP KA 8796 and XQ 536, produce these zinc methacrylate modified HNBRs for improved toughness and durability.

The higher levels of unsaturation in some grades, while sacrificing some high-temperature performance, give the lowest heat build-up and best resilience for dynamic applications.

TABLE 3.4
Test Formulae for Figures 3.9 and 3.10

Ingredient	phr
NBR Polymer (20, 35, and 50 mole% Acrylonitrile)	100
N660 carbon black	50
Zinc oxide	2
Stearic acid	1
Varox DCP-40C	1.00, 2.00, and 3.00
HNBR Polymer (31, 38, and 45 mole% Acrylonitrile)	100
Naugard 445 antioxidant	2
N660 carbon black	50
Zinc oxide	5
Stearic acid	1
Varox 802-40KE	7
Vanax MBM	0.0, 1.25, and 2.50

Table 3.2 shows that there are various low- and high-bulk viscosity materials. The low viscosity polymers are designed to closely match the properties of their high-viscosity variants while providing improved flow for high shear molding such as transfer or injection molding.

Some of the specialty alloys and blends are also given in Table 3.2 such as Therban XT VP KA 8889 carboxylated HNBR for extreme abrasion resistance and Zetpol PBZ-123, which is a high Acrylonitrile HNBR/PVC blend for flex fuel resistance and low fuel permeation in fuel hose applications.

3.4 COMPOUNDING—VULCANIZATION SYSTEMS

The key to understanding the requirements for vulcanization of HNBR is to understand the chemistry as shown in Figures 3.1 and 3.2. While conventional Nitrile rubber, NBR, is easily reactive to active sulfur and sulfur donor chemicals due to its high level of unsaturation, HNBRs are much less reactive since most of the highly reactive unsaturation sites have been saturated with hydrogen. Likewise, it takes fairly low levels of organic peroxides to vulcanize conventional Nitriles, NBR, while it takes much higher levels of organic peroxides to vulcanize HNBRs and often coagents are needed to produce the desired properties of organic peroxide vulcanized HNBR.

Sulfur vulcanization can be done with, mostly, the polymers with higher levels of unsaturation listed in Table 3.2 such as those with unsaturation levels above roughly 4%. Generally, sulfur vulcanization is most effective with low levels of free sulfur and with aggressive thiuram disulfide and thiocarbamate type accelerators. Some typical, effective sulfur vulcanization systems are shown in Table 3.5 [6]. These compounds provide a good balance of relatively high ultimate elongation and fair compression set resistance. However, for improved compression set resistance and heat resistance for extended periods at 150°C, organic peroxide vulcanization is better. However, due to the relatively high elongation and high tensile strength of many sulfur vulcanized HNBRs, their dynamic flex life may be superior to peroxide vulcanized HNBRs for applications such as reinforced drive belts, oil-resistant flexible couplings or rolls.

For organic peroxide vulcanization, a variety of organic peroxides have been shown to be effective [7]. Such peroxides would include di-t-butyl peroxide; dicumyl peroxide; t-butyl cumyl peroxide; 1,1-di(t-butylperoxy)-3,3,5-trimethyl cyclohexane; 2,5-dimethyl-2,5-di(t-butylperoxy)hexane; 2,5-dimethyl-2,5-di(t-butylperoxy)hexyne-3; α,α-bis(t-butylperoxy)diisopropylbenzene; t-butyl perbenzoate; and t-butylperoxy isopropylcarbonate. In contrast to NBR elastomers where 1–3 phr of peroxide is effective, HNBRs generally require peroxide levels of 5–8 phr for effective vulcanization. As discussed earlier, this is due to the very low levels of unsaturation present in the HNBR polymers. The specific peroxide chosen will depend on the process safety desired and the vulcanization temperature to be used. Generally, the most common organic peroxides used in HNBR elastomers are Dicumyl peroxide, such as Varox Dicup 40C from R.T. Vanderbilt, on an inert carrier for vulcanization below 177°C and α,α-bis(t-butylperoxy)diisopropylbenzene, such as Varox VC40KE from Vanderbilt, on an inert carrier for vulcanization above 150°C.

The effectiveness of organic peroxide vulcanization can be enhanced by coagents. The most common coagents used are N-N'-m-phenylenedimaleimide (Vanax

TABLE 3.5
Semi-EV and EV Sulfur Cure Systems for HNBR

Ingredient	phr in Recipe					
	1	2	8	9	10	11
Zetpol 1020	100.00	100.00	100.00	100.00	100.00	100.00
Zinc oxide	5.00	5.00	5.00	5.00	5.00	5.00
Stearic acid	1.00	1.00	1.00	1.00	1.00	1.00
N550 carbon black	40.00	40.00	40.00	40.00	40.00	40.00
Permanox OD	2.00	2.00	2.00	2.00	2.00	2.00
Sulfur (325 mesh)					0.50	0.30
4,4'-Dithiomorpholine (DTDM)	0.50	0.50	2.00	2.00	1.00	1.00
Dipentamethylenethiuram hexasulfide (DPTH)			0.70	0.70		0.70
Tetramethyl thiuram disulfide (TMTD)	2.00	2.00	0.70	0.70	0.70	0.70
N-cylclohexyl-2-benzothiazolesulfenamide (CBS)	1.00					
2-Mercaptobenzothiazole (MBT)		0.50				
Zinc dimethyl dithiocarbamate (ZMDC)					0.50	0.50
Zinc dibutyl dithiocarbamate (ZBDC)			0.50	0.50	0.50	0.50
Tellurium diethyl dithiocarbamate (TEDC)				0.50		

Properties after vulcanization for 20 min at 160°C

Property	Recipe from above					
	1	2	8	9	10	11
Shore A hardness	74	74	75	75	75	75
Tensile strength, MPa	23.8	24.4	23	23.1	23.1	22.5
Ultimate elongation, %	460	450	410	400	420	390
100% Modulus, MPa	4.1	4.4	4.6	4.8	4.3	4.9
300% Modulus, MPa	16.6	16.9	18.3	18.6	17.8	18.6

Compression set, ASTM D395 Method B, 25% compression

	Recipe from above					
	1	2	8	9	10	11
70 h at 100°C	35	34	37	41	42	41
70 h at 120°C	54	46	46	56	57	57

MBM), triallyl isocyanurate (TAIC), and triallyl cyanurate (TAC). These coagents provide a more stable free radical source by reacting with the free radicals formed from peroxide decomposition. These more stable free radicals formed by the combination of the decomposed organic peroxide and the coagent are then more likely to react with the HNBR polymer to form effective cross-links. An excellent study of the interaction of peroxides and coagents was conducted by Zeon Chemicals and shows the utility of statistical experimental design and response surface methods to provide a comprehensive picture of this interaction [8]. The compounds used in this

TABLE 3.6

Peroxide and Coagent Interaction Study in HNBR

Ingredient	phr in Recipe											
	1	2	3	4	5	6	7	8	9	10	11	12
Zetpol 2010	100.00	100.00	100.00	100.00	100.00	100.00	100.00	100.00	100.00	100.00	100.00	100.00
Zinc oxide	5.00	5.00	5.00	5.00	5.00	5.00	5.00	5.00	5.00	5.00	5.00	5.00
Stearic acid	0.50	0.50	0.50	0.50	0.50	0.50	0.50	0.50	0.50	0.50	0.50	0.50
N330 carbon black	40.00	40.00	40.00	40.00	40.00	40.00	40.00	40.00	40.00	40.00	40.00	40.00
Naugard 445	1.50	1.50	1.50	1.50	1.50	1.50	1.50	1.50	1.50	1.50	1.50	1.50
Zinc salt of 4 and 5 methylmercaptobenzimidazole	1.50	1.50	1.50	1.50	1.50	1.50	1.50	1.50	1.50	1.50	1.50	1.50
α,α-bis(t-butylperoxy)diisopropylbenzene	5.00	5.00	5.00	5.00	7.50	7.50	7.50	7.50	10.00	10.00	10.00	10.00
N,N-m-phenylene dimaleimide	0.00	2.50	5.00	7.50	0.00	2.50	5.00	7.50	0.00	2.50	5.00	7.50

Properties after 15 min at 170°C vulcanization

Property	Recipe											
	1	2	3	4	5	6	7	8	9	10	11	12
Shore A hardness	70	73	76	78	71	73	76	78	71	75	77	80
Tensile strength, MPa	28.0	30.1	28.1	28.0	32.4	30.2	29.3	27.1	32.0	28.2	28.6	27.7
Ultimate elongation, %	560	340	270	240	400	270	230	190	310	220	190	160
50% Modulus, MPa	1.6	2.0	2.5	3.3	1.7	2.2	2.8	3.8	1.8	2.5	3.2	4.4
100% Modulus, MPa	2.2	4.1	6.4	9.0	2.9	5.0	7.7	11.0	3.7	6.6	9.9	13.4
Compression set, ASTM D395 Method B, 70 h at 150°C	61	32	26	22	42	29	26	23	34	25	24	22

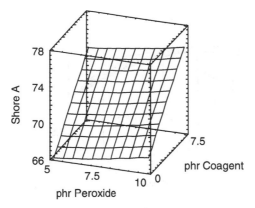

FIGURE 3.11 Study of peroxide and coagent influence on hardness in HNBR. (From Zeon Chemicals L.P., *Study of Level and Ratio of Peroxide and Co-Agent in Zetpol 2010* (Z5.1) [Brochure], Louisville, KY. With permission.)

study are shown in Table 3.6 with the response surfaces plotted in Figures 3.11 through 3.14. Again, this shows the utility of response surface methods and experimental design in rubber compound formulation and optimization.

Unlike conventional NBRs, HNBRs benefit from oven post-curing processes. Post-curing is particularly effective in improving compression set resistance. These improvements in compression set resistance without significant impact on original physical properties generally hold true for both sulfur and peroxide cured materials. One study conducted by Zeon Chemicals is illustrative in the effects of oven post-curing [9]. Their data are shown in Table 3.7 and Figures 3.15 and 3.16. Generally, they found that an oven post-cure of 4 h at 150°C was effective for sulfur vulcanized materials and an oven post-cure of 2 h at 150°C was effective for peroxide vulcanized materials. Where maximizing compression set is necessary, these types of studies with particular compounds will help discern which oven post-cure is optimal for specific compounds. Table 3.7 also shows the difference in

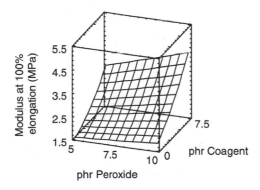

FIGURE 3.12 Study of influence of peroxide and coagent on modulus in HNBR. (From Zeon Chemicals L.P., *Study of Level and Ratio of Peroxide and Co-Agent in Zetpol 2010* (Z5.1) [Brochure], Louisville, KY. With permission.)

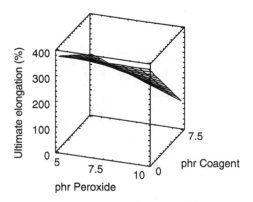

FIGURE 3.13 Influence of peroxide and coagent on elongation in HNBR. (From Zeon Chemicals L.P., *Study of Level and Ratio of Peroxide and Co-Agent in Zetpol 2010* (Z5.1) [Brochure], Louisville, KY. With permission.)

compression set behavior of sulfur vulcanized and peroxide vulcanized HNBR compounds of similar original physical properties. Particularly at temperatures of 150°C, the peroxide vulcanized material shows better compression set resistance and better cross-link stability than the sulfur vulcanized material.

Currently, both Bayer and Zeon Chemicals have developed additives to give slight improvements in upper temperature operation with claims that this technology will extend HNBRs to 160°C operating temperatures.

3.5 COMPOUNDING—CARBON BLACK FILLERS

Carbon black is still the most important commercial reinforcing and extending filler for use in the rubber industry (Zeon studies [10]). Carbon blacks covering the

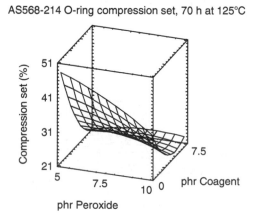

FIGURE 3.14 Influence of peroxide and coagent on compression set in HNBR. (From Zeon Chemicals L.P., *Study of Level and Ratio of Peroxide and Co-Agent in Zetpol 2010* (Z5.1) [Brochure], Louisville, KY. With permission.)

TABLE 3.7
Study of Post-Cure Effects in HNBR

Post-Cure Effects Study

Ingredient	phr in Recipe							
	1	2	3	4	5	6	7	8
Zetpol 1020	100.0	100.0	100.0	100.0	100.0	100.0	100.0	100.0
Zinc oxide	5.0	5.0	5.0	5.0	5.0	5.0	5.0	5.0
Stearic acid	1.0	1.0	1.0	1.0	1.0	1.0	1.0	1.0
N770 carbon black	40.0	40.0	40.0	40.0	40.0	40.0	40.0	40.0
Dibutoxyethoxyethyl adipate	5.0	5.0	5.0	5.0	5.0	5.0	5.0	5.0
Sulfur (325 mesh)	0.5	0.5	0.5	0.5				
Tetramethyl thiuram disulfide	2.5	2.5	2.5	2.5				
2-Mercaptobenzothiazole	0.5	0.5	0.5	0.5				
40% Dicumyl peroxide on an inert carrier					8.0	8.0	8.0	8.0

	Recipe							
	1	2	3	4	5	6	7	8
Press-cure conditions	30 min at 160°C	30 min at 160°C	30 min at 160°C	30 min at 150°C	30 min at 160°C	30 min at 160°C	30 min at 160°C	30 min at 160°C
Oven post-cure conditions, h at 150°C	0	2	4	8	0	2	4	8

Property	Recipe							
	1	2	3	4	5	6	7	8
Shore A hardness	65	66	66	67	65	67	67	67
Tensile strength, Mpa	27.5	23.5	25.1	25.6	24.3	26.8	26.4	25.6
Ultimate elongation, %	510	460	480	460	450	440	430	410
100% Modulus, Mpa	2.4	2.7	2.7	2.7	2.4	2.6	2.6	2.9
200% Modulus, Mpa	6.4	7.7	7.7	8.1	7.8	8.4	8.7	10
Compression set, ASTM D395 Method B, 70 h at 120°C	50	36	31	24	27	15	14	13
Compression set, ASTM D395 Method B, 70 h at 150°C	75	67	60	50	35	25	24	23

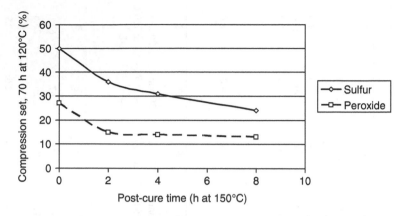

FIGURE 3.15 Post-cure effects on HNBR compression set. (From Zeon Chemicals L.P., *Effects of Post Curing on Zetpol 1020* (Z5.1.) [Brochure], Louisville, KY. With permission.)

spectrum from N110 carbon black to N990 carbon black were studied. A simple set of rules was developed for estimating the Shore A hardness of the finished compound. In addition, work at Wynn's-Precision for basic compounding rules gave similar simple equations for predicting the reinforcing effects of various carbon black grades and those results are placed in Table 3.8 as well. Both models give approximately the same final Shore A hardness for the HNBR compound within the usual +/−5 point Shore A hardness specification range.

More interesting is the detailed data [10] on carbon black reinforcement of Zetpol 2020. As had been found to be the case by the author, the key determinant of the reinforcement of the carbon black is the DBP number of the carbon black. The DBP measures the ability of the combined effects of particle size and structure and gives a measure of how much polymeric material will be absorbed onto the surface

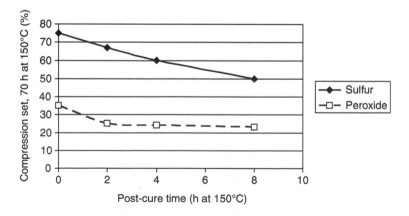

FIGURE 3.16 Effects of post-cure on HNBR compression set. (From Zeon Chemicals L.P., *Effects of Post Curing on Zetpol 1020* (Z5.1.) [Brochure], Louisville, KY. With permission.)

TABLE 3.8
Carbon Black Reinforcement of HNBR Predictive Equations to Determine Shore A

Carbon Black	Wynn's-Precision Model	Zeon Model (10)
N330		$53 + (0.333 \times \text{phr CB}) + 10$
N550	$55 + (0.500 \times \text{phr CB})$	$53 + (0.333 \times \text{phr CB}) + 7$
N660	$55 + (0.450 \times \text{phr CB})$	
N762 or N774	$55 + (0.377 \times \text{phr CB})$	$53 + (0.333 \times \text{phr CB}) + 5$
N990	$55 + (0.238 \times \text{phr CB})$	$53 + (0.25 \times \text{phr CB})$

Comparison of models with different levels of carbon blacks

Carbon Black	phr	Wynn's-Precision Model	Zeon Model (10)
N550	30	70.0	73.0
	40	75.0	76.3
	50	80.0	79.7
	60	85.0	83.0
	80	95.0	89.6
N762	30	66.3	68.0
	40	70.1	71.3
	50	73.9	74.7
	60	77.6	78.0
	80	85.2	84.6
N990	30	62.1	60.5
	40	64.5	63.0
	50	66.9	65.5
	60	69.3	68.0
	80	74.0	73.0
	100	78.8	78.0

Note: phr CB = phr carbon black.

of the carbon black. Plotting the results from Ref. 10 in terms of phr black, between 40 and 80 phr, used and the DBP of the black used gives the response surface shown in Figures 3.17 and 3.18. Either the simple calculations above, or the response surfaces in Figures 3.17 and 3.18 could be used to determine the effects of carbon black reinforcement of HNBR polymers. These types of experimental designs can be used to give expert systems, which can predict the effects of different carbon black fillers on key properties.

3.6 COMPOUNDING—NON-BLACK FILLERS

The wide variety of non-black fillers make for a wide variety of finished compound properties. Studies done by Zeon [11,12] show very different reinforcing effects, which do not correlate with particle size of the fillers suggesting some very strong interactions with particular non-black fillers. Information from that study is shown in

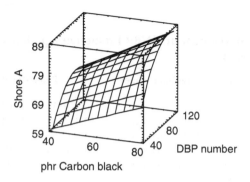

FIGURE 3.17 Influence of amount and DBP number of carbon black on Shore A in HNBR. (From Zeon Chemicals L.P., *Carbon Black Reinforcement of Zetpol* (Z5.2.1 Ed.) [Brochure], Louisville, KY. With permission.)

Table 3.9. The reinforcing effects are the change in Shore A hardness per phr of filler and the change in Modulus at 100% per phr of filler. A good starting point with any filler system is to determine the amount of hardness or modulus gain per amount of filler used. High alkalinity fillers, such as Carplex 1120 and Zeolex 23, have been shown to improve heat resistance in terms of physical property losses, but they have little effect on compression set.

3.7 COMPOUNDING—AGING STABILIZERS

While HNBRs have much better stability to oxygen and ozone attack than NBRs, there is still the need to stabilize the compounds for oxygen attack and, possibly depending on the application, ozone attack. Studies by Zeon Chemicals [13] have shown that polymerized 1,2-dihydro-2,2,4-trimethyl quinoline (trade names such as Agerite Resin D, Flectol H, and Naugard Q) at levels from 1.5 to 3.0 phr were very effective in both oxygen and dynamic ozone protection of peroxide vulcanized HNBR. Addition of 0.5 phr protective wax with the polymerized 1,2-dihydro-2,2,4-trimethyl

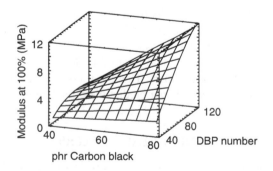

FIGURE 3.18 Influence of Amount and DBP number on 100% modulus in HNBR. (From Zeon Chemicals L.P., *Carbon Black Reinforcement of Zetpol* (Z5.2.1 Ed.) [Brochure], Louisville, KY. With permission.)

TABLE 3.9
Non-Black Filler Systems in HNBR

Filler	phr	Relative Particle Size	Shore A	100% Modulus	Delta Shore A per phr	Delta Modulus at 100% per phr
None	0		50	1.1		
Ultra-Pflex	90	Small	65	1.6	0.167	0.0056
Roy-Cal L	90	Medium	65	1.8	0.167	0.0078
Zeospheres 200	90	Medium	66	2.4	0.178	0.0144
Imsil A-8	90	Medium	68	3.4	0.200	0.0256
Nulok 321	90	Medium	73	5.0	0.256	0.0433
Nulok 390	90	Medium	73	7.0	0.256	0.0656
Pyrax B	90	Large	74	7.5	0.267	0.0711
Mistron Cyprubond	90	Medium	76	8.1	0.289	0.0778
Translink 77	90	Medium	77	9.2	0.300	0.0900
Celite 350	90	Large	80	9.2	0.333	0.0900
Silene 732D	90	Small	83	10.2	0.367	0.1011
Zeolex 23	40	Medium	65	2.7	0.375	0.0400
HiSil 532EP	40	Small	68	3.3	0.450	0.0550
Cab-O-Sil TS-720	40	Small	71	2.5	0.525	0.0350
HiSil 233	40	Small	73	2.9	0.575	0.0450
Ultrasil VN-3 SP	40	Small	74	3.0	0.600	0.0475
Aerosil 300	40	Small	76	2.8	0.650	0.0425
Cab-O-Sil M-7D	40	Small	78	3.1	0.700	0.0500

Base formula without filler

Ingredient	phr
Zetpol 2010	100.0
Plasthall TOTM	5.0
Titanium dioxide	5.0
Zinc oxide	5.0
Stearic acid	0.5
Naugard 445	1.5
Vanox ZMTI	1.0
Carbowax 3350	2.0
A-172 Silane	1.0
Vul-Cup 40KE	8.0

Source: From Zeon Chemicals L.P., *Evaluation of Larger Particle Non-Black Fillers in Peroxide Cured Zetpol 2010* (Z5.3.3.) [Brochure], Louisville, KY.

quinoline gave excellent dynamic ozone resistance. This study also showed that another effective antioxidant combination was a combination of 4,4'-bis(α,α'-dimethylbenzyl) diphenylamine, trade name Naugard 445, with the zinc salt blend of 4 and 5 methylmercaptobenzimidazole, trade name Vanox ZMTI, at 1.5 phr each was a very effective antioxidant system for long-term air aging. It is important to note that some of these systems inhibit the peroxide cure mechanism and may require more coagent for efficient vulcanization [13].

Both Zeon and Bayer have recently introduced proprietary masterbatch aging stabilizers, which extend HNBR compounds to 160°C long-term life, measured either by elongation loss or by long-term compression set [14].

3.8 COMPOUNDING—PLASTICIZERS

While HNBR and NBR can be compounded with many of the same plasticizers and softeners, care must be taken with HNBR to assure that the plasticizer used is not so volatile that it detracts from the overall heat resistance of the finished compound. Common plasticizers used at a 20 phr level are summarized in Table 3.10 [15]. While plasticizers such as Dioctyl phthalate and dibutoxy ethoxy ethyl adipate are effective in reducing hardness and viscosity, they are also volatile at high temperatures and can give finished part shrinkage in prolonged heat aging. For an overall balance of low-temperature flexibility improvement, viscosity reduction, and stability to heat aging, the trioctyl trimellitate and the triisononyl trimellitate plasticizers are excellent choices.

TABLE 3.10
Plasticizers in HNBR

			All Plasticizers at 15 phr in Base Recipe			
Property	**No. Plasticizer**	**Dioctyl Phthalate**	**Dibutoxy Ethoxyethyl Adipate**	**Ether Ester**	**Trioctyl Trimellitate**	**Triisononyl Trimellitate**
Compound Mooney viscosity, ML1 + 4 (100°C)	133	86	78	81	91	87
Oscillating disc rheometer at 170°C, Vmin	11.4	6.5	7.1	7.4	8.0	8.3
Oscillating disc rheometer at 170°C, Vmax	51.6	40.4	46.2	41.4	53.1	56.3
Oscillating disc rheometer at 170°C, T95, min	19.5	18.3	15.1	16.5	17.2	17.0
Shore A hardness	76	70	69	69	70	69
Ultimate elongation, %	310	360	450	430	380	370
100% Modulus, MPa	6.2	4.1	3.4	3.3	4.2	4.3
Weight loss after 168 h at 150°C air aging, %	−1.4	−9.7	−8.6	−6.1	−3.4	−1.7

TABLE 3.10 (Continued)
Plasticizers in HNBR

All Plasticizers at 15 phr in Base Recipe

Property	No. Plasticizer	Dioctyl Phthalate	Dibutoxy Ethoxyethyl Adipate	Ether Ester	Trioctyl Trimellitate	Triisononyl Trimellitate
Temperature retraction per ASTM D1329						
TR-10°C	−22.6	−28.7	−25.1	−25.0	−28.1	−27.9
TR-30°C	−16.5	−20.6	−12.7	−12.5	−20.0	−19.6
TR-70°C	−4.5	−8.5	−3.3	−3.8	−6.8	−6.8
Heat resistance weight loss, 168 h at 150°C	−1.4	−9.7	−8.6	−6.1	−3.4	−1.7

Base Recipe

Ingredient	phr
Zetpol 2010	100
N550 carbon black	50
α,α9-bis(*t*-butylperoxy)-diisopropylbenzene, 40% on inert carrier	6

Source: From Campomizzi, E.C., *A new master batch to extend the heat resistance of Therban HNBR compounds*, Bensenville, IL, January 27, 2000.

3.9 COMPOUNDING—OTHER INGREDIENTS (TACKIFIERS, INTERNAL LUBRICANTS)

Tackifiers may be required for building plies of HNBR compounds to other HNBR compounds or to other polymer formulations for composite constructions. Coumarone-Indene resins, Phenol-acetylene resins, Aromatic hydrocarbon resins, and Styrene-vinyl toluene resins work well at levels of up to15 phr to tackify the resulting HNBR compound.

Sometimes, such as in O-Rings and seals for valves, internal lubrication is desirable to prevent valve sticking and squeaking. Oleamides are, by far, the most efficient internal lubricants for HNBR. Typical trade names for such Oleamides are Armoslip CP and Kenamide E.

3.10 COMPOUNDING—BLENDING WITH EPDM

This is a well-known technique with NBRs to improve heat and ozone resistance and works quite well with HNBRs as well. As one might expect, the EPDM improves low-temperature flexibility and weather resistance while giving some sacrifice in resistance to swelling in hydrocarbon fluids. Generally, 10–20 phr of EPDM elastomer with medium to high third monomer content to give efficient vulcanization will give improvements in low-temperature flexibility and weather resistance without great sacrifices in hydrocarbon fluid resistance. The ethylene segments in HNBR as a result of the hydrogenation process provide better overall compatibility in blending HNBRs and EPDMs.

3.11 APPLICATIONS

In this section, we will combine the basic properties of the HNBR polymer material with the compounding information presented to give starting points for compounds for specific applications. In each case, the application will be defined, starting compounds will be suggested, and recommendations for further optimization will be given. HNBR is a viable rubber material because of the combination of fluid resistance, toughness and abrasion resistance, higher temperature resistance compared with NBR, and its overall value in many applications, which experience temperatures in the range of 125°C–150°C.

3.11.1 APPLICATION—HIGH-TEMPERATURE OIL COOLER O-RING SEAL

Our customer is a diesel engine manufacturer who uses circulating water plus ethylene glycol coolant in a small heat exchanger to lower the temperature and extend the life of diesel engine lubricating oil. The O-Ring seal will see exposure to both engine oil and to coolant at 125°C maximum temperature with typical operation in the range of 100°C–120°C. Conventional NBRs do not have sufficient durability at 125°C to survive in the application, EPDMs have great coolant resistance but have poor resistance to engine oils, and FKMs are quite expensive and have poor resistance to high-temperature water-based systems. Thus, an HNBR material may be a good choice for this application. Being an O-Ring, mostly static seal, resistance to the application fluids and application temperatures combined with excellent compression set resistance are the critical properties in compound development. Generally, the most universal hardness for O-Rings is 75 Shore A, so we will target this in the development. Standard compression molding will be the production process, so low compound unvulcanized viscosity will not be a requirement. A starting recipe and recommendations for optimization are shown in Table 3.11. This compound will give the best overall compression set and fluid resistance based on the compounding parameters described earlier in this chapter. The optimization factors should then be combined in an experimental design with response surface techniques for final compound development.

3.11.2 APPLICATION—HIGH-TEMPERATURE, ABRASION-RESISTANT JOINT BOOT

In this application, our customer is a manufacturer of drive train components and has a particularly demanding application for a protective joint boot. The boot is for the take-off shaft of the all-wheel drive assembly, the joint is located very near a catalytic converter so it sees high temperatures from 100°C to 125°C, and the vehicle is intended for off-road capability so the boot needs to be tough and abrasion resistant to tolerate brush, weeds, mud, and dirt. High tear strength Silicones, VMQ, have sufficient resistance to joint grease, sufficient high-temperature resistance and sufficient resistance to the external environment, but are still not tough enough with sufficient abrasion resistance to survive in the off-road environment. Chloroprene rubber, CR, does not have sufficient high-temperature capability for the environment near the catalytic converter. Thermoplastic materials used for the majority of the joint boots, likewise,

TABLE 3.11
Diesel Engine Oil Cooler O-Ring

Compound Development and Optimization

Compound Basics Ingredient	phr	Logic/Comments	Experimental Design Optimization
34%–36% Acrylonitrile HNBR polymer with 4%–5.5% residual unsaturation (variable)	100.0	Best overall balance of fluid resistance and low-temperature properties; the higher levels of unsaturation give the most efficient cure and balance temperature resistance and compression-set resistance	Evaluate 2–3 polymers for best overall balance of properties
Zinc oxide	2.0		
Stearic acid	1.0		
Naugard 445 antioxidant	0.7	Best overall heat resistance stabilizer package	
Vanox ZMTI antioxidant	0.7	Best overall heat resistance stabilizer package	
Carbon black (variable)	50–75	N762 black as low level in design at 75 phr, N550 black as high level in design at 50 phr	Evaluate two carbon black types for best overall balance of properties
Varox VC40KE	10.0	Peroxide for compression molding at 180°C	
Vanax MBM (variable)	4.5–7.5	Coagent for increased efficiency of cure and best compression set resistance	Optimize coagent levels for best balance of properties

Note: Optimization – 3 factor, 2,2 and 3-level full factorial design—12 compounds and experiments. *Responses (properties) to be measured:* Resistance to diesel engine oil at 125°C–150°C; resistance to water/glycol coolant at 125°C–150°C; O-Ring compression-set resistance at 125°C–150°C.

do not have sufficient high-temperature resistance for this application. The critical properties for this application are heat resistance to 125°C, joint grease resistance, weather and ozone resistance, good low-temperature flexibility, and good dynamic flex life. For an overall balance of toughness with flexibility, a target Shore A hardness of 65 will work. Due to the complexity of the part, injection molding will be the required process so low viscosity will be necessary. A possible starting recipe with parameters for experimental design, response surface methods, and final optimization is shown in Table 3.12. The recipe in Table 3.12 uses a blend of HNBR and EPDM polymers to improve low-temperature flexibility and improve weathering while still maintaining good resistance to joint grease, which is not a terribly aggressive hydrocarbon fluid. From the previous discussion on protective materials, the blend of polymerized 1,2-dihydro-2,2,4-trimethyl quinoline antioxidant with a wax protectant is used for the best ozone resistance. TOTM is used as the plasticizer for the best balance of low-temperature improvements, raw compound viscosity reduction, and high-temperature stability.

TABLE 3.12

High Temperature, Abrasion-Resistant Joint Boot

Compound Development and Optimization

Compound Basics Ingredient	phr	Logic/Comments	Experimental Design Optimization
21%–25% Acrylonitrile HNBR polymer with 1%–4% residual unsaturation (variable)	70–90	Good combination of grease resistance and low-temperature flex properties	Vary level from 70 to 90 phr with blending of EPDM polymer to achieve total 100 phr of polymer
EPDM polymer—low ethylene, low viscosity, 3%–5% third monomer unsaturation level (variable)	30–10	Low viscosity for injection mold flow; medium to high third monomer level for efficient vulcanization	Vary level from 30 to 10 phr with blending of HNBR polymer to achieve total 100 phr of polymer
Zinc oxide	2.0		
Stearic acid	1.0		
Agerite Resin D	1.0	Best antioxidant and antiozonant balance with wax	
Protective wax	2.0	Best antioxidant and antiozonant balance	
N762 carbon black	55.0	Reasonable reinforcement while maintaining high elongation for flex resistance	
Plasthall TOTM (trioctyl trimellitate)	10.0	Reduction of compound viscosity for injection molding; improvement of low-temperature flex; good heat stability	
Varox DCP-40C (variable)	3.0–5.0	Fast curing in injection molding where machine cycle time is critical	Optimize level of peroxide for best balance of properties

Note: Optimization = 2 factor, 3 and 3-level full factorial design with center point—10 compounds and experiments.

Responses (properties) to be measured: 125°C heat resistance; 125°C joint grease resistance; low-temperature flexibility; dynamic flex life; static and dynamic ozone resistance; mold flow and mold cleanliness in injection molding.

3.11.3 APPLICATION—OIL FIELD HIGH PRESSURE WELL PACKER

Our customer here is a manufacturer of oil well tools and requires a rubber packer with excellent resistance to crude oil, hydrogen sulfide, aggressive amines, and water. Fluorocarbon elastomers have excellent high-temperature resistance, but their poor resistance to aggressive amines makes them not practical for this application. Conventional NBRs have the oil, amine, and water resistance, but do not have suitable high-temperature resistance of 125°C continuous with excursions to 150°C. When the packer element seals the well, pressure differentials of several thousand psi can be experienced, so we will need to formulate a high hardness, tough material. The packer element is a relatively simple cylindrical shape, so compression molding will be a viable production method. The critical properties for our packer compound are heat resistance of 125°C continuous with excursions to 150°C, crude oil resistance, hydrogen sulfide resistance, steam resistance, aggressive amine resistance, and pressure resistance. HNBRs show less drop in modulus of elasticity, stiffness, with increasing temperatures, so our HNBR packer material will retain its pressure resistance better than other materials at high temperatures. Our starting compound is shown in Table 3.13 with variables for further study and optimization using experimental design and response surface methods.

TABLE 3.13
Oil Field Packer

Compound Development and Optimization

Compound Basics Ingredient	phr	Logic/Comments	Experimental Design Optimization
34%–36% Acrylonitrile HNBR polymer with less than 2% residual unsaturation (variable)	100.0	Best overall balance of heat and chemical resistance	Two or three different polymers optimized for balance of properties
Zinc oxide	2.0		
Stearic acid	1.0		
Naugard 445	1.5	Best antioxidant package for long-term heat	
Vanox ZMTI	1.5	Best antioxidant package for long-term heat	
N330 or N550 carbon black (variable)	55.0	To obtain 90 + Shore A durometer for pressure resistance	Evaluate two different carbon blacks for best overall balance of properties
Varox VC40KE	10.0	High-temperature peroxide for compression molding at 180°C	
Coagent: Vanax MBM at 7.5 phr or TAIC at 5.0 phr (variable)	VARIABLE	High level of coagent for efficient vulcanization	Optimize compound based on evaluation of two chemical types of coagents

Note: Optimization = 3 factor, 2,2 and 2-level full factorial design with center point—8 compounds and experiments.

Responses (properties) to be measured: 125°C–150°C heat resistance; crude oil resistance; hydrogen sulfide resistance; steam resistance; aggressive amine resistance; pressure extrusion resistance.

3.11.4 APPLICATION—HIGH-TEMPERATURE, LONG LIFE AUTOMOTIVE SERPENTINE BELT

Our customer is a manufacturer of automotive engines and has noticed an unacceptable warranty on Chloroprene rubber serpentine belts. While Chloroprene rubber has excellent flex resistance and excellent weather resistance in this application, the reduced air flow through the engine compartment required to reduce drag on the car has resulted in continuous operating conditions for the serpentine belt in the range of 100°C–125°C. Previously, the serpentine belt saw temperatures in the range of 70°C–90°C and reduced life from heat aging was not an issue in warranties. The belt is fabric reinforced to prevent creep and elongation, so creep or compression set are not extremely critical. We do want a relatively high elongation compound to give excellent dynamic and static flex life. Thus, the critical properties for this application are heat resistance from 125°C to 150°C, static and dynamic ozone resistance, and dynamic flex life. The better retention of stiffness at high temperatures of HNBR compounds will result in the meshing features of the belt being retained better at high temperatures. A basic recipe for this application is shown in Table 3.14 with compounding variables for experimental design study and optimization.

3.11.5 APPLICATION—ORANGE, WATER PUMP MECHANICAL SEAL PROTECTIVE BELLOWS

Our customer is a manufacturer of mechanical face seals for automotive water pump applications. They had been using conventional NBRs for the protective bellows around the water pump seal, but the engine compartment heat has risen to 125°C continuous with stop-and-go traffic excursions to 150°C and the NBR bellows does not meet the requirements for 100,000 mile warranties. For overall toughness, weather resistance, splash fluid resistance, and cost, HNBR is a good choice. The customer would like an orange color to distinguish, for the mechanics and assemblers, the new HNBR water pump bellows from the previous black NBR. While there is some flexing of the bellows requiring some flex-fatigue resistance, the flexing is fairly small. The initial target hardness for the material is 65 Shore A. Key properties for this application are color, heat resistance continuous at 125°C with excursions to 150°C, weather resistance, water and coolant resistance, low-temperature flexibility for cold starts, and dynamic flex resistance. For high volume production of this fairly complex shape, injection molding is the desired production process. A starting recipe for this application is shown in Table 3.15 with additional variables for statistical experimental design and final compound optimization. Organic pigments are, generally, not good choices for peroxide cured materials since the pigments can act as free radical scavengers and interfere with the vulcanization reaction. In addition, the oxidizing effects of peroxides may completely destroy the color of organic pigments.

3.11.6 APPLICATION—GREEN HNBR FOR AUTOMOTIVE AIR CONDITIONING O-RINGS

HNBRs show excellent resistance to HFC-134a refrigerant and the polyalkalene glycol, PAG, lubricants used with these refrigerants. Our customer is looking for

TABLE 3.14

High-Temperature Serpentine Belt

Compound Development and Optimization

Compound Basics Ingredient	phr	Logic/Comments	Experimental Design Optimization
34%–36% Acrylonitrile HNBR polymer with 1%–4% residual unsaturation (variable)	70–90	Best overall balance of heat resistance and efficient vulcanization	Vary levels of HNBR with Zinc methacrylate modified HNBR to balance properties; total polymer phr = 100
Zinc methacrylate modified HNBR	30–10	Added tear and tensile strength	Vary levels of HNBR with Zinc methacrylate modified HNBR to balance properties; total polymer phr = 100
Zinc oxide	2.0		
Stearic acid	1.0		
Agerite Resin D	1.0	With wax, gives best overall heat and ozone resistance	
Protective wax	2.0	With Agerite Resin D, gives best overall heat and ozone resistance	
N762 carbon black (variable)	10–20	Reinforcement without loss of flex resistance	Vary level between 10 and 20 phr to obtain best balance of properties
Varox VC40KE (variable)	4–8	High-temperature vulcanization system	Vary level between 4 and 8 phr to obtain best balance of properties

Note: Optimization = 3 factor, 3,3 and 3-level full factorial design with center point—27 compounds and experiments.

Responses (properties) to be measured: Heat resistance from 125°C to 150°C; static and dynamic ozone resistance; dynamic flex life.

O-Ring seals for this environment as engine compartment temperatures have increased. The current black Chloroprene rubber, CR, seals are marginal at the 125°C continuous and 150°C intermittent operating temperatures that the O-Rings will see. The green color is to easily track the change for original assembly, warranty claims, and aftermarket parts supply. Moreover, there is a need for excellent low-temperature flexibility to prevent refrigerant leakage in winter. Because of the high volumes involved and the small O-Ring inner diameters less than 1 in., injection molding is the most cost-effective process. For a balance of compliance along with pressure tolerance and initial hardness of 75 Shore A is the target. The critical properties will be resistance to HFC-134a and PAG blends at 125°C continuous and 150°C intermittent, good long-term compression set resistance at 150°C, low-temperature flexibility, and good mold flow and rapid curing for injection molding.

TABLE 3.15
Orange Water Pump Seal Protective Bellows
Compound Development and Optimization

Compound Basics Ingredient	phr	Logic/Comments	Experimental Design Optimization
21%–25% Acrylonitrile HNBR polymer, 1%–4% residual unsaturation (variable)	100.0	Best low-temperature flexibility for cold starts	Evaluate two different polymers
Zinc oxide	1.0		
Stearic acid	2.0		
Agerite Resin D	1.0	With wax, gives best overall heat and ozone resistance	
Protective wax	2.0	With Agerite Resin D, gives best overall heat and ozone resistance	
Plasthall TOTM	10.0	Reduction of compound viscosity for injection molding; improvement of low-temperature flex; good heat stability	
Nulok 321	50.0	Extending clay filler	
HiSil 532EP (variable)	0–20	Reinforcing filler	Level 1 will be this grade
HiSil 233 (variable)	0–20	Reinforcing filler	Level 2 will be this grade
Vinyl-tris(β-methoxyethoxy) silane (Silane A-172)	1.0	Coupling agent for filler dispersion	
Red iron oxide	4.0	Stable pigment	
Varox DCP-40C (variable)	4–8	Rapid vulcanization in injection molding at 170°C	Three levels to optimize properties

Note: Optimization = 3 factor, 2,2 and 3-level full factorial design—12 compounds and experiments.
Responses (properties) to be measured: Heat resistance, 1008 h at 125°C, 504 h at 150°C; ozone resistance; coolant resistance, 1008 h at 125°C; low-temperature flexibility; dynamic flex-fatigue resistance; molding performance, flow and cure rates.

A starting compound is given in Table 3.16 with variables for experimental design and compound optimization.

3.11.7 APPLICATION—HIGH-TEMPERATURE DIFFERENTIAL SHAFT SEAL

In this application, our customer requires a higher temperature material for the output shaft of the drive differential joint of an automobile. The customer had been using NBR based shaft seals for many years, but relocation of the exhaust and catalytic converter system with vehicle redesign has created an environment near the shaft seal where the seal elastomer lip sees temperatures continuously in the range of 100°C to −125°C and premature hardening of the NBR seal is causing shortened seal life and warranty issues. The extreme pressure lubricant oils used in the differential

TABLE 3.16
Green, O-Ring Compound for Automotive Air Conditioning Seals

Compound Development and Optimization

Compound Basics Ingredient	phr	Logic/Comments	Experimental Design Optimization
21%–25% Acrylonitrile HNBR polymer, 1%–4% residual unsaturation (variable)	80–100	Best low-temperature flexibility for cold starts	Vary level of HNBR and level of EPDM to give optimum properties; total polymer phr = 100
EPDM polymer—low ethylene, low viscosity, 3%–5% third monomer unsaturation level (variable)	20–0	Low viscosity for injection mold flow; medium to high third monomer level for efficient vulcanization	Vary level from 20 to 0 phr with blending of HNBR polymer to achieve total 100 phr of polymer
Zinc oxide	2.0		
Stearic acid	1.0		
Naugard 445	0.7	Best aging performance	
Vanox ZMTI	0.7	Best aging performance	
Nulok 321 (variable 2, level 1)	0 or 50	Extending filler	Evaluate two types of extending fillers for optimum properties
Zeolex 23 (variable 2, level 2)	0 or 50	Extending filler	Evaluate two types of extending fillers for optimum properties
Triisononyl mellitate plasticizer	10.0	Viscosity reduction and low-temperature flex improvement	
Chrome Green pigment	7.0	Temperature stable pigment	
Varox DCP-40C	6.0	Fast vulcanization at 170°C	
Vanax MBM coagent (variable)	5.0–7.5	Efficient vulcanization for low compression set	Evaluate two levels of coagent for optimum properties

Note: Optimization = 3 factor, 3,2 and 2-level full factorial design—12 compounds and experiments.
Responses (properties) to be measured: Resistance to HFC-134a, PAG oil, and HFC-134a/PAG blend at 150°C; O-Ring compression set, long term, at 150°C; low-temperature flexibility; molding flow and vulcanization rate.

have active sulfur and phosphorous chemicals, which act as vulcanizing agents for the highly unsaturated NBR material. The highly saturated nature of HNBRs should give the desired resistance to the differential gear lubricants at the temperatures involved and the toughness and abrasion resistance of the HNBR materials should provide the long-term life necessary in the potentially abrasive environment due to dust and mud. For the high volumes involved, transfer molding is the desired process

TABLE 3.17

High-Temperature Differential Shaft Seal

Compound Development and Optimization

Compound Basics Ingredient	phr	Logic/Comments	Experimental Design Optimization
34%–36% Acrylonitrile HNBR, 1%–4% residual unsaturation (variable)	100.0		Evaluate two polymers for best overall properties
Zinc oxide	2.0		
Stearic acid	1.0		
Naugard 445	0.7	With ZMTI, gives best antioxidant performance	
Vanox ZMTI	0.7	With Naugard 445, gives best antioxidant performance	
Celite 350 or Mistron Cyprubond (variable)	75.0	Mineral fillers show best rotating shaft seal performance	Evaluate two different fillers for best overall properties
N550 carbon black	10.0	Black pigment	
Varox VC40KE	5–10		Evaluate three levels of peroxide for best overall properties

Note: Optimization = 3 factor, 2,2 and 3-level full factorial design—12 compounds and experiments.
Responses (properties) to be measured: Resistance to gear lubricant at 125°C for 1008 h; molding evaluations = flow, cleanliness, bonding to steel case; shaft seal life tests at extremes of operating environment.

and the HNBR material must bond well to the metal casing of the shaft seal. Optimal performance in this application typically dictates an 80 Shore A hardness material. The critical properties of this material are long-term resistance to the gear lubricants at 125°C, molding performance in terms of flow and cleanliness, bonding to the steel case material, and shaft seal life in the simulated application extremes test program. The starting recipe and variables for experimental design and optimization are shown in Table 3.17.

3.11.8 APPLICATION—SHORT POWER STEERING SYSTEM HOSE

For this application, the customer needs a short hose segment for power steering hydraulic fluid. Because of tight space requirements, the fabric reinforced hose needs to have one compound that will resist heat, weathering, and ozone on the outside and mineral based hydraulic fluid inside the hose. The hose will be plied around the braided fabric reinforcement and the plies will be formed by extrusion. Because of the location of an exhaust header near the short section of hose, the operating temperature of the hose will be continuous at 125°C with occasional temperature

TABLE 3.18
Short Power Steering Fluid Hose

Compound Development and Optimization

Compound Basics Ingredient	phr	Logic/Comments	Experimental Design Optimization
34%–36% ACN, 1% and 5% unsaturation HNBR polymer	100.0	Best overall balance of fluid resistance	Low level at 1% unsaturation, high level at 5% unsaturation
Zinc oxide	2.0		
Stearic acid	1.0		
N550 carbon black	60.0		
Plasthall TOTM	10.0	Viscosity reduction without sacrifice of heat aging	
Agerite Resin D	1.0	With protective wax, gives best ozone resistance	
Protective wax	2.0	With Agerite Resin D, gives best ozone resistance	
Varox DCP-40C	3.0–5.0	Fast and efficient curing	Variable level to optimize properties

Note: Optimization = 2 factor, 2 and 3-level full factorial design—12 compounds and experiments.
Responses (properties) to be measured: Static ozone resistance; dynamic ozone resistance; flex-fatigue resistance; extrusion processing; high temperature, pressure pulse simulated application test of hose sample.

excursions to 150°C at idle in heavy traffic. These requirements dictate the use of HNBR for its temperature resistance, overall toughness, retention of modulus at high temperatures, and ozone resistance. The critical properties of the compound and hose assembly are the static and dynamic ozone resistance, flex-fatigue resistance, smooth processing in extrusion, adhesion of the plies, and a pressure pulse test established by the customer on the finished hose section. A 75 Shore A hardness compound offers the best balance of pressure resistance and conformity in the fittings. The starting recipe and variables for Statistical Experimental Design optimization are given in Table 3.18.

3.11.9 APPLICATION—HIGH-PRESSURE RESISTANT O-RING FOR POWER STEERING PUMP OUTPUT

The customer needs a high modulus, pressure-resistant O-Ring for the output fitting on a power steering pump. Because the engine compartment temperatures have risen from 100°C continuous to 125°C continuous with excursions to 150°C, NBR materials do not have sufficient long-term life in this application. The output fitting of the pump typically sees pressures of 1500 psi with pulses to 2000 psi. The maximum extrusion gap of the O-Ring gland is such that an 80 Shore A rubber compound is desirable. Due to the high volumes and relatively small inner diameter of the O-Ring, injection molding is the most cost-effective process. The key properties of the rubber

TABLE 3.19
Power Steering Pump O-Ring

Compound Development and Optimization

Compound Basics Ingredient	phr	Logic/Comments	Experimental Design Optimization
21% and 25% ACN, 1%–4% unsaturation HNBR polymer, low viscosity	100.0	Best low-temperature flexibility, low viscosity for injection molding	Evaluate two polymers, 21% and 25% ACN
Zinc oxide	2.0		
Stearic acid	1.0		
Naugard 445	1.0	With Vanox ZMTI, gives the best antioxidant performance	
Vanox ZMTI	1.0	With Naugard 445, gives the best antioxidant performance	
N550 carbon black	60.0		
Plasthall TOTM	5.0	Viscosity reduction without sacrifice of heat aging	
Varox DCP-40C	4.0–6.0	Fast curing in injection molding	Vary levels for optimized properties
Vanax MBM	4.5–7.5	Efficient curing and low compression set	Vary levels for optimized properties

Note: Optimization = 3 factor, 2,2 and 3-level full factorial design—12 compounds and experiments.
Responses (properties) to be measured: Heat resistance 1008 h at 125°C, 504 h at 150°C; power steering fluid resistance, 1008 h at 150°C; low-temperature flexibility; compression-set resistance of O-Rings, 1008 h at 125°C; injection molding processing.

compound are resistance to temperatures of 125°C continuous with excursions to 150°C, resistance to mineral-based power steering fluids at 150°C, low-temperature flexibility to −40°C to prevent air aspiration during cold weather starts, injection molding processing, and long-term compression set resistance. A starting compound and variables for Experimental Design optimization are given in Table 3.19.

3.11.10 APPLICATION—OZONE RESISTANT AND ABRASION-RESISTANT HOSE COVER

Our customer needs a cover for a hydraulic hose, which resists temperatures to 125°C, is abrasion resistant for long life rubbing against cement floors in the application, and has excellent long-term ozone resistance. The high-temperature requirements dictate that NBR materials will not work. The combination of abrasion resistance, resistance to a wide variety of fluids that may be spilled on the hose cover, temperature resistance, and ozone resistance makes HNBR blends with Zinc Methacrylate modified HNBR the best choice. The use of Zinc Methacrylate

TABLE 3.20
Ozone-Resistant Hose Cover

Compound Development and Optimization

Compound Basics Ingredient	phr	Logic/Comments	Experimental Design Optimization
34%–36% ACN, 1%–4% unsaturation HNBR polymer	50–40	Best overall chemical resistance	Vary level between 50 and 40 phr with total polymer at 100 phr
Zinc methacrylate modified HNBR	50–60	Contributes toughness, abrasion resistance, and ozone resistance	Vary level between 50 and 60 phr with total polymer at 100 phr
Naugard 445	1.0	With Vanox ZMTI, gives good heat aging	Package 1
Vanox ZMTI	1.0	With Naugard 445, gives good heat aging	Package 1
Agerite Resin D	1.0	With protective wax, gives best ozone resistance	Package 2
Protective wax	2.0	With Agerite Resin D, gives best ozone resistance	Package 2
Zinc oxide	2.0		
Stearic acid	1.0		
N330 carbon black	5.0	Pigment	
Plasthall DBEEA	5.0	Smooth extrusion without loss of heat resistance	
Phenol-acetylene resin	10–15	Tackifier for ply building	Vary level for best properties
Varox DCP-40C	5.0–7.0	Fast curing	Vary level for best properties

Note: Optimization = 4 factor, 2,2,2 and 2-level full factorial design—16 compounds and experiments.
Responses (properties) to be measured: Abrasion resistance; static ozone resistance; resistance to spill fluids, 70 h at 125°C; tack for ply build-up; extrusion processing; heat resistance 1008 h at 125°C.

modified HNBR requires a peroxide to react the acrylate as well as vulcanizing the compound. The hose plies will be built from extruded profiles of the various layers. The critical properties for the compound are relative abrasion resistance, static ozone resistance, resistance to potential spilled fluids at 125°C, successful adhesion on plying the hose construction, extrusion processing, and long-term temperature resistance to 125°C. The starting compound and variables for Experimental Design optimization are given in Table 3.20.

3.11.11 APPLICATION—CHEMICALLY RESISTANT, LOW HYSTERESIS ROLLER FOR PAPER MILLS

The customer for this application is a paper mill with requirements for longer running calender gloss rolls at temperatures slightly above 120°C. The variety of

TABLE 3.21
Paper Mill Calender Gloss Roll

Compound Development and Optimization

Compound Basics Ingredient	phr	Logic/Comments	Experimental Design Optimization
34%–36% ACN, 5% unsaturation HNBR polymer	50.0	Basic HNBR for chemical resistance	Level 1 evaluation of HNBR
34%–36% ACN, 15% unsaturation HNBR polymer	50.0	Higher unsaturation for less hysteresis	Level 2 evaluation of HNBR
Zinc methacrylate modified HNBR polymer	50.0	Adds toughness, abrasion resistance, and high hardness	
Silane treated, calcined clay	70.0	Mineral filler	Level 1 filler evaluation
Mistron Cyprubond	70.0	Mineral filler	Level 2 filler evaluation
Varox VC40KE	7.0		
Vanax MBM	4.5–7.5	Efficient curing	Variable levels to optimize properties

Note: Optimization = 3 factor, 2,2 and 2-level full factorial design—8 compounds and experiments.
Responses (properties) to be measured: Dynamic hysteresis; heat resistance, 1008 h at 120°C; resistance to application fluids, 70 h at 120°C; extrusion processing.

chemicals used dictates that the roll compound must have good resistance to water-based and mineral-based fluids. Current urethane rolls are not lasting long enough due to the combination of chemicals and temperatures. Low hysteresis is required to prevent excess heat build-up in the rollers, which will upset the process. The desired hardness of the roller compound is 60 Shore D. Because the rolls will be used with brightly colored papers, a black filled compound is not acceptable. The abrasion resistance, temperature resistance, chemical resistance, and retention of modulus at high temperatures make a HNBR + Zinc Methacrylate modified HNBR a good choice for this application. The rolls will be built from smooth calendered stock plied onto steel cores and the rolls will be ground to size after molding. Critical properties for the roll compound are low hysteresis, temperature resistance to 120°C, resistance to application chemicals, and smooth calendering. A starting compound with variables for Experimental Design optimization is shown in Table 3.21.

3.11.12 APPLICATION—HIGH-PRESSURE RESISTANT O-RING FOR OIL DRILL BITS

This customer has a need for a material resistant to temperatures of 150°C at high pressure differentials up to 10,000 psi with crude oil, natural gas, and hydrogen sulfide resistance. Even with tight control of the O-Ring gland extrusion gap, a 90 Shore A material is required. The combination of resistance to temperature, resistance to crude oil, resistance to pressure, and the good retention of modulus at high temperatures makes HNBR a good choice for this application. Critical properties are resistant to crude oil (we will use ASTM reference oils as the test media), resistance

TABLE 3.22

High Pressure O-Ring for Oil Drill Bits

Compound Development and Optimization

Compound Basics Ingredient	phr	Logic/Comments	Experimental Design Optimization
34%–36% ACN, 1% unsaturation HNBR polymer	100.0	Best overall balance of properties	−1 level of polymer
34%–36% ACN, 5% unsaturation HNBR polymer	100.0	Best overall balance of properties	+1 level of polymer
Zinc oxide	2.0		
Stearic acid	1.0		
Naugard 445	1.0	With Vanox ZMTI, gives good heat resistance	
Vanox ZMTI	1.0	With Naugard 445, gives good heat resistance	
N330 or N550 carbon black	70.0	Reinforcing filler	−1 level using N550, +1 level using N330
Varox VC40KE	7.0		
Vanax MBM	4.5–7.5	Efficient curing	Vary level between 4.5 and 7.5 phr for optimization of properties

Note: Optimization = 3 factor, 2,2 and 2-level full factorial design—8 compounds and experiments.
Responses (properties) to be measured: ASTM reference oil resistance, 70 h at 150°C; hexane resistance to simulate light hydrocarbons, 70 h at room temperature; O-Ring compression set, 1008 h at 150°C; pressure extrusion test in simulated application gland.

to light hydrocarbons such as methane, and resistance to pressure extrusion in a simulated gland. The starting compound with Experimental Design variables for optimization is shown in Table 3.22.

REFERENCES

1. American Society for Testing and Materials (2002). D2000–01. In *Rubber* (Vol. 9.02). Washington, DC: American Society of Testing and Materials.
2. Zeon Chemicals L.P. *A Comparison of Oil Resistant Elastomers in Engine Oil and ATF* (Z7.3.16). Louisville, KY.
3. Zeon Chemicals L.P. (1999). *Zetpol Product Guide* [Brochure]. Louisville, KY.
4. McCormick, T. and Keller, R.W. (February 5, 1992). *Raw Material Test Audit Program for NBR Polymers* (R&D Test Report 979). Lebanon, TN: Wynn's-Precision, Inc.
5. Baird, W. and Keller, R.W. (February 26, 1992). *Raw Material Test Audit Program for HSN Polymers* (R&D Test Report 980). Lebanon, TN: Wynn's-Precision, Inc.
6. Zeon Chemicals L.P. *Semi-EV and EV Curing Systems for Zetpol 1020 (Z5.1.2)* [Brochure]. Louisville, KY.

7. Zeon Chemicals L.P. *A Study of Organic Peroxides in Zetpol 2010* (Z5.1) [Brochure].
    Louisville, KY.
8. Zeon Chemicals L.P. *Study of Level and Ratio of Peroxide and Co-Agent in Zetpol 2010*
    (Z5.1) [Brochure]. Louisville, KY.
9. Zeon Chemicals L.P. *Effects of Post Curing on Zetpol 1020* (Z5.1) [Brochure]. Louisville,
    KY.
10. Zeon Chemicals L.P. *Carbon Black Reinforcement of Zetpol* (Z5.2.1 Ed.) [Brochure].
    Louisville, KY.
11. Zeon Chemicals L.P. *Evaluation of Larger Particle Non-Black Fillers in Peroxide Cured
    Zetpol 2010* (Z5.3.3) [Brochure]. Louisville, KY.
12. Zeon Chemicals L.P. *Evaluation of Silica and Silicate Fillers in Peroxide Cured Zetpol
    2010* (Z5.3.2) [Brochure]. Louisville, KY.
13. Zeon Chemicals L.P. *Protection Systems for Zetpol* (Z5.5 Ed.) [Brochure]. Louisville, KY.
14. Campomizzi, E.C. (2000 January 27). *A new master batch to extend the heat resistance of
    Therban HNBR compounds*. Paper presented at the meeting of the Chicago Rubber Group
    Technical Meeting. Bensenville, IL.
15. Zeon Chemicals L.P. *Low Temperature Properties of Zetpol and Influence of Plasticizers
    and Tackifiers* (Z5.4 Ed.) [Brochure]. Louisville, KY.

4 Fluoroelastomers, FKM, and FEPM

Pascal Ferrandez

CONTENTS

The information set forth herein is furnished free of charge and is based on technical data that DuPont Performance Elastomers believes to be reliable. It is intended for use by persons with technical skill, at their own discretion and risk. Handling precaution information is given with the understanding that those using it will satisfy themselves that their particular conditions of use present no health hazards. Because conditions of product use and disposal are outside our control, we make no warranties, expressed or implied, and assume no liability in connection with any use of this information. As with any material, evaluation of any compound under end-use conditions before specification is essential. Nothing herein is to be taken as a license to operate under or a recommendation to infringe on any patents. While the information presented here is accurate at the time of publication, specifications can change. Check www.dupontelastomers.com for the most up-to-date information.

CAUTION: Do not use in medical applications involving permanent implantation in the human body. For other medical applications, discuss with your DuPont Performance Elastomers customer service representative and read Medical Caution Statement H-69237.

Copyright© DuPont Performance Elastomers. All Rights Reserved.

Viton® is a registered trademark of DuPont Performance Elastomers.

4.1   FLUOROELASTOMER OVERVIEW

4.1.1   INTRODUCTION

Fluoroelastomers are typically used when other elastomers fail in harsh environments. Chemical resistance and heat resistance are the two main attributes that gave fluoroelastomers successful growth since their introduction in 1957 by the DuPont Company. Applications range from niche applications in aerospace to large series applications with O-rings and shaft seals in the automotive industry. Use of fluoroelastomers continues to grow due to ever more demanding applications, in automotive to more stringent regulations, and in the chemical industry to support environmental protection. There are a multitude of fluoroelastomer products, different in composition and end-use performance, but also different in viscosity or incorporated level of curatives and process aids. The compounder has to be in close contact with his supplier to understand the product line and follow new product introductions.

4.1.2   HISTORY

Fluoroelastomers were introduced in the late 1950s and have been developed over the years with various compositions from dipolymers [1,2] to pentapolymers [3–7], to offer different levels of chemical resistance, and many cure systems to either improve compression set resistance or steam and acid resistance.

4.1.3 Suppliers

There are four major suppliers in the world: DuPont Performance Elastomers (USA) [8], Dyneon (USA) [9], Solvay Solexis (Europe) [10], and Daikin (Japan) [11]. Asahi Glass in Japan also supplies some specialty fluoroelastomers while some local suppliers also exist in China and Russia.

4.2 WHY USE A FLUOROELASTOMER

Fluoroelastomers have excellent heat resistance with continuous service temperature of 205°C. They are referenced as FKM and rated as HK with 250°C heat resistance and excellent oil resistance, and less than 10% swell at 150°C as per the ASTM D2000 classification.

Another category of fluoroelastomers, called FEPM has been created recently for base resistant fluoroelastomers. For higher temperature and chemical resistance fully fluorinated materials, called perfluoroelastomers, are available in the form of molded seals and parts, such as Kalrez® perfluoroelastomers parts and Chemraz® Perfluoroelastomer Compounds. They bear the classification of JK or HK depending on supplier and compound offered, JK offering a continuous service temperature of 315°C. FFKM compounds are available from most FKM suppliers but gums are rarely available and in this chapter we focus on fluoroelastomers FKM and FEPM.

Fluoroelastomers can withstand limited exposure to 300°C. However, they should not be exposed to temperatures higher than 315°C as HF emissions could occur. See handling precautions bulletins for more information [12].

Fluoroelastomers are used extensively in O-rings and gaskets because they have very good compression set resistance as shown in Figure 4.1. Another way to look at sealing performance is to measure retained sealing force and Figure 4.2 shows the performance of fluoroelastomers versus other commonly used elastomers in sealing applications.

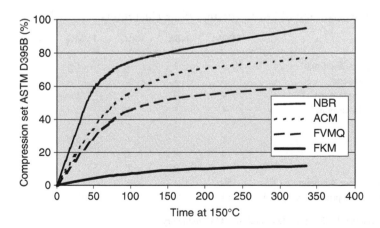

FIGURE 4.1 Compression set resistance of fluoroelastomers versus other elastomers.

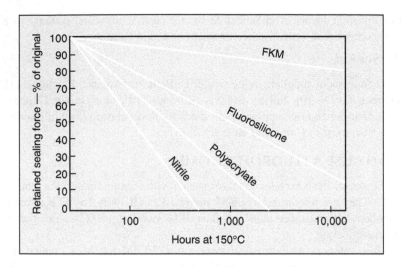

FIGURE 4.2 Retained sealing force of FKM versus other elastomer O-rings.

Chemical resistance is the other main attribute of fluoroelastomers with excellent resistance to fuels, solvents, and oils. Steam and acid resistance is satisfactory with bisphenol cured FKM, and very good with peroxide cured FKM (Table 4.1). Strong alkalis require the use of FEPM, namely TFE/P and ETP types (Table 4.2).

Fluoroelastomer compounds, although more costly than most mid-performance elastomers, provide additional value with heat, chemical resistance, and longer seal life. Figure 4.3 expresses the value in use versus the cost per liter (more adapted as part prices are volume based, whereas material prices are mass based).

TABLE 4.1
Chemical Resistance of FKMs versus Other Elastomers

Elastomer	EPDM	NBR	VMQ	FVMQ	FKM	FFKM
Lubricating oils, Fuel oils	4	1	4	1	1	1
Gasoline, Kerosene	4	1	4	1	1	1
Aromatic hydrocarbons	4	3	4	3	1	1
Aliphatic hydrocarbons	4	1	4	2	1	1
Alcohols	2	1	2	2	1	1
Ketones	1	4	4	4	4	1
Acids	4	4	4	3	2	1
Alkalis	1	2	2	2	4	1
Maximum service temperature (°C)	150	121	230	175	205	315
Low temperature properties, T_g (°C)	−54	−30	−85	−65	−17, −30	−8

1 = Satisfactory; 2 = Fair; 3 = Doubtful; 4 = Unsatisfactory.

TABLE 4.2
Chemical Resistance of Major FKM Families

	Family, Type of Viton Fluoroelastomer							
	A	B	F	GBL	GF	GLT	GFLT	ETP
	Cure System							
	Bisphenol					Peroxide		
Aliphatic hydrocarbons, process fluids, chemicals	1	1	1	1	1	1	1	1
Aromatic hydrocarbons (toluene, etc.), process fluids, chemicals	2	1	1	1	1	2	1	1
Automotive and aviation fuels (pure hydrocarbons— no alcohol)	1	1	1	1	1	1	1	1
Automotive fuels containing legal levels (5%–15%) of alcohols and ethers (methanol, ethanol, MTBE, TAME)	2	1	1	1	1	2	1	1
Automotive/methanol fuels blends up to 100% methanol (flex fuels)	NR	2	1	2	1	NR	1	1
Engine lubricating oils (SE-SF grades)	2	1	1	1	1	1	1	1
Engine lubricating oils (SG-SH grades)	3	2	2	1	1	1	1	1
Acid (H_2SO_4, HNO_3), hot water, and steam	3	2	2	1	1	1	1	1
Strong base, high pH, caustic, amines	NR	NR	NR	NR	NR	NR	NR	1–2
Low molecular weight carbonyls—100% concentration (MTBE, MEK, MIBK, etc.)	NR	NR	NR	NR	NR	NR	NR	1–2
Low temperature sealing capability TR-10 test results	−17°C	−14°C	−7°C	−15°C	−6°C	−30°C	−24°C	−11°C

Note: 1 = Excellent, minimal volume swell; 2 = Very good, small volume swell; 3 = Good, moderate volume swell.
NR = Not Recommended, excessive volume swell or change in physical properties.

4.3 FLUOROELASTOMERS MOLECULAR STRUCTURE

4.3.1 MONOMERS

Fluoroelastomers are made of fluorinated monomers. Dipolymers made from VF2 (vinylidene fluoride) and CTFE (chlorotrifluoroethylene) were first developed in

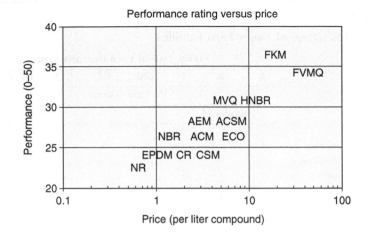

FIGURE 4.3 Performance versus cost for various elastomers. Performance is a combination of end-use performance criteria such as heat and chemical resistance, and price is taking into account the volume of the part rather than the mass as parts are sold by the piece and not by weight.

the mid-1950s, but were quickly replaced by dipolymers of VF2 and HFP (hexa-fluoropropylene) with the introduction of Viton®, a fluoroelastomer from DuPont in 1957. Dipolymers are still the main compositions used in the industry.

TFE (tetrafluoroethylene), a third monomer was soon introduced to increase the fluorine level of fluoroelastomers and increase their chemical resistance, particularly to alcohols. TFE is usually added at the expense of VF2, reducing hydrogen content, whereas HFP is maintained at sufficient levels to break crystallinity.

The addition of a fourth, cure site monomer is used for peroxide curing and to improve the acid and steam resistance.

Unfortunately, while fluorine level is increased from typically 66% to 70%, low-temperature flexibility is reduced with glass transition temperatures increasing from $-19°C$ to $-7°C$. This drop can be compensated by the substitution of HFP by PMVE (perfluoromethylvinyl ether) a more expensive monomer that improves low-temperature performance by about 15°C.

4.3.2 POLYMER STRUCTURE—MOLECULAR WEIGHT

Most fluoroelastomers have a Mooney viscosity ranging from 20 to 70 (ML $1 + 10$ at 121°C), usually with a molecular weight average from 100,000 to 1,000,000 and a polydispersity ranging from 2 to 5. Polydispersity is the ratio of weight average to number average molecular weight. The higher the ratio, the broader the molecular weight distribution. There are some exceptions with very low molecular weight average products, very high weight average products, and some bi-modal products. These products were developed for special processing needs.

4.3.3 Vulcanization

First, fluoroelastomers were cured with amines. However, the scorchy nature of those cure systems as well as a rather poor compression set resistance led to the development of bisphenol cure systems [13,14]. Peroxide curing is performed thanks to the addition of a cure site monomer [4,15–18] for compositions that cannot be cured with bisphenol, such as high fluorine compositions, low-temperature polymers where HFP has been substituted by PMVE, and non-VF2 containing polymers such as FEPMs (TFE/P and ETP polymers), unless a bisphenol cure site has been added to the polymer [19,20].

4.4 CHOOSING THE RIGHT FLUOROELASTOMER

4.4.1 Chemical Resistance and Low Temperature

The two major differentiators for fluoroelastomers are chemical resistance (specifically, base resistance and oxygenated fuel resistance) and low-temperature flexibility.

Once the right fluoroelastomer has been selected, there is little that can be done to improve these properties. For example, the wide use of plasticizers for improving the low-temperature performance of other elastomers is not acceptable in fluoroelastomers: even ester plasticizers are not compatible and tend to exude in the post-cure step or in service.

4.4.2 Processes

Most common rubber processes are also used with fluoroelastomers. Compression molding is still the preferred way for small production quantities. However, compression transfer, injection transfer, and injection molding are now commonly used. Fluoroelastomers can be extruded, calendered, and dissolved for mastic and coating applications.

Typical mold cycle times range from less than a minute for injection molding to up to 10 min for thick parts made by compression molding. Molding temperatures are typically between 175°C and 205°C. Most formulations and end-use will require a post-cure cycle ranging from 1 to 24 h, at temperatures between 200°C and 240°C.

For hose applications an autoclave forming stage will be performed at 150°C–170°C for 20–60 min.

Fluoroelastomers require processing under pressure; hence ultrasound, salt bath, or hot air curing after extrusion is not effective as the part will foam with the cure system generating off-gas.

Processing areas need good ventilation. See handling precautions bulletins for more information [12].

4.4.3 Selection Tree

The first step is to select the right composition for the fluoroelastomer since little can be achieved through compounding. Then, the right viscosity and the right type must be selected. Finally, the formulation will tackle properties such as hardness, modulus, color or mold release, and extrudability. A comprehensive tool, The Chemical Resistance Guide, is available on DuPont Performance Elastomers' Web site [21] www.dupontelastomers.com, which guides the compounder to the right composition and type, depending on end-use application and process.

4.4.3.1   Base Resistance Grades of FKM

While only TFE/P [22] and ETP [23] elastomers were available a few years ago, there is now a wide range of products marketed as base-resistant fluoroelastomers. The trade off is usually on low temperature and oil resistance. Figure 4.4 shows the current offering of base-resistant fluoroelastomers. Only Aflas® 100 series and Viton® Extreme™ TBR-S [24], and ETP-S [25] series have the designation of FEPM, whereas the others, containing VF2 still have the FKM designation. The absence of VF2 in the polymer is key in achieving true base resistance, as the sequence of VF2 and HFP is particularly prone to basic attack and dehydrofluorination [24]. Aflas 100 and TBR have a glass transition T_g above freezing point, whereas the T_g for Dyneon® 7132 [26] and Tecnoflon® BR-9151 [27] is slightly below 0°C. Only ETP, Aflas® MZ series, and GFLT have relatively low T_g, respectively, of -10°C, -13°C, and -26°C, but ETP is the only FKM and FEPM that combines base resistance, fuel resistance, and adequate low temperature flexibility.

4.4.3.2   Fuel Resistance or Permeation

Most applications call for either base resistance or fuel resistance. In the case where both are needed, the choice becomes very limited. Using GFLT for fuel

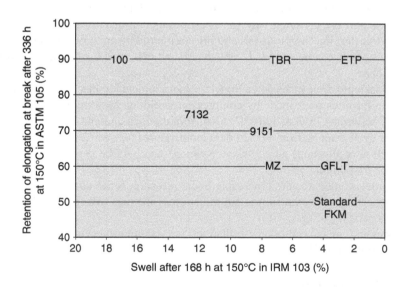

FIGURE 4.4 Relative oil resistance of FEPM and FKM Materials. Aflas is a registered trademark from Asahi Glass; Viton is a registered trademark from DuPont Performance Elastomers; Extreme is a trademark from DuPont Performance Elastomers; Dynamine is a registered trademark from Dyneon; Tecnoflon is a registered trademark from Solvay Solexis.

and moderate base resistance, and ETP for extreme chemical resistance, ETP can handle fuels, solvents, acids, steam, and strong bases. ETP is actually the only fluoroelastomer that meets both FKM and FEPM designation.

If base resistance is not needed, a breadth of fluoroelastomers can be used. We use DuPont Performance Elastomers' designation for families to differentiate fuel resistance, from an A type to design standard dipolymers to a B type commonly used in shaft seals, to an F type for the highest fluorine level while maintaining adequate rubber properties (higher fluorine level can be achieved but require compositions into the elasto-plastic region that do not deliver the typical FKM compression set resistance). Another family can be designed by replacing HFP monomer by PMVE. These types are designated by an LT suffix. These compositions are preferably cured with peroxide, and a G suffix designates this feature, which can also be applied to other HFP compositions and achieve better chemical resistance, particularly steam and acid resistance.

Since low-temperature flexibility is one of the weak points for fluoroelastomers, Figure 4.5 is using TR-10 retraction temperature to differentiate the different families.

A more distinct way to look at fluoroelastomers is to compare their fuel permeation resistance, particularly when the fuel has been modified by an oxidizing agent [28]. Figure 4.6 illustrates typical permeation rates achieved with fluoroelastomers compared with other elastomers.

Table 4.2 provides additional information on fluid resistance and a more comprehensive tool exists on DuPont Performance Elastomers Web site [29].

4.4.3.3 Low-Temperature Flexibility

In view of the premium associated with specialty fluoroelastomers, the compounder must decide what degree of low-temperature flexibility is really required. The glass

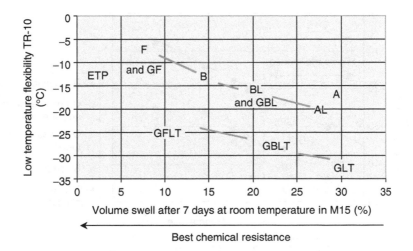

FIGURE 4.5 Relative fuel resistance versus low temperature flexibility of major FKM families (M15 fuel is a blend of 15% methanol in 85% Fuel C. Fuel C is an aromatic hydrocarbon mixture of 50% isooctane/50% toluene).

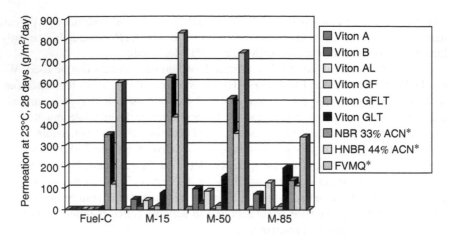

FIGURE 4.6 Permeation of various FKMs versus NBR, HNBR, and FVMQ.

transition temperature and temperature of retraction TR-10, are very close for most FKM, and for dry applications would certainly qualify for lowest dynamic sealing temperature. On the other hand, when soaked or in static applications, seals can perform well below their glass transition temperature [30] (Figure 4.7).

4.4.3.4 End Use and Processing

The fluoroelastomer market has a product class that was developed in the 1970s, with precompounds or cure-incorporated products. FKM suppliers developed this

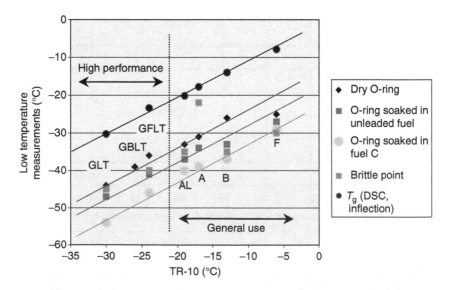

FIGURE 4.7 Various low temperature measurement methods on FKM compositions.

offering to help the customers handle difficult to mix and handle bisphenol cure systems. Most products are available in precompounded form but peroxide gums are usually not precompounded because of bin stability. Separate gum and chemical offerings are available from some FKM suppliers to help the compounder tailor compounds for a specific use.

Once considerations involving chemical resistance and low-temperature flexibility have narrowed down the selection of fluoroelastomers, there are still quite a few parameters to consider such as finer aspects of chemical resistance, complexity of the part to make, and the forming and vulcanization processes envisioned to make the part.

1. Resistance to acid and steam will direct the compounder to peroxide curing systems.
2. Difficult to mold parts with undercuts require lower level of curatives.
3. Bonding to a substrate will require the use of a bonding promoter, which might be already incorporated.
4. Process will, in part, dictate the viscosity in terms of Mooney viscosity, molecular weight distribution, level of end groups, and degree of branching. Although these are the major parameters governing rheological parameters, fluoroelastomer suppliers usually supply their products without much rheological information, and prefer to guide the compounder with the Mooney viscosity and the process for which the product has been designed.

4.5 COMPOUNDING

One basic rule the compounder must remember about fluoroelastomers is that these products are extremely resistant to chemicals and heat. Therefore, any compounding ingredient is likely to adversely affect FKM performance. Plasticizers are not tolerated and most chemicals commonly used in rubber compounding are not recommended.

Typical properties for fluoroelastomers are

1. Hardness, Shore A, pts 55–95
2. Tensile Strength, 7–20 MPa
3. Elongation at break, % 100–500
4. Compression Set, 22 h at 200°C, % 10–50

4.5.1 CURING CHEMICALS

The preferred way to cure fluoroelastomers is with Bisphenol AF and accelerator salts. They are usually incorporated in the commercial product but can be purchased separately (e.g., Viton® Curative 20, 30, and 50) to adjust the cure state or cure rate.

Amine curatives such as DIAK™ 1 and DIAK™ 3 were first introduced in the late 1950s but were rapidly abandoned when the bisphenol cure system was introduced as it could give much better scorch safety and superior compression set resistance.

Peroxide curing systems can be used only with those fluoroelastomers containing a peroxide cure site, for example the low temperature, PMVE containing types, the highly fluorinated types, and other specialty type such as BR 9151 or ETP. Peroxides such as Luperco® 101XL are preferred, in conjunction with TAIC coagent DIAK™ 7.

The entire cure system is used in conjunction with metal oxides, which act as acid acceptors in all systems, capturing HF formed during vulcanization. In the case of bisphenol cure systems, they act as bases to dehydrofluorinate the backbone, and their concentration is essential to the cure response [14].

4.5.2 METAL OXIDES

For bisphenol curing, a combination of high activity magnesium oxide and calcium hydroxide is preferred, typically 6 phr of $Ca(OH)_2$ and 3 parts of MgO. Different ratios can be used with 3 phr of $Ca(OH)_2$ and 6 parts of MgO for better bonding to metal inserts, and 9 phr of MgO only for shorter post-cure cycles.

For peroxide types, lower levels of oxides can be used from 1 to 3 phr. Zinc Oxide is preferred for better acid and steam resistance.

4.5.3 FILLERS

4.5.3.1 Black Fillers

Fluoroelastomers exhibit a rather high viscosity before curing and low hardness is difficult to achieve. Without filler, most FKM exhibit Shore A hardness of 50–60. Therefore, mainly MT and SRF black fillers can be used. MT N990 Black is the best multipurpose black for fluoroelastomers. More reinforcing blacks like SRF N762 and HAF N330 can be used alone or blended with MT at low levels, to achieve higher modulus and higher tensile strength at the expense of elongation at break. Such blends can be used in hose applications to reduce collapse. Other specialty blacks can be considered, such as conductive blacks (XC-72, N472) for electrostatic or conductive properties and Austin black for better compression set resistance.

4.5.3.2 Other Fillers

Many types of white fillers are available. Barium sulfate, or $BaSO_4$ (also referenced as Blanc Fixe in formulations) is a versatile filler with good acid and water resistance. It has a specific gravity of 4 compared to 1.8 for most fluoroelastomers.

Calcium metasilicate or Wollastonite® (Nyad® 400) offers more reinforcement with high hot tear resistance, necessary when the part is undercut, for example, in a shaft seal. It is often used in conjunction with barium sulfate.

Calcium carbonate, $CaCO_3$ provides superior stress-strain properties and Diatomaceous Silica (Celite® 350) high tensile strength and low elongation.

4.5.3.3 Specialties

Aerosil® R-972 can be used when there is need for extreme reinforcement.

Titanium dioxide or TiO_2 (TiPure® R960) can be used as a filler as well as a white pigment.

Fluoropolymer micropowder such as Zonyl® MP, can be used to reduce the coefficient of friction (10–20 phr of MP 1000), or to reinforce due to the fibrillating effect, with as low as 3 phr of MP 1500.

Surface treated fillers are used more often as they enable the compounder to achieve properties approaching those obtained in black formulations. This particularly is true in the case of peroxide formulations containing recently introduced types [31]. Wollastocoat® 10,022 (methacrylate treated), Wollastocoat 10,014 (vinyl silane treated), and Wollastocoat 10,012 (amino silane treated) are examples of treated calcium metasilicates that are preferred to enhance physical properties in non-black peroxide curing formulations.

4.5.4 Pigments

The choice of pigments is limited for fluoroelastomers as most would not withstand high temperatures. The most typical are red iron oxide and chrome oxide, which are often used at 1–2 phr level with 2 phr of TiO_2 to get consistent white background, and 1 phr of MT black to give less vivid colors. More pigments are now available, which exhibit adequate heat resistance, such as blue compounds (Stan-Tone® D-4005 and D-4006, or Cromophtal® Blue 4GNP).

4.5.5 Process Aids

Table 4.3 below lists the most commonly used process aids for fluoroelastomers. Formulations typically contain a few tenths of a percent to a few percent of process aids, very often in combinations of two or more. The compounder will have to check

TABLE 4.3
Major Process Aids for FKMs

| | Features | | | |
Type	Mill Calender Release	Improves Extrusion Smoothness	Mold Release	Compression Set Resistance
VPA #1[a]	+	+	+	0/−
VPA #2[a]	++	++	0/+	−
VPA #3[a]	0/+	0/+	+	0
Carnauba Wax	++	++	0/+	−
Polyethylene	0/+	0/+	0/+	−
Struktol WS280	+	+	+	−
Armeen 18D	0/+	0/+	++[b]	−

[a] DuPont Performance Elastomers supplier.

[b] Peroxy cures:

++ = very positive effect;

+ = positive effect;

0 = no effect;

− = negative effect.

with his supplier if process aids are already incorporated in the precompounds, along with cure chemicals, to avoid excessive process aid levels.

For peroxide curing formulations, Struktol® HT 290 and PAT 777 are also recommended for best mold release [31].

4.5.6 PLASTICIZERS

There is only one fluoroelastomer plasticizer, Dai-El® G-101 from Daikin [32]. No plasticizer, including ester plasticizers, can withstand post-cure cycles and service temperatures typically used with fluoroelastomers.

4.6 COMPOUNDING FOR SPECIAL NEEDS

4.6.1 HARDNESS

Hardness can be adjusted with filler level, as shown in Figure 4.8. Mooney viscosity of the polymer is the only other variable, ranging from 20 for Viton® A-201C, to 65 for A-601C. Mooney viscosity impacts hardness only at low filler levels. Another way to reduce hardness is to reduce crosslink density (Figure 4.9). That can be achieved by purchasing gums or curative masterbatches or by blending a precompound with a pure gum.

The use of the plasticizer Dai-El® G-101 from Daikin can provide hardness values substantially lower, typically 6–10 points for 15–33 phr of G-101 [32].

Achieving lower hardness through the use of low levels of fillers, low Mooney polymers, and low curative levels is illustrated in Figure 4.9.

4.6.2 TENSILE STRENGTH AND ELONGATION AT BREAK

Most black fluoroelastomer formulations achieve a tensile strength above 10 MPa and 150% elongation at break at room temperature. With colored stocks, usually

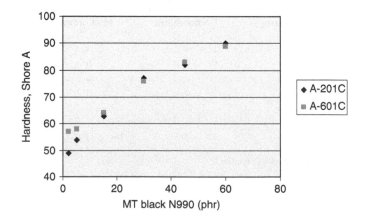

FIGURE 4.8 Hardness versus MT black level in type A FKMs. (Base formulation polymer 100; Ca(OH)$_2$ 6; MgO 3; MT variable), (cured 10 min at 177°C, post-cured 24 h at 232°C).

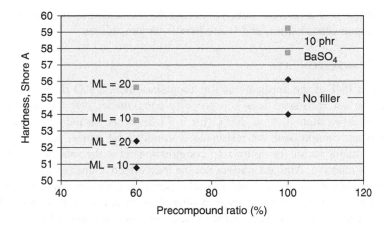

FIGURE 4.9 Lower hardness using low viscosity FKM and low filler and curative levels (cured 10 min at 177°C, post-cured 24 h at 232°C). Formulations are based on precompound made of A-100 or A-200, with 2.5 phr of Viton Curative 50 and 1 phr of VPA 3. 60% precompound means 60% of above precompound blended with 40% of the gum, either A-200 or A-100. All contain, 2 phr of TiPure R960, 1 phr of Chrome Oxide Green, 0.5 phr Carnauba Wax on a 100 phr level.

more than one kind of filler is needed to achieve the required tensile properties with adequate hardness.

Table 4.4 shows an example of properties achieved with various white fillers used to formulate a multipurpose A-type, in comparison with the same level of MT

TABLE 4.4

Comparison of Various White Fillers with MT Black

	A-1	A-2	A-3	A-4	A-5
Viton A-331C	100	100	100	100	100
Maglite D	3	3	3	3	3
Calcium hydroxide	6	6	6	6	6
MT Black (N990)	30	—	—	—	—
Albagloss	—	30	—	—	—
Nyad 400	—	—	30	—	—
Celite 350	—	—	—	30	—
Blanc Fixe	—	—	—	—	30
Stress/strain at room temperature (press cure 10' at 177°C, post-cured 24 h at 232°C)					
M_{100} (MPa)	4.5	3.5	5.8	9.8	2.3
T_b (MPa)	13.5	13.7	11.4	16.1	12.0
E_b (%)	235	240	240	175	325
Hardness, A (pts)	77	68	69	77	65
Compression set, method B, O-rings					
70 h at 200°C	21	20	18	22	17

black. Albagloss is a calcium carbonate, Nyad 400 is a calcium metasilicate, Celite 350 is a diatomaceous silica, and Blanc Fixe is barium sulfate.

4.6.3 Compression Set

Since about half of the applications for fluoroelastomers are O-rings and gaskets, compression set resistance is a key property compounder that typically aim for a value of 20% after 70 h at 200°C for a 75 Shore A formulation. For higher fluorine types, compression set resistance is usually a bit worse, in the 30%–40% range. The major factor for optimizing compression set is the selection of the type and level of curative. Amine curatives are the least effective and led to the development of bisphenol curatives. Until recently, peroxide curatives could not provide the excellent compression set resistance offered by bisphenol cure systems. However, recent advances in development of new cure site monomers have addressed these issues, even offering the capability to reduce the post-cure cycles to only 1 h [15–18].

Figure 4.10 shows the influence of level of curatives and viscosity in the same compounds as those used in Figure 4.9. Only compounds with 10 phr of $BaSO_4$ are represented.

Once the polymer has been selected, fillers can play an important role and Tables 4.5 and 4.6 show the effect of filler level and filler type on compression set resistance. Generally, for a given filler, the lower the level, the better the compression set resistance.

4.7 PROCESSING

Fluoroelastomers can be processed like most elastomers. The major difference to consider is high specific gravity of gum polymer, that is, approximately 1.8 g/cc. Most compounds have specific gravities between 1.8 and 2.0 g/cc. Quantities to be

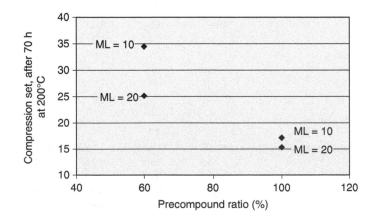

FIGURE 4.10 Influence of curative levels and viscosity on compression set. Same formulations as in Figure 4.9, with 10 phr of $BaSO_4$.

TABLE 4.5
Effect of Filler Level on Compression Set and Physical Properties

	B-1	B-2	B-3	B-4	B-5
Viton A-331C	100	100	100	100	100
Maglite D	3	3	3	3	3
Calcium hydroxide	6	6	6	6	6
MT Black (N990)	60	45	30	15	5
Stress/strain at room temperature (press cure 10' at 177°C, post-cured 24 h at 232°C)					
M_{100} (MPa)	9.9	6.9	4.5	2.3	1.5
T_b (MPa)	15.1	14.4	13.5	12.6	10.0
E_b (%)	165	205	235	305	315
Hardness, A (Pts)	88	83	77	67	59
Compression set, method B, O-rings					
70 h at 23°C	23	20	18	14	13
70 h at 200°C	29	23	21	17	21

processed in internal mixers must be adjusted accordingly. The other major difference is the higher thermal expansion of fluoroelastomers than most rubbers, leading to shrinkage of 2%–4% depending on formulation and mold temperature [33]. Therefore, special molds are usually required for fluoroelastomers.

We refer the reader to literature from certain suppliers, in particular the Processing Guide for Viton from DuPont Performance Elastomers [34], for further information.

TABLE 4.6
Influence of Black Fillers on Compression Set and Physical Properties

	B-6	B-7	B-8	B-9	B-10	B-11
Viton A-331C	100	100	100	100	100	100
Maglite D	3	3	3	3	3	3
Calcium hydroxide	6	6	6	6	6	6
N990—MT Black	30	—	—	—	—	—
N762—SRF Black	—	20	—	—	—	—
N787—SRF HS Black	—	—	20	—	—	—
N550—FEF Black	—	—	—	20	—	—
N330—HAF Black	—	—	—	—	15	—
N110—SAF Black	—	—	—	—	—	15
Stress/strain at room temperature (press cure 10' at 177°C, post-cured 24 h at 232°C)						
M_{10} (MPa)	0.9	1.0	1.2	1.3	1.3	1.5
T_b (MPa)	14.2	17.4	17.7	17.7	18.0	18.1
E_b (%)	255	260	255	245	290	310
Hardness, A (pts)	74	74	78	82	79	82
Compression set, method B, O-rings						
70 h at 200°C	19	19	27	29	33	48

4.8 TROUBLESHOOTING

This quick guide is intended to aid the compounder in finding the cause of unexpected problems.

4.8.1 VULCANIZATION SPEED

Slow cure: may result from sulfur contamination. Add accelerator Viton® Curative 20 to speed up cure, and eliminate contamination areas. If slow cure is not due to sulfur contamination, add $Ca(OH)_2$.

Fast cure: check for condensation on preforms, or high humidity at mixing and other production sites. Check for wet oxides possibly from open containers. Oxides should always be stored in metal containers that one can seal properly after use. Add crosslinker Bisphenol AF through Viton® Curative 30 or calcium oxide.

4.8.2 MOLDING DEFECTS

Flaws: possible contamination from oil leaks from internal mixer, or from other elastomers remaining from previous batches.

Flow lines: often observed in injection molding, they result from use of excessive amounts of process aids. As for knit lines, reducing the level of incompatible process aids is often a solution.

Fissures: often observed in production of thick parts. A post-cure with step rise in oven temperature is often the solution. Start a post-cure cycle at 90°C, below the boiling point of water, for 2–4 h, and increase temperature to post-cure temperature with increments of 25°C/h.

Back rinding: this defect appears as a rough edge at the parting line, and occurs when polymer cures at the surface while the core is still uncured and cold. Thermal expansion forces partially cured rubber out of the mold and scorched, foamy rubber on the mold line. If molding is done through compression, preforms could be warmed before use. Additional bumps or a difference sequence for bumping might help. For all molding processes, one can achieve greater scorch safety and slow cure rate by reducing mold temperature, adding bisphenol AF, replacing some $Ca(OH)_2$ with MgO or adding CaO.

Tear when demolding: hot tear is always a problem with parts designed with an undercut. Using lower mold temperatures, higher average molecular weight material, lower cure levels, or to a lesser extent adding fillers such as fibrillating PTFE (MP 1500) will help improve hot tear resistance. Hot tear improvement should be tested at molding temperatures, with a goal of over 100% for elongation at break. Hot tear caused by excessive mold sticking can be eliminated in many cases by applying additional external mold release to the mold surface.

Knit lines: usually appear when excessive process aids are present in the formulation. Reduce levels, particularly when using highly incompatible process aids such as VPA 2 or carnauba wax.

Blisters: may come from undispersed ingredients, trapped water, and contaminants from other stocks or trapped air before molding:

1. Undispersed ingredients can readily be seen by cutting through mixed stock or preform with sharp knife. Check the particle size distribution of fillers and chemicals, review mixing process, and ensure a refining mixing cycle is performed after overnight stand.
2. Trapped water: Some condensation may occur when compound is stored in a cool place and brought to the molding shop that is warmer and more humid. The compound should be protected by a plastic bag, and the top removed only when the stock has reached the temperature of the molding shop.
3. Contamination from other stock: More difficult to find, these defects may appear as a crack in the molded part or as a particle near the surface. Make sure that all equipment is thoroughly cleaned before handling fluoroelastomers.
4. Trapped air: Preforms should be dense. If an extruder is used for preforming, use a screen or a breaker plate to build back pressure and eliminate trapped air.

Shrinkage: if there is too much shrinkage, not much can be done, except to reduce mold temperature. On the other hand, shrinkage can be increased by the use of fugitive process aids such as VPA 3.

Non-fills: appear as depressed areas on the surface of the molded part. This may happen if the preform or the shot volume is too light. The operator has to adjust and optimize the volume of rubber to mold. Non-fills also occur when the rubber begins to cure before the cavity is full. In that case, cure of the product must be slowed, either by lowering mold temperature or slowing cure, by addition of bisphenol AF, reduction of the accelerator or $Ca(OH)_2$, or eventually the addition of CaO.

ACKNOWLEDGMENTS

The author would like to greatly thank Ronald D. Stevens (Stow, Ohio), John G. Bauerle (Wilmington, Delaware), and Dr Stephen Bowers (Geneva, Switzerland), for their previous advices and comments in composing this document.

REFERENCES

1. Dittman, A.L., Passino, A.J., Whrightson, J.M., US patent 2 689 241 (1954), to M.W. Kellog.
2. Rexford D.R., US patent 3 051 677 (1962), to DuPont.
3. Pailthorp, J.R., Schroeder, H.E., US patent 2 968 649 (1961), to DuPont.
4. Apotheker, D., Krusic, P.J., US patent 4 035 565 (1977), to DuPont.
5. Apotheker, D., Krusic, P.J., US patent 4 214 060 (1977), to DuPont.
6. Brasen, W.R., Cleaver, C.S., US patent 3 467 635 (1969), to DuPont.
7. Albano, M., Arcella, V., Brinati, G., Chiodini, G., Minutillo, A., US patent 5 264 509 (1993), to Ausimont.
8. www.dupontelastomers.com.
9. www.dyneon.com.
10. www.solvaysolexis.com.
11. www.daikin.com.

12. Handling Precautions for Viton® and Related Chemicals—Technical Bulletin from DuPont Performance Elastomers.

13. Moran, A.L., Pattison, D.B., *Rubber World* 103 (1971) 37.

14. Schmiegel, W.W., *Angew. Makromol. Chemie* 76/77 (1979) 39.

15. Tatemoto, M., Suzuki, T., Tomoda, M., Furusaka, Y., Ueta, Y., US patent 4 243 770 (1981), to Daikin.

16. Arcella, V., Brinati, G., Albano, M., Tortelli, V., US patent 5 585 449 (1996), to Ausimont.

17. Stevens, R., Lyons, D.F., Paper #29, "New, Improved Processing HFP-Peroxide Cured Types of Viton®," 160th meeting of the Rubber Div./ACS, October 2001.

18. Lyons, D.F., Stevens, R.D., Paper #30, "New, Improved Processing PMVE-Peroxide Cured Types of Viton®," 160th meeting of the Rubber Div./ACS, October 2001.

19. Bowers, S., Thomas, E.W., Paper #52, "Improved Processing Fluorohydrocarbon Elastomers," ACS Rubber Div., October 2000.

20. Bauerle, J., Tang, P.L., SAE Paper # 2002-01-636, "A New Development in Base-Resistant Fluoroelastomers," (2002).

21. http://www.dupontelastomers.com/Products/Viton/selectionGuide.asp.

22. Kojima, G., Wachi, H., *Rubber Chem. Technol.* 51 (1978) 940.

23. Stevens, R.S., Moore, A.L., Paper #32, "A New, UniqueViton® Fluoroelastomer With Expanded Fluids Resistance," ACS Rubber Div., October 1997.

24. Bauerle, J., Paper #9, "A New, Improved Processing Base-Resistant Fluoroelastomer Based on APA Technology, Viton® TBR-S," ACS Rubber Div., October 2003.

25. Dobel, T.M., Stevens, R.S., Paper #8, "A New Broadly Fluid Resitant Fluoroelastomer Based on APA Technology, Viton® Extreme™ ETP-S," ACS Rubber Div., October 2003.

26. Dyneon® 7132X data sheet from Dyneon.

27. Tecnoflon® BR-9151 data sheet from Solvay Solexis.

28. Stahl, W.M., Stevens, R.D., SAE Paper # 920163, "Fuel-Alcohol Permeation Rates of Fluoroelastomers, Fluoroplastics, and Other Fuel Resistant Materials," (1992).

29. www.dupontelastomers.com/Tech_Info/chemical.asp.

30. Ferrandez, P., Bowers, S., *Kautsch Gummi Kunstst* 6 (1999) 429.

31. Compounding, Mixing and Processing Peroxide-Cured Viton® made with Advanced Polymer Architecture, DuPont Performance Elastomers Technical Bulletin (06/04 VTE-A10123-00-B0604).

32. Dai-El G-101 Technical Information Bulletin from Daikin, www.daikin.cc/.

33. MacLachlan, J.D., Fogiel, A.W., *Rubber Chem. Technol.* 49 (1976) 43.

34. Processing Guide for Viton® Fluoroelastomers, DuPont Performance Elastomers Technical Bulletin (07/03 VTE H90171-00-A0703).

TRADEMARKS

1. Viton® and Viton® Extreme™, Kalrez®, and DIAK™ trademarks or registered trademarks of DuPont Performance Elastomers

2. Chemraz® is a registered trademark of Greene Tweed

3. Aflas® is a registered trademark of Asahi Glass

4. Dyneon® is a registered trademark of Dyneon

5. Tecnoflon® is a registered trademark of Solvay Solexis

6. Dai-El® is a registered trademark of Daikin

7. Luperco® is a registered trademark of Atofina

8. Celite® is a registered trademark of Manville
9. Nyad® is a registered trademark of Nyco
10. Struktol® is a registered trademark of Struktol
11. PAT® 777 is a registered trademark from IDE Processes International Sales
12. Aerosil® is a registered trademark of Degussa
13. Wollastocoat® is a registered trademark from Nyco Minerals
14. Cromophtal® is a registered trademark of Ciba Specialty Chemicals
15. TiPure® and Zonyl are registered trademarks of the DuPont Company
16. Stan-Tone® is a registered trademark of PolyOne
17. Armeen® is a registered trademark of Akzo Nobel Polymer Chemicals

5 Polyacrylate Elastomers—Properties and Applications

Robert C. Klingender

CONTENTS

5.1 INTRODUCTION

The U.S. Department of Agriculture Eastern Region Laboratory first developed polyacrylate elastomers in the 1940s [1]. The first commercial introduction of these rubbers was in 1947 by B.F. Goodrich. Since then there have been many upgrades in the quality and technical capabilities of these polymers. There were several curing systems developed to achieve improved properties [2–5].

Polyacrylate, designated ACM by ASTM, is a cost-effective family of heat and oil resistant polymers with a long history of use by the automotive industry, and to a much lesser extent for mechanical goods. These polymers operate very well at 150°C to 175°C in oil, hence are used in applications operating above the capability of nitrile rubber, (NBR). Polyacrylate is also more cost-effective than hydrogenated nitrile (HNBR) and fluoroelastomer (FKM), thus may be used where the properties of the ACM meet all of the requirements of a given application. ACM competes more directly with AEM ethylene acrylic elastomer. The ASTM D2000/SAE J200 designations for polyacrylate compounds are in the DH (150°C) up to the EH (175°C) categories.

The excellent heat, oxidation, ozone, lubricating, and transmission fluid resistance of ACM make it a natural selection for automotive dynamic and static sealing applications, transmission oil cooler hose, transmission signal wire cover, flexible magnets, and heavy duty air duct tubing. Nonautomotive applications include seals, gaskets, O-rings, rolls, sponge, and sealants. It is also used as a binder for explosives that generate nontoxic gas to deploy air bags for automobiles and aircraft jettison mechanisms.

5.1.1 STRUCTURE AND TYPES

Polyacrylate elastomers may consist of one or more of the following monomers: ethyl acrylate, EA; butyl acrylate, BA; and methoxy ethyl acrylate MEA, the structures of which are given in Figure 5.1. These are coupled with cure sites, which may be reactive halogens, epoxy or carboxyl groups, or combinations thereof, as seen in Figure 5.2. [1,6,11].

The ethyl acrylate group provides the best heat and oil resistance but is deficient in low-temperature flexibility. Butyl acrylate will give low-temperature properties down to −40°C, but the oil resistance is poorer. The methoxy ethyl acrylate (MEA), when combined with EA and BA, results in a much-improved balance of oil and low-temperature resistance. Figure 5.3 demonstrates the property balance obtained with the various combinations of EA, BA, and MEA [6].

The original, and currently extensively used, cure-site group is chlorine based. A number of cure systems may be used, but the most common is the "soap-sulfur" one using sodium and potassium stearate with sulfur or a sulfur donor. Very good physical properties and heat resistance may be obtained with the chlorine cure sites, however compression set resistance is only fair.

Epoxy side cure sites offer a reduction in mold and metal corrosion and are better for water resistance [7]. Two main cure systems are available for these products, Zeonet A, BF, and HyTemp SR50 combination as well as ammonium benzoate.

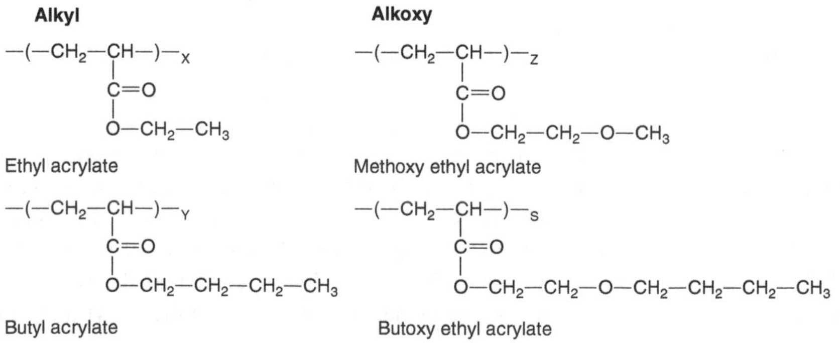

Alkyl

$-(-CH_2-CH-)-_x$
|
C=O
|
O$-CH_2-CH_3$

Ethyl acrylate

$-(-CH_2-CH-)-_y$
|
C=O
|
O$-CH_2-CH_2-CH_2-CH_3$

Butyl acrylate

Alkoxy

$-(-CH_2-CH-)-_z$
|
C=O
|
O$-CH_2-CH_2-O-CH_3$

Methoxy ethyl acrylate

$-(-CH_2-CH-)-_s$
|
C=O
|
O$-CH_2-CH_2-O-CH_2-CH_2-CH_2-CH_3$

Butoxy ethyl acrylate

FIGURE 5.1 Base structure of polyacrylate.

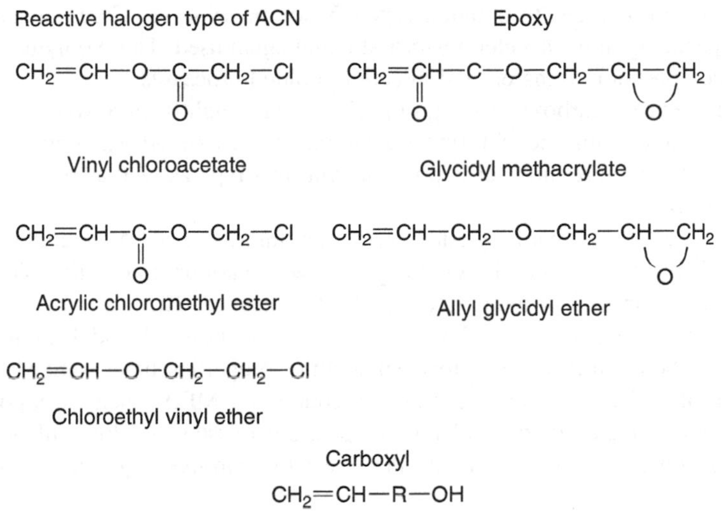

Reactive halogen type of ACN

$CH_2{=}CH{-}O{-}\underset{\underset{O}{\|}}{C}{-}CH_2{-}Cl$

Vinyl chloroacetate

$CH_2{=}CH{-}\underset{\underset{O}{\|}}{C}{-}O{-}CH_2{-}Cl$

Acrylic chloromethyl ester

$CH_2{=}CH{-}O{-}CH_2{-}CH_2{-}Cl$

Chloroethyl vinyl ether

Epoxy

$CH_2{=}CH{-}\underset{\underset{O}{\|}}{C}{-}O{-}CH_2{-}CH{-}CH_2$

Glycidyl methacrylate

$CH_2{=}CH{-}CH_2{-}O{-}CH_2{-}CH{-}CH_2$

Allyl glycidyl ether

Carboxyl

$CH_2{=}CH{-}R{-}OH$

FIGURE 5.2 Polyacrylate typical reactive cure sites.

This type of ACM offers good to excellent scorch resistance and bin stability. The heat resistance is good to fair.

The carboxyl cure sites offer fast cures and excellent resistance to heat and automotive fluids and low compression set but result in compounds that have shorter scorch with certain curatives.

The dual chlorine/carboxyl cure sites have the capability of using a very wide variety of cure systems and will produce compounds with excellent heat resistance

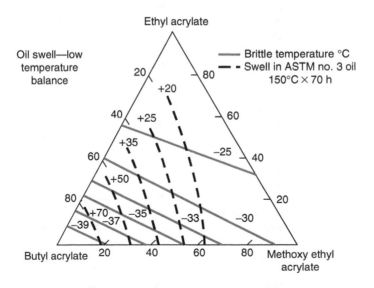

FIGURE 5.3 Influence of ACM composition on properties.

and can be compounded to obtain a very low compression set. Cure rate and scorch times depends upon the accelerator/crosslinking agent used. These polymers may be used in compression, transfer, and injection molded processes.

A dual epoxy/carboxyl side group also offers slightly improved water resistance compared with the chlorine containing grades of ACM. Very good heat resistance and compression set is possible with this type as well as better corrosion resistance [8].

Commercially available grades from one family of ACM manufacturers is given in Table 5.1, which shows the grades with various cure active side groups and an indication of low temperature and oil resistance. The elastomers with low-temperature resistances of −15°C would offer the best oil and heat resistance; conversely those with the best low-temperature properties have somewhat poorer oil and heat resistance. Those elastomers containing MEA, such as Nipol AR53, have a better balance of oil and low-temperature resistance. This will provide an initial screening of grades available that meet the end-use requirements regarding

TABLE 5.1
HyTemp and Nipol ACM Grades

HyTemp	Nipol	Reactive Cure Site	Volume Swell %	Gehman T_{100} °C	Sp. Gr.	Mooney Viscosity
4051		Chlorine/carboxyl	11	−18	1.1	46–58
4051EP		Chlorine/carboxyl	11	−18	1.1	35–47
4051CG		Chlorine/carboxyl	11	−18	1.1	25–37
AR71	AR71	Chlorine	11	−18	1.1	42–54
AR715		Chlorine	15	−24	1.1	30–42
	AR31	Epoxy	16	−17	1.1	35–45
	AR51	Epoxy	11	−14	1.1	50–60
4052		Chlorine/carboxyl	17	−32	1.1	32–40
4052EP		Chlorine/carboxyl	17	−32	1.1	20–35
4062		Chlorine/carboxyl	17	−32	1.1	32–40
4065		Chlorine/carboxyl	15	−32	1.1	32–40
AR72LF		Chlorine	22	−28	1.1	28–36
	AR72LS	Chlorine	22	−28	1.1	28–38
	AR72HF	Chlorine	20	−28	1.1	45–55
	AR32	Epoxy	18	−26	1.1	30–40
	AR42W	Epoxy	20	−26	1.1	30–40
	AR12	Carboxyl	30	−30	1.1	30–36
	AR14	Carboxyl	27	−40	1.1	28–38
	AR22	Carboxyl	24	−25	1.1	44–54
4053EP		Chlorine/carboxyl	24	−42	1.1	23–31
	AR53L	Epoxy	15	−32	1.1	31–37
4054		Chlorine/carboxyl	63	−41	1.1	22–30
	AR54	Epoxy	29	−37	1.1	24–34
	AR74	Chlorine	28	−40	1.1	25–35
PV-04		Proprietary	45	−30	1.1	25–40

low-temperature and oil-resistance properties. Further selection will be needed to determine the appropriate viscosity and cure-site type for the required processing and physical properties [8–11].

Nipol AR12, AR14, and AR22, special grades of polyacrylate, have been developed to provide properties similar to or better than an AEM elastomer, although they are not chemically equivalent [12–14].

A peroxide curable grade, HyTemp PV-04, is available from Zeon Chemicals for O-rings, seals, gaskets as well as binders, adhesives, caulks, and plastic modification. This grade offers good compression set without the need for a post-cure, but is sensitive to exact curative levels [15]. NOK also has a polyacrylate that may be cured with peroxides, but information on it is not readily available.

A comparison by cure sites and low-temperature properties of the various grades available from known manufacturers is given in Table 5.2. Remember that new types and grades are constantly being developed, so keep in close touch with the various manufacturers.

The product number has some significance with a few of the manufacturers. The AR70 series from the Zeon family all have chlorine side groups and the last digit indicates roughly the low-temperature resistance in °C, with 1 corresponding to −15°C with excellent oil resistance; 2 being −28°C with excellent oil resistance; 3 at −40°C with very good oil resistance; and 4 relating to −40°C with good oil resistance. Zeon Chemicals' 4000 number series of elastomers usually have chlorine/carboxyl cure sites with the third digit indicating the current family, 4050 and the newer low-temperature polymerized 4060 types [16]. The last digit also relates to the low-temperature properties as with the AR70 series, except for HyTemp 4065, which is a special grade with a low-temperature flexibility of −30°C. Zeon Corporation's Nipol AR 30, 40, and 50 series have epoxy cure sites, with the last digit indicating the low-temperature properties. TOA Acron's products 800 series contain chlorine side groups, 600 and 700 groups have an epoxy cure site and the 500 polymers contain chlorine/carboxyl groups. JSR products use the first digit to indicate the low-temperature properties, 1 for −15°C, 2 for −28°C, 3 for −30°C to −40°C, and 4 for −40°C. The second digit indicates the cure active side groups with 1 being chlorine and 2 indicating epoxy types.

5.2 COMPOUNDING

5.2.1 POLYMER SELECTION

Complete information on the specific application is essential before considering the appropriate grade or grades of polyacrylate to choose. The design engineer will need to determine the maximum and minimum operating temperatures, any short-term excursions above or below the expected range, any and all fluids or gases to be encountered, with their full descriptions. The design of the part and the mold is important for the design of the rheological properties of the compound. Dynamic properties should be considered for rotating or reciprocating seals, rollers, and hose.

The cure-site selection should be based upon manufacturing and process considerations as well as performance requirements such as dynamic properties,

TABLE 5.2
Available Grades of ACM

Reactive Cure Site	Low Temperature	Zeon Chemicals	Zeon Corporation	TOA Paint	NOK	JSR	Nissin
Chlorine	−15	AR71	AR71	AR801	A1095, PA401, A1165	AR110	RV1220
	−25	AR72LF, AR715	AR72LF, AR72LS, AR72HF	AR840, AR825		AR210	RV1240
Epoxy	−40		AR74	AR860		AR310, AR411	RV1260
	−15		AR31, AR51	AR601		AR120	RV1020
	−25		AR32, AR42W	AR740		AR220	RV1040
	−40		AR53L, AR54	AR760		AR320	RV1060
Chlorine/carboxyl	−15	4051, 4051EP, 4051CG		AR501, AR501L			
	−25	4052, 4052EP, 4062, 4065		AR540, AR540L			
	−40	4053EP, 4054					
Carboxyl	−15				PA501		
	−25		AR22		PA502		
	−30		AR12				
	−40		AR14				
Vinyl	−15						RV2520
	−25						RV2540
	−40						RV2560

compression set, high retention of compressive stress, coupling retention, and so on. One must use the cure system designed for each elastomer family. Table 5.2 lists many of the types and grades available.

Chlorine cure sites in ACM polymers exhibit excellent heat resistance and were the original type used for many years. The typical cure systems are based on either the soap/sulfur or triazine. A diamine such as Diak #1 or #3 also may be used, but are more difficult to process in the factory. All of these will require a post-cure for optimum results, the triazine system giving much better compression set resistance and shelf life. The families of these elastomers include the Zeon Group's HyTemp or Nipol AR71, AR715, AR72, AR72LF; Nipol AR72LS, AR72HF, AR74; TOA Acron AR801, AR840, AR825, AR860; JSR AR110, AR210, AR310, AR411; and NOK Noxtite A1095, PA401, A1165.

A comparison of one family of these ACMs is given in Table 5.3. This will provide an indication of their original and aged properties with and without a post-cure [8].

Polymers with Epoxy cure sites include Zeon Corporation's Nipol AR31, AR51, AR32, AR42, AR42W, AR53L, AR54; TOA Paint Acron AR601, AR740, AR760; Japan Synthetic Rubber AR120, AR220, and AR320. This family of elastomers may be cured with either ammonium benzoate or a diacid/amine combination. The ammonium benzoate curative will require a post-cure, whereas the diacid/amine combination will provide low compression set without a post-cure. Odor may be objectionable in the factory with ammonium benzoate, which may limit its use. The ratio of Zeonet A (diacid), Zeonet BF (amine), and HyTemp SR50 (urea) retarder may be adjusted to achieve the desired cure rate, scorch, and shelf life, as well as compression set resistance. A comparison of Nipol ARs with an epoxy cure site is provided in Table 5.4 [7].

Dual chlorine/carboxyl active cure site offers the greatest versatility in available cure systems and resultant process and performance properties. Zeon Chemicals has an NPC (no post-cure) curative system that offers the chemist the ability to design a compound that does not require a post-cure and that has lower compression set than available with other systems. Even better compression set is obtainable with a short post-cure. Other curatives that may be used, but are not recommended, are soap/sulfur (or sulfur donor), triazine, t-amines, or thiadiazole. Available grades are Zeon Chemicals HyTemp 4051, 4051EP, 4051CG, 4052, 4052EP, 4062, 4065, 4053EP, 4054; TOA Acron AR501, AR501L, AR540, and AR540L. The HyTemp 4060 series from Zeon Chemicals, which is covered in more detail, offer improved processing characteristics and lower compression set properties. The HyTemp family of ACMs with the dual cure sites is compared in Table 5.5 [17].

Low-temperature polymerized ACMs, which are available from Zeon, exhibit improved processing properties and slightly better compression set resistance. HyTemp 4062 with the improved factory handling characteristics and better compression set resistance, is a good replacement for HyTemp 4052/4052EP. HyTemp 4065, in addition to being polymerized at low temperatures and is believed be a more linear polymer for very good processing properties and much improved compression set resistance. A comparison is given in Table 5.6 [16].

TABLE 5.3

Comparison of a Chlorine Cure-Site Polyacrylate Series

HyTemp Grade	AR71	AR715	AR72LF	AR74
AR71	100	—	—	—
AR715	—	100	—	—
AR72LF	—	—	100	—
AR74	—	—	—	100
N55O	65	75	80	80
Stearic acid	1	1	1	1
TE80 Paste	2	2	2	2
AgeRite Stalite S	2	2	2	2
HyTemp ZC50	5.2	5.2	5.2	5.2
Total	172.5	185.2	190.2	190.2
Processing properties				
Mooney viscosity, ML 1 + 4 at 100°C				
Viscosity	76	79	87	75
Mooney scorch, ML at 125°C				
Minimum viscosity	73	77	94	73
t_5, minutes	5.6	10.9	10	12.7
Rheometer, 100 cpm, 3° arc at 190°C				
ML, N · m	1.7	1.5	1.6	1.9
MH, N · m	7.5	7.6	8.6	8.9
t_{s2}, minutes	1.1	1.4	1.3	1.4
t_{90}, minutes	5.2	5.2	5.8	6.1

Physical properties, cured 4 min at 190°C—post-cured 4 h at 177°C

	Original	Post-cured	Original	Post-cured	Original	Post-cured	Original	Post-cured
Shore A, pts	66	70	69	77	68	80	70	79
Modulus at 100%, MPa	6.7	8.5	7.2	9.2	8.8	10.5	8.5	10.2
Tensile, MPa	11.8	12.5	9.7	11.5	10.1	11.5	9.4	10.9
Elongation, %	230	160	195	180	140	135	125	120
Tear C,	20.1	18.4	20.0	20.1	17.5	16.6	13.1	14.0
Gehman torsional stiffness								
T_{100}, °C		−17		—		−28		−40
Compression set B, 70 h at 150°C, buttons cured 6 min at 190°C								
Set, %	21	14	22	15	23	16	26	16
Aged in air oven 70 h at 190°C-change								
Shore A, pts	14	13	16	4	20	9	17	9
Tensile, %	−18	−29	−3	−16	−5	−16	−18	−20
Elongation, %	−28	−3	−3	−3	−32	−22	−60	−46
Aged in ASTM #1 Oil 70 h at 150°C-change								
Shore A, pts	5	3	−2	3	7	−3	6	1
Tensile, %	17	14	21	7	18	9	3	2
Elongation, %	−24	13	−23	−3	−14	4	−24	4
Volume swell, %	−3.5	−2	−3.2	−1	−3	−1	−2.2	−0.5
Aged in ASTM #3 Oil 70 h at 150°C-change								
Shore A, pts	−6	−10	−7	−15	−7	−19	−15	−24
Tensile, %	6	0	7	−9	1	−11	−8	−20
Elongation, %	−7	34	−3	6	7	11	12	17
Volume swell, %	9.5	11.5	14.5	17.5	15.5	18	22	25

TABLE 5.4
Comparison of an Epoxy Cure-Site Polyacrylate Series

Nipol	AR31	AR51	AR32	AR42	AR53	AR54
Nipol AR31	100	—	—	—	—	—
Nipol AR51	—	100	—	—	—	—
Nipol AR32	—	—	100	—	—	—
Nipol AR42	—	—	—	100	—	—
Nipol AR53	—	—	—	—	100	—
Nipol AR54	—	—	—	—	—	100
Stearic acid	1	1	1	1	0.5	1
FEF N550	50	50	50	55	—	—
MAF	—	—	—	—	65	60
Interstab G8205	—	—	—	2	1	—
Naugard 445	—	—	—	1	2	—
ODPA	—	—	—	—	—	1
Ammonium benzoate	1	1	1	1.5	—	1.5
Zeonet A	—	—	—	—	0.5	—
Zeonet BF	—	—	—	—	1.8	—
Zeonet U	—	—	—	—	1.3	—
Total	152	152	152	160.5	172.1	163.5
Physical properties, cured	20 min at 155°C	20 min at 155°C	20 min at 155°C	20 min at 170°C	20 min at 170°C	30 min at 155°C
Hardness, JIS, pts	63	65	53	64	69	64
Modulus at 100%, MPa	3.1	2.9	2.1	5.2	4.6	3.6
Tensile, MPa	10.3	10.8	10.3	10.7	9.7	9.0
Elongation, %	330	450	400	230	250	230
Physical properties, cured	30 min at 155°C, PC 8 h at 150°C	30 min at 155°C, PC 8 h at 150°C	30 min at 155°C, PC 8 h at 150°C	20 min at 170°C, PC 2 h at 170°C	20 min at 170°C, PC 4 h at 150°C	30 min at 155°C, PC 16 h at 150°C
Hardness, JIS, pts	66	68	55	68	70	69
Modulus at 100%, MPa	5.1	4.7	3.1	7.6	6.0	4.7
Tensile, MPa	11.9	13.7	11.5	11.5	11.0	9.7
Elongation, %	230	310	300	170	200	210
Low-temperature properties						
Brittle point, °C	NT	NT	NT	−22.6	−23.4	−32.2
Gehman T_{10}, °C	−11	−9	−22	NT	NT	−29
Gehman T_{100}, °C	−17	−16	−28	NT	NT	−37.5
Compression set B 70 h at 150°C, set %						
Cure 20 min at 170°C	NT	NT	NT	NT	24.6	NT
Cure 30 min at 155°C, PC 4 h at 150°C	54	35	62	NT	15.2	39.3
Cured 30 min at 155°C, PC 16 h at 150°C	43	23	51	NT	14.2	29.6
Cured 20 min at 170°C, PC 2 h at 170°C	NT	NT	NT	29.6	NT	NT

TABLE 5.4   (Continued)
Comparison of an Epoxy Cure-Site Polyacrylate Series

Nipol	AR31	AR51	AR32	AR42	AR53	AR54
Aged in air 21 days at 150°C-change						
Hardness, pts	9	10	12	NT	NT	13
Tensile, %	−8	−22	−2	NT	NT	5
Elongation, %	−48	−48	−46	NT	NT	−35
Aged in air 3 days at 175°C-change						
Hardness, pts	7	7	10	4	9	12
Tensile, %	−2	−27	5	5	19	−5
Elongation, %	−43	−21	−47	0	−44	−30
Aged in #1 Oil 3 days at 150°C-change						
Hardness, pts	0	0	−2	0	1	1
Tensile, %	−17	−15	−24	7	8	6
Elongation, %	−29	−41	−32	−18	−16	−10
Volume swell, %	0	0	1	−1.5	−2.3	0.5
Aged in #3 Oil 3 days at 150°C-change						
Hardness, pts	−15	12	−14	−12	−16	−21
Tensile, %	−23	−18	−27	−18	−3	−29
Elongation, %	−5	7	−7	−23	−8	−25
Volume swell, %	17	14	19	18.6	13.6	29.8

TABLE 5.5
Comparison of Chlorine/Carboxyl Cure-Site Polyacrylate Series

HyTemp	4051	4051EP	4052	4052EP	4053EP	4054
HyTemp 4051	100					
HyTemp 4051EP	—	100	—	—	—	—
HyTemp 4052	—	—	100	—	—	—
HyTemp 4052EP	—	—	—	100	—	—
HyTemp 4053EP	—	—	—	—	100	—
HyTemp 4054	—	—	—	—	—	100
Stearic acid	1	1	1	1	1	1
N550	65	65	80	80	80	80
Struktol WB222	2	2	2	2	2	2
AgeRite Stalite S	2	2	2	2	2	2
Sodium stearate	4	4	4	4	4	4
HyTemp NPC-50	2	2	2	2.5	2	3
Total	176	176	191	191.5	191	192
Mooney viscosity, ML 4 at 100°C						
Viscosity	63	55	64	53	60	52
Rheometer at 190°C, 3° arc, 100 cpm						
ML, N · m	1.8	1.2	1.5	1.1	1.3	1.1
MH, N · m	8.0	7.9	6.7	5.7	6.4	6.1

(continued)

TABLE 5.5   (Continued)
Comparison of Chlorine/Carboxyl Cure-Site Polyacrylate Series

HyTemp	4051	4051EP	4052	4052EP	4053EP	4054
t_{s2}, minutes	0.5	0.5	0.6	0.6	0.5	0.5
t_{90}', minutes	3.8	3.5	3.9	3.0	3.3	3.3
Physical properties, cured 4 min at 190°C						
Shore A, pts	66	66	66	67	65	66
Modulus at 100%, MPa	6.3	6.5	5.6	5.6	6.3	6.0
Tensile, MPa	13.1	11.9	10.3	9.5	9.3	9.0
Elongation, %	210	200	160	170	140	150
Physical properties, cured 4 min at 190°C, PC 4 h at 177°C						
Shore A, pts	65	67	66	67	64	64
Modulus at 100%, MPa	7.0	6.5	5.9	5.0	6.2	5.7
Tensile, MPa	13.2	12.3	10.0	8.8	9.0	8.8
Elongation, %	200	210	170	180	140	160
Low-temperature properties						
Brittle point, °C	−12	−12	−26	−26	−34	−32
Gehman T_{10}, °C	−12.2	−11.8	−25.5	−25.3	−33.4	−32.3
Gehman T_{100}, °C	−17.7	−17.9	−33.1	−32.4	−41.5	−40.7
TR10, °C	−16.5	−16.0	−32.1	−31.2	−37.1	−35.1
Aged in air 168 h at 177°C-change, samples cured 4 min at 190°C						
Shore A, pts	6	7	12	12	14	8
Tensile, %	−17	−12	−22	−27	−25	−16
Elongation, %	−38	−35	−44	−47	−43	−33
Aged in ASTM #1 Oil 168 h at 150°C-change, samples cured 4 min at 190°C						
Shore A, pts	2	5	13	3	3	−2
Tensile, %	10	12	−2	−1	−2	3
Elongation, %	−29	−25	−19	−6	−14	−20
Volume swell, %	−4	−5	−3	−4	−4	2
Aged in ASTM #3 Oil 168 h at 150°C-change, samples cured 4 min at 190°C						
Shore A, pts	−7	−9	−11	−16	−16	−25
Tensile, %	−3	−1	−2	−8	−21	−47
Elongation, %	−19	−20	−12	6	−21	−40
Volume swell, %	11	10	16	16	25	65
Aged in Dexron II ATF 168 h at 150°C-change, samples cured 4 min at 190°C						
Shore A, pts	−3	−2	−4	−7	−8	−21
Tensile, %	7	10	1	−1	−3	−18
Elongation, %	−19	−15	−6	−6	−7	−13
Volume swell, %	2	1	4	3	6	27
Aged in GM rear axel fluid 168 h at 150°C-change, samples cured 4 min at 190°C						
Shore A, pts	0	−1	−2	−4	−5	−14
Tensile, %	8	2	1	−4	3	−10
Elongation, %	−5	0	0	0	7	0
Volume swell, %	1	1	2	2	3	12

TABLE 5.6

Comparison of Cold Polymerized HyTemp 4062 and 4065 with Standard HyTemp 4052

HyTemp 4065	100	—	—
HyTemp 4062	—	100	—
HyTemp 4052	—	—	100
Stearic acid	1.5	1.5	1.5
Naugard 445	2	2	2
Struktol WB222	2	2	2
N550	65	65	65
HyTemp NS-70	5.7	5.7	5.7
HyTemp NPC-25	4	4	4
HyTemp SR-50	2	2	2
Total	182.2	182.2	182.2
Compound properties			
ML(1 + 4) at 100°C	47	47	48
Mooney scorch, T_5 at 125°C	2.4	2.9	2.4
Injection on Rep M36, Spider Flow at 200°C, 200 Bars, Screw 50°C, Inject 60°C			
Injection time, seconds	5.2	5.0	5.1
Physical properties, cured 11 min at 190°C, post-cured 4 h at 180°C			
Shore A, pts	56	56	62
Modulus at 200%, MPa	10.8	10	10
Tensile, MPa	11.4	11.6	12.1
Elongation, %	235	260	270
Low-temperature properties			
Gehman T_{10}, °C	−30	−30	−30
Aged in air 70 h at 150°C-change			
Tensile, %	−11	−16	−36
Elongation, %	−9	−19	−30
Compression set, VW Method, 22 h at 150°C			
Set, %	37	42	42
Aged 70 h in ASTM #1 Oil at 150°C-change			
Tensile, %	7	3	4
Elongation, %	−2	−4	−4
Aged 70 h in SF-105G at 150°C-change			
Tensile, %	4	1	2
Elongation, %	−9	−12	−10
Aged 70 h in Dexron III at 150°C-change			
Tensile, %	4	2	0
Elongation, %	−2	−7	−4

Peroxide-curable grade from Zeon Chemical is HyTemp PV-04. It has an oil swell of 45% and a low-temperature resistance of −30°C. This polymer may also be used for plastic modification and for sealants or caulks. The formulation given in Table 5.7 contains the recommended level of peroxide, which is critical with this elastomer for

TABLE 5.7
Peroxide Curable ACM Compared with Standard ACM and Lower Swell VMQ

	HyTemp PV-04	HyTemp 4052	Proprietary VMQ
HyTemp PV-04	100	—	—
HyTemp 4052	—	100	—
Proprietary improved VMQ compound	—	—	100
N550	70	35	—
Stearic acid	1	1	—
Naugard 445	2	2	—
TE-80 paste	2	2	—
HyTemp NPC-25	—	4	—
HyTemp SR-50	—	2	—
HyTemp NS-70	—	5.7	—
Vulcup 40KE	2	—	—
HVA #2	2	—	—
Total	179	186.7	100
Physical properties, cured at 190°C			
Shore A, pts	44	45	41
Modulus at 100%, MPa	1.3	3.6	0.9
Tensile, MPa	5.7	12.8	8.4
Elongation, %	350	205	850
Gehman low temperature			
T_{100}, °C	−29	−31	−72
Aged in air 168 h at 150°C-change			
Shore A, pts	13	8	5
Tensile, %	−8	−5	9
Elongation, %	−10	−2	−5
Compressive stress relaxation, retained force, %	75	44	49
Compression set B, 168 h at 150°C			
Set, %	28	12	29
Aged in IRM 903 168 h at 150°C-change			
Shore A, pts	−25	−9	−25
Tensile, %	−30	−23	−59
Elongation, %	−20	−18	−90
Volume swell, %	45	19	67
Compressive stress relaxation, retained force, %	43	68	66
Aged in SAE 5W30 Oil 168 h at 150°C-change			
Shore A, pts	−7	−3	−22
Tensile, %	2	−15	−62
Elongation, %	0	−18	−46
Volume swell, %	13	3	53
Compressive stress relaxation, retained force, %	54	64	40
Aged in ATF 168 h at 150°C-change			
Shore A, pts	−14	−4	NT
Tensile, %	−28	−19	NT
Elongation, %	10	−18	NT
Volume swell, %	6	18	NT

optimum properties. A comparison of properties and sealing characteristics is made of HyTemp PV-04 with a chlorine/carboxyl ACM and a propriety improved VMQ [16]. Nippon Oil Seal also has development polymers that are peroxide curable, Mektron AY1120 and AY1122; however, please check on current availability of these.

Carboxyl ACM with much improved compression set, compression stress relaxation resistance, and heat resistance while maintaining −30°C low-temperature flexibility, has been developed and named Nipol AR-12 by Zeon Corporation. A comparison of this grade with a standard ACM as well as a low swell AEM-G in Table 5.8 indicates that it will function very well as an improved replacement for AEM-G or standard ACMs [15].

5.2.2 FILLER SELECTION

Carbon black, such as N550 or N774, is widely used with polyacrylates and is a more effective reinforcing agent than mineral fillers. The only caution with carbon

TABLE 5.8
Nipol AR12 Compared with Standard ACM and AEM-G

Nipol AR12	100	—	—
HyTemp 4062	—	100	—
AEM-G	—	—	100
Stearic acid	2	1	2
Armeen 18D	—	—	0.5
Vanfre VAM	—	—	0.5
Struktol WB222	2	2	—
Naugard 445	2	—	2
Proprietary antioxidant	—	3	—
N550	50	65	—
N774	—	—	50
DOTG	—	2	4
Diak #1	—	0.6	1.5
HyTemp NPC-25	4	—	—
HyTemp NS-70	6	—	—
HyTemp SR50	2	—	—
Total	168	173.6	160.5
Mooney scorch ML at 125°C			
Minimum viscosity	32	35	—
T_5, minutes	4.5	2.9	—
Rheometer at 190°C			
t_{s2}, minutes	0.8	0.7	—
T'_{90}, minutes	9.7	11	—
Physical properties, cured 4 min at 190°C, PC 4 h at 180°C			
Shore A, pts	56	56	62
Modulus at 100%, MPa	3.6	2.9	3.8
Tensile, MPa	10.5	11.1	19.1
Elongation, %	270	285	340

(continued)

TABLE 5.8   (Continued)
Nipol AR12 Compared with Standard ACM and AEM-G

Low-temperature properties

Gehman T_{100}, °C	−29	−30	−29
TR10, °C	−29	−27	−26

Aged in air 3024 h at 150°C-change

Shore A, pts	8	18	11
Tensile, %	−11	−14	−41
Elongation, %	−28	−63	−59

Aged in air 1008 h at 175°C-change

Shore A, pts	17	17	20
Tensile, %	−30	−23	−40
Elongation, %	−43	−43	−88

Aged in SF105 Oil 1512 h at 150°C-change

Shore A, pts	−3	3	−6
Tensile, %	−1	0	−14
Elongation, %	−30	−41	−44
Volume swell, %	10	7	30

Aged in ATF fluid 1512 h at 150°C-change

Shore A, pts	10	19	−4
Tensile, %	30	13	−10
Elongation, %	−40	−54	−32
Volume swell, %	6	3	20

Compression set Daimler Benz Method at 150°C

1008 h, Set, %	20.5	40	17
2016 h, Set, %	30	81	24
3024 h, Set, %	30	96	28

Compressive stress relaxation in SF105 Oil at °C—% force retained

1008 h	66	22	70
2016 h	45	5	50
3024 h	30	1	37

black is to be sure that the brand used has low sulfur content and does not have a low pH since acidic materials will retard the cure rate. This also applies to graphite and ground coal dust. The amount of various carbon blacks to increase hardness by one point Shore A is given in Table 5.9.

TABLE 5.9
Carbon Black Loading to Increase Shore A by One Point Unfilled Elastomer Base Hardness Shore A24

Carbon Black Type	N774	N650	N550	N330
High-temperature ACM	2.00	1.75	1.65	1.50
Low-temperature ACM	3.65	2.60	2.20	1.80

TABLE 5.10

Non-Black Filler Loading to Increase Shore A by One Point Unfilled Base Elastomer Hardness Shore A24

Filler Type	Calcined Clay with Amino Silane	Precipitated Silica, Low Reinforcement	Talc
High-temperature ACM	3.80	1.60	3.70

Non-black fillers may also be employed, but it is important to use neutral or medium high pH grades such as silane-treated calcined clay, synthetic sodium aluminum silicate, platy talc, neutral pH silicas, diatomaceous earth, hydrated aluminum silicate, calcium meta silicate, precipitated calcium carbonate, and so on. An indication of non-black filler requirements for one point of hardness increase is shown in Table 5.10.

Polyacrylates with carbon black, which gives the appearance of scorch upon bin aging, should be retested for Mooney scorch and if the minimum viscosity is 10 or more points higher than the original figures, it is scorched. If, however, the change in viscosity is minimal, then remilling the compound on a cool, tight mill usually refreshes the compound and it may then be processed in the normal fashion.

When designing a dynamic seal compound, it is often required to have a low coefficient of friction and a "pumping action" on the lip of the seal for lubrication. Hence a combination of reinforcing filler and a coarse one, such as a coarse-ground silica, needle-like talc, graphite, or ground coal dust may be considered to provide a pumping action on the surface.

Static seals are best formulated with larger particle size carbon blacks such as N774 for good compression set, combined with a more reinforcing one if necessary to achieve a higher modulus.

5.3 PLASTICIZERS

ACM elastomers are expected to operate at very high temperatures; hence the use of plasticizers is quite limited. It is obviously necessary to select ones with very low volatility at operating temperatures and extractability in the fluids to be encountered. The polarity of the plasticizer needs to be compatible with ACM to minimize or prevent bleeding.

If low-compound viscosity is needed, a low-viscosity polymer selection should be the primary consideration as well as the cure system, with lower reinforcing fillers considered next followed by the addition of minimum amounts of plasticizer, generally less than 5 pphr.

Low-temperature flexibility is dependent on the base elastomer(s) as the prime consideration with the addition of small amounts of a good low-temperature plasticizer as the secondary selection. However, adding plasticizer to an ACM will result

TABLE 5.11

Plasticizer Selection Guide

Plasticizer	Volatility	Oil Solubility	Fuel Solubility	Low Temperature
Plasthall P670	Low	P. S.	IS	F-G
Plasthall P7046	Low	P.S.	IS	P
Paraplex G25	Low	P.S.	S	P
TOTM	Low	S	S	F-P
Hercoflex 600	Low	P.S.	—	G
Vulkanol OT	Low	S	—	VG
Plasthall 7050	High	S	IS	VG
Thiokol TP759	High	S	IS	VG

in poorer aging properties. Attention must also be given to the allowed swell limits in the fluids in which the part must function, which again depends mainly on the grade of ACM selected as well as the extractability of the plasticizer.

Plasticizers to consider are given in Table 5.11. Equivalent grades from other manufacturers would be very acceptable.

5.4 PROCESS AIDS

Process aids, as the name implies, will assist in mixing, milling, and molding of ACM compounds. The selection of a given process aid will depend on the process improvement desired, taking into consideration its influence on other properties. Always keep in mind that the Mooney viscosity of the base polymer selected will also be a major factor in the processability of an ACM compound.

Stearic acid acts not only as an effective process aid, but also as a retarder with most cure systems used with ACM. Keep in mind that process aids may contain fatty acids, which will retard the cure rate; hence an overall balance of stearic acid with certain process aid needs to obtain for minimum influence on cure rate.

The best overall process aid with the NPC cure system has been found to be Struktol WB222 followed by TE80 paste and Struktol WS-280 (not powder form). Some plants have found good results with Axel Int 216.

5.5 ANTIOXIDANTS

Polyacrylate compounds are very stable at high temperatures in oxygen, ozone, and mineral oils and fluids. However, there is some benefit to be gained by the inclusion of 1–2 pphr of a good antioxidant, stable at high temperature and with limited solubility in oil. Recommended antioxidants, or equivalents from other suppliers, are AgeRite Stalite S or Naugard 445.

5.6 CURE SYSTEMS

There is a very broad range of cure systems available for ACM, however, these are specific to the cure-site groups on the various families of polymers selected, and are often not interchangeable with one another. The cure system and family of ACM selected depends on the process and application properties desired.

ACM polymers have lower viscosities; hence dispersion of ingredients may require more care than with other elastomers. For this reason, it is highly recommended that dispersed forms of various curatives or cure packages be used. In addition, soaps may vary in moisture content, making them difficult to disperse if the water content is low. The moisture content will also influence the cure rate, another reason to use dispersed forms. Relative humidity, thus moisture content of the compound, will influence cure rate, thus changes may occur with different seasons of the year.

Chlorine containing ACM elastomers have been traditionally cured with a soap/sulfur mechanism; however, the development of a triazine system has offered the compounder greater versatility and better process and vulcanizate properties. There is a slight sacrifice in elongation and aged hardness change with the triazine system. These curatives are available as individual basic ingredients or as dispersions and as combinations in a cure package, and are illustrated in Table 5.12.

TABLE 5.12
Cure Systems for Chlorine Cure-Site ACM

Criteria	Soap/Sulfur	Soap Dispersion	Triazine	Triazine Package
Sodium stearate, pphr	2–4	—	—	—
Potassium stearate, pphr	1–0	—	—	—
Sulfur, pphr	0.2–0.6	0.2–0.6	—	—
HyTemp NS-70, pphr	—	5–6	—	—
Zisnet ZSC	—	—	0.3–1.0	—
Ethyl Zimate, pphr	—	—	1–2	—
Vulcalent E/C, pphr	—	—	0.1–0.5	0.1–0.5
HyTemp ZC-50, pphr	—	—	—	5–7
Black fillers	Yes	Yes	Yes	Yes
Non-black fillers	Yes	Yes	Yes	Yes
Scorch safety	Excellent	Excellent	Good	Good
Shelf stability	Fair	Fair	Excellent	Excellent
Cure rate	Variable	Good	Fast	Fast
Post-cure processes	Normal compression molding	Normal compression molding	Optional molding and extrusion	Optional molding and extrusion
Tensile strength	Fair	Good	Good	Good
Elongation	Good	Good	Good	Good
Hardness	Normal	Normal	Higher	Higher
Compression set	Fair	Fair	Excellent	Excellent
Air/Oil aging	Excellent	Excellent	Excellent	Excellent

A comparison of properties obtained with the various curing systems used in chlorine cure-site polymers is given in Table 5.13. The soap/sulfur system has fair shelf stability and gives excellent heat resistance, but requires a post-cure to provide reasonable compression set resistance. HyTemp NS-70 dispersion of sodium stearate is recommended for more consistent cure rates and scorch resistance. The triazine system also gives very good heat resistance and low compression set and good shelf

TABLE 5.13
Cure System Data for Chlorine Cure-Site ACM

	Soap/Sulfur	Triazine	Triazine Package
HyTemp AR71	100	100	100
Stearic acid	1	1	1
Struktol WB-222	—	2	—
TE 80 Paste	2	—	2
AgeRite Stalite S	2	2	2
N550	65	65	65
Spider Sulfur	0.3	—	—
Sodium stearate	2.25	—	—
Potassium stearate	0.8	—	—
Zisnet F-PT	—	0.5	—
Butyl Zimate	—	1.5	—
Santogard PVI	—	0.2	—
Thiate H	—	0.3	—
HyTemp ZC-50	—	—	5.2
Total	173.3	172.5	175.3
Mooney scorch ML at 125°C			
Minimum viscosity	47	46.3	58
T_5, minutes	14.4	13.6	8
Rheometer, 100 cps, 3° arc at 190°C			
ML, N·m	1.3	1.0	1.4
MH, N·m	4.7	5.7	7.1
t_{s2}, minutes	1.8	1.1	1.2
t'_{90}, minutes	11.9	3	5.0
Physical properties, cured 4 min at 190°C, PC 4 h at 177°C			
Shore A, pts	69	70	67
Modulus at 100%, MPa	5.2	4.5	7.2
Tensile, MPa	12.6	10.7	12.8
Elongation, %	295	260	235
Compression set B 70 h at 150°C (cured 6 min at 190°C, PC 4 h at 177°C)			
Set, %	46	30.5	12
Aged 70 h at 177°C in air-change			
Shore A, pts	3	7	—
Tensile, %	−27	−23	—
Elongation, %	−18	−33	—

stability. HyTemp ZFC-65 may be accelerated with a butyl or ethyl dithiocarbamates and retarded with PVI (ODOR) or Thiate H or Vulkalent E/C. The triazine cure package, HyTemp ZC-50 provides very good cure characteristics, physical properties, and excellent compression set.

Chlorine/carboxyl cure-site ACM provides a wide variety of cure systems depending on the process and functional properties required. Table 5.14 gives a summary of the various systems used with the expected characteristics for each. Of these, there are four cure systems used in common practice, the Diuron, HyTemp SC-75, HyTemp NPC-25, and Butyl Tuads. A comparison of these cure systems is shown in Table 5.15.

The Diuron cure system provides excellent scorch safety, good shelf life, but moderate cure rates. The use of up to 6 pphr will improve cure rates at some sacrifice in scorch safety. This product is not recommended with non-black fillers. The physical properties obtained with Diuron are excellent, except the elongation is lowered with higher loadings. The compression set is good and heat aging is excellent with Diuron compounds. Rhenocure Diuron is an 80% dispersion that is recommended in place of the pure Diuron.

HyTemp NPC-25 is the cure system of choice as it contributes a very fast rate of cure; excellent shelf stability, and fair scorch safety. Processing conditions must be carefully controlled to maintain a maximum temperature of 85°C in the compound to prevent scorch. The very fast cure rate limits the use of the NPC cure system for injection or transfer molding. The NPC cure system does work well for injection or transfer molding operations provided that care is taken to process the compound under 85°C and control runner temperatures. Very good compression set with out post-cure offers a major advantage and if even lower compression set is required, a post-cure will give values of less than 15%, 70 h at 150°C.

The HyTemp SC-75 cure package also will provide very fast cure rates with excellent shelf stability and fair scorch safety, but does require a post-cure to obtain desired physical properties. This cure system is recommended for injection and transfer as well as compression molding. The physical properties are good with a higher elongation and lower hardness than with other mechanisms. The heat aging and compression set are excellent, making this a very useful product if process conditions are hard to control and a post-cure can be tolerated.

A combination of Vanax 829 and TBTD in an ACM gives excellent scorch safety and fast cure rates but the shelf stability is sensitive to moisture. The physical properties after a post-cure are very good, but the compression set is only fair. Aging properties with this cure combination are excellent.

One would expect that soap/sulfur, or sulfur donor or triazine cure systems may be used because of the presence of the chlorine cure site, however, the shelf stability is only fair and the cure rates are slow. The physical properties, heat aging and compression set obtained are good with the soap/sulfur combination but better properties may be obtained with other curatives such as NPC, hence these are not recommended. Diak #3 gives properties similar to those of the soap/sulfur system, but has the added disadvantage of a poor compression set and a higher hardness.

Epoxy containing ACM may also offer the opportunity to bypass the post-cure step when the Zeonet A (0.6 pphr), BF (1.75–2.25), and HyTemp SR50 (1.0–2.0) cure combination is employed. This type of elastomer also may be cured with

TABLE 5.14

Cure System Summary for Chlorine/Carboxyl Cure-Site ACM

	NPC-50	Diuron	Vanax 829	Soap	Diak#3	Triazine	SC-75
Sodium stearate	4–6	4	—	3–4	—	—	—
Stearic acid	0–3	0–3	0–2	0–2	0–2	0–2	0–2
HyTemp NPC-25	4–6	—	—	—	—	—	—
HyTemp SR-50	2–6	—	—	—	—	—	—
HyTemp SC-75	—	—	—	—	—	—	—
Polydex HA80	—	2–6	—	—	—	—	—
Vanax 829	—	—	0.9	—	—	—	—
TBTD 72% Disp.	—	—	3.5	—	—	—	—
TMTD	—	—	—	0.7	—	—	—
Diak #3	—	—	—	—	3	—	—
Zisnet F-PT	—	—	—	—	—	0.75–1.2	—
ZDEC	—	—	—	—	—	2	—
HyTemp SC-75	—	—	—	—	—	—	7–8
Black compounds	Yes	Yes	Yes	Yes	Yes	Yes	Yes
Non-black compounds	Yes	No	Yes	Yes	Yes	Yes	Yes
Scorch safety	Fair	Excellent	Excellent	Good	Good	Fair	Fair
Shelf stability	Excellent	Good	Good	Fair	Fair	Good	Excellent
Cure rate	Very fast	Moderate	Fast	Slow	Slow	Fast	Fast
Post-cure	Optional	Yes	Yes	Yes	Yes	Yes	Yes
Tensile strength	Good	Excellent	Good	Good	Good	Good	Good
Elongation	Low	Fair	Good	Good	Good	Low	Good
Hardness	Lower	Normal	Normal	Normal	Higher	High	Lower
Compression set	Excellent	Good	Fair	Good	Poor	Good	Excellent
Air/Oil aging	Excellent	Excellent	Good	Good	Good	Good	Excellent
Suggested process	Compression molding	Molding extrusion	Molding extrusion	Molding extrusion	Molding extrude	Compression molding	Injection; transfer molding

TABLE 5.15
Cure System Study in Chlorine/Carboxyl ACM

HyTemp 4052	100	100	100	100
Stearic acid	1	1	1	1
Struktol WB-222	2	2	2	2
Naugard 445	2	2	2	2
N550	80	80	80	80
Sodium stearate	4	—	4	—
Diuron-80	2	—	—	—
HyTemp SC-75	—	7	—	—
HyTemp NPC-25	—	—	4	—
HyTemp SAR-50	—	—	3	—
Vanax 829	—	—	—	0.9
TBTD 72% Drimix	—	—	—	3.5
Total	191	192	194	189.4
Mooney scorch, ML at 125°C				
Minimum viscosity, original	40	45	40	39
Minimum viscosity, shelf aged 2 wks at 23°C	40	50	40	37
T_5, Original	18.8	7.3	3.6	8.9
T_5, Shelf aged 2 wks at 23°C	12.6	5.6	3.7	22.1
Rheometer at 190°C, 3° arc, 100 cpm				
ML, N·m	0.8	1.0	0.9	0.8
MH, N·m	4.1	4.2	4.7	5.0
t_{s2}, minutes	1.8	1.2	0.8	0.9
t_{c90}, minutes	15.7	8.4	5.0	4.2
Physical properties, cured 4 min at 190°C				
Shore A, pts	75	64	69	76
100% Modulus, MPa	9.4	5.2	4.8	9.3
Tensile, MPa	13.3	11.3	10.9	11.1
Elongation, %	145	210	210	185
Physical properties, cured 4 min at 190°C, post-cured 4 h at 177°C				
Shore A, pts	73	69	68	80
100% Modulus, MPa	6.7	5.8	5.2	5.1
Tensile, MPa	11.1	11.5	10.6	12.1
Elongation, %	175	205	205	150
Compression set B, 70 h at 150°C—cured 6 min at 190°C				
Set, %	50	33	18	61
Compression set B, 70 h at 150°C—cured 4 min at 190°C, PC 4 h at 177°C				
Set, %	26	19	15	42
Aged in air 70 h at 175°C—cured 4 min at 190°C, PC 4 h at 177°C-change				
Shore A, pts	8	9	7	8
Tensile, %	−19	−27	−20	−11
Elongation, %	−17	−29	−22	−13

ammonium benzoate (0.5–2.5) as well as zinc dimethyl dithiocarbamate (0.5–2.5) with ferric dimethyl dithiocarbamate (0.25–1.0) and Vulkalent E (0.5–1.0) retarder. The presence of ammonia fumes during the vulcanization process may not be acceptable in all factories, which limits the use of ammonium benzoate. A comparison of these three cure systems is shown in Table 5.16.

The Zeonet system does provide very good compression set without post-cure, good bin stability, reasonably fast cure rates, and it may be cured in open steam. The vulcanizates have good water and corrosion resistance.

The ammonium benzoate cured compounds are safe processing, but slow curing. Even with a long post-cure, the compression set obtained is only fair to poor. The odor of ammonia given off during vulcanization is often objectionable.

TABLE 5.16
Cure System Study for Epoxy Cure-Site ACM

Nipol AR31	100	100	100	100
Stearic acid	1	1	1	1
Interstab G-8205	—	—	1	1
MAF carbon black	—	—	50	50
FEF carbon black	50	50	—	—
Naugard 445	—	—	2	2
Ammonium benzoate	1	—	1.3	—
ZMDC (Zinc dimethyldithiocarbamate)	—	1	—	—
FMDC (Ferric dimethyldithiocarbamate)	—	0.5	—	—
Zeonet A	—	—	—	0.8
Zeonet BF	—	—	—	1.8
Zeonet U	—	—	—	1.3
Total	152	152.5	155.3	157.9
Physical properties—cured 20 min at 170°C				
Hardness, JIS	63	66	67	69
100% modulus, MPa	3.1	3.3	3.3	4.3
Tensile, MPa	10.3	10.3	9.4	10.2
Elongation, %	330	340	360	300
Physical properties—cured 20 min at °C, post-cure 4 h at 170°C				
Hardness, JIS	66	71	69	70
100% modulus, MPa	5.1	6.9	4.4	4.9
Tensile, MPa	11.9	12.8	11.8	10.9
Elongation, %	230	200	290	260
Compression set B, 70 h at 150°C—cured 20 min at 170°C, PC 4 h at 170°C				
Set, %	43	59	64.8	20.5
Compression set B, 70 h at 150°C—cured 20 min at 170°C, PC 4 h at 150°C				
	54	67	45.3	16.4
Aged in air 70 h at 175°C—cured 20 min at 170°C, PC 2 h at 170°C-change				
Hardness, pts	7	3	7	9
Tensile, MPa	−2	−37	5	−24
Elongation, %	−43	−39	−28	−47

The dithiocarbamates cure system has a long scorch time and cure time, which may not be practical for high-speed production. The heat aging properties obtained are good, but the compression set is poor, making this cure combination a poor candidate for seal applications, but acceptable for extruded, steam-cured parts.

Nipol AR12, AR14, and AR22 with a carboxyl cure site use approximately 0.6 pphr of Diak #1, hexamethylene diaminecarbamate with 2 pphr of Vanax DOTG, di-ortho-tolylguanidine as optimum levels. A basic formulation is given in Table 5.8 [15,18].

Peroxide Curable HyTemp PV-04 is sensitive to the amount of peroxide used, hence it is necessary to use 2.0 pphr of Vulcup 40KE with 2.0 pphr of HVA #2 coagent for optimum properties. Information on this combination is provided in Table 5.7.

5.7 PROCESSING PROPERTIES

Polyacrylate elastomers have lower viscosities than most other elastomers and also are more thermoplastic, hence it is essential when mixing to maintain full cooling on all mixing surfaces and incorporate reinforcing fillers very early in the mixing cycle in order to obtain good dispersion. This applies to both internal and mill mixing.

Premature scorch, whether severe, rendering the compound unfit for further use, or as a very mild condition, which may be smoothed out with milling through a tight, cool mill, needs to be addressed carefully. Caution: remilling may be deceiving since the compound, when further processed into a finished part, may have poorer physical properties and compression set than expected. It is important to re-test the compound for Mooney scorch and Rheometer properties to ascertain if processing and physical properties meet established standards.

ACM polymers tend to be scorchy and slow curing, thus requiring a post-cure to obtain the optimum physical properties. While the NPC (No Post-Cure) system available from Zeon Chemicals is very fast curing, it is also very sensitive to process temperatures; hence the compound must be kept under 90°C after the curative has been incorporated to prevent scorch. If the compound is handled properly, the bin stability and scorch resistance are excellent.

Mixing may be done preferably in an internal mixer, but also on a mill, bearing in mind that there may be some mill-sticking problems. A two-step mix is strongly recommended to control scorch and process temperature of the compound. If the compound has a long scorch time, a single pass mix may be used if the process temperatures can be controlled to prevent scorch [19].

5.7.1 INTERNAL MIXING PROCEDURES

An ACM clean-out should be used before starting to mix a series of ACM compounds. Start on slower speed with full cooling. Be sure there is no condensed moisture in mixer before starting.

Conventional	**Upside Down**
First Pass at Slow Speed	*First Pass at Highest Speed Obtainable*
Fill factor 74%	Fill factor 65%–70%
Add all polymer and 2/3 fillers	Add all ingredients, except curatives, polymer last

At 85°C add remaining filler
  and other ingredients, except
  curatives, then sweep well	

At 105°C lift ram and sweep	At 1.5 min or 105°C lift ram and sweep
Dump at 120°C	Dump at 165°C
Dip, cool, and dry	Dip, cool, and dry

REST MINIMUM OF 24 h

Second Pass at Slow Speed	*Second Pass at Slow Speed*
Start slow speed, full cooling, being sure there is no condensate in the mixer before starting	

Add 1/2 of the masterbatch, the curatives,
  then the rest of the masterbatch
	Add 1/2 of the masterbatch, the
	  curatives, then the rest of the
	  masterbatch

Dump at 107°C maximum, 80°C with
  the NPC cure system as measured on
  the sheet-off mill
	Dump at 107°C maximum, 80°C
	  with the NPC cure system as
	  measured on the sheet-off mill

Dip, cool, and dry	Dip, cool, and dry

5.7.2 MILL MIXING PROCEDURE

Start with cool rolls, cooling full on.
Band polymer with nip adjusted to provide a rolling bank.
Add filler, stearic acid, antioxidant, and sweep.
Adjust nip to maintain a rolling bank.
Add process aid.
Add plasticizer.
Slab off, dip, cool, and dry.

REST A MINIMUM OF 24 h.

Re-band on a cool mill with full cooling.
Add curatives and disperse well, keeping batch temperature below 100°C, or
    80°C for the NPC cure system.
Slab off, dip, cool, and dry.

Do not use zinc-containing dusting agents such as zinc stearate as these will act
    as a pro-degradent.

The shelf life of mixed compounds depends primarily upon the choice of curative,
but is also very dependent upon the process temperature control and storage

conditions. If refrigerated storage is available, this will dramatically extend the shelf life of the compound. Aged compounds can be freshened on the mill before subsequent processing, as long as the material has not scorched. The normal bin life expected is approximately 2 weeks, but this is reduced as storage temperature and humidity increases.

Extrusion and calendering of black filled ACM compounds may be accomplished easily; however, non-black filled compounds tend to be nervy and difficult to extrude or calender. Normal conditions to use are as follows:

Extrusion	Calender
Feed section temperature, 55°C	Top roll temperature, 60°C–70°C
Zone 1 temperature, 70°C	Middle roll temperature, 70°C–80°C
Zone 2 temperature, 85°C	Bottom roll temperature, 80°C–100°C
Head temperature, 85°C	If using NPC curative, keep below, 85°C
Screw temperature, 80°C	

Cool well with water then blow dry. *Do not use zinc stearate* as a dusting agent. When calendering, cool well before winding up with a polyethylene liner.

Continuous vulcanization may be accomplished in a salt bath or in open steam. It is best if the extrusion is first wrapped with a Mylar tape to exclude moisture, followed by a nylon tape.

Compression molding may be done with any cure system, but HyTemp NPC is the most efficient one; transfer or injection molding is best using HyTemp SC-75, Vanax 829, Zisnet F-PC, or NPC if care is used in processing to keep the compound below a measured temperature of 85°C.

Typical cure cycle used in industry for ACM compounds are as follows:

Compression molding	2 min at 190°C to 1 min at 205°C
Transfer molding	8 min at 165°C to 4 min at 177°C
Injection molding	90 s at 190°C to 45 s at 204°C
CV cure in steam	2 min at 190°C to 1 min at 204°C
Open steam cure	30–60 min at 163°C

These cycles vary depending on the size and shape of the parts and the presence of metal inserts. It is good practice with compression molding to allow the pre-forms to rest 24 h and structure for better back pressure. Pre-form shapes should allow for good flow to push out the air ahead of the compound. The use of freshly mixed and prepared compounds is recommended for transfer or injection molding.

Post curing is best done in a circulating air oven. Typical cycles are in the range of 4–8 h at 177°C, depending upon the heat transfer characteristics of the oven and the mass of the parts. Shorter cycles may be used at higher temperatures, but it is recommended that the maximum post-cure temperature not exceed 190°C.

Bonding to metal and other substrates may be accomplished with a variety of commercial adhesives, with proper metal or fabric preparation recommended by the adhesive supplier. Suggested adhesives include Chemlok 250, 8500, 8560, TyPly BN, and Thixon 718/719. With the increased use of water-based adhesives, it is suggested that you ask your supplier for their latest recommendations.

5.8 APPLICATIONS

The most important consideration when designing a compound for any given application is complete information on the expected performance of the part. This may require education of the design engineer and repeated questions about the environment to be encountered. Too often assumptions are made based on limited information, which invariably leads to failure and belated finding of all the facts.

Dynamic or Shaft Seals are designed with a garter spring to hold the lip against the rotating shaft to maintain a seal. A low coefficient of friction and good abrasion resistance is essential for this application. In order to maintain a low coefficient of friction and low abrasion loss, coarse fillers such as coarse-ground mineral silica, Pyrax B, Celite 350, and so on are used to provide a rough or dimpled surface, which acts as a pump between the lip and shaft to maintain a film of oil on the rubber/metal interface and thus lubricate the sealing surface. If the lip runs dry there will be very high frictional heat generated and the sealing surface will crack and result in leakage. In the past fillers such as PTFE powder, graphite, ground coal dust, were commonly used, and may even be still used for older applications. Excellent heat resistance is also needed and low, but positive swell in the lubricant to maintain dimensional stability. Table 5.17 provides a commercial compound for either compression or injection/transfer molding.

Lip Seal starting point compound to meet GM 8645003-C specification is given in Table 5.18. If an injection molding process were used, the cure system would need to be changed to HyTemp SC-75. It is also suggested that the sodium stearate be replaced with a dispersion such as HyTemp NS-70.

A Commercial Colorable Bearing Seal compound based on HyTemp AR-72LF is shown in Table 5.19. The low compression set is obtained by using Zisnet F-PT cure package HyTemp ZC-50.

A Rocker Cover Gasket with HyTemp 4051EP and 4054 to meet Nissan Specification 13270 is given in Table 5.20. This compound is designed to avoid "weeping" of oil through the gasket as often occurs with silicone.

Oil Pan Gasket using HyTemp AR72LF is demonstrated in Table 5.21 using the Zisnet F-PT cure system. This compound, like the rocker cover gasket, was formulated to provide good resistance to "weeping" of oil through the gasket.

Table 5.22 lists O-ring compounds using the NPC and SC-75 cure systems for a choice of compression or injection molding. For many specifications a post-cure

TABLE 5.17
Commercial Toacron 840 Shaft Seal Compound, ASTM D2000 M5DH706A26B16EO16EO36F13

Toacron 840	100
Stearic acid	1
ODPA	2
N762	25
Cabosil M-7D	12
Celite 350	55
TL 102 PTFE micropowder	20
Silane A1100 DLC	0.5
Plasticizer TP 759	4
Sodium stearate	4
Spider Sulfur	0.4
Total	223.9
Physical properties, cured 5 min at 177°C, post-cured 3 h at 204°C	
Shore A, pts	72
Modulus at 50%, MPa	4.1
Modulus at 100%, MPa	7.7
Tensile, MPa	8.5
Elongation, %	140
Tear, die C, KN/m	15
Low temperature	
Brittle point, Pass T_P, °C	−20
Gehman T_5, °C	−18
Compression set B, 22 h at 150°C	
Set, %	31
Aged in air 70 h at 150°C-change	
Shore A, pts	1
Tensile, %	4
Elongation, %	0
Aged in ASTM #1 Oil 70 h at 300°C-change	
Shore A, pts	0
Tensile, %	−2
Elongation, %	−7
Volume, %	−1
Aged in IRM 903 Oil 70 h at 150°C-change	
Shore A, pts	−8
Tensile, %	−6
Elongation, %	0
Volume, %	15

would not be required with the NPC curative, but for the more critical requirements, a post-cure will improve the set to 15%. The un-post-cured NPC compound has a compression set equal to the post-cured SC-75 compound.

TABLE 5.18
Lip Seal for GM 8645003-C

HyTemp 4051 CG	67	
HyTemp 4052 EP	33	
Stearic acid	3	
N550	95	
AgeRite Stalite S	1	
HyTemp SR-50	5	
Sodium stearate	6	
HyTemp NPC-25	6	
Maglite Y	0.5	
Total	216.5	
Specific gravity	1.35	

Mooney viscosity at 100°C

ML 1 + 4	66	

Rheometer at 210°C, 3° arc, 100 cpm

ML, N·m	0.8	
MH, N·m	7.2	
T_{s2}, minutes	0.4	
T_{c90}, minutes	1.6	

Physical properties—cured 1.5 min at 210°C

	Results	Specification
Shore A, pts	84	80–90
Modulus at 50%, MPa	4.6	—
Modulus at 100%, MPa	9.7	4.5 min.
Tensile, MPa	11.1	10.3 min.
Elongation, %	120	100 min.
Tear, Die B, ASTM D624, kN m	33.4	21.9 min.

Low-temperature brittleness at −23°C

Result	Pass	Pass

Aged in air 70 h at 149°C-change

Shore A, pts	0	0 to +7
Tensile, %	−6	−30 max.
Elongation, %	−17	−20 max.

Compression set B, 70 h at 149°C, cured 1.5 min at 210°C, button 2.25 min at 210°C

Plied, %	38	—
Button, %	33	50 max.

Aged in ASTM # 3 Oil 70 h at 149°C-change

Shore A, pts	−15	0 to −15
Tensile, %	−6	−25 max.
Elongation, %	−8	−15 max.
Volume, %	17	0 to +20

Aged in Dexron II ATF 70 h at 149°C-change

Shore A, pts	−10	0 to −10
Tensile, %	+2	−30 max.
Elongation, %	−8	−30 max.
Volume, %	+6.6	0 to +10

TABLE 5.19
Commercial HyTemp AR-72LF Non-Black Bearing Seal, ASTM D2000 M5DH806A26B16EO16EO36F13

HyTemp AR-72LF	100
Stearic acid	2
Struktol WB 222	2
HiSil 233	40
Celite Superfloss	60
DSC-18	1.5
Naugard 445	2
Plasticizer TP 759	5
HyTemp ZC-50	5
Thiate H	0.3
Total	217.8
Physical properties—cured 5 min at 177°C, PC 3 h at 204°C	
Shore A, pts	79
Modulus at 50%, MPa	6.0
Modulus at 100%, MPa	8.1
Tensile strength, MPa	9.3
Elongation, %	135
Tear, Die C, KN/m	13.9
Aged in air 70 h at 150°C-change	
Shore A, pts	2
Tensile, %	−4
Elongation, %	4
Aged in ASTM #1 Oil 70 h at 150°C-change	
Shore A, pts	2
Tensile, %	7
Elongation, %	−2
Volume, %	−0.5
Aged in IRM 903 Oil 70 h at 150°C-change	
Shore A, pts	−6
Tensile, %	−4
Elongation, %	−9
Volume, %	15
Compression set B, 22 h at 150°C	
Set, %	28
Low temperature	
Brittle point, Pass, °C	−16
Gehman T_5, °C	−11

Anti-Drain Back Gasket Compound in Table 5.23 is designed to be very heat resistant as well as being resilient to effect proper closure. The Vanax 829 and Butyl Tuads curative were chosen for the resilience that it provides.

TABLE 5.20
Rocker Cover Gasket to Meet Nissan Specification 13270

HyTemp 4051EP	85	
HyTemp 4054	15	
HyTemp SR-50	4	
Stearic acid	2	
AgeRite Stalite S	2	
Struktol WB-222	2	
N550	75	
Thiokol TP-759	8	
Sodium stearate	5	
HyTemp NPC-25	5	
Total	203	
Mooney viscosity, ML 4 at 100°C		
Viscosity	36	
Rheometer at 190°C, 3° arc, 100 cpm		
ML, N · m	0.5	
MH, N · m	4.2	
t_{s2}, minutes	0.8	
t'_{90}, minutes	4.0	
Physical properties—cured 3 min at 190°C		Specification
Shore A, pts	61	55–65
Modulus at 100%, MPa	3.5	—
Tensile, MPa	8.3	7.9 min.
Elongation, %	260	150 min.
Specific gravity	1.29	—
Compression set B, plied discs, 70 h at 150°C		
Set, %	30	50 max.
Low-temperature brittleness, ASTM D2137		
Pass, °C	−20	−20 max.
Aged in air 70 h at 150°C-change		
Shore A, pts	2	±15
Tensile, %	−5	±30
Elongation, %	−12	−50 max.
180° Bend	Pass	—
Aged in Shell 5W30 Oil 70 h at 150°C-change		
Shore A, pts	0	±5
Tensile, %	11	−30 max.
Elongation, %	−12	−30 max.
Volume, %	−2.2	±5
180° Bend	Pass	—

TABLE 5.21
Oil Pan Gasket Compound

HyTemp AR 72LF	100
Stearic acid	1
Struktol WB-222	2
AgeRite Stalite S	2
N550	48
Zisnet F-PT	0.55
Butyl Zimate	1.0
Santogard PVI	0.3
Thiate H	0.5
Total	155.35
Mooney scorch, ML at 125°C	
Minimum viscosity	28
t_5, minutes	21.6
Rheometer at 190°C, 3° arc, 100 cpm	
ML, N · m	0.6
MH, N · m	4.2
t_{s2}, minutes	1.6
t'_{90}, minutes	5.0
Physical properties, cured 4 min at 190°C, PC 4 h at 177°C	
Shore A, pts	54
Modulus at 100%, MPa	3.5
Tensile, MPa	9.6
Elongation, %	270
Tear strength, kN/m	28.9
Compression set B, 70 h at 150°C, cured 6 min at 190°C, PC 4 h at 177°C	
Set, %	26
Aged in air 46 h at 177°C-change	
Shore A, pts	7
Tensile, %	−9
Elongation, %	−17
Aged in ASTM #3 Oil 46 h at 150°C-change	
Shore A, pts	7
Tensile, %	−9
Elongation, %	−17
Volume, %	15.5
Aged in BP Oil 10W30 46 h at 150°C-change	
Shore A, pts	0
Tensile, %	5
Elongation, %	−4
Volume, %	2.6

TABLE 5.22
70 Shore A O-Ring Compounds

HyTemp 4052	100	100
Stearic acid	1	1
Struktol WB-222	2	2
Naugard 445	2	2
N550	80	80
Sodium stearate	4	—
HyTemp SC-75	—	7
HyTemp SR-50	3	—
HyTemp NPC-25	4	—
Total	196	192
Mooney scorch, ML at 125°C		
Minimum viscosity	40	45
T_5, minutes	3.6	7.3
Rheometer at 190°C, 3° arc, 100 cpm		
ML, N·m	0.9	1.0
MH, N·m	4.7	4.2
t_{s2}, minutes	0.8	1.2
t_{c90}, minutes	5	8.4
Physical properties, cured 4 min at 190°C		
Shore A, pts	69	64
Modulus at 100%, MPa	4.8	5.2
Tensile, MPa	10.9	11.3
Elongation, %	210	210
Physical properties, cured 4 min at 190°C, post-cured 4 h at 177°C		
Shore A, pts	68	69
Modulus at 100%, MPa	5.2	5.8
Tensile, MPa	10.6	11.5
Elongation, %	205	205
Compression set B, 70 h at 150°C		
Original set, %	18	33
Post-cured set, %	15	19
Aged in air 70 h at 175°C, cured 4 min at 190°C, PC 4 h at 177°C-change		
Shore A, pts	7	9
Tensile, MPa	−20	−27
Elongation, %	−22	−29
Aged in 10W40 Oil 70 h at 150°C, cured 4 min at 190°C, PC 4 h at 177°C-change		
Shore A, pts	3	3
Tensile, %	12	7
Elongation, %	−24	−22
Volume, %	3	3

TABLE 5.23
Anti-Drain Back Valve

HyTemp 4053EP	100
Stearic acid	1
Struktol WB-222	2
N990	120
Thiokol TP-759	5
Vanax 829	0.9
Butyl Tuads, 72% Drimix	3.5
Synpro-Ware G (DPS) D-80	0.2
Total	232.6

Mooney viscosity, ML 4 at 100°C

ML 1 + 4	45

Rheometer at 177°C, 3° arc, 100 cpm

ML, N·m	5.7
MH, N·m	25.3
t_{s2}, minutes	2.0
t_{c90}, minutes	6.1

Physical properties, cured 10 min at 177 C, post-cured 4 h at 177°C

Shore A, pts	58
Modulus at 100%, MPa	5.9
Tensile, MPa	7.2
Elongation, %	130
Specific gravity	1.40

Low-temperature properties

Brittleness, ASTM D2137, pass °C	−32
Gehman, ASTM D1053, T_{100}, °C	−43

Compression set B, 22 h at 150°C, cured 15 at 177°C, PC 4 h at 177°C

Set, %	19

Aged in air in test tube 70 h at 150°C-change

Shore A, pts	2
Tensile, %	5
Elongation, %	−15
180° Bend	Pass

Aged in ASTM #1 Oil 70 h at 150°C-change

Shore A, pts	4
Tensile, %	14
Elongation, %	−15
Volume, %	−3.6

Aged in ASTM #3 Oil 70 h at 150°C-change

Shore A, pts	−12
Tensile, %	−13
Elongation, %	−8
Volume, %	17

TABLE 5.24
Transmission Oil Cooler Hose

HyTemp AR72HF	20	—
HyTemp AR74	80	—
Nipol AR-53	—	100
Stearic acid	1	1
Struktol WB-222	2	—
TE-80 paste	—	2
Naugard 445	2	2
Inhibitor OABH	2	2
N550	90	80
Thiokol TP-759	7	8
Sodium stearate	—	4
Zisnet F-PT	0.5	—
Ethyl Zimate	1.5	—
Vulkalent E/C	0.5	—
Thiate H	0.3	—
Diuron 80	—	2
Total	206.8	201

Mooney viscosity, ML 1 + 4 at 100°C

Viscosity	56	—

Rheometer, 3° arc, 100 cpm at 190°C

ML, N·m	1.0	—
MH, N·m	7.3	—
T_2, minutes	1	—
t_{s90}, minutes	5.8	—

Physical properties, cured	60 min at 166°C, PC 4 h at 177°C	45 min at 170°C, PC 6 h at 166°C
Shore A, pts	82	68
Modulus at 100%, MPa	8.2	6.0
Tensile, MPa	8.9	11.4
Elongation, %	126	210

Aged in air-change 70 h at 175°C, 168 h at 150°C

Tensile, %	−2	−5
Elongation, %	4	−37

Aged in ASTM IMR 903 Oil 70 h at 150°C-change

Tensile, %	3	15
Elongation, %	22	−5
Volume, %	15	11

Aged in Dexron III at 150°C-change 168 h and 70 h

Tensile, %	13	9
Elongation, %	1	−7
Volume, %	3	2

TABLE 5.24   (Continued)
Transmission Oil Cooler Hose

Aged in Motor Oil 70 h at 150°C-change		
Tensile, %	—	13
Elongation, %	—	−10
Compression set B, 70 h at 150°C		
Set, %	33	23
Low-temperature properties		
Bend at −40°C ASTM D736	—	Pass
Gehman T_{100}, °C	−46	—

Transmission Oil Cooler Hose compounds are shown in Table 5.24 in which two non-post-cure systems are given. Both the NPC-25 and Zeonet system provides low compression set without post-cure. The temperature of the NPC-25 compound during processing must NOT exceed 85°C, as measured in the stock. During vulcanization, a moisture barrier such as Mylar film should be wrapped on under the nylon tape. The Zeonet cure mechanism also does not require a post-cure and it has the added advantage that it may be used in open steam without a moisture barrier.

ACKNOWLEDGMENTS

The Zeon group of companies provided most of the data for this chapter, for which I would like to extend my appreciation. In addition, the recommendations from Dr Paul Manley of Zeon Chemicals have been very helpful.

REFERENCES

1. D.S. Seil and F.R. Wolf, *Rubber Technology*, chapter 11, third edition, Maurice Morton, Van Nostrand Reinhold Company, New York.
2. R.M. Montague, Advances in Acrylic Cure Technology, Detroit Rubber Group Meeting, March 10, 1988.
3. C.T. Smith and P.E. Manley, New Polyacrylate Curing Systems, Chicago Rubber Group, January 28, 1999.
4. K. Umetsu, T. Kawanaka, K. Hosoyo, K. Hashimoto, and S. Hayashi, Crosslinking of Polyacrylate Elastomer with Iso-Cyanuric Acid and Ammonium Salt, 148th Meeting, Rubber Division, ACS, Cleveland, October 17–20, 1995.
5. K. Hosoya, T. Nakagawa, and S. Yagishita, A New Polyacrylic Elastomer Cure System for Automatic Transmission Fluid Hose Application, SAE Meeting, Detroit, February 29–March 4, 1988.
6. R.C. Klingender, Polyacrylate Properties and Uses, Fort Wayne Rubber & Plastics Group, April 20, 1989.
7. Zeon Corporation, Nipol AR-52 Nipol Acrylic Rubber, brochure.

8. Zeon Corporation, Nipol-AR Nipol Acrylic Rubber, brochure.

9. Zeon Chemicals, 2001 Product Guide, E2-6/00.

10. TOA Acron, AR Technical Information Bulletin.

11. TOA Acron, AR-501, 501L, AR540, AR540L Heat and Oil Resistant Polyacrylate Elastomer Bulletin.

12. I. Kubota and H. Kanno, Improved Heat & Compression Set Resistance Polyacrylate Elastomers—Part A: Polymer Development, 158th Meeting, Rubber Division, ACS, Cincinnati, October 17–20, 2000.

13. P.E. Manley and C.T. Smith, Improved Heat & Compression Set Resistant Polyacrylate Elastomers—Part B: Applications in Engine and Automatic Transmission Fluids Requiring Low Compressive Stress Relaxation, 158th Meeting, Rubber Division, ACS, October 17–20, 2000.

14. I. Kubota and H. Kanno, Improved Low-Temperature Properties Polyacrylate Elastomers, 160th Meeting, Rubber Division, ACS, Cleveland, October 16–19, 2001.

15. P.E. Manley, New Technology for Polyacrylic Elastomers in Automotive Sealing Applications, SAE Meeting, Detroit, February 26–29, 1996.

16. Zeon Chemicals, Introducing HyTemp 4065 Improved Processing and Compression Set, brochure.

17. Zeon Chemicals Technical Bulletin HPA-1A, HyTemp.

18. Zeon Corporation, Business and Products Guide, www.zeoncorporation.com.

19. A. Anderson and P.E. Manley, Improved Mixing of Polyacrylate, 160th Meeting, Rubber Division, ACS, Cleveland, October 16–19, 2001.

6 Ethylene/Acrylic (AEM) Elastomer Formulation Design

Lawrence C. Muschiatti, Yun-Tai Wu, Edward McBride, and Klaus Kammerer

CONTENTS

Vamac ethylene/acrylic (AEM) elastomers were introduced commercially in 1975 to meet the escalating temperature and fluid resistance requirements of the automobile industry. This high-performance elastomer was designed to combine high heat and service fluid resistance with good low-temperature properties. Good physical strength of vulcanizates, excellent ozone and weather resistance, and excellent damping characteristics—over a broad temperature range, are a consequence of the chemical composition of AEM.

6.1 CHEMICAL COMPOSITION

Vamac ethylene/acrylic elastomers are noncrystalline copolymers of ethylene and methyl acrylate. Vamac terpolymers contain a small quantity of a polar cure site monomer with carboxylic acid functionality. Both ethylene and methyl acrylate monomers contribute to the high-temperature stability of the AEM polymers. Good low-temperature properties are derived from the nonpolar ethylene monomer, while the polar methyl acrylate monomer provides the oil and fluid resistance, including mineral oil-based synthetic fluids. Ethylene-methyl acrylate ratios have been optimized and form the basis for the different grades of Vamac. The completely saturated halogen-free polymer is noncorrosive and the backbone imparts excellent resistance to ozone, oxidation, UV radiation, and weathering. The basic structure of AEM is shown schematically in Figure 6.1 and an overview of the benefits provided by this composition is given in Table 6.1.

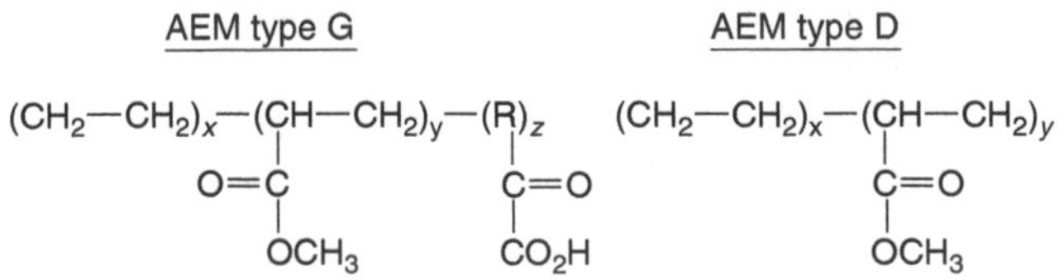

FIGURE 6.1 Structures of Vamac ethylene acrylic elastomers.

TABLE 6.1
Primary Benefits of AEM

High temperature durability
Good oil and automotive fluid resistance
Excellent water resistance
Good low temperature flexibility
Outstanding ozone and weather resistance
Good mechanical strength
Very good compression set
Good flex fatigue resistance
Good abrasion and wear resistance
High and consistent vibration damping
Low permeability to gases
Non-halogen, low smoke emissions
Colorability
Moderate cost

6.2   COMMERCIAL GRADES

Vamac elastomers are commercially available as copolymers of ethylene and methyl acrylate (Vamac D grades) or terpolymers of ethylene, methyl acrylate, and an acidic cure site monomer (Vamac G grades). In both the copolymer and terpolymer classes there are polymers with increased levels of methyl acrylate for lower volume swell in automotive service fluid environments. These grades have an "LS" designation, that is, Vamac DLS and Vamac GLS. A list of the present commercial grades of Vamac is shown in Table 6.2. The Vamac D grades are cured with peroxides and may not require a post-cure to optimize properties. The terpolymer VamacG grades are vulcanized with diamines and do require a post-cure to optimize vulcanizate properties, especially modulus and compression set. Vamac copolymers are not sensitive to ionic impurities or additives in fluids and fillers as are the terpolymers.

TABLE 6.2
Vamac Product Line

Terpolymer Grades	Form	ML (1 + 4) at 100°C
Vamac G	Translucent gum	16 ± 3
Vamac HG	Gum—high green strength	35 ± 10
Vamac GLS	Gum—low fluid swell	16 ± 3
Copolymer grades		
Vamac D	Translucent gum	22 ± 3
Vamac DLS	Gum—low fluid swell	16 ± 3

However, copolymers must be cured with peroxides and are subject to the constraints normally associated with peroxide cures.

The shelf stability of AEM gum polymers is excellent, being measured in years.

6.3 OVERVIEW OF VULCANIZATE PROPERTIES

6.3.1 DRY HEAT RESISTANCE

Vamac compounds can be formulated to perform capably over a broad temperature range. The approximate useful life, that is, time to embrittlement, varies with continuous exposure temperature as shown in Table 6.3. The useful life with continuous exposure at 121°C is approximately 2 years, decreasing to 5 days at 204°C. As per the ASTM D2000 and SAE J200 classification systems, the heat resistance rating for AEM is Type "E," that is, vulcanizates of AEM polymers are capable of maintaining 50% elongation at break after 1008 h continuous exposure to 171°C. Vamac is not recommended for applications involving continuous exposure to high-pressure steam.

6.3.2 LOW-TEMPERATURE PERFORMANCE

Vulcanizates of unextended Vamac elastomers have brittle points below −60°C with good low-temperature flexibility. Although the addition of fillers causes the brittle point to increase, selected low-volatility plasticizers can be incorporated to improve low-temperature performance. Practical Vamac vulcanizates can be formulated to meet low-temperature brittleness requirements to temperatures lower than −40°C. Because of their slightly higher glass-transition temperatures, compounds based on the "LS" grades are generally 5°C to 7°C higher in their low-temperature limits than comparable "non-LS" based compounds.

TABLE 6.3

Dry Heat Resistance of Vamac. Approximate Useful Life—Time to Embrittlement (50% Elongation)

Continuous Air Oven Exposure (°C)	Approximate Useful Life
121	2 years
149	6 months
171	1000 h (6 weeks)
177	4 weeks
191	10 days
204	5 days

Note: Approximate useful life is increased when immersed in most fluids.

Vamac is not recommended for applications involving continuous exposure to high pressure steam.

6.3.3 General Fluid and Chemical Resistance

Vamac is a polar polymer and is resistant to aliphatic and paraffinic-based fluids. In general, Vamac is serviceable in fluids such as

1. Lubricating oils and greases
2. Automatic transmission fluids
3. Mineral-based engine oils
4. Synthetic engine oils
5. Diesel fuel
6. Kerosene
7. Mineral oil-based hydraulic fluids
8. Water to 100°C
9. Diluted acids and bases
10. Wet and dry sour gas

In most fluids, the "LS" grades of AEM offer lower swell than the polymers with lower levels of methyl acrylate, nominally lower by half. The fluids to avoid with Vamac are

1. Gasoline
2. Aromatic hydrocarbons
3. Esters
4. Ketones
5. Concentrated acids

Some typical fluid resistance values (percent volume change) of compounds of Vamac G, in a variety of fluids [1], are shown in Table 6.4.

After 6 weeks immersion in highly aromatic ASTM #3 oil or IRM 903 oil at 150°C, compounds of Vamac G, Vamac HG, and Vamac D show a nominal volume increase of 50%. The volume swell of the "LS" grades in a similar environment is nominally 25%–30%. This performance results in ASTM D2000 (SAE J200) ratings of "EE" to "EH" for Vamac compounds.

6.3.4 Vibration Damping

AEM polymers are excellent damping materials, characterized by high loss tangent values, and are unique in that they maintain the same degree of damping over a broad range of temperature and frequency [2] (Figure 6.2). This unique property allows the use of Vamac for vibration damping in high-temperature environments.

6.3.5 Physical Properties

AEM polymers can be compounded to meet a broad range of physical property requirements. Typical physical properties for AEM vulcanizates are shown in Table 6.5. AEM compounds combine high physical strength with excellent compression set. The high physical strength of compounds from Vamac is retained at elevated temperatures.

TABLE 6.4
Fluid Resistance of Vamac G

Type of Fluid	Immersion Conditions		
	Temperature °C	Time Days	Volume Change %
ASTM Standard oils and reference fuels			
ASTM Oil no. 1	177	3	3.5
ASTM Oil no. 2	177	3	23
ASTM Oil no. 3	177	3	53
ASTM Oil no. 3	149	3	50
Service fluid 105G	150	7	17
ASTM Reference fuel A	24	7	20
ASTM Reference fuel B	24	7	73
ASTM Reference fuel C	24	7	153
Hydraulic fluids and lubricants			
SAE 5W30 Motor oil	24	7	1.7
SAE 5W30 Motor oil	150	7	20
SAE 5W30 Motor oil	150	42	24
Mobil one synthetic motor oil	150	7	10
Mobil one synthetic motor oil	150	42	13
Dexron III ATF	150	7	25
Dexron III ATF	150	42	29
Sunamatic 141 ATF	177	3	21
Texaco ATF	149	7	16
ELCO EP Axle Lube M2C-105A	149	7	16
Grease, Lithium additive	121	1	11
Silicone grease	24	7	1.4
Skydrol 500A	121	7	214
Wagner brake fluid (DOT 3)	121	7	166
Texaco power steering fluid	163	3	21
Pentosin CHF 11S PSF	125	7	12
Pentosin CHF 11S PSF	125	42	13
Water, coolants, acids and bases			
Water	100	14	24
Dowtherm 209 (46% in Water)	98	14	14
Ethylene glycol	100	7	3.5
Sulfuric acid, 20%	24	7	0.3
Hydrocloric acid, 20%	24	7	2.0
Sodium hydroxide, 20%	24	7	4.5
Dry sour gas at 14.5 MPa	150	10	4.6
Wet sour gas at 3.0 MPa	205	10	−2.6
Solvents			
Cotton seed oil	24	7	3.2
Ethyl alcohol	24	7	40
Freon 12	24	7	30
Kerosene	24	7	35
Methylene chloride	24	7	175
Methyl isobutyl ketone	24	7	244

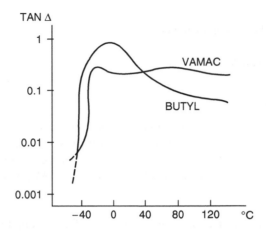

FIGURE 6.2 Vibration damping of Vamac compared with BUTYL rubber.

TABLE 6.5
Typical Physical Properties

Compound Property	Typical Range of Results
Hardness, durometer A, Pts	40–90
Tensile strength, MPa	6.9–17.2
Elongation at break, %	100 600
Die C Tear strength, kN/m	17–44
Compression set B, 70 h at 150°C	
Set, % as molded	25–65
Set, % post-cured	10–30

6.3.6 MISCELLANEOUS PROPERTIES

Vamac polymers are halogen-free and therefore do not promote corrosion and are suitable for use in noncorrosive, flame-retardant applications. The permeability of Vamac to gases is low, compared with that of butyl rubber. In addition, AEM compounds have good adhesion to a variety of metals, fibers, and other elastomers. Mineral-filled compounds of Vamac can be brightly colored and are capable of maintaining their color quality after thermal aging.

6.4 COMPOUND DESIGN

The basic principles and procedures used to compound Vamac are straightforward. The components of most Vamac compounds are

1. Polymer
2. Release system—process aids

3. Stabilizer(s)
4. Fillers—carbon black or mineral fillers
5. Plasticizer(s)
6. Vulcanization system

6.4.1 Polymer Selection

Because of the broad balance of properties and their relative ease of processing, AEM terpolymers should be considered as the polymers of choice unless their use is precluded by mitigating circumstances. Considering the responses to a few simple questions can make a determination of the optimum AEM polymer for a specific application

1. What are the applications and methods of fabrication; that is, molded gasket, extruded hose, flame-retardant cable, and so on? Will the compound be highly extended? Choices: Vamac G versus Vamac HG.
2. Which performance properties are critical? Choices: Vamac G versus Vamac GLS (Vamac D versus Vamac DLS).
3. Is a post-cure impractical? Choices: Vamac D versus Vamac G (Vamac DLS versus Vamac GLS).
4. Does the polymer interact with any compounding ingredients? Choices: Vamac D versus Vamac G (Vamac DLS versus Vamac GLS).

High Mooney Viscosity Vamac HG is the polymer of choice for compounds that are used in fabrication processes, which require high green strength and for compositions that are highly filled. Because of its high viscosity, Vamac HG is used to enhance processing, extension with fillers and plasticizers, collapse resistance, and sponging. The low Mooney viscosity grades are favored for fabrication processes such as injection and transfer molding. Both viscosity grades are applicable for compression molding, except that the higher viscosity grades may be needed to avoid trapped air.

Fluid resistance and low-temperature requirements are the principle considerations that determine if "LS" grades are to be considered. Requirements for low swell in aromatic fluids necessitate the use of the "LS" grades or blends of "LS" grades with their lower methyl acrylate content counterparts. Compounds based entirely on the "LS" grades sacrifice 5°C–7°C in low-temperature flexibility. If low-temperature flexibility is critical, Vamac G (Vamac D) is preferred. With Vamac GLS it is sometimes possible to regain low-temperature flexibility through the addition of a plasticizer. As is explained later, this is generally not true for Vamac DLS. A good balance of fluid swell and low-temperature flexibility can often be obtained by blending Vamac GLS (Vamac DLS) with Vamac G (Vamac D) at the ratios required to optimize both properties.

AEM terpolymers, which contain acidic cure sites, are susceptible to the formation of ionic crosslinks and should not be used in compositions with fillers containing soluble divalent metal ions. Ionic crosslinks can affect the stock by significantly increasing stock viscosity, making the material difficult to process. Ionic crosslinks significantly increase the compression set and variability of the physical properties of vulcanizates. In cases where the use of fillers containing soluble divalent metal ions is required, flame-retardant compositions, for example, AEM copolymers, should be used. In addition, if the application requires continuous contact with fluids that contain additives, which are reactive toward the acidic cure sites in the AEM terpolymer, Vamac D copolymer grades must be used. Examples of reactive additives are amine stabilizers with primary amine impurities.

Terpolymers must be post-cured to attain optimum properties. If a post-cure is either not practical or not desired, and excellent compression set is a requirement, the copolymer Vamac D grades are the preferred choice.

A brief overview of the polymer selection criteria discussed earlier is given in Table 6.6, entitled Polymer Selection Guide. Table 6.6 lists the polymers that are recommended as the preferred alternatives to Vamac G in applications where special processing or properties need to be considered. This table should be used as a general guide and it is not intended to be an all-inclusive listing of alternatives.

6.4.2 RELEASE SYSTEM

As with many elastomers, AEM polymers require the addition of internal release systems. Release agents and process aids must be incorporated to facilitate the release of AEM compounds from the mixer and from the mold. The standard release packages recommended for almost all AEM compounds are shown in Table 6.7. Note that the recommended release systems differ for terpolymer and copolymer. Optimization studies have confirmed that all of the components of the release system are essential for good productivity. Balancing release properties with vulcanizate performance derived the suggested quantities of each component. The alkyl phosphate is a very effective release agent for both the terpolymer and copolymer. Unfortunately it interferes with the peroxide cure of the copolymer and is not recommended for use with copolymers unless necessary for processing. Increasing the level of Vanfre VAM or Vanfre UN to two (2) phr significantly improves mold release, usually at the sacrifice of vulcanizate physical properties and compression set. Although melt fracture during extrusion is usually not a problem with Vamac, one (1) phr of Struktol WS180, a fatty acid/silicone condensation product, can improve extrusion performance. Higher levels of Struktol WS180 are effective but will tend to bloom.

6.4.3 STABILIZER(S)

Vamac requires the use of stabilizers for long service life. The types of stabilizers used in Vamac include antioxidants, metal deactivators, and zinc deactivators. The most effective antioxidants for Vamac are diphenylamines. Table 6.8 contains the recommended stabilizers and suggested levels for various requirements.

The recommended level of diphenylamine stabilizer for terpolymers is 2 phr. Antioxidants are required in copolymer compositions but their level must be reduced to 1 phr because of the interference of the diphenylamine with the peroxide cure. Diphenylamines are discoloring antioxidants and cannot be used where high temperature color stability is required. The addition of 4–7 phr of titanium dioxide

TABLE 6.6
Polymer Selection Guide

Polymer	Property/Process								
	High Viscosity	High Green Strength	Collapse Resistance	High Extension	CV Cure	No Post Cure	Plastic Inserts	Metal Ion Contamination	Higher Fluid Resistance
Vamac HG	X	X	X	X	—	—	—	—	—
Vamac HG/Vamac G	X	X	X	X	—	—	—	—	—
Vamac GLS	—	—	—	—	—	—	—	—	X
Vamac GLS/Vamac G	—	—	—	—	—	—	—	—	X
Vamac D	—	—	—	—	X	X	X	X	—
Vamac DLS	—	—	—	—	X	X	X	X	X
Vamac DLS/Vamac D	—	—	—	—	X	X	X	X	X

Note: In general, vamac g should be considered as the polymer of choice for most applications unless there are special considerations, the special considerations and the AEM polymers that should be considered as alternatives to Vamac G are listed.

TABLE 6.7
Release Package

Vamac terpolymers require the standard release package
Armeen 18D (Octadecylamine)	0.5 phr
Stearic acid	1–2 phr
Vanfre VAM (Alkyl Phosphate Paste)	1 phr

Vamac copolymers use the standard release package
Armeen 18D (Octadecylamine)	0.5 phr
Stearic acid	0.5–1.5 phr

Other release agents can be used to further improve processing characteristics at the expense of affecting other physical properties.
Two (2) phr of Vanfre VAM or Vanfre UN (fatty acid phosphate) significantly improves mold release at the sacrifice of physical properties and compression set properties.
One (1) phr of Struktol WS 180 can improve extrusion performance, but at higher levels it will bloom.

(TiPure 103) to color stable compounds is recommended to produce bright whites and colors. The recommended antioxidant package for optimum high temperature color stability in terpolymer-based mineral filled compounds includes

1. Naugard XL1 0.5–1 phr
2. Weston 600 1–3 phr
3. Naugard 10 0.5 phr
4. Irganox MD-1024 2 phr (if necessary for metal deactivation)

TABLE 6.8
Stabilizers

Standard antioxidant systems
Vamac terpolymers	2 phr Naugard 445 or Dustanox 86
Vamac dipolymers	1 phr Naugard 445 or Dustanox 86

Specialty systems
Optimum high temperature color stability for terpolymer compounds
> 0.5–1.0 phr Naugard XL1
> 1 phr Weston 600
> 0.5 phr Naugard 10
> Irganox MD-1024

Optimum high temperature color stability for copolymer compounds
> 0.5 phr Naugard XL1
> 0.5 phr Naugard 10

Zinc resistance (terpolymers and copolymers)
> 2 phr Eastman Inhibitor OABH

Metal deactivation is usually needed in pigmented compounds. When absolutely no discoloration can be tolerated, copolymer compositions stabilized with 0.5 phr Naugard XL1 and 0.5 phr Naugard 10 are recommended.

Vulcanizates that are in contact with galvanized metals, hose couplings, for example, need to be protected from degradation with zinc inhibitors. The zinc inhibitor that has been found effective for Vamac is Eastman Inhibitor OABH.

6.4.4 FILLERS—CARBON BLACK AND MINERAL FILLERS

As with most elastomers, various fillers have different effects on the properties of the compound. Because of the potential for the formation of ionic crosslinks with the terpolymer, some fillers are not suitable for use with terpolymers. Table 6.9 is a brief overview of filler performance when incorporated into Vamac.

The expected relationships between carbon black particle size and structure, and the properties of vulcanizates, can be extended to Vamac. Tables 6.10 and 6.11 show the effect of carbon black particle size on the physical properties of Vamac terpolymers and copolymers, respectively.

As expected, as the particle size of the carbon black decreases, the hardness, modulus and tensile strength of Vamac vulcanizates increases, while the elongation at break decreases. There is essentially no difference in the behavior of the terpolymer

---

TABLE 6.9
Performance of Fillers in Vamac

Reinforcing fillers

Carbon black (N762, N774, N550)	Optimum toughness and heat resistance
Graphite	Reduced coefficient of friction, better abrasion resistance
Fumed silica (Cab-O-Sil MS-7DS)	High tensile strength and elongation
	High stock viscosity
	High (poor) compression set
Hydrated alumina (Hydral 710)	Used for flame-retardant, low-smoke compounds
Treated magnesium silicate (Mistron CB, Cyprubond)	Maximum modulus with minimum hardness increase

Nonreinforcing fillers
Carbon black, N990 (MT)
Uncoated calcium carbonate
Barium sulfate (Blanc Fixe)

Unsatisfactory fillers for terpolymers
Divalent metal oxides
Metal stearate coated fillers
Clays
Hydrated/precipitated silicas

Fillers permissible in small amounts in terpolymers

Titanium dioxide	Up to 7 phr
Chrome oxide	Up to 5 phr
Antimony trioxide	Up to 5 phr
Iron oxide	Up to 3 phr

TABLE 6.10
Carbon Black Fillers in Vamac Terpolymers

Carbon Blacks	Compounds[a]					
MT (N990)	60	—	—	—	—	—
SRF (N762)	—	60	—	—	—	—
GPF (N660)	—	—	60	—	—	—
FEF (N550)	—	—	—	60	—	—
HAF (N330)	—	—	—	—	60	—
ISAF (N220)	—	—	—	—	—	60
Physical properties[b]—stress/strain at 23°C						
Hardness, durometer A (Pts)	51	63	68	72	76	77
100% Modulus, Mpa	1.69	3.21	3.90	4.62	5.14	5.17
Tensile strength, Mpa	10.0	10.4	11.3	11.6	14.8	15.5
Elongation at break, %	415	355	345	320	280	280
Die C tear strength, kN/m	18.2	24.2	26.4	27.1	25.9	27.3

[a] Basic compound: Vamac G—100 parts, Naugard 445—2 phr; Armeen 18D—0.5 phr; Vanfre VAM—1 phr; stearic acid—1.5 phr; DOS—7 phr; 40% TETA dispersion—5 phr; DOTG—4 phr.
[b] Test slabs press-cured for 30 min at 166°C.

and the copolymer. Because of the broad property range attainable and the minimal effect on processing, the carbon blacks most widely used in Vamac are N762, N774 (SRF), and N550 (FEF) carbon blacks. For high hardness, abrasion resistant Vamac compounds, very small particle size carbon blacks and graphite are required.

TABLE 6.11
Carbon Black Fillers in Vamac Copolymers

Carbon Blacks	Compounds[a]					
MT (N990)	60	—	—	—	—	—
SRF (N762)	—	60	—	—	—	—
GPF (N660)	—	—	60	—	—	—
FEF (N550)	—	—	—	60	—	—
HAF (N330)	—	—	—	—	60	—
ISAF (N220)	—	—	—	—	—	60
Physical properties[b]—stress/strain at 23°C						
Hardness, durometer A (Pts)	54	67	69	74	76	75
100% Modulus, MPa	1.83	2.86	3.41	5.10	4.28	5.83
Tensile strength, Mpa	9.79	11.8	12.6	13.7	15.0	14.4
Elongation at break, %	320	275	265	230	260	215
Die C tear strength, kN/m	22.1	22.9	23.3	22.6	25.9	25.4

[a] Compound: Vamac D (100 parts); Naugard 445 (1 phr); stearic acid (1.5 phr); Armeen 18D (0.5 phr); DiCup 40C (8 phr); HVA-2 (2 phr); carbon black (as shown).
[b] Test slabs press-cured 10 min at 177°C.

TABLE 6.12
Mineral Fillers in Vamac Terpolymers

Mineral Fillers	Compounds[a]					
Cyprubond	60	—	—	—	—	—
Blanc Fixe	—	60	—	—	—	—
Atomite Whiting	—	—	60	—	—	—
Nyad 400	—	—	—	60	—	—
Cab-O-Sil MS-7SD	—	—	—	—	60	—
Minusil (5u)	—	—	—	—	—	60
Physical properties[b]—stress/strain at 23°C						
Hardness, durometer A (Pts)	58	40	42	47	84	46
100% Modulus, MPa	4.31	0.83	0.86	1.34	3.62	1.24
Tensile strength, MPa	7.31	2.59	3.00	2.07	12.3	4.93
Elongation at Break, %	325	330	400	215	415	405
Die C tear strength, kN/m	20.0	8.1	7.9	8.8	36.6	11.2

[a] Basic compound: Vamac G—100 parts; Naugard 445—2 phr; Armeen 18D—0.5 phr; Vanfre VAM—1 phr; stearic acid—1.5 phr; DOS—7 phr; 40% TETA dispersion—5 phr; DOTG—4 phr.
[b] Test slabs press-cured for 30 min at 166°C.

Some mineral fillers also reinforce Vamac. Magnesium silicates and fumed silica are reinforcing while barium sulfate, calcium carbonate, calcium metasilicate, and silicon dioxide are nonreinforcing (Table 6.12). Hydrated alumina is also a reinforcing filler in Vamac. Fumed silica is a very effective filler for Vamac. Compounds containing fumed silica have high strength and tear properties but suffer from high compression set and high compound viscosity.

The potential for the formation of ionic crosslinks with Vamac terpolymers precludes the use of fillers that may contain soluble divalent metal ions. Metal stearates, metal stearate-coated fillers, clays, and precipitated silicas are not recommended for use in Vamac terpolymers. Precipitated fillers can contain impurities, which can lead to ionic bonding. Another potential source of soluble divalent metal ions is the pigment used to color mineral-filled compounds. Ionic bonds can significantly increase the viscosity of the uncured stock and increase the compression set of vulcanizates. The effect of metal ions and ionic bonds on stock and vulcanizate properties can vary considerably from batch to batch and is dependent on the impurity concentration in the lot of filler as well as the mixing/processing procedures.

Vamac can accept high filler levels but very high filler levels may require the use of the higher viscosity Vamac HG. N550 (FEF) carbon black is recommended for extrusion applications.

6.4.5 PLASTICIZERS

Effective plasticizers for Vamac include monomeric esters, polymeric esters, and polymeric ether–ester types. Table 6.13 lists the plasticizers typically used for Vamac. The petroleum-based plasticizers commonly used in most elastomers are

TABLE 6.13
Plasticizers for Vamac[a]

Plasticizer Type	Examples	Comments
Monomeric esters	DOS, DOP	Imparts optimum low-temperature flexibility but are only functional up to continuous use temperatures of 125°C
		DOS has the highest boiling point in this group
Polyesters	Drapex 409	Best permanence
	Plasthall P550	Recommended for continuous use temperatures >150°C
	Plasthall 670	Polyesters do not lower the T_g of the compound
Ether/ester	TP 759	Optimum balance of high- and low-temperature properties
	Nycoflex ADB30	Usable for continuous use temperatures up to 150°C
	Merrol 4426	Reduce compound T_g

[a] Most plasticizers interfere with the peroxide cure of dipolymers and as a result have a significant negative impact on dipolymer compound physical properties and compression set. Trimellitates like Plasthall 810TM, ADK Cizer C-9N and UL 100, or the polyester plasticizer Plasthall P670, have been found to have the least negative impact on the properties of peroxide cured dipolymer compounds. Use of these plasticizers should be limited to 5–10 phr.

generally not suitable for Vamac. Because of the polarity of AEM elastomers and the high end-use temperatures that Vamac parts are exposed to, only specific classes of plasticizers are acceptable. In general, aromatic plasticizers are compatible with Vamac but are too volatile; naphthenic plasticizers are less volatile than aromatics but only marginally compatible; paraffinic plasticizers are thermally stable at end-use temperatures but incompatible.

Monomeric esters are the most effective plasticizers for Vamac, significantly enhancing the low-temperature flexibility by reducing compound T_g, while reducing compound viscosity and hardness. Monomeric ester plasticizers can reduce TR-10 values of highly filled Vamac compounds to −50°C and lower, but they are low boiling and should not be used in applications requiring continuous use temperatures higher than 125°C. Dioctylsebacate (DOS) is the least volatile of the monomeric esters and is very effective in Vamac. Monomeric plasticizers are about twice as effective in reducing compound hardness as either of the polymeric types.

Polyester plasticizers are the least volatile, most permanent of the three plasticizer types, and should be used for optimum high-temperature performance. Unfortunately, polyester plasticizers do not lower the compound T_g and as a result will not improve the low-temperature flexibility. Blends of monomeric and polymeric plasticizers can provide a favorable blend of low-temperature and high-temperature performance.

Mixed ether–ester plasticizers offer the best balance of high- and low-temperature properties. Ether–ester plasticizers improve the low-temperature flexibility of Vamac compounds and are functional for continuous use up to temperatures of 150°C. The low-temperature properties of the three plasticizer types are compared in Table 6.14.

TABLE 6.14
Low-Temperature Properties of Plasticizers in Vamac G[a]

Variation	1	2	3	4
Monomeric ester (DOS)	—	40	—	—
Polyester (Drapex 409)	—	—	40	—
Ether/ester (TP 759)	—	—	—	40
Hardness, durometer A, Pts	72	66	71	67
Low temperature properties °C[b]				
Brittle point	−37	−57	−41	−50
Clash-Berg	−24	−52	−31	−41
Gehman T-100	−29	−52	−31	—
Compound Heat Resistance,°C	170	125	170	150

[a] Compound: Vamac G (100 parts); Naugard 445 (2 phr); stearic acid (2 phr); Armeen 18D (0.5 phr); Vanfre VAM (1 phr); N650 carbon black (65 phr); N775 carbon black (20 phr); Diak #1 (1.25 phr); DOTG (4 phr); plasticizer (as shown). Test slabs press-cured 15 min at 177°C and post-cured 4 h at 177°C.

[b] These data are based on Vamac G. The low fluid swell "LS" types, Vamac GLS and Vamac DLS, lose 5°C to 7°C in low-temperature properties compared with Vamac G.

Most plasticizers have a significant negative impact on the physical properties and compression set of copolymer compounds. Trimellitates, like Plasthall 810TM, ADK Cizer C-9N, and UL-100, or polyester plasticizer Plasthall P670, have performed best in copolymers. The use of these plasticizers in copolymer compounds should be limited to 5–10 phr to minimize interference with peroxide curing systems.

Table 6.15 is a collation of general rules for the adjustment of carbon black and plasticizer levels to control the hardness of Vamac compounds.

TABLE 6.15
Rules of Thumb for Carbon Black/Plasticizer Adjustment in AEM Compounds

1. Hardness will increase six (6) points with the addition of each 10 phr of FEF, N550, carbon black
2. Hardness will increase four (4) points with the addition of each 10 phr of SRF, N762 or N774, carbon black
3. Hardness will decrease six (6) points with the addition of each 10 phr of monomeric plasticizer and decrease about three (3) points with the addition of each 10 phr of either the polyester or ether/ester plasticizers
4. Extending with fillers and plasticizers can lower cost, improve smoothness, and reduce "nerve," while maintaining compound hardness and modulus. However, compound tensile strength, elongation at break, tear strength, and compression set could be lowered
5. To maintain a constant hardness value during compound extension
   Add nine (9) phr FEF carbon black or twelve (12) phr SRF carbon black for each 10 phr of monomeric plasticizer added
   Add four (4) phr of FEF carbon black or six (6) phr SRF carbon black for each 10 phr of either polyester or ether/ester plasticizer added

6.5   VULCANIZATION: TERPOLYMERS (VAMAC G, HG, AND GLS)

AEM terpolymers contain an organic acid cure site monomer and are usually vulcanized with primary diamines. The terpolymer structure and vulcanization mechanism are shown schematically in Figure 6.3.

During press cure the primary diamine reacts with the acid cure site producing an amide crosslink. In the chemical environment of the AEM terpolymer backbone the amide crosslinks are reactive and are converted with time and temperature to imide crosslinks. Since the reduction of the amide to the imide needs to occur before exposing a functional part to high temperatures and stress, a post-cure is required. A separate post-cure step can minimize press-cure times, increases vulcanizate modulus, tensile strength and hardness, and generally results in compression set values <20% (70 h at 150°C). Alternative post-cure cycles are

1. 1 h at 200°C
2. 4 h at 175°C
3. 10 h at 150°C
4. 24 h at 125°C

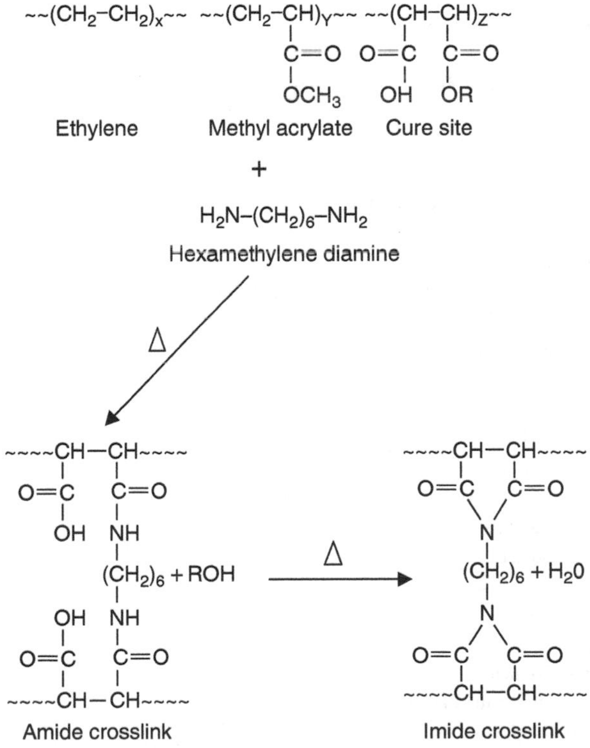

FIGURE 6.3 Terpolymer structure and curing mechanism.

Post-cures at temperatures around 175°C are the most common.

Primary diamine cure systems require the incorporation of accelerators, soluble organic bases to control pH, and accelerate the closing of the imide ring. The most widely used accelerators are guanidines; diphenylguanidine (DPG) and diorthotolylguanidine (DOTG). A partial list of acceptable diamine curing agents and accelerators, including comments around their applicability, is given in Table 6.16.

TABLE 6.16
Cure Systems for Terpolymers

Vamac terpolymers are usually cured with a combination of primary diamines and guanidine accelerators

Curatives

Hexamethylenediamine carbamate (HMDC)
- Diak No.1, Vulcofac HDC, pure powder form or as a dispersion in polymer or plasticizer
- Good general purpose curative
- Yields excellent compression set and aging properties

Ethylenediamine (ED)
- Significantly less costly than HMDC
- Liquid
- 3/0/3 F/R/H (Fire/Reactivity/Health) rating

Ketimine
- Epicure 3502 (contains some free ethylenediamine)
- Less costly than HMDC
- More scorch safety than HMDC
- Liquid
- 2/0/3 F/R/H (Fire/Reactivity/Health) rating

Triethylenetetramine (TETA)
- Less costly than HMDC
- Slightly poorer scorch safety and thermal aging than HMDC
- Liquid, but dispersions available

Vulcofac CA-64 (Proprietary diamine)
- Lower cost than HMDC
- Improved scorch safety over HMDC
- Slower curing than HMDC
- Physical properties sacrificed

Accelerators

DOTG (diorthotolyl guanidine)
- Optimum compression set
- Optimum modulus

DPG (diphenyl guanidine)
- High tear strength
- High flex-fatigue resistance
- Higher elongation at break than DOTG
- Lower modulus and hardness than DOTG

The selection of a preferred diamine cure system depends on the priority of properties and processing constraints. Hexamethylenediamine carbamate, HMDC, is the best general-purpose curing agent and the most widely used. Table 6.17 shows the property trends expected as a function of HMDC level and the accelerator preferred for the optimization of specific properties.

Combinations of HMDC with DOTG yield properties consistent with those that would be expected from vulcanizates with high states of cure. Strength properties, abrasion resistance, and compression set are all enhanced. Increasing the HMDC level to 1.5 phr or higher also optimizes compression set and resistance to abrasion. Lower levels of HMDC result in vulcanizates with improved flexural properties, and in general, higher elongation than with higher levels of HMDC. Tear strength initially increases with increasing HMDC level, attains a maximum value, and then decreases as the level of HMDC continues to increase. In many practical applications the maximum tear strength occurs at an HMDC level lower than that necessary to optimize other required performance properties.

The combination of HMDC with DPG yields properties typical of AEM vulcanizates with a lower state of cure. In contrast to DOTG at the same level, compounds accelerated with DPG generally have higher tear strength and elongation and better flexural properties. High levels of diamine minimize these differences between DOTG and DPG. For both accelerators the recommended level is 4 phr.

Conversion of the hexamethylenediamine to the carbamate transforms the diamine liquid to a more process-friendly solid. The carbamate also blocks the amine and adds scorch safety to the fast curing HMDC system. Bin stability is not usually a problem with HMDC but there are cases where the scorch safety may not be acceptable. In general, if the Mooney Scorch minimum is low, a compound

TABLE 6.17

Characteristics of HMDC/Accelerator Cure Systems

Property	0.75 phr HMDC	⟵——⟶	1.50 phr HMDC
Property trends as a function of HMDC level (relative comparison)			
Cure rate	Slower		Faster
100% Modulus	Lower		Higher
Tensile strength	Lower		Higher
Elongation at break	Higher		Lower
Tear strength	Lower	Highest	Lower
Flex resistance	Better		Poorer
Compression set	Poorer		Better

Property	Preferred Accelerator
Preferred accelerator to optimize specific properties	
100% Modulus	DOTG
Elongation at break	DPG
Tear strength	DPG
Flex resistance	DPG
Compression set	DOTG

with a Mooney Scorch t_{10} value of around 10 min would have adequate bin stability. If the scorch safety of an AEM compound is inadequate, it can be improved by

1. Adding plasticizer to reduce compound viscosity
2. Changing the diamine cure system
3. Changing the compound
4. Adding retarders

Two materials, which have been shown to effectively retard the premature cure of AEM compounds, are octadodecylamine (Armeen 18D) and salicylic acid. Armeen 18D is usually present in AEM compounds as part of the internal release system at 0.5 phr. Increasing the Armeen 18D level to 1.5 phr can result in a nominal 50% improvement in scorch safety, t_{10}, a nominal 20% reduction in cure rate, and approximately 10% reduction in cure state. Compression set and tensile strength will be negatively affected out of the mold but these properties are normally recovered after post-cure. When Armeen 18D is used to address scorch safety, the recommendation is to add this retarder in 0.25 phr increments.

Salicylic acid, although not as effective as Armeen 18D, is also an acceptable retarder for AEM compounds. The effect of salicylic acid on cure rate and cure state is significantly less than that of Armeen 18D. Salicylic acid does not appear to affect the physical properties and compression set of the post-cured vulcanizates. The recommended starting level of salicylic acid is 1.5 phr.

Vulcanizing thick sections of Vamac terpolymer compounds, using the standard cure system, can result in the formation of fissures. The fissures are the result of the large temperature gradient that can be generated when press-curing a thick part and the concurrent differences in cure rate within the part. The fissures are created by the compressed gasses that are trapped below the surface of the cured outer skin of the part. Cured Vamac has very low permeability to most gasses and will not allow the vulcanization by-products to escape in the time frame allotted. Fissures in these thick parts can be eliminated by replacing the 4 phr of the DOTG or DPG accelerator with 5–6 phr of Armeen 2C (dicocoamine).

AEM polymers, both terpolymers and copolymers, must be cured under pressure to avoid "sponging." Pressures of 413.7 kPa and higher, preferably higher, are recommended for curing Vamac compounds without sponging. High viscosity compounds with fast cure systems can be molded at the lower end of the pressure range. If compression set is not a consideration, the viscosity of a terpolymer-based compound can be increased sufficiently with ionic crosslinks to be cured at atmospheric pressure [3].

6.6 VULCANIZATION: TERPOLYMERS—FAST DIAMINE/PEROXIDE CURE SYSTEMS

Very fast diamine cure systems with good bin stability have been developed for AEM terpolymers. These fast, safe (scorch) cure systems have the potential to improve productivity by reducing press cure time and post-cure time, thus reducing compound cost. The basic components of the fast cure systems are shown in Table 6.18.

TABLE 6.18

Basic Components of the Fast, Safe Vulcanization System for Terpolymers

Component	Comments	Recommended Starting Level (phr)
Diamine		
HMDC	Preferred	0.75–1
Ketimine	Less costly, safer	3
Ethylenediamine	Much less costly	0.75
Peroxide		
Dicumyl peroxide	Preferred	5
Triganox 17–40B	Lower activation temperature	3
Vulcup 40KE	Higher activation temperature	3
Accelerator		
DOTG	Fastest	4
DPG	Higher elongation	4
Coagent		
Ricon 152	Preferred	2
HVA-2	Fast, scorchy	2
Retarder		
Salicylic acid	Preferred	1.5–3
Armeen 18D	Most efficient, can affect properties.	Add in 0.25 phr increments.

The Fast, Safe Cure System Consists of: a Diamine, a Peroxide, a Coagent, an Accelerator and Optionally, a Scorch Retarder.

Components. Diamine: HMDC Ethylenediamine/methylisobutylketimine (Epicure 3502) ethylenediamine. Peroxide: Dicumyl peroxide a,a-bis(*t*-butyl peroxy) diisopropylbenzene (Vulcup). Accelerator: Diphenyl guanidine (DPG) Diorthotolyl guanidine (DOTG). Coagent: Ricon 152 (low molecular weight polybutadiene polymer). Retarder (Optional): Salicylic acid Armeen 18D (Octadecylamine). Recommended Starting Levels.

With these cure packages it is possible to formulate bin stable AEM compounds with greater than 2X the rate of cure as the standard 1.5 phr HMDC system. The compression sets of the press-cured materials, using the fast HMDC/peroxide cure system, are about 30% lower (better) than the equivalent standard HMDC press-cured vulcanizates.

6.7 VULCANIZATION: COPOLYMERS (VAMAC D, VAMAC DLS)

AEM copolymers do not contain acid cure sites and were developed initially to eliminate the need to post-cure. In the absence of cure sites, copolymers must be cured with peroxides. The proposed mechanism for the peroxide cure is shown in Figure 6.4.

The peroxide cure of Vamac copolymers requires a coagent. The two preferred peroxide/coagent packages are

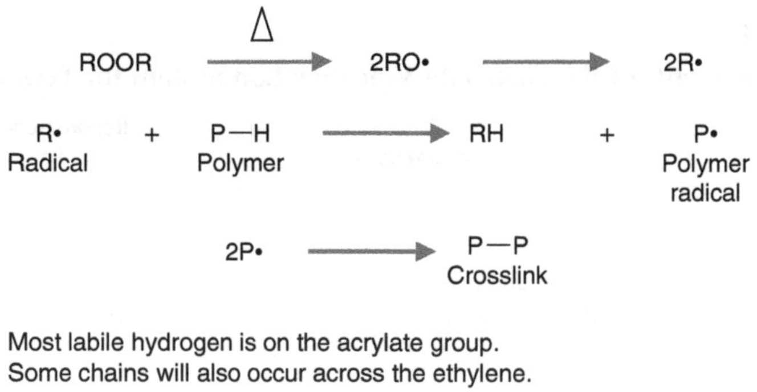

$$ROOR \xrightarrow{\Delta} 2RO\cdot \longrightarrow 2R\cdot$$

$$R\cdot + P{-}H \longrightarrow RH + P\cdot$$
Radical   Polymer   Polymer
                                           radical

$$2P\cdot \longrightarrow P{-}P$$
Crosslink

Most labile hydrogen is on the acrylate group.
Some chains will also occur across the ethylene.

FIGURE 6.4 Peroxide cure mechanism for Vamac.

1. DiCup/HVA-2
2. Vulcup/HVA-2

Although these are the preferred systems a number of other peroxides and coagents
are functional. A list of acceptable peroxides and coagents is given in Table 6.19.

As with all peroxide cure systems, there are limitations on the type and concentration of fillers and additives that can be used. In the case of AEM copolymer,
interference with the peroxide cure can come from some essential components
including

TABLE 6.19
Vulcanization Systems for Dipolymers

Components	Comments
Preferred peroxide/coagent combinations	
DiCup 40C/HVA-2	Optimum compression set with short cycle times
Vulcup/HVA-2	Higher 100% modulus
	Higher tensile strength
	Good compression set
Other peroxides	
Varox	Bloom and odor
Triganox 29 (Luperco 231XL)/Dicup	Faster curing
Other coagents	
Triallylisocyanurate (TAIC)	Fast cure—lower elongation
Triallylcyanurate (TAC)	Fast cure—lower elongation
Polybutadiene, Ricon 152	Fast, safe cure system
Trimethylolpropane trimethacrylate, Sartomer 351	Good heat aging
	Less efficient

Note: Vamac dipolymers are cured with a combination of peroxide and coagent.

1. Stabilizers—Naugard 445
2. Plasticizers—all three types
3. Armeen 18D
4. Vanfre VAM
5. Acidic materials

As a result, the way a copolymer compound is formulated differs considerably from that of the terpolymer. Since stabilization is necessary, Naugard 445 is added, but at a 1 phr level. The Armeen 18D level remains at 0.5 phr, the stearic acid level is held around 1 phr, and if possible, the Vanfre VAM is eliminated. Vanfre should be added to a peroxide-cured copolymer if release is a problem, and then at the minimum level necessary.

Most plasticizers have a significant negative impact on physical properties and compression set properties of peroxide-cured AEM copolymer. Unless required for processing, plasticizers should be eliminated from copolymer compounds. Trimellitates like Plasthall 810TM, ADK Cizer C-9N, and UL-100, or the polyester plasticizer, Plasthall P670, have the least negative effect on copolymer compounds and can be used if needed. Use of these plasticizers in copolymer compounds should be limited to 5–10 phr.

6.8   MIXING OF AEM COMPOUNDS

AEM compounds can be mixed in internal mixers, intermeshing and tangential, or on a mill, with relatively short mix cycles. The guiding principle when mixing AEM compounds is to keep the polymer temperature as low as possible initially to increase mixing times.

6.8.1   INTERNAL MIXERS

If possible, single pass, upside-down mixing is preferred to control overheating. Upside-down mixing allows the fillers and additives to be blended by the rotor without any temperature increase. The polymer is added last maximizing the time of mixing. Table 6.20 outlines the recommended internal mixing procedure for AEM compounds.

If the scorch safety of the compound is marginal, the curing agents and accelerators can be added at the sweep or on a second pass. The mixing process can be modified to accommodate highly filled compounds or compounds with difficult to disperse fillers or curative packages (fine particle size or agglomerated fillers). When incorporating fillers that are difficult to disperse, adding half of the problem filler as part of an upside-down mix with the remaining filler added at the sweep is usually sufficient to attain good mixing. If adding the filler in two or three stages still does not allow for complete mixing before the drop, a two pass mixing process is required with the curative package added on the second pass, either in an internal mixer or on a mill.

Because of the relatively short mixing cycles, the dispersion of solid HMDC could be a problem. Poor dispersion of HMDC usually manifests itself in the formation of bubbles and blisters. Solid white particles of HMDC are often visible in compounds in which the HMDC is poorly dispersed. Concentrates of pre-dispersed

TABLE 6.20

Recommended Internal Mixing Procedure

Suggested conditions
- Use smoked sheet natural rubber or NBR for clean-out
- One pass upside-down mix
- Slow rotor speed
- Full cooling water on chamber and rotors
- Release agents necessary for efficient drop
- 70% load factor
- Dump terpolymers at or below 100°C
- Dump dipolymers between 105°C and 110°C
- Add curatives at the final sweep with highly loaded compounds

Typical cycle
Start with a clean internal mixer
  1. "0" min
     Add fillers (sandwich plasticizers)
     Add stabilizers, release agents and curatives
  2. 30 s
     Add Vamac
  3. 1–1.5 min
     Sweep
  4. 2 min
     Sweep
     Add curatives if not added at the start
  5. 2.5–3 min (100°C)

DUMP
Blend on sheet-off mill.
Slab-off
Air cool or water dip tank
     Acceptable dusting agents: soapstone, talc and Crystal 2000.

HMDC are recommended to facilitate proper dispersion of HMDC in Vamac compounds. Pre-dispersions of HMDC in plasticizer and polymer are commercially available and should be used in place of solid HMDC if dispersion is a problem.

Sheet-off mill rolls must be kept cool to eliminate sticking to the rolls. Zinc and calcium stearate, and metal oxide dusting agents should be avoided with terpolymers because of the potential for forming ionic crosslinks. Soapstone, talc, Crystal 2000, or other commercially available dusting agents that do not contain divalent metal ions can be used. AEM copolymer compounds can be dropped from the internal mixer at 5°C to 10°C higher than terpolymer compounds.

6.8.2 ROLL MILL MIXING

AEM compounds can be mixed on a mill. The recommended mixing procedure is given in Table 6.21. Care must be taken to add the release package first and to avoid excessive mill mixing and overheating.

TABLE 6.21
Recommended Mill Mixing Procedure

Place full cooling water on the rolls
Open the nip and pass the cut polymer through one (1) time
Close the nip and band the polymer
Immediately add the release package
Remove the mixed polymer/release blend after a few cuts and folds (remove blend before the polymer
  gets warm)
Reintroduce and band the polymer/release blend (this procedure insures that the release agents contact the
  steel rolls)
Add the plasticizer/filler mixture
After the filler is adsorbed, add the curatives
Cut and fold three (3) times from each side
Cigar roll at least three (3) times
Sheet off
Cool in air of dip in a water bath of soapstone, talc or Crystal 2000

6.9   PROCESSING OF AEM COMPOUNDS

AEM compounds can be easily extruded, calendered, and molded (injection, transfer, and compression).

6.9.1   EXTRUSION

AEM compounds can be extruded using equipment commonly used to extrude thermoset elastomers. Short or moderate L/D extruders are suitable. Most AEM compounds are not susceptible to melt fracture under normal production conditions and compounds can be successfully extruded over a broad range of draw-down ratios. General extrusion guidelines for AEM compounds are shown in Table 6.22.

The melt strength and collapse resistance of extrudates of Vamac G compounds can be increased by replacing a portion of the Vamac G with the higher viscosity Vamac HG and by reducing the temperature of the extrudate.

6.9.2   CALENDERING

Although very little AEM is calendered, it is possible to produce excellent quality sheet using this process. The calender rolls are kept cool to optimize melt ("green") strength and to maintain operating temperatures between 30°C and 50°C. A small rolling "bank" of warmed stock is preferred. High quality, blister-free sheet can be calendered up to approximately 2 mm. Sheets must be plied to build thicker 3.2 or 6.4 mm materials.

6.9.3   MOLDING

Injection and transfer molding are favored for AEM compounds because of their low viscosity. In addition, because of their low viscosity, care must be taken to

TABLE 6.22
Extrusion Guidelines

	Cold Feed	Hot Feed	Ram
Preheat stock	No	Yes	Yes
Stock temperature, C	*Ambient*	70–90	70–90
Extruder zone temperature, °C			
Feed	40	40	—
Barrel zone near feed	50	50	—
Barrel zone near head	60	50	—
Head	70	70	—
Screw temperature, °C	50	—	—

- Purge the extruder with smoked sheet or NBR
- Turn the cooling water on to facilitate the clean-out
- Pre-treat dies, mandrels, and metal-curing mandrels with release agents
- Lubricating solution
- Lower the set temperatures if necessary to maintain maximum melt
- Strength for dimensional stability

Curing

Steam autoclave
- 30–45 min at 551.6 kPa steam
- Purge (remove air) for peroxide cures
- Both wrapped and open steam cures acceptable
- Vamac will "sponge" if cured without pressure

Air autoclave
- 170°C to 190°C
- 413.7–551.6 kPa minimum pressure to prevent "sponging"
- Slow curing systems require the higher minimum pressure
- Purge and use N_2 for peroxide cures

CV curing (wire and cable)
- 1.5–2.5 min at 1551–1724 kPa steam
- Use the peroxide or peroxide/diamine fast curing system

avoid trapped air and blisters. Correct mold design, with the proper number and placement of vents, and adequate vacuum on the mold, are essential to eliminate these potential problems. Ram barrel temperatures can be lowered to minimize air entrapment. Mold surfaces should be treated with external release agents to reduce the wetting of the metal by the compound. "Bumping" the mold during the cycle is sometimes helpful.

Typical injection molding stock temperatures for Vamac compounds at the nozzle should be in the range of 70°C–85°C. The cold runner systems should be set in the same range of 70°C–85°C. Scorch can be a serious problem if the temperature exceeds 100°C–110°C. Mold cavity temperatures are typically 175°C–190°C for injection molding cycle times of 1–3 min.

Acceptable and ideal injection molding conditions vary with compound composition primarily: the cure system—the level of curing agents and accelerators, filler

---

TABLE 6.23

Lubricants

Recommended External Release Agents (Alphabetical Order)

Aqualine R-150	Dexter Corporation
Crystal 1063	TSE Industries
DiamondKote W3052 & W3489	Franklynn Industries, Inc.
McLube 1733L/1711L	McGee Industries Inc.
TM-967	Release Coatings
TRA 9825	DuPont Performance L

---

type, and filler level. The number of shots in the extruder or transfer chamber and attainable inject pressures must be taken into account when selecting operating temperatures. Control of compound temperature is essential to balance scorch safety and cure rate, and to eliminate trapped air. Low viscosity (usually low hardness) compounds of Vamac may need to be molded at lower than recommended inject and runner temperatures to increase compound viscosity during fill. Replacing a portion of Vamac G with higher viscosity Vamac HG is also an option to increase compound viscosity and maintain higher injection temperatures.

AEM compounds have a strong affinity to metals and steps must be taken to prevent mold fouling and part sticking, as well as trapped air at the mold /compound interface. It is important to incorporate the recommended internal release package into the compound and to apply external release agents to the mold surface. Hardened steel molds may be acceptable but chrome plated or stainless steel molds are preferred to minimize mold fouling and sticking of the compound. External mold release is generally required as are secondary over-sprays. External mold release is necessary to reduce the coefficient of friction at the compound/mold interface sufficiently to allow trapped air to escape, reducing the tendency for the development of sink marks and blisters. Care must be taken in the application of external mold release since excess mold release is one of the principle causes of mold fouling and weld lines. A list of external release agents suitable for Vamac is shown in Table 6.23 and a summary of typical molding cycles is given in Table 6.24. Typical injection molding conditions for Vamac are given in Table 6.25.

TABLE 6.24

Typical Molding Cycles

Operation	Mold Temperature (°C)	Cure Time (min)
Injection molding	175–190	1–3
Transfer molding	165–185	5–10
Compression molding	160–180	10–20

TABLE 6.25

Typical Injection Molding Conditions

Position	Temperature, °C
Strip	Ambient
Barrel	70–85
Nozzle	70–85
Screw	Ambient—70
Inject (compound out of extruder)	70–85
Cold runners[a]	70–85
Mold cavity	175–190

Cycle times: 60–180 s.

[a] Scorch can be a serious problem if the temperature exceeds 100°C to 110°C.

Other possible causes of internal blisters and voids within the part are

1. Insufficient external pressure during vulcanization
2. Insufficient stock to fill the mold
3. Poor mold design
4. Poor dispersion of curatives and accelerators
5. Excess of volatile plasticizers and release agents
6. Under-cure
7. Moisture

Typical shrinkage values for Vamac compounds are around 2.5% after press-cure, increasing to 3.5%–4% after post-cure. Part shrinkage can be reduced by increasing the amount of nonvolatile filler, decreasing the level of plasticizer, reducing mold temperatures, and reducing the temperature and duration of post-cure.

6.10   DEVELOPMENT PROGRAMS—NEW POLYMERS

Development programs are in progress, which are expected to result in commercial products designed to meet the changing requirements of the automotive industry. These include

1. AEM polymers suitable for continuous use at 175°C and higher
2. AEM polymers with T_g in the range of −40°C to −50°C
3. AEM polymer with a lower loss tangent than present grades and a more constant loss tangent over the temperature range of −50°C to 150°C
4. Faster curing copolymers that develop optimum properties without a post-cure

6.11   SUGGESTED STARTING FORMULATIONS
FOR SPECIFIC APPLICATIONS

This section describes suggested starting compound formulations based on both Vamac terpolymers and copolymers. Most have proven satisfactory in prototype or commercial applications. Where available, compound properties critical to the specific applications are included.

6.11.1 SEALS, GASKETS, AND GENERAL MOLDED APPLICATIONS

Vamac is widely used in a broad range of molded seal and gasket applications. Starting formulations listed below vary primarily in hardness and resistance to selected automotive fluids (Tables 6.26 and 6.27).

TABLE 6.26
Seals, Gaskets, and General Molded Applications—Terpolymer

Ingredients	Parts						
	A	B	C	D	E	F	G
Nominal hardness	30/35	30/35	55/60	55/60	65/70	65/70	65/70
Vamac G	—	—	100	—	100	—	—
Vamac GLS	—	100	—	100	—	100	100
Vamac HG	100	—	—	—	—	—	—
Naugard 445	2	2	2	2	2	2	2
Armeen 18D	0.5	0.5	0.5	0.5	0.5	0.5	0.5
Stearic acid	1.5	1.5	1.5	1.5	1.5	1.5	1.5
Vanfre VAM	1	1	1	1	1	1	1
SRF, N774 carbon black	—	—	70	60	90	82	72
FEF, N550 carbon black	10	10	—	—	—	—	—
Cab-O-Sil M7D	10	10	—	—	—	—	—
TP 759	20	20	10	10	10	10	10
HMDC	1.5	1.5	1.5	1.5	1.5	1.5	1.5
DOTG	4	4	4	4	4	4	4

Selected vulcanizate properties[a]

Original properties at 23°C

Hardness, durometer A, Pts	30	32	57	55	66	68	67
100% Modulus, MPa	0.72	0.75	4.2	3.7	6.9	6.4	6.4
Tensile strength, MPa	4.5	5.0	13.9	14.4	13.8	15.0	15.5
Elongation at break, %	373	371	285	297	228	242	251
T_g (DSC), °C	−41.5	−35.7	−35.4	−29.7	−35.5	−28.8	−24.5

Compression set B—Plied, 70 h at 150°C

Set, %	32	27	20	16	17	18	16

Properties after air oven aging 1008 h at 150°C

Hardness, durometer A, Pts	37	46	70	70	82	83	79
Change from original, Pts	7	14	13	15	16	15	12
100% Modulus, Mpa	1.2	1.9	8.2	8.2	11.2	12.6	10.3
Change from original, %	66.7	153	95.2	122	62.3	97	61
Tensile strength, Mpa	6.7	7.8	13.9	15.4	14.1	15.1	15.5
Change from original, %	48.8	56.0	0.0	6.9	2.2	0.7	−1.9
Elongation at break, %	334	286	171	186	150	137	173
Change from original, %	−10.4	−22.9	−40.0	−37.4	−34.2	−40.9	−31.1

(*continued*)

TABLE 6.26   (Continued)
Seals, Gaskets, and General Molded Applications—Terpolymer

Ingredients				Parts			
	A	B	C	D	E	F	G

Properties after aging 1008 h in Mobil 5W30 motor oil at 150°C (no oil changes)

Volume change, %	24.1	5.6	24.9	8.8	24.1	8.2	10.7
Hardness, durometer A, Pts	33	40	53	57	60	69	67
Change from original, Pts	3	8	−4	2	−6	1	0
100% Modulus, Mpa	0.80	1.3	4.8	4.6	7.0	8.2	7.3
Change from original, %	11.1	76.3	14.3	24.3	1.4	28.1	14.1
Tensile strength, Mpa	3.6	6.1	10.6	12.8	10.9	14.3	14.4
Change from original, %	−20.0	22.0	−23.7	−11.1	−21.0	−4.7	−8.9
Elongation at break, %	239	263	191	215	145	166	188
Change from original, %	−35.9	−29.1	−33.0	−31.0	−36.4	−31.4	−25.1

Compressive stress relaxation[b], percent retained force after

504 h	55	54	77	56	46	43	49
1008 h	48	49	78	53	41	38	43

Properties after aging 1008 h in Mobil One "New Trisynthetic Formula" motor oil at 150°C
(no oil changes unless specified)

Volume change, %	7.1	−1.6	13.3	2.7	13.4	2.4	4.4
Hardness, durometer A, Pts	35	44	63	60	69	76	70
Change from original, Pts	5	12	6	5	3	8	3
100% Modulus, MPa	1.0	1.5	5.4	5.7	7.9	10.1	9.0
Change from original, %	38.9	100	28.6	54.0	14.5	57.8	40.6
Tensile strength, MPa	5.2	7.1	12.8	13.0	14.1	16.0	16.2
Change from original, %	15.5	42.0	−7.9	−9.7	2.2	6.7	2.5
Elongation at break, %	286	275	200	185	160	155	169
Change from original, %	−23.3	−25.9	−29.8	−37.7	−29.8	−35.9	−32.7

Compressive stress relaxation[b]—weekly oil changes, percent retained force after

504 h	35	28	41	41	14	29	32
672 h	50	32	30	38	22	15	37
1008 h	20	26	23	42	20	30	31

Properties after aging 1008 h in Dexron III ATF at 150°C (no fluid changes)

Volume change, %	31.7	11.0	30.5	13.9	29.4	13.7	16.0
Hardness, durometer A, Pts	32	42	52	62	60	73	69
Change from original, Pts	2	10	−5	7	−6	5	2
100% Modulus, MPa	0.9	1.4	5.1	5.9	7.6	10.1	8.9
Change from original, %	25.0	86.7	21.4	59.4	10.1	57.8	39.1
Tensile strength, MPa	4.1	6.3	10.9	14.6	13.2	16.3	16.6
Change from original, %	−8.9	26.0	−21.5	1.4	−4.3	8.7	5.1
Elongation at break, %	252	264	162	196	155	153	176
Change from original, %	−32.4	−28.8	−43.1	−34.0	−32.0	−36.8	−29.9

[a] Test slabs post-cured four (4) hours at 175°C.
[b] Akron Polymer Laboratory. Compressive stress relaxation according to ISO 3384, Method B. Samples: 12.7 mm (diameter) × 6.35 mm (nominal thickness)—Plied.

TABLE 6.27
Seals, Gaskets, and General Molded Applications—Copolymer

Ingredients	Parts		
	H	I	J
Nominal hardness	45	65	70
Vamac D/Vamac DLS[a]	100	100	100
Naugard 445	1	1	1
Armeen 18D	0.5	0.5	0.5
Stearic acid	1.5	0.5	1.5
Vanfre VAM	0.5	0.5	0.5
SRF, N774 carbon black	—	—	60
FEF, N550 carbon black	45	50	—
Plasticizer	—[b]	—[b]	—[b]
Vulcup R (Vulcup 40KE)[c]	2 (5)	2 (5)	3.2 (8)
HVA-2	2.0	1.0	1.0
Selected vulcanizate properties[d,e]			
Original properties at 23°C			
Hardness, durometer A, Pts	61	66	68
100% Modulus, MPa	3.3	3.9	3.6
Tensile strength, MPa	12.2	13.7	14.5
Elongation at break, %	260	280	255
Compression set B—Plied, 70 h at 150°C			
Set, %	17	16	14
Properties after air oven aging 1008 h at 150°C			
Hardness, durometer A, Pts	—	—	68
Change from original, Pts	—	—	0
100% Modulus, MPa	—	—	3.9
Change from original, %	—	—	8.3
Tensile strength, MPa	—	—	13.0
Change from original, %	—	—	−10.3
Elongation at break, %	—	—	255
Change from original, %	—	—	0
Properties after aging 1008 h in Mobil 5W30 motor oil at 150°C (no oil changes)			
Hardness, durometer A, Pts	—	—	50
Change from original, Pts	—	—	−13
100% Modulus, MPa	—	—	2.9
Change from original, %	—	—	−19.4
Tensile strength, MPa	—	—	11.2
Change from original, %	—	—	−22.7
Elongation at break, %	—	—	240
Change from original, %	—	—	−5.9
Volume change, %	—	—	(Vamac D 22)
			(Vamac DLS 8)
Properties after aging 1008 h in Mobil Infilrex factory fill motor oil at 150°C (oil changed weekly)			
Hardness, durometer A, Pts	—	—	58
Change from original, Pts	—	—	−10

(*continued*)

TABLE 6.27   (Continued)

Seals, Gaskets, and General Molded Applications—Copolymer

Ingredients	Parts		
	H	**I**	**J**
100% Modulus, MPa	—	—	3.5
Change from original, %	—	—	−2.8
Tensile strength, MPa	—	—	9.9
Change from original, %	—	—	−31.7
Elongation at break, %	—	—	185
Change from original, %	—	—	−27.4
Volume change, %	—	—	(Vamac D 20)
			(Vamac DLS 7)
Compression stress relaxation[f], percent retained force after			
500 h	—	—	38
1008 h	—	—	32
1500 h	—	—	36
Properties after aging 1008 h in Ford ATF at 150°C (no fluid changes)			
Volume change, %	—	—	(Vamac D 21),
			(Vamac DLS 9)
Hardness, durometer A, Pts	—	—	52
Change from original, Pts	—	—	−16
100% Modulus, MPa	—	—	3.0
Change from original, %	—	—	−16.7
Tensile strength, MPa	—	—	11
Change from original, %	—	—	−24.1
Elongation at break, %	—	—	210
Change from original, %	—	—	−17.6
Compression stress relaxation[f], percent retained force after			
504 h	—	—	41
1008 h	—	—	39
1500 h	—	—	35

Source: D.A. Kotz, T.M. Dobel. Stress relaxation of ethylene/acrylic and other gasket materials in automotive fluids. Society of Automotive Engineers International Congress and Exposition, Detroit, Michigan, February/March 1994, Paper No. 940959; Basic principles for the compounding and processing of Vamac® dipolymers. Dufont Dow Elastomers Technical Information Bulletin H-68560.

[a]   Vamac D and Vamac DLS can be blended at various ratios to adjust fluid resistance and volume increase. The T_g of the polymer blend increases with increasing levels of Vamac DLS.

[b]   Trimellitate plasticizers can be added if necessary for processing. Plasticizers should be used at low levels and the carbon black level adjusted to maintain hardness. Refer to the text for guidelines on plasticizer addition to dipolymers.

[c]   Vulcup and DiCup can be used interchangeably. DiCup gives faster cures than Vucup. See text for more detailed information.

[d]   All vulcanizate data in this Table are based on compounds of Vamac D.

[e]   Test slabs press-cured: "H," 3 min at 193°C. "I," 5 min at 193°C. "J," 10 min at 177°C.

[f]   Shawbury-Wallace [4].

6.11.2 HIGH PERFORMANCE AUTOMOTIVE HOSES

Vamac is used in many automotive hose applications including turbo charger hoses, oil coolant hoses, air conditioning hoses, and power steering hoses (Table 6.28). The

TABLE 6.28
Automotive Hoses—Terpolymer

Ingredients[a]	Parts			
	K	L	M	N
Vamac G	100	100	100	75
Vamac GLS	—	—	—	25
Naugard 445	2	2	2	2
Stearic acid	1.5	1.5	1.5	1.5
Vanfre VAM	0.3	0.75	0.3	0.3
Armeen 18D	0.5	0.5	0.5	0.5
FEF, N550 carbon black	30	60	—	30
Cab-O-Sil M7D	—	15	—	—
SRF, N762 carbon black	65	—	—	65
Spheron 6000	—	—	95	—
TP 759	10	10	10	10
HMDC	1.25	1.25	1.25	1.25
DPG	4	4	4	4
Selected vulcanizate properties[b]				
Original properties at 23°C				
Hardness, durometer A, Pts	78	72	74	77
100% Modulus, MPa	6.9	4.7	6.4	7.2
Tensile strength, MPa	14.1	13.9	11.9	14.1
Elongation at break, %	247	358	258	248
Compression set B—94 h at 23°C				
Set, %	47	45	49	49
Compression set B—94 h at 120°C				
Set, %	64	67	84	87
Die C tear strength at 125°C				
N/mm	9.0	12.3	8.6	8.8
Properties after air oven aging 504 h at 160°C				
Hardness, durometer A, Pts	89	85	83	87
Change from original, Pts	11	13	9	10
100% Modulus, MPa	8.5	8.5	7.7	9.2
Change from original, %	23.2	80.8	20.3	27.8
Tensile strength, MPa	12.0	14.1	10.7	12.8
Change from original, %	−14.9	1.4	−10.1	−9.2
Elongation at break, %	219	233	257	214
Change from original, %	−11.3	−34.9	−0.4	−13.7

(*continued*)

TABLE 6.28 (Continued)
Automotive Hoses—Terpolymer

Ingredients[a]	Parts			
	K	L	M	N
Properties after aging 94 h at 150°C in Mobil Cecilia 20 synthetic motor oil				
Hardness, durometer A, Pts	66	63	63	69
Change from original, Pts	−12	−9	−11	−8
100% Modulus, MPa	6.4	5.1	5.3	6.5
Change from original, %	−7.2	8.5	−17.2	−9.7
Tensile strength, MPa	13.3	13.1	11.9	13.5
Change from original, %	−5.7	−5.7	0	−4.2
Elongation at break, %	203	297	235	214
Change from original, %	−17.8	−17.0	−8.9	−13.7
Volume change, %	15	16	15	13
Properties after aging 94 h at 23°C in Diesel Fuel AP20/NP II				
Hardness, durometer A, Pts	65	59	62	69
Change from original, Pts	−13	−13	−12	−8
100% Modulus, MPa	5.6	4.0	5.0	5.9
Change from original, %	−23.2	−14.9	−21.9	−18.0
Tensile strength, MPa	11.7	11.8	10.6	11.9
Change from original, %	−17.0	−15.1	−10.9	−15.6
Elongation at break, %	207	300	228	200
Change from Original, %	−16.2	−16.2	−11.6	−19.3
Volume change, %	13	15	13	10

Source: Vamac® compounds for high performance hoses. DuPont Product Data Sheet.

[a] Zinc inhibitor can be added at the 1 phr level to eliminate degradation due to galvanized clamps.

[b] Test slabs press-cured 20 min at 170°C and post-cured 4 h at 175°C.

following formulations can serve as starting formulations for pure Vamac hoses. Most existing Vamac hose specifications call for a Shore A hardness of around 70. For best collapse resistance and to achieve the lowest volume increase values, it is recommended to design the compound at the upper limits of the hardness specification. The optimum ether–ester plasticizer level to optimize flexibility and total performance of the terpolymer is 10 phr (Tables 6.29 and 6.30).

6.11.3 DAMPERS—TORSIONAL DAMPERS

Vamac compounds are characterized by high loss tangents and are therefore excellent choices for applications requiring a combination of damping and heat resistance. In addition, the high levels of damping remain reasonably constant over a broad range of temperature and frequency [2]. In most high-temperature damping applications the quantity of plasticizer is minimized to avoid changes in damping characteristics over time. Some typical formulations are shown below. Within limits, the

TABLE 6.29
Automotive Hoses—Terpolymer
Low Fluid Swell with Zinc Inhibitor

Ingredients	Parts P
Vamac G	50
Vamac GLS	50
Naugard 445	2
Stearic acid	1.5
Vanfre VAM	1
Armeen 18D	0.5
FEF, N550 Carbon Black	20
SRF, N764 Carbon Black	60
Eastman Inhibitor OABH	1
TP 759	10
HMDC	1.5
DOTG	4
Nominal hardness	75

TABLE 6.30
Automotive Hoses—Copolymer

Ingredients	Turbo Hose/Air Ducts Parts	TOC
	Q	R
Vamac D	100	100
Naugard 445	1	1
Stearic acid	1.5	0.5
Vanfre VAM	1	—
Armeen 18D	1	0.5
Struktol WS 180	1	—
FEF, N550 Carbon Black	35	82.5
SRF, N764 Carbon Black	30	—
Cab-O-Sil M7D	15	—
Plasthall 810TM	5	7.5
Eastman Inhibitor OABH	—	2
Vulcup 40KE	4	—
DiCup 40C	—	8.8
HVA-2	1.5	1.75
Nominal hardness	72	75

Source: Basic principles for the compounding and processing of Vamac®
dipolymers. DuPont Dow Elastomers Technical Information Bulletin
H-68560.

TABLE 6.31
Torsional Dampers

Ingredients	Parts		
	S	T	U
Vamac G	100	100	100
Naugard 445	2	2	2
Stearic acid	1.5	1.5	1.5
Vanfre Vam	1	1	1
Armeen 18D	0.5	0.5	0.5
Carbon Black FEF N550	60	60	—
Carbon Black SRF N774	—	—	95
HMDC	1.5	1.25	1.25
DOTG	4	4	4
Selected vulcanizate properties[a]			
Original properties at 23°C			
Hardness, durometer A, Pts	66	66	59
100% Modulus, MPa	7.7	6.3	3.9
Tensile Strength, MPa	19.1	18.2	10.8
Elongation at Break, %	250	301	262
Compression set B—168 h at 150°C, Plied			
Set, %	9.6	19.4	22.1
T_g by DMA, °C	−15.0	−14.8	−14.7
Stress/strain properties at 150°C			
100% Modulus, MPa	5.1	4.3	2.8
Tensile strength, MPa	5.3	6.0	3.5
Elongation at break, %	103	130	130
Stress/strain properties at −40°C			
Tensile strength, MPa	51.9	49.1	59.4
Elongation at break, %	23	28	3

[a] Test slabs press-cured 5 min at 177°C and post-cured 4 h at 175°C.

levels and types of carbon black can be varied (and plasticizer added) to modify the damping characteristics of these compounds (Table 6.31).

6.11.4 WIRE AND CABLE

The primary use for Vamac in wire and cable is in ATH filled, flame resistant, non-halogen jackets, usually peroxide cured (Table 6.32). The following typical jacket compounds are based on Vamac D and Vamac G. A color concentrate can be used in place of the carbon black. Armeen 18D and Vanfre VAM can be added to improve release properties but at the high filler levels associated with flame-retardant jacketing compounds, release is usually not a problem.

TABLE 6.32
Flame Retardant Non-Halogen Jacket Compounds

Ingredients	Parts	
	V	W
Nominal hardness	75	75
Vamac D	100	—
Vamac G	—	100
Hydral 710	150	150
SRF, N774 Carbon Black	3	3
Stearic acid	1.5	1.5
Silquest A-172	1	1
Irganox 1010	2	2
DLTDP	1	1
HVA-2	2	2
Vulcup R	2.5	2.5
Selected vulcanizate properties[a]		
Original properties at 23°C		
Hardness, durometer A, Pts	75	73
100% Modulus, MPa	4.8	6.4
Tensile strength, MPa	5.5	8.1
Elongation at break, %	266	227
Trouser tear, kN/m	5.4	6.3
Limiting oxygen index[b]	33.4	37.5
Properties after air oven aging 7 days at 150°C		
Hardness, durometer A, Pts	81	75
Change from original, Pts	6	2
100% Modulus, MPa	6.8	8.0
Change from original, %	41.7	25.0
Tensile strength, MPa	7.2	9.6
Change from original, %	30.9	18.5
Elongation at break, %	73	80
Change from original, %	−72.5	−64.7
Properties after air oven aging 7 days at 175°C		
Hardness, durometer A, Pts	86	77
Change from original, Pts	11	4
100% Modulus, MPa	8.3	11.3
Change from original, %	72.9	76.6
Tensile strength, MPa	8.5	11.4
Change from original, %	54.5	40.7
Elongation at break, %	143	106
Change from original, %	−46.2	−53.3

(*continued*)

TABLE 6.32   (Continued)
Flame Retardant Non-Halogen Jacket Compounds

Ingredients	Parts	
	V	W
Volume change after 70 h in IRM 903 oil at 150°C		
Percent volume change	44	46
Moisture absorption after 7 days in 70°C water		
Percent absorbed	2.6	4.6

ᵃ Tests run on press-cured slabs.
ᵇ Limiting Oxygen Index (LOI) or Oxygen Index (OI) is the minimum
   concentration of oxygen, expressed as volume percent, in a mixture of
   O_2 and N_2 that will just support flaming combustion of a material
   initially at 23°C under the conditions specified in ASTM D 2863.

Radiation curing is also feasible. In such cases the peroxide can be replaced with a coagent to enhance crosslinking by radiation (Table 6.33).

Vamac also functions well as a strippable semi-conductive insulation shield, a wire cover in automotive transmission wire, and a heat resistant ignition wire cover (Tables 6.34 through 6.36).

Other specialty applications for Vamac include: o-rings, CVJ boots, low smoke flooring and extruded edging, roll covers, sponge, and oil field applications. The starting formulations for Vamac compounds designed for these applications have proved satisfactory in laboratory, prototype, or commercial evaluations.

TABLE 6.33
Flame Retardant, Non-Halogen
Jacket for Radiation Curing

Ingredients	Parts X
Vamac G	100
Cab-O-Sil MS-7	15
Nipol 1511	10
Naugard 445	2
Stearic acid	1.5
Armeen 18D	0.5
Vanfre VAM	1
Vanox ZMTI	2
Aminox	1
Atomite Whiting	65
Sartomer SR-350 (TMPTM)	4

TABLE 6.34
Strippable Insulation Shield

Ingredients	Parts	
	Y	Z
Vamac G	100	—
Vamac D	—	100
Natural Whiting	20	—
XCF, N472 Black	50	—
Black Pearls 3200	—	75
Stearic acid	1.5	1
Santowhite Powder	2	—
Agerite Resin D	—	1.5
Armeen 18D	0.5	—
Vanfre VAM	0.5	—
Vulcup R	2	1.5
Selected vulcanizate properties[a]		
Original properties at 23°C		
Hardness, durometer A, Pts	—	84
100% Modulus, MPa	—	4.4
200% Modulus, MPa	7.6	—
Tensile strength, MPa	9.7	6.2
Elongation at break, %	365	374
Volume resistivity		
at 23°C, ohm-meters	1.34	0.048–0.069
at 90°C, ohm-meters	1.0	—
at 110°C, ohm-meters	0.88	—
Properties after air oven aging 7 days at 121°C		
Tensile strength, MPa	12.2	—
Change from original, %	25.8	—
Elongation at break, %	325	—
Change from original, %	−11.0	—
Properties after air oven aging 7 days at 150°C		
Hardness, durometer A, Pts	—	84
Change from original, Pts	—	0
100% Modulus, MPa	—	6.5
Change from original, %	—	47.7
Tensile strength, MPa	—	7.9
Change from original, %	—	27.4
Elongation at break, %	—	316
Change from original, %	—	−15.5
Properties after air oven aging 7 days at 175°C		
Hardness, durometer A, Pts	—	83
Change from original, Pts	—	−1
100% Modulus, MPa	—	5.6

(*continued*)

TABLE 6.34 (Continued)
Strippable Insulation Shield

Ingredients	Parts	
	Y	Z
Change from original, %	—	27.3
Tensile strength, MPa	—	6.0
Change from original, %	—	−3.2
Elongation at break, %	—	255
Change from original, %	—	−31.8
Volume change after aging 70 h in IRM 903 Oil at 150°C		
Percent volume change	—	70

[a] Tests run on slabs press-cured 15 min at 177°C.

TABLE 6.35
Automotive Transmission Wire Cover

Ingredients	Parts AA
Vamac G	100
Natural Ground Whiting	100
Cab-O-Sil MS-7	5
Hydral 710	15
Naugard 445	1.5
Stearic acid	1.5
Vanfre UN	0.5
Ricon 152	2
Dibasic lead phosphite	2.5
HMDC	0.5
Vulcup R	2.5
Selected vulcanizate properties[a]	
Original stress/strain properties at 23°C	
Hardness, durometer A, Pts	68
100% Modulus, MPa	3.0
200% Modulus, MPa	3.5
Tensile strength, MPa	7.4
Elongation at break, %	415
Electrical properties	
Dielectric strength, kV/mm	37

TABLE 6.35   (Continued)
Automotive Transmission Wire Cover

Ingredients	Parts AA
Properties after air oven aging 7 days at 150°C	
Tensile strength, MPa	6.7
Change from original, %	−9.4
Elongation at break, %	280
Change from original, %	−32.5
Properties after aging in Dexron II ATF for 1008 h at 150°C	
Hardness, durometer A, Pts	57
Change from original, Pts	−11
Tensile strength, MPa	5.7
Change from original, %	−22.9
Elongation at break, %	175
Change from original, %	−57.8
Dielectric strength—withstand 1 kV	Pass
Flexibility—Bend Around 12.7 mm Mandrel	No cracks

[a] Tests run on slabs cured 20 min at 177°C.

TABLE 6.36
Heat Resistant Ignition Wire Cover

Ingredients	Parts BB
Vamac G	100
Elvax 40	20
Natural Ground Whiting	150
Cab-O-Sil MS-7	15
TiPure R960	5
Naugard XL1	1.5
Weston 600	1
Irganox MD-1024	2
Armeen 18D	0.5
Stearic acid	1.5
Vanfre VAM	1
Santicizer 409	5
Multiwax 180M	3
Selected vulcanizate properties[a]	
Original properties at 23°C	
100% Modulus, MPa	2.1

(*continued*)

TABLE 6.36   (Continued)
Heat Resistant Ignition Wire Cover

Ingredients	Parts BB
Tensile strength, MPa	6.6
Elongation at break, %	440
Dielectric strength, kV/mm	16.2
Properties after air oven aging 7 days at 177°C	
100% Modulus, MPa	4.5
Change from original, %	114
Tensile strength, MPa	7.4
Change from original, %	12.1
Elongation at break, %	320
Change from original, %	−27.2
180° Bend Test (ASTM D470)	No cracking
Properties after air oven aging 70 h at 191°C	
100% Modulus, MPa	4.3
Change from original, %	105
Tensile strength, MPa	6.7
Change from original, %	1.5
Elongation at break, %	300
Change from original, %	−31.8
180° Bend Test (ASTM D470)	No cracking
Properties after air oven aging 70 h at 200°C	
100% Modulus, MPa	5.0
Change from original, %	138
Tensile strength, MPa	5.5
Change from original, %	−16.7
Elongation at break, %	160
Change from original, %	−63.6
180° Bend Test (ASTM D470)	No cracking

[a] Tests run on slabs cured 20 min at 177°C.

6.11.5 O-Rings

Vamac can be used to manufacture O-rings, lathe-cut or machined gaskets with a broad range of diameters and thickness. Because of the excellent compression set of Vamac, it is the material of choice for high-temperature applications in contact with automotive fluids. O-rings and gaskets from Vamac can be molded or extruded and lathe-cut or machined (Table 6.37).

6.11.6 CVJ Boots

Depending on the location of the vehicle, Vamac can be used as the base polymer in CVJ boots in environments where good fluid resistance is required. When they

TABLE 6.37
O-Rings

Ingredients	Parts			
	CC	DD	EE	FF
Nominal Hardness	75	75	75	75
Vamac G	100	100	100	100
Naugard 445	2	2	2	2
Armeen 18D	0.5	0.5	0.5	0.5
Stearic acid	2	2	2	2
Vanfre VAM	1	1	1	1
FEF, N550 Carbon Black	65	—	—	60
SRF, N762 Carbon Black	—	75	100	—
TP 759	10	—	20	5
HMDC	1.5	1.5	1.5	1.5
DOTG	4	4	4	4
Selected vulcanizate properties[a]				
Original properties at 23°C				
Hardness, durometer A, Pts	76	78	77	74
100% Modulus, MPa	5.5	6.6	5.8	5.0
Tensile strength, MPa	15.3	17.4	13.5	15.2
Elongation at break, %	285	265	225	285
Die C tear strength, kN/m	31.7	29.1	27.5	33.5
Compression set B, Plied Slabs				
Set, % after 70 h at 150°C	14	11	18	11
Set, % after 336 h at 150°C	28	23	33	24
Set, % after 1008 h at 150°C	44	36	48	38
Compression set, O-Rings				
Set, % after 70 h at 150°C	28	26	40	26
Set, % after 336 h at 150°C	59	50	73	56
Set, % after 1008 h at 150°C	87	76	94	81
T_g (DSC Inflection), °C	−36	−29	−42	−32
Brittle point (average), °C				
No breaks	−40	−34	−34	−40
All breaks	−48	−40	−40	−48
Properties after air oven aging 1008 h at 150°C				
Hardness, durometer A, Pts	80	81	85	80
100% Modulus, MPa	7.9	7.9	10.1	6.9
Tensile strength, MPa	13.1	16.6	12.6	15.1
Elongation at break, %	180	195	145	240
Properties after aging 1008 h in Mopar 5W30 motor oil at 150°C				
Hardness, durometer A, Pts	64	61	66	61
100% Modulus, MPa	5.5	6.3	6.1	4.8
Tensile strength, MPa	13.4	14.6	13.9	13.5
Elongation at break, %	215	215	200	250

[a] Tests run on slabs press-cured 10 min at 177°C and post-cured 4 h at 177°C.

become available, lower T_g terpolymers can be substituted for Vamac G if improved low-temperature flexibility is critical (Table 6.38).

6.11.7 CELLULAR VAMAC—CLOSED CELL

Cellular Vamac can be prepared with a wide range of final densities (Tables 6.39 and 6.40). The load-bearing capacities and other physical properties of laboratory samples meet the ASTM specifications for medium oil resistance cellular rubber.

TABLE 6.38
CVJ Boots

Ingredients	Parts			
	GG	HH	JJ	KK
Nominal hardness	55/60	65	65	65
Vamac G/HG	100	100	100	100
Naugard 445	2	2	2	2
Armeen 18D	0.5	0.5	0.5	0.5
Stearic acid	2	1.5	1.5	1.5
Vanfre VAM	1	1	1	1
APF, N683 Carbon Black	70	—	—	—
FEF, N550 Carbon Black	—	80	90	100
TP 759	37.5	30	40	50
HMDC	1.25	1.5	1.25	1.25
DPG	—	4	—	—
DOTG	4	—	4	4
Selected vulcanizate properties[a]				
Original properties at 23°C				
Hardness, durometer A, Pts	56	65	65	66
100% Modulus, MPa	2.3	4.5	3.6	3.5
Tensile strength, MPa	9.4	10.9	9.8	8.5
Elongation at break, %	380	258	299	276
Die C tear strength, N/mm	29.1	23.8	22.4	20.1
T_g (DSC), °C	—	−48	−53	−57
Compression set B, Plied				
Set, % after 22 h at 125°C	—	13	15	18
Set, % after 70 h at 150°C	16	—	—	—
Brittle point, °C	−52	—	—	—
TR-10, °C	−39	—	—	—
DeMattia Flex—not Pierced				
Kcycles to cracks	171	—	—	—
Kcycles to failure	990	—	—	—

TABLE 6.38 (Continued)
CVJ Boots

Ingredients	Parts			
	GG	HH	JJ	KK
Properties after air oven aging 168 h at 125°C				
Hardness, durometer A, Pts	—	65	68	69
Change from original, Pts	—	0	3	3
Tensile strength, MPa	—	11.1	9.5	8.2
Change from original, %	—	1.8	−3.1	−3.5
Elongation at break, %	—	249	286	267
Change from original, %	—	−3.5	−4.3	−3.3
Properties after air oven aging 70 h at 150°C				
Hardness, durometer A, Pts	60	—	—	—
Change from original, Pts	4	—	—	—
100% Modulus, MPa	2.6	—	—	—
Change from original, %	13.0	—	—	—
Tensile strength, MPa	9.1	—	—	—
Change from original, %	−3.2	—	—	—
Elongation at break, %	380	—	—	—
Change from original, %	0	—	—	—
TR-10, °C	−37	—	—	—
Properties after aging 168 h in ASTM #1 Oil at 125°C				
Hardness, durometer A, Pts	—	78	80	81
Change from original, Pts	—	13	15	15
Tensile strength, MPa	—	10.4	12.0	10.8
Change from original, %	—	−4.6	22.4	27.0
Elongation at break, %	—	262	267	260
Change from original, %	—	1.5	−10.7	−5.8
Percent volume change	—	−9	−11	13

[a] Test slabs for "GG" were press-cured 15 min at 180°C and post-cured 4 h at 180°C. Test slabs for the remaining compounds were press-cured 5 min at 180°C and post-cured 4 h at 175°C. Data based on Vamac G.

Very fast curing is preferred for expanded Vamac. Although the HMDC/tetra-methylguanidene system is fast, the HMDC/peroxide fast cure system may be advantageous for producing low density cellular Vamac. The curing of cellular Vamac requires two steps—a press-cure (pre-cure) where the blowing agent is decomposed, and a post-cure to complete the vulcanization.

6.11.8 ROLL COVERS

Vamac can be calendered and fabricated into roll covers for specialty applications. Compositions for roll covers should be formulated to minimize heat buildup (Table 6.41).

TABLE 6.39
Black Closed Cell Cellular Vamac

Ingredients	Parts			
	LL	MM	NN	PP
Vamac G	100	100	100	100
Naugard 445	2	2	2	2
Armeen 18D	0.5	0.5	0.5	0.5
Stearic acid	2	2	2	2
Vanfre VAM	0.5	0.5	0.5	0.5
SRF, N774 carbon black	20	20	20	20
FEF, N550 carbon black	—	—	10	20
HMDC	2	2	2	2
Tetramethylguanidine	1	1	1	1
Unicel ND	7.5	10	10	10
Activator DN	2.4	3.2	3.2	3.2
Cured sponge density, g/cc	0.217	0.156	0.154	0.177
Compression/deflection				
25% Deflection, kPa	50	30	34	56

Source: Cellular Vamac®. DuPont Technical Bulletin IF-EA-560.F06.

TABLE 6.40
Non-Black Closed Cell Cellular Vamac

Ingredients	Parts		
	QQ	RR	SS
Vamac G	100	100	100
Naugard 445	2	2	2
Stearic acid	2	2	2
Vanfre VAM	0.5	0.5	0.5
Nipol 1511	10	10	10
Cab-O-Sil MS-7	10	10	20
Hydral 710	—	—	50
DiCup 40C	7	7	7
Triallylisocyanurate (TAIC)	—	2	2
HVA-2	2	—	—
Unicel ND	5	5	5
Activator DN	1.6	1.6	1.6
Cured sponge density, g/cc	0.26	0.284	0.263
Compression/deflection			
25% Deflection, kPA	64	88	82

Source: Cellular Vamac®. DuPont Technical Bulletin IF-EA-560.F06.

TABLE 6.41
Roll Covers

Ingredients	Parts			
	TT	UU	VV	WW
Vamac G[a]	100	100	100	100
Naugard 445	2	2	2	2
Stearic acid	1.5	1.5	1.5	1.5
Vanfre VAM	1	1	1	1
Armeen 18D	0.5	0.5	0.5	0.5
CaCO$_3$	80	90	80	90
Cab-O-Sil MS7D	35	40	35	40
Plasthall P670	10	10	10	10
HMDC	1.5	1.5	1	1
DOTG	4	4	4	4
DiCup 40C	—	—	5	5
Ricon 152	—	—	2	2

[a] Blends with other polymers may be required to control solvent resistance and flexibility.

6.11.9 Oil Field Applications

The resistance to degradation at high operating temperatures in the wide variety of fluids encountered in the oil industry, coupled with the resistance of AEM to wet and dry sour gas, make Vamac a candidate in elastomeric oil field applications. Some starting formulations for oil field applications are shown below, but because of the widely differing requirements, especially fluid resistance, from site to site, polymer blending may be necessary. To reduce the susceptibility of Vamac to polar fluid environments and enhance low-temperature flex life, Vamac can be blended with nonpolar polymers such as polybutadiene.

Duplication of vaguely defined oil field conditions in the laboratory is not possible. The range of chemical compositions, temperatures, pressures, and performance requirements encountered are so broad that it is impossible to select a general composition for oil field service. Table 6.42 is an attempt to illustrate how Vamac can perform in certain specified conditions. It is suggested that prototype parts be evaluated and modifications made to optimize the compound for the specific application or site.

6.11.10 Low Smoke Flooring and Extruded Edging

Vamac D is used as the primary polymeric component in low smoke flooring. Vamac flooring also has sound deadening capabilities (Table 6.43).

TABLE 6.42
Oil Field Applications

Ingredients	Parts			
	XX	YY	ZZ	AAA
Nominal hardness, durometer A	80	85	90	90
Vamac G/GLS	100	100	100	100
Naugard 445	2	2	2	2
Armeen 18D	0.5	0.5	0.5	0.5
Stearic acid	1.5	1.5	2	2
Vanfre UN	—	—	2	2
HAF Carbon Black	—	—	—	80
SRF, N762 Carbon Black	—	—	20	20
FEF, N550 Carbon Black	60	80	80	—
Graphite	20	20	—	10
TP 759	5	10	—	—
HMDC	1.25	1.25	1.25	1.25
DOTG	4	4	4	4
Selected vulcanizate properties[a,b]				
Original properties at 23°C				
Hardness, durometer A, Pts	—	—	86	88
100% Modulus, MPa	—	—	10.4	11.0
Tensile strength, MPa	—	—	13.8	15.3
Elongation at break, %	—	—	165	165
Die C tear strength, N/mm	—	—	—	31.5
Compression set B—70 h at 150°C				
Set, %	—	—	14	25
NBS Abrasion Index	—	—	90	—
Taber Abrasion, H-18 Wheel, 500 g Load				
Weight loss, mg/revolution	—	—	0.20	—
Properties at 149°C				
Hardness, durometer A, Pts	—	—	79	—
Tensile strength, MPa	—	—	6.7	—
Elongation at break, %	—	—	90	—
Properties after exposure to dry and wet sour gas[c]				
Dry sour gas exposure[d] *−150°C, 13.8 MPa, 3 days*				
100% Modulus, MPa	—	—	—	11.5
Change from original, %	—	—	—	4.5
Tensile strength, MPa	—	—	—	13.0
Change from original, %	—	—	—	−15.0
Elongation at break, %	—	—	—	125
Change from original, %	—	—	—	−24.2
Dry sour gas exposure[d] *−205°C, 14.5 MPa, 3 days*				
100% Modulus, MPa	—	—	—	14.2

TABLE 6.42   (Continued)
Oil Field Applications

Ingredients	Parts			
	XX	**YY**	**ZZ**	**AAA**
Change from original, %	—	—	—	29.0
Tensile strength, MPa	—	—	—	14.6
Change from original, %	—	—	—	−4.6
Elongation at break, %	—	—	—	105
Change from original, %	—	—	—	−36.4
Dry sour gas exposure[d] *−235°C, 15.9 MPa, 3 days*				
Hardness, durometer A, Pts	—	—	—	60
Change from original, Pts	—	—	—	−28
Tensile strength, MPa	—	—	—	11.7
Change from original, %	—	—	—	−23.5
Elongation at break, %	—	—	—	25
Change from original, %	—	—	—	−84.8
Wet sour gas exposure[e] *−150°C, 9.7 MPa, 3 days—liquid phase*				
100% Modulus, MPa	—	—	—	7.3
Change from original, %	—	—	—	−33.6
Tensile strength, MPa	—	—	—	10.7
Change from original, %	—	—	—	−30.1
Elongation at break, %	—	—	—	145
Change from original, %	—	—	—	−12.1
Wet sour gas exposure[e] *−150°C, 9.7 MPa, 3 days—vapor phase*				
100% Modulus, MPa	—	—	—	6.9
Change from original, %	—	—	—	−37.3
Tensile strength, MPa	—	—	—	9.5
Change from original, %	—	—	—	−37.9
Elongation at break, %	—	—	—	155
Change from original, %	—	—	—	−6.1
Wet sour gas exposure[e] *−205°C, 14.5 MPa, 3 days—liquid phase*				
Tensile strength, MPa	—	—	—	12.4
Change from original, %	—	—	—	−18.9
Elongation at break, %	—	—	—	60
Change from original, %	—	—	—	−63.6
Wet sour gas exposure[e] *−205°C, 14.5 MPa, 3 days—vapor phase*				
100% Modulus, MPa	—	—	—	9.4
Change from original, %	—	—	—	−14.5
Tensile strength, MPa	—	—	—	12.7
Change from original, %	—	—	—	−17.0
Elongation at break, %	—	—	—	135
Change from original, %	—	—	—	−18.2

(*continued*)

TABLE 6.42 (Continued)
Oil Field Applications

Ingredients	Parts			
	XX	YY	ZZ	AAA
Wet sour gas exposure[e] *−205°C, 3 MPa, 10 days—vapor phase*				
Tensile strength, MPa	—	—	—	22.2
Change from original, %	—	—	—	45.1
Elongation at break, %	—	—	—	40
Change from original, %	—	—	—	−75.7
Volume change, %	—	—	—	−2.6

Source: Vamac® for oil well applications. DuPont Technical Bulletin VA-D-8.

[a] Test slabs post-cured 4 h at 177°C.

[b] Data based on Vamac G.

[c] Testing—Dumbells from 1.9 mm slabs were exposed to the various conditions outlined in the table. Exposures were carried out in high pressure autoclaves (bombs) under the conditions shown. Properties were measured after equilibrating the specimens for 3 days at 23°C and 50% RH.

[d] *Dry sour gas:* 35 weight percent H_2S, 15 weight percent CO_2, and 50 weight percent CH_4. All ingredients charged to a 1.2L bomb.

[e] *Wet sour gas:* 11.5 weight percent H_2S, 11.5 weight percent CO_2, 38.5 weight percent #10 oil, and 38.5 weight percent H_2O. All ingredients charged to a 1.2L bomb.

TABLE 6.43
Low Smoke Flooring and Extruded Edging

Ingredients	Parts BBB
Vamac D	65
Elvax 265	25
Elvax 40LO3	10
Stearic acid	2
Armeen 18D	1
Vanfre VAM	0.5
Pigment	To choice
Vinyl silane	2.5
Aluminum hydroxide	200
Magnesium hydroxide	50
Varox DBPH50	3.5
TAC	1.75
Nominal hardness	95

Source: Basic principles for the compounding and processing of Vamac® dipolymers. DuPont Dow Elastomers Technical Information Bulletin H-68560.

APPENDIX—LIST OF CHEMICALS

Sources of compounding ingredients used in the text and formulations shown in this chapter are listed below. Their presence in this list does not imply that ingredients from other sources might not be equally satisfactory.

Chemical	Composition	Source
AKD-Cizer C-9N, C-810, UL-100	Polyester plasticizer	Asahi Denka Kogyo K.K.
Armeen 2C	Dicocoamine	Akzo Nobel Chemicals
Armeen 18D	1-Octadecanamine	Akzo Nobel Chemicals
Atomite Whiting	Calcium carbonate	Thompson, Weinman
Blanc Fixe	Precipitated barium sulfate	BASF
Cab-O-Sil	Fumed silica	Cabot Corporation
Cyprubond	Magnesium silicate	Luzenac America Inc.
DiCup 40C	Dicumyl peroxide on calcium carbonate	Hercules, Inc.
DOTG	Diorthotolylguanidine	Akron Chemical
DPG	Diphenyl guanidine	Akron Chemical
Dustanox 86	4,4-Dicumyldiphenylamine	Duslo, Slovak Republic
Elvax 265, 40L03	Ethylene/vinyl acetate copolymer	DuPont Company
Epicure 3501	Ketimine	Shell Chemical Company
HMDC	Hexamethylenediamine carbamate	DuPont Dow Elastomers, L.L.C.
[Diak #1]	100% solid HMDC	Alcan Rubber and Chemical Co.
[Vulcafac HDC]	100% solid HMDC	Flow Polymer
[PAB-1685]	66.7/33.3 HMDC/Unflex 330	Elastochem (Rheinchemie)
[Vulcofac HDS 66]	53/47 HMDC/dibutyl sebacate paste	Alcan Rubber and Chemical Co.
	66/34 HMDC/glycol ether–ester plasticizer	
HVA-2	N,N'-m-phenylene dimaleimide	DuPont Dow Elastomers, L.L.C.
Hydral 710	Hydrated aluminum oxide	Alcoa
Irganox MD-1024	1,2-bis(3,5-di-t-butyl-4-hydroxy-hydrocinnamoyl) hydrazide	Ciba-Geigy
Naugard 10	Tetrakis[methylene(3,5-di-t-butyl-4-hydroxylhydrocinnamate)]methane	UniRoyl Chemicals
Naugard 445	4,4-bis(a,a-dimethylbenzyl) diphenylamine	UniRoyl Chemicals
Naugard XL1	2,2-Oxamido bis[ethyl 3-(3,5-di-t-butyl-4-hydroxyphenyl)]proprionate	UniRoyl Chemicals
OABH	Oxalyl bis(benzylidenehydrazide)	Eastman Chemical Company
Plasthall 810TM	Linear trimellitate	C.P. Hall Company
Plasthall P670	Polyester plasticizer	C.P. Hall Company
Ricon 152	1,2-Polybutadiene liquid	Colorado Chemicals
Salicyclic acid	Salicyclic acid	Bayer Corporation
Stearic acid	Stearic acid	C.P. Hall Company

(*continued*)

(Continued)

Chemical	Composition	Source
Struktol WS180	Fatty acid-silicone condensation product	Struktol Company of America
TAC	Triallyl cyanurate	Cytec
TAIC (Diak 7)	Triallyl isocyanurate	DuPont Dow Elastomers, L.L.C.
Tetramethylguanidine	Tetramethylguanidine	American Cyanamid
TP-759	Ether/ester plasticizer	Morton Chemical
Trigonox 17–40B	40% Butyl 4,4-di(t-butylperoxy) valerate	Elf Atochem
Unicell MD	Azodicarbonamide	Dong Jin (USA), Inc.
Vanfre UN	Fatty alcohol phosphate, unneutralized	R.T. Vanderbilt Co., Inc.
Vanfre VAM	Alkyl phosphate	R.T. Vanderbilt Co., Inc.
Vulcup R	a-a-bis(t-butyl peroxy) diisopropylbenzene	Hercules, Inc.
Vulcup 40KE	Vulcup R on Burgess KE clay	Hercules, Inc.
Varox	2,5-Dimethyl-2,5-di(t-butylperoxy)hexane	R.T. Vanderbilt Co., Inc.
Varox DBPH-50	50% Varox powder	R.T. Vanderbilt Co., Inc.
Weston 600	Phosphite stabilizer	GE Specialty Chemicals

REFERENCES

1. W.M. Stahl, Fluid resistance of Vamac®. DuPont Technical Bulletin EA-510.1 (R2).
2. A.E. Hirsch and R.J. Boyce. Dynamic properties of ethylene/acrylic elastomer: A new heat resistant rubber. Proceedings of the International Rubber Conference, Brighton, England, May 1997. DuPont Technical Bulletin C-EA-53C.G04.
3. Compounding Vamac® for continuous vulcanization without external pressure. DuPont Technical Bulletin HHSD 141.
4. D.A. Kotz and T.M. Dobel. Stress relaxation of ethylene/acrylic and other gasket materials in automotive fluids. Society of Automotive Engineers International Congress and Exposition, Detroit, Michigan, February/March 1994, Paper No. 940959.
5. Basic principles for the compounding and processing of Vamac® dipolymers. DuPont Dow Elastomers Technical Information Bulletin H-68560.
6. Vamac® compounds for high performance hoses. DuPont Product Data Sheet.
7. Cellular Vamac®. DuPont Technical Bulletin IF-EA-560.F06.
8. Vamac® for oil well applications. DuPont Technical Bulletin VA-D-8.

NOTE: Since this chapter was written there have been improvements in Vamac products and cure systems, such as Advancure. As of 2007 available grades are Vamac G, Vamac HVG, Vamac GLS, Vamac GXF, Vamac DP, Ulta IP and Ulta LT. Please contact Douglas D. King, DuPont Performance Elastomers at 302 792 4063 or web site www.dupontelastomers.com/vamac.

7 Polyepichlorohydrin Elastomer

Robert C. Klingender

CONTENTS

Elastomers containing polyepichlorohydrin, also known as ECO, CO, or GECO according to ASTM, offer an excellent balance of properties, combining certain desired dynamic properties of natural rubber (NR), with much of the fuel, oil, and chemical resistance of other specialty elastomers such as nitrile (NBR), polyacrylate (ACM), and neoprene (CR) rubbers. The combination of the basic properties of oil, fuel, heat, low-temperature flexibility, and ozone resistance imparted by the saturated main chain and the chlorine groups, coupled with low permeability, makes polyepichlorohydrin a very useful elastomer for automotive applications. Specific applications include fuel hoses, emission tubing, air ducts, seals, and diaphragms.

This elastomer is also very useful in certain mechanical goods applications such as oil field and refinery seals and pressure accumulator bladders. Specialty rollers for both printing and copying applications make use of the excellent static conductivity of the GECO elastomer.

There are three classes of polyepichlorohydrins, the homopolymer CO and GCO, the copolymer ECO, and the terpolymer GECO, as shown in Table 7.1. The structure of these is given in Figure 7.1. Each family of these elastomers has its own characteristics as illustrated in Table 7.2.

The homopolymer has unusually good fuel and gas permeation resistance and very good fuel resistance, making it very suitable for air-conditioning hose and seals as well as oilfield and refinery applications. A version of polyepichlorohydrin is available with a small amount of allyl glycidyl ether (AGE) as a side group for improved heat resistance and greater versatility in cure systems. This elastomer also provides better dampening properties than the copolymer or terpolymer.

The copolymer has ethylene oxide in the backbone of the polymer, which, coupled with less of the bulky chlorine in the side group, provides good mobility and thus excellent low-temperature properties and good fuel, weathering, and aging resistance, thus making it the polymer of choice for most automotive applications.

The terpolymer has excellent low temperature and very good fuel, ozone, and weathering resistance, and in addition, has better reversion resistance. The terpolymer contains a small amount of AGE in the form of an allyl side group, which allows for the utilization of a broader range of vulcanization systems. In addition, polyepichlorohydrin

---

TABLE 7.1
ASTM D1418 Designations

Homopolymer of epichlorohydrin (ECH)	CO
Copolymer of epichlorohydrin/allyl glycidyl ether (ECH/AGE)	GCO
Copolymer of epichlorohydrin/ethylene oxide (ECH/EO)	ECO
Terpolymer of epichlorohydrin/ethylene oxide/allyl glycidyl ether (ECH/EO/AGE)	GECO

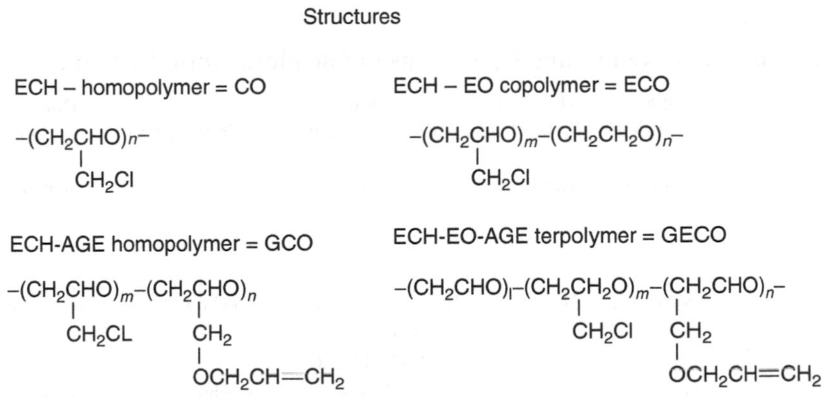

Structures

ECH – homopolymer = CO

$-(CH_2CHO)_n-$
        |
     CH_2Cl

ECH – EO copolymer = ECO

$-(CH_2CHO)_m-(CH_2CH_2O)_n-$
        |
     CH_2Cl

ECH-AGE homopolymer = GCO

$-(CH_2CHO)_m-(CH_2CHO)_n$
        | |
     CH_2CL CH_2
                       |
                  $OCH_2CH=CH_2$

ECH-EO-AGE terpolymer = GECO

$-(CH_2CHO)_l-(CH_2CH_2O)_m-(CH_2CHO)_n-$
                              |
                           CH_2Cl CH_2
                                          |
                                     $OCH_2CH=CH_2$

FIGURE 7.1 Structures.

terpolymer elastomers are electrostatically dissipative, which makes it well suited for electro-assist printing and copy machine roller applications.

The SAE J200/ASTM D2000 designation for compounds based on the copolymer of epichlorohydrin and ethylene oxide is "CH," indicating a service temperature of 125°C and a maximum volume swell in ASTM No. 3 oil of 30%.

7.1 HISTORY

E.J. Vandenberg discovered, in 1957, that organometallic catalysts gave high molecular weight polymers from epoxides [1]. The commercially important elastomers developed as a result of this work are polyepichlorohydrin (CO), epichlorhydrin-ethylene oxide copolymer (ECO), and epichlorhydrin-ethylene oxide-allyl glycidyl ether terpolymer (GECO). The polyepichlorohydrin-allyl glycicyl ether copolymer (GCO) was developed at B.F. Goodrich at a later date.

B.F. Goodrich first commercialized polyepichlorohydrin in 1965, under a license agreement from Hercules, Inc., using the trade name Hydrin. In 1966, Hercules, Inc. began marketing similar polymers under the trade name Herclor. (Table 7.3 has a historical as well as a current list of the trade names and their equivalents.)

TABLE 7.2
Polyether Elastomers Information

Polymers	ECH, wt%	Chlorine, wt%	Ethylene Oxide, wt%	Allyl Glycidyl Ether, wt%	Specific Gravity	ML Mooney at 100°C	T_g, °C
ECH	100	38	0	0	1.36	40–80	−22
GCO	92	35	0	7	1.24	60	−25
ECO	68	26	32	0	1.27	40–130	−40
GECO	65–76	24–29	13–31	3–11	1.27	50–100	−38

TABLE 7.3

History of Trade Names and Equivalents of Epichlorohydrin Elastomers

	B.F. Goodrich[a]	Hercules, Inc.[a]	Zeon Chemicals L.P.	Nippon Zeon	Daiso Co., Ltd.
Homopolymer (CO)	Hydrin 100	Herclor H	Hydrin H 45–75	Gechron 1000	Epichlomer H
Copolymer of CO (GCO)			Hydrin H1100	Gechron 1100	
Copolymer (ECO)	Hydrin 200	Herclor C	Hydrin C, C2000XL C2000LL C2000L, C2000	Gechron 2000	Epichlomer C
Terpolymer (GECO)	Hydrin 400	Herclor T	Hydrin T, T3000	Gechron 3000	Epichlomer CG
Terpolymer (GECO) different combinations			Hydrin T3100, T3102, T3105, T3106	Gechron 3100, 3101, 3102, 3103, 3105, 3106	

[a] Obsolete, no longer in business.

B.F. Goodrich purchased the Herclor business from Hercules, Inc. in 1986 and the two families of elastomers were consolidated under the Hydrin name and production was moved to the former Hercules plant. In 1989, Zeon Chemicals bought the elastomer division of B.F. Goodrich, including the Hydrin business. Nippon Zeon, the parent company of Zeon Chemicals, has its own line of epichlorohydrin elastomers and continues to make these under the trade name of Gechron. In addition, Daiso in Japan produces polyepichlorohydrins under the Epichlomer trade name.

7.2   COMMERCIAL POLYEPICHLOROHYDRIN

Polyepichlorohydrin is manufactured by a continuous solution polymerization process using aluminum-based catalysts with base monomers of epichlorohydrin (ECH), ethylene oxide (EO), and AGE. A cationic coordination mechanism is suggested as the chain propagation mechanism. The commercial polymer is 97%–99% head-to-tail and has been shown to be stereo random and atactic. The CO, GCO, ECO, and GECO polymers are amorphous and linear [1].

The production system in making ECO, as shown in Figure 7.2, is polymerization, steam coagulation, drying, and packaging. Typical composition and polymer data are given in Table 7.2.

There are two families of Hydrin polymers available from Zeon Chemicals, the original being the consolidated grades of Hydrin H, C, and T. The original Hydrin C and T grades had a tendency to stick to cold metal surfaces and were somewhat nervy if lightly loaded. The newer versions of Hydrin C2000 and T3000 series now

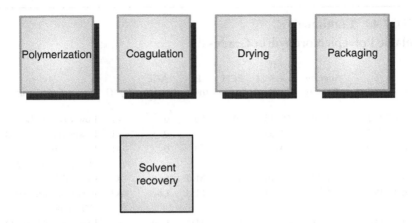

FIGURE 7.2 Plant process.

have a reduced low-molecular weight tail, which slightly reduces the tendency to stick on cold metal surfaces.

The range of currently available grades of CO, GCO, ECO, and GECO from the Zeon Group is given in Table 7.4. It is advised that you consult your suppliers to be sure that these are still available, or if newer versions have been introduced.

7.3 PHYSICAL PROPERTIES

Polyepichlorohydrin terpolymer compounds can be formulated to withstand for a very short term, 70 h at 150°C in air, while being flexible at -40°C. A more practical operating temperature of polyepichlorohydrin elastomers, for longer-term aging in

TABLE 7.4
Available Polyepichlorohydrin Grades

Grade	Specific Gravity	ML 1 + 4 at 100°C	ECH mol%	EO mol%	AGE mol%	T_g, °C	Comments
Hydrin H, Gechron 1000, Epichlomer H	1.37	40–80	100	0	0	-24	Original CO
Hydrin H1100, Gechron 1100	1.35	48–68	93	0	7	-26	GCO for better O_3 resistance and curing versatility
Hydrin C, Epichlomer C	1.28	50–90	50	50	0	-41	Original ECO
Hydrin C2000, Gechron 2000	1.28	90–102	50	50	0	-41	Better processing than "C" type

(*continued*)

TABLE 7.4   (Continued)
Available Polyepichlorohydrin Grades

Grade	Specific Gravity	ML 1 + 4 at 100°C	ECH mol%	EO mol%	AGE mol%	T_g, °C	Comments
Hydrin C2000L	1.28	65–75	50	50	0	−41	Low viscosity ECO
Hydrin C2000LL	1.28	53–65	50	50	0	−41	Lower viscosity ECO
Hydrin C2000XL	1.28	40–52	50	50	0	−41	Lowest viscosity ECO for injection
Hydrin T	1.28	50–100	Med.	Med.	Low	−43	Original GECO
Hydrin T3000	1.28	80–94	Med.	Med.	Low	−43	Improved processing "T" type
Hydrin T3000L	1.28	65–79	Med.	Med.	Low	−43	Low viscosity GECO
Hydrin T3000LL	1.28	50–64	Med.	Med.	Low	−43	Lowest viscosity GECO for injection
Hydrin T3100, Gechron 3100	1.30	63–77	High	Low	High	−36	High AGE for O_3 and sour gas resistance
Gechron 3101	1.28	75–85	Med.	Med.	Low	−37	Basic GECO
Hydrin T3102, Gechron 3102	1.30	80–100	Med.	Med.	Low	−38	Highest ECH for fuel and permeation resistance
Gechron 3103	1.29	80–90	High	Low	Med.	−41	Better scorch resistance
Hydrin T3105, Gechron 3105	1.29	70–80	High	Low	High	−41	High ECH and AGE for fuel and O_3 resistance
Hydrin T3106, Gechron 3106	1.28	53–67	Low	High	Med.	−48	High EO for low temperature and static dissipation

air, is 125°C. An ECO compound can be designed to meet a typical ASTM line call out D2000 5CH 712 A_{25} B_{34} C_{12} EF_{31} EO_{16} EO_{36} F17. The range of physicals for typical formulations is given in Table 7.5.

TABLE 7.5
Typical Formula for an Epichlorohydrin Elastomer

Ingredients	Phr	Range, Phr
Polymer	100	
Carbon black (e.g., N762)	70	0–130
Plasticizer (e.g., DBEEA)	10	0–30
Antioxidant	1	0–2
Processing aid	1	0–3
Acid acceptor	5	3–10
Cure chemicals	1	0.3–2.0

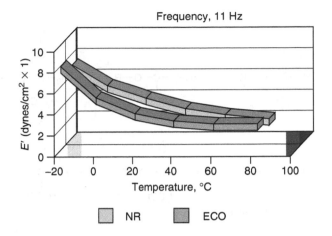

FIGURE 7.3 Dynamic modulus versus temperature.

7.4 DYNAMIC PROPERTIES

Polyepichlorohydrin copolymer or terpolymer compounds can provide vibration dampening comparable to natural rubber (NR), but at an extended temperature range. This characteristic makes polyepichlorohydrin compounds a good choice for suspension mounts and impact absorbers, which must operate at higher temperatures than practical for natural rubber. Typical data are shown in Figures 7.3 and 7.4.

The homopolymer, CO or GCO, lacks resilience, hence may be used where good energy absorption is desired.

7.5 HEAT AND OIL RESISTANCE

Tensile strength change is the best stress/strain indicator of service life of an epichlorohydrin compound [2]. CO/GCO/ECO/GECO ultimately degrade by reversion or chain

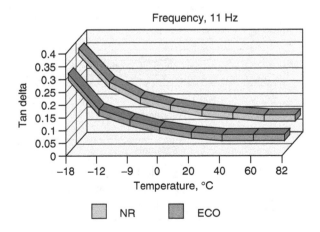

FIGURE 7.4 Damping versus temperature.

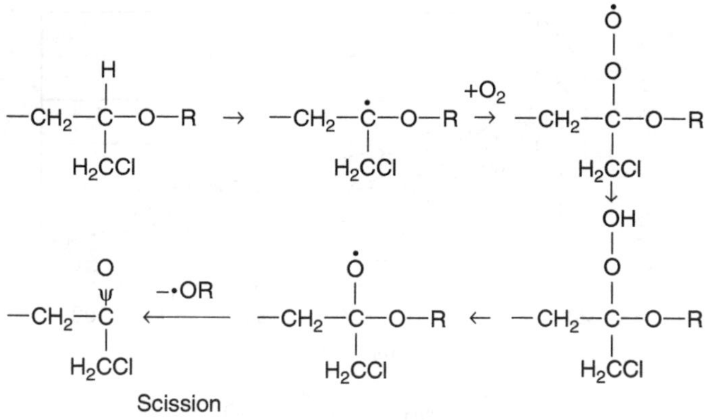

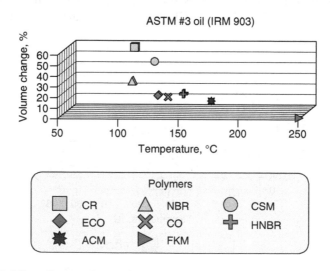

FIGURE 7.5 Autoxidation mechanism of epichlorohydrin rubber.

scission [3], the mechanism of which is shown in Figure 7.5. The elongation of epichlorohydrin initially decreases on aging in air; then it stabilizes for a time before ultimately reducing to zero. The tensile strength decreases steadily to zero. A drastic decrease in the tensile strength of ECO should not occur until after 1000 h at 120°C in air. The relation of polyepichlorohydrin versus other elastomers is shown in Figure 7.6.

The effect of reversion on polyepichlorohydrin compounds cured with ETU/red lead is most noticeable; however, this may be decreased quite significantly by using a Zisnet F-PT cure system and a high AGE content GECO polymer. The effect of AGE content on heat aging may be seen in Table 7.6. In addition, the homopolymers CO or GCO exhibit better heat resistance because of the absence of ethylene oxide. After aging 144 h at 150°C the hardness increase of a CO-based compound is

FIGURE 7.6 Oil swell versus heat resistance.

TABLE 7.6
Effect of AGE Content on Aging

AGE Content (mol%)	Hardness Change (pts)
Aged in air 288 h at 150°C	
0	−21
2.5	−10
6	−5
10.5	−1
15	2

5 points, whereas the hardness change of a 50% EO containing ECO compound was −10 points and with 70% EO in ECO the change was −19 points.

7.6 OZONE RESISTANCE

The ozone resistance of polyepichlorohydrin compounds is very good because of the saturated backbone of the polymer. This may be further enhanced by the inclusion of AGE, which further protects the backbone of the elastomer by providing some unsaturation in the side chain, which scavenges the ozone. This influence is shown in Table 7.7.

7.7 LOW-TEMPERATURE PROPERTIES

As shown in Figure 7.7, ECO also has excellent low-temperature flexibility to −40°C. This attribute leads to ECO and GECO compounds being used for hoses, ducts, vibration isolators, and diaphragms in vehicles. Further improvement in low-temperature properties may be obtained by the addition of a plasticizer such as DOA, TP-95, TP-795, DBP, or Vulcanol OT.

TABLE 7.7
Effect of AGE Content on Ozone Resistance

AGE Content (mol%)	Time to Initial Cracking (h)
Aged sequentially in Fuel C 48 h at 40°C; air 72 h at 100°C; 50 pphm ozone; 0%–30% elongation at 40°C	
0	50
4	98
5	125
6	150
8	200
10	300

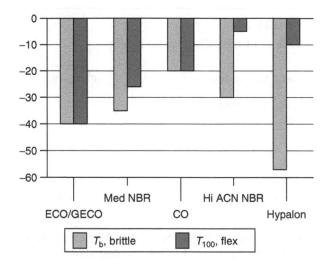

FIGURE 7.7 Temperature balance.

The higher the ethylene oxide content of the polymer the better the low-temperature flexibility. The homopolymer CO has a Gehman T_{100} of $-18°C$ as compared with the copolymer ECO that has 30% EO where the Gehman T_{100} is $-38°C$ and with 70% EO it is $-48°C$.

7.8 CORROSION RESISTANCE

As with most halogen-containing elastomers, polyepichlorohydrin polymers can corrode metal surfaces, but only noticeable under extreme conditions, such as being left in the mold at 200°C for 2 h. The Zisnet F-PT cure system is much better than the ETU/red lead curative package as given in Table 7.8, for lower mold fouling and better corrosion resistance.

TABLE 7.8
Metal Corrosion Rating

	ECO/ETU	GECO/Triazine F-PT
Corrosion test rating[a]		
Steel, SAE 1020	3	1
Zn-Cr coated	2	1
Zinc plate	4	1
Aluminum plate	1	0

0 No corrosion
1–2 Light corrosion
3–4 Dramatic corrosion
5 Severe corrosion

[a] Rating, GM 9003-P 72 h, 100% humidity at 50°C.

Most other compounding ingredients have little influence on the corrosion properties of the vulcanizate. Basic materials such as sodium bicarbonate reduce corrosion as do Linde 4A molecular sieves, Aerosil R-792 (a hydrophobic silica), epoxy resins, Duomene (tallow propylene diamine), and Teiton B (b-benzyltrimethylammonium hydroxide). Acidic ingredients as well as those containing chlorine, such as chlorinated paraffins, increase corrosion.

If high volume, repeated runs are expected for a given application, the use of stainless steel, or other corrosion-resistant materials, for the molds is recommended.

7.9 FUEL RESISTANCE

FKM has been the classic choice for best fuel permeation resistance. The high cost of FKM has resulted in its use in a veneer-type construction with less expensive, yet fuel-resistant material used to make a tie layer or a cover stock. ECO is commonly used as both a tie layer as well as a cover stock material.

Fuel hoses are now made that use thermoplastic types of fluoropolymer materials, for example, THV[5], to provide the high degree of permeation resistance. In these hoses, ECO compounds are also used as a cover stock material. Polyepichlorohydrin compounds, being fuel resistant, provide another layer of protection in case the thin inner layer leaks.

The homopolymer CO has as good a fuel resistance as ECO, but if low-temperature properties need to be considered, either the copolymer ECO or the terpolymer GECO would be suggested. All tend to swell about 40% in Fuel C. The inclusion of 10% methanol or ethanol in Fuel C results in volume swells of 85% or 70%, respectively, for a typical ECO compound. The lower the ethylene oxide content of polyepichlorohydrin ECO compounds, the more resistant they are to fuels containing alcohols, with a CO-based one being the best.

7.10 PERMEABILITY

The homopolymer of epichlorohydrin, CO, has excellent permeation resistance. This superior permeation resistance to gases provides a characteristic that is desirable in hoses for air conditioners and refrigerators. The air permeation of CO and ECO versus other elastomers is given in Figure 7.8.

Fuel permeation is also important to meet SHED requirements. The permeation through a CO compound of 90 Fuel D/10 methanol is 290 $g/m^2/day$, which is comparable with 310 for NBR and 40 for FKM. A comparison using 90 Fuel D/10 ethanol shows the CO compound to lose 100 $g/m^2/day$ versus 315 for NBR and 15 for FKM.

7.11 ELECTROSTATIC PROPERTIES

Most elastomers do not conduct electricity or static well. In contrast, ECO and GECO elastomers characteristically have volume resistivity values as low as

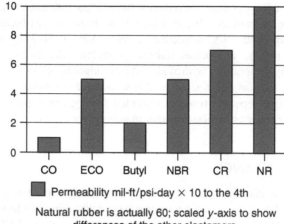

FIGURE 7.8 Air permeability.

10^8 ohm-cm (Figure 7.9). The electrical properties of ECO and GECO can be used advantageously in a situation where the buildup of a static charge is a problem. Certain non-black floor mats, printer/copier rolls, and computer static dissipation pads are examples.

Including a conductive carbon black, N472, in a CO compound can give volume resistivity values in the conductive range, as low as 1.6×10^3 ohm-cm, as shown in Figure 7.10.

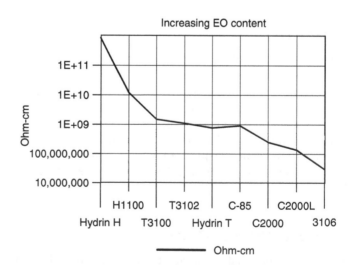

FIGURE 7.9 Volume resistivity values as low as 10^8 ohm-cm.

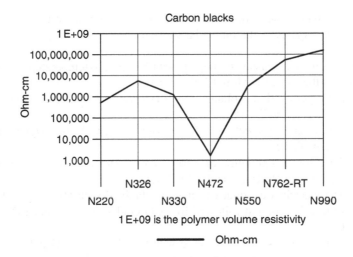

FIGURE 7.10 Volume resistivity values as low as 1.6×10^3 ohm-cm.

7.12 FORMULATION DESIGN: GENERAL

The design of CO/ECO/GECO compounds is very similar to some of the more common elastomers such as CR, CM, NBR, and so on. Common carbon blacks, white fillers, ester plasticizers, and process aids can be used, along with various antioxidants (metal dithiocarbamates, amines, and imidazoles) and curatives appropriate to the class of elastomer and application.

Plasticizers, which have a polarity similar to that of the polymer, should be used to a maximum of approximately 30 pphr, depending on the filler level. These plasticizers are normally of the diester and ether families, examples of which are DOP (dioctyl phthalate), DIDP (diisodecyl phthalate), TOTM (trioctyl trimellitate), and DBEEA (di (butoxy-ethoxy-ethyl) adipate). Plasticizers with low volatility, such as Struktol WB300 and KW400, Paraplex G25 and TP795 provide better aging properties as well as better low-temperature properties. The best low-temperature properties, as measured by Gehman T_{10} in a Gechron 3102 compound with 20 pphr of plasticizer, may be obtained with DBP ($-42°C$), DOA ($-43°C$), TP-95 ($-44°C$), TP-795 ($-44°C$), and Vulkanol OT ($-42°C$). Paraplex G25 and Struktol WB300 have the lowest extraction, and thus the highest volume swell in Fuel C, whereas DOA, DOP, DOS, Hercoflex 600, and Vulkanol OT have a higher extraction and thus lower volume swell. All of these characteristics of plasticizers must be considered when selecting one, or a combination, to achieve the desired properties for any given application.

7.13 POLYMER SELECTION

The choice of the elastomer depends entirely on all of the application requirements, provided that the customer supplies complete information. Often a compromise is necessary to balance the various needs and priorities that have to be given to the more important aspects of the particular application.

1. Very good fuel resistance coupled with better low-temperature resistance indicates the use of Hydrin T3102 or T3100.
2. Minimum permeation is obtained with a CO or GCO elastomer followed by Hydrin T3100 and T3105.
3. Maximum ozone protection is found with GCO, CO, then Hydrin T3100 and Hydrin T3105.
4. The best low-temperature properties are obtained with Hydrin T3106 followed by the Hydrin T and T3000 series.
5. The best electrical conductivity may be obtained with Hydrin T3106 with the highest EO content followed by ECO grades and medium EO content GECOs.
6. Maximum heat resistance is found with the GCO and CO polymers. The very high AGE content elastomers with the Zisnet F-PT curative provide the better reversion resistance based on elongation change.
7. High resilience similar to that of natural rubber may be obtained with ECO or GECO elastomers with medium or low ECH contents, Hydrin T3106 being the best. Conversely, very good dampening is achieved with CO, GCO, and GECO polymers with high ECH contents.

7.14 CURATIVES

The homopolymer (CO), and the copolymer of (ECO), are fully saturated, so that a cure system compatible with chlorine-containing polymers must be employed. The GCO and GECO polymers, containing an unsaturated allyl side chain, can be readily cured by sulfur or a peroxide system as well as the halogen compatible cure systems.

For CO/ECO elastomers, unique cure packages are often required. Ethylene thiourea (ETU) and lead oxide was the cure system that replaced the original Diak 1; however, changing environmental regulations has led to the use of nonlead cure systems. Ethylene thiourea is a goitrogen, teratogen, and carcinogen. Moreover, the effect of lead on mental deterioration is not an acceptable condition, which eliminates its use in most factories. Work begun in the early 1990s has led to the development of nonlead and nontoxic cure systems, which are very effective alternatives [3].

Three viable cure systems for the CO and ECO elastomers are: triazine (Zisnet F-PT), thiadiazole (Echo MPS), and bis-phenol (Dynamar 5157/5166). An overview of the systems is given in Table 7.9 [5].

The cure rate of the triazine system is readily adjustable as seen in Table 7.10. The triazine, (Zisnet F-PT), curative may be changed by adding varying amounts of certain key ingredients. When using 0.8 pphr of the triazine, 0.2–1.0 pphr of DPG acts as an accelerator and 0.2–1.0 pphr of either Santogard PVI or Vulkalent E/C retards the cure and gives better scorch safety.

The accelerator DPG (diphenylguanidine) is very effective at curing temperatures. Santogard PVI (N-(cyclohexylthio) phthalimide) or CTP is very efficient at

TABLE 7.9
Alternative Nonlead/ETU Cure Systems

Characteristic	Triazine	Thiadiazole	Bis-phenol
Disadvantages	Slightly lower tensile	Less scorch time—w/o barium carbonate (heavy metal)	Expensive
Advantages	Less toxic	Less toxic	Less toxic (using Dynamar RC-5251Q)
Adhesion to FKM	Fair to good	Fair to good	Excellent

processing temperatures, and for prolonging shelf life, hence the two ingredients are used together to gain a good balance of scorch safety versus cure rate. Although 1 phr of CTP provides more scorch safety and increased shelf life, levels above 0.5 phr are not recommended because of the offensive odor CTP emits during and after curing. Vulkalent E/C (sulfonamide derivative with 5% $CaCO_3$ and 1%–2% oil) is a very effective replacement for the CTP, and does not have the odor problem.

DTDM (4,4′ dithio-bis-morpholine or Sulfasan R) is the best retarder for the less desirable ETU cure system. If this mechanism is employed, a level of 0.5 pphr provides sufficient scorch safety for most processes.

The versatility of curing systems that may be employed with a GECO such as Hydrin 3105 is illustrated in Table 7.11. The Zisnet F-PT curative provides the best overall balance of properties for both processing and aging of the vulcanizate. The Echo system is also good and is fast curing. The fastest cure system is with Vulcup 40KE, but the oil and fuel aged properties are not desirable. Imadacure KE provides good aged properties but is slow curing and gives lower tear strength.

7.15 ACID ACCEPTORS

Chlorine-containing elastomers require an acid acceptor such as a metal oxide to scavenge any free chlorine created during vulcanization. Three to five parts of a precipitated calcium carbonate, sometimes coupled with three parts of magnesium

TABLE 7.10
Cure Rate Adjustment

Ingredients, phr	Function	Normal Cure Rate	Faster Cure Rate	Slower Cure Rate	Combination Cure Rate
Calcium carbonate	Acid acceptor	5.0	5.0	5.0	5.0
Zisnet F-PT	Curative	0.5–1.0	0.8	0.8	1.1
DPG (diphenylguanidine)	Accelerator	0	0.2–1.0	0	0.3
CTP (Santogard PVI) Vulkalent E/C	Retarders	0	0	0.2–1.0; 0.2–1.0	0.2–0.6; 0.5–1.0

TABLE 7.11
Hydrin Terpolymer 3105 Cure Systems Study

Curative	ETU/Pb	ETU/MgO	Zisnet F-PT	Echo	Diak	Imadazole	Peroxide	Sulfur
Formulations								
Hydrin 3105	100	100	100	100	100	100	100	100
N330	50	50	50	50	50	50	50	50
Stearic acid	1	1	1	1	1	1	1	—
NBC, 70%	3	3	3	3	3	3	3	3
Dynamar PPA 790	1	1	1	1	1	1	1	1
Calcium carbonate	—	—	5	—	—	—	—	—
Red lead, 90%	6	—	—	—	6	6	—	—
Magnesium oxide	—	5	3	—	—	—	—	—
Barium carbonate, 80%	—	—	—	9	—	—	—	4
Potassium stearate	—	—	—	—	—	—	3	—
Calcium oxide	—	—	—	—	—	—	2.5	—
Sartomer SR350	—	—	—	—	—	—	3	—
Vulcup 40KE	—	—	—	—	—	—	1.2	—
Imadacure KE	—	—	—	—	—	2	—	—
Echo AP	—	—	—	2.8	—	—	—	—
Diak #1	—	—	—	—	1	—	—	—
Ethylene thiourea, 75%	1	1	—	—	—	—	—	—
Zisnet F-PT	—	—	1	—	—	—	—	—
Sulfur	—	—	—	—	—	—	—	0.8
MBT	—	—	—	—	—	—	—	2
TMTM	—	—	—	—	—	—	—	1
TOTAL	162	161	164	166.8	162	163	164.7	161.8

Rheometer, 100 cpm, 3° arc, 60 min at 175°C								
MH, N·m	11	9	13	15	15	15	10	9
ML, N·m	2	2	3	3	3	2	2	2
ts₂, minutes	1.3	1.9	1.4	1.1	0.9	3	1.1	2.6
t₉₀, minutes	32	31	20	14	30	30	7	48
Physical properties, cured 40 min at 175°C								
Shore A	72	76	76	75	76	79	70	74
Modulus at 100%, MPa	5.2	6.3	6	7.1	8.3	8.3	5.0	5.6
Tensile strength, MPa	18.7	16.2	18.6	17.3	16.7	19.4	14.9	19.0
Elongation, %	280	215	255	255	185	235	265	340
Tear, Die C, kNm	48.2	40.8	34.7	36.8	33.4	27.3	23.1	35.9
Compression set B, Plied disks, 22 h at 125°C								
Set, %	19	44	17	18	20	14	20	60
Aged in air 72 h at 150°C-change								
Hardness, pts	11	6	6	7	4	4	4	10
Modulus at 100%, %	58	27	37	37	-26	38	42	134
Tensile strength, %	-20	-33	-12	3	-43	-2	1	-19
Elongation, %	-39	-40	-33	-31	-16	-32	-34	-62
Aged in ASTM #3 Oil 72 h at 150°C-change								
Hardness, pts	-8	-11	-9	-2	-3	-6	-13	-3
Modulus at 100%, %	7	-11	6	2	-1	17	-7	28
Tensile strength, %	-13	-16	-10	-26	4	-9	-13	-18
Elongation, %	-23	-14	-16	-37	24	-30	-19	-38
Volume, %	15	15	14	13	11	13	13	8
Aged in Fuel C 48 h at 23°C-change								
Hardness, pts	-16	-16	-17	-16	-14	-15	-23	-18
Modulus at 100%, %	-2	3	1	4	-100	-4	2	3
Tensile strength, %	-61	-60	-51	-46	-51	-49	-59	-49
Elongation, %	-55	-53	-45	-53	-49	-51	-57	-54
Volume, %	45	44	44	53	25	42	45	38

oxide, work well with the triazine curing mechanism. The formerly used ETU system, as well as Diak #1, normally employed red lead as an acid acceptor, or magnesium oxide, which results in much higher compression set. Barium carbonate is used with Echo AP. Peroxide cured compounds require calcium oxide with potassium stearate to counteract the halogen.

7.16   PROTECTIVE AGENTS

As with most elastomers, the inclusion of an antioxidant in epichlorhydrin compounds enhances the aging resistance of the vulcanizates. Table 7.12 presents a comparison of various antioxidants in a Gechron 2000 compound. There is considerable deterioration with no antioxidant, but in order of effectiveness, NBC is best followed by PBNA, TMQ, and MBI. Although this study was with an ETU/red lead system, it is expected that similar results would be obtained using Zisnet F-PT as the curative.

Antiozonants may also be included in CO, ECO, and GECO compounds for improved resistance to ozone. The selection of the proper antiozonant is important as some will actually reduce ozone resistance, as may be seen in Table 7.13. NBC and MBI are the only effective ones tested, although methyl niclate would also be good.

TABLE 7.12
Antioxidants in Gechron 2000 ECO

Antioxidant	None	TMDQ	PBNA	MBI	NBC
Formulations					
Gechron 2000	100	100	100	100	100
Tin stearate	2	2	2	2	2
N550	40	40	40	40	40
Red lead	5	5	5	5	5
ETU	1.2	1.2	1.2	1.2	1.2
AgeRite Resin D	—	2	—	—	—
PBNA	—	—	2	—	—
Antioxidant MBI	—	—	—	2	—
NBC	—	—	—	—	2
TOTAL	148.2	150.2	150.2	150.2	150.2
Physical properties—cured 30 min at 155°C					
Shore A, pts	65	63	66	67	65
Modulus at 100%, MPa	2.8	2.4	2.4	2.9	2.6
Tensile, MPa	12.9	12.2	12.4	11.9	12.3
Elongation, %	500	540	530	440	450
Aged in air 192 h at 140°C-change					
Hardness, pts	−20	+2	−7	+13	−1
Modulus at 100%, %	−72	+29	−29	+196	+18
Tensile, %	−81	−36	−34	−26	−13
Elongation, %	−36	−59	−35	−77	−35

TABLE 7.13
Antiozonants in Gechron 2000/1100

Formulation	Gechron 2000	70						
	Gechron 1100	30						
	Stearic acid	1						
	Red lead	8						
	N550	40						
	ETU	1.2						
	Antiozonant	2						
Antiozonant	None	PBNA	IPPD	NBC	EMDQ	MBI	Wax

Ozone resistance in 80 pphm O$_3$, 120 h at 40°C

20% extension	NC	C1	C3	NC	C2	NC	NC
40% extension	NC	C2	C3	NC	C3	NC	NC
60% extension	A1	C2	C3	NC	C3	NC	A1
Rating	No. of cracks	Rating	Crack size				
NC	None	0					
A	Few	1	Tiny				
B	Several	2	Small				
C	Many	3	Large				

7.17 PROCESSING SECTION—MIXING

Polyepichlorohydrin compounds can be mixed either on a mill or in an internal mixer. The mixing procedure for either method is virtually the same as for SBR, NBR, and EPDM compounds. Because of shelf-life concerns, a two-pass mix may be preferable.

7.17.1 MILLING

Mill roll surface preparation is extremely important. Polyepichlorohydrin type ECO and GECO compounds tend to stick to mill roll surfaces when certain residues from previously processed compounds, based on different polymer types, remain on the rolls and if the mill roll is too cold. One of the worst residues seems to come from an EPDM compound cured with a peroxide. The mill rolls must be consistently thoroughly cleaned to avoid problems with sticking. Millathane Glob has been an effective, safe method of cleaning the rolls.

If more mill release is wanted, then coating the mill with stearic acid before milling may improve the release of the material from the roll surface, initially. Adding MSC to the compound also should help the mill release of the material.

Polyepichlorohydrin compounds process best if the mill rolls are residue free and hot, 60°C–80°C. Both suggestions alleviate the problem of the compound sticking to the mill roll surfaces.

If the compounder has the freedom to make changes to the formulation, then substituting one of the Hydrin C2000 series for Hydrin C or Hydrin T3000 series for

TABLE 7.14
Banbury Mix Procedure

Temperature (°C)	Approximate Time, Minutes	Actions
60 max.	0	Add polymer, carbon black, and all masterbatch and cure chemicals
85	1[a]	Add all the plasticizer (if any)
90	2	Raise ram, scrape, lower ram
105	3[b]	Drop batch onto the mill with clean roll surfaces. Regulate rotor speed to control to a maximum drop temperature 115°C

[a] Just before peak power draw, or max. amperage.
[b] When the batch has "come together."

Hydrin T will also provide much better release from the metal surfaces of the mixer and mill rolls.

7.17.2 PROCEDURE FOR INTERNAL MIXERS

Polyepichlorohydrin compounds are typically mixed in a two-stage process. Upside-down and regular mix procedures have been used to make quality compounds, with good dispersion of the other ingredients. One-pass mixing is also feasible with proper control of the chamber and rotor temperatures and rotor speed. A one-pass mix procedure is given in Table 7.14.

7.18 EXTRUSION

Epichlorohydrin compounds extrude very well, especially if the amount of the filler is large as this has an impact on the extrusion characteristics, as it does with any elastomer. The use of relatively high loadings of carbon black also helps to give the extruded compound a smooth surface. The high loadings also help to provide sufficient green strength, which allows a tube to hold its extruded shape and provide enough stiffness for subsequent braiding of hose.

Laboratory batches based on the formulas in Table 7.15, with relatively high loadings of different carbon black types (N330, N550, N762, and N990) were extruded into tubes using a Davis-Standard 38 mm cold feed extruder, with a 20 to 1 L/D screw ratio. Almost all of the compounds were processed well, with smooth surfaces, good feed rates, and acceptable head pressures (Table 7.16). All of the tubes maintained their extruded shape well.

Laboratory extrusion results do not always predict compound performance on production extruders; however, good correlation of laboratory extrusions versus those performed on a full-scale facility has been obtained. A recommended temperature profile is given in Table 7.17.

TABLE 7.15
Compounds for Laboratory Extrusion Trial

Compound	1	2	3	4	5
Hydrin T3102	100	100	100	100	100
N330	55	—	—	30	—
N550	—	60	35	—	60
N990	—	—	130	50	—
Polyester adipate	5	5	5	5	5
DBEEA	5	5	5	5	5
2-Mercapto-4(5)-methylbenzimidiazol	1	1	1	1	1
Potassium stearate	3	3	3	3	3
Calcium oxide	4	4	4	4	4
Dicumyl peroxide	1.3	1.3	1.3	1.3	1.3
Inhibited TMPTMA	—	—	—	—	5

7.19 MOLDING

CO/ECO/GECO compounds can be successfully compression, transfer, injection, or injection/transfer molded. Injection molding may be the most commonly used method. When molding with metal inserts, Ty-Ply BN, Chemlok 250, and Chemlok 855 are some of the recommended adhesives for bonding ECO to metal.

Compounds for compression molding tend to require a higher viscosity, typically a ML $1+4$ Mooney at 100°C of 90–120. This is usually accomplished with a more highly reinforcing filler and a higher loading. As an example, 50 phr of N330 or 90 phr of N762 carbon black coupled with 2–5 phr of a plasticizer, such as TOTM or DBEEA, should help the compound flow easier to fill the cavity without being too soft, which could lead to air entrapment.

Compounds for transfer molding normally have lower viscosities, depending on the sprue, gate, and runner sizes and locations, as well as the cavity shape. Process aids such as Struktol WA48 or WB222 or occasionally waxes are typically used in

TABLE 7.16
Results of Lab Extrusion

Compound	1	2	3	4	5
Die swell, %	27	36	14	9	3
Linear shrinkage, %	0.5	1.0	0.8	2.7	3.3
Rate, m/min	1.98	2.1	2.52	1.52	2.22
Head pressure, MPa	13.1	12.8	18.6	13.2	13.4
Motor current, amps	26	30	41	26	30
Extruder temperature, °C	110	110	127	104	110
Visual appearance	Smooth	Smooth	Smooth	Smooth/dull	Lumpy

TABLE 7.17

Recommended Extruder Profile Temperatures, °C

Zone 1	Zone 2	Zone 3	Screw	Head	Die
70–90	75–95	80–100	70–90	80–100	90–120

an attempt to allow the compound to flow even easier; however, there is the danger of poor knitting if too much is used.

Compounds for injection molding can vary substantially in filler to plasticizer loading ratios, and hence have a wide range of viscosity values. Here again, gate, sprue, and runner sizes and cavity shapes are important factors in determining the requirements to successfully fill the mold cavities. A lower viscosity than with transfer compounds is normally desired. The heating and mastication of the rubber compound through the barrel and screw and the runner system is often sufficient as to not require a low viscosity, and it may not mold as well as a higher viscosity compound.

7.20 APPLICATIONS

Applications for epichlorohydrin are primarily automotive due to the key properties including heat, fuel, ozone, and permeation resistance coupled with very good low temperature and dynamic properties. It is also used in the roller and mechanical goods industries where the electrical properties and chemical resistance of ECO are of advantage. Basic properties of epichlorohydrin compounds given in Table 7.18 will help in determining if this elastomer is suitable for a given application.

TABLE 7.18

General Properties

	Typical Properties	Range
Tensile strength, MPa	10	8–18
Elongation, %	300	200–800
Fuel C swell, %	35	20–40
IMR 903 Oil swell, %	10	0–12
	Properties	Description
Heat resistance-change,		
Tensile and	−30	125°C–150°C
Elongation, %	−50	
Low temperature flex	−40°C	Down to −50°C
Ozone resistance	Excellent	Passes 70 h at 100 pphm
Resistivity, ohm-cm (naturally static-dissipative)	1×10^4 (1×10^8)	$1 \times 10^{1\ to\ 9}$
Tear strength, kNm	49	26.3–61.3
Dynamic and flex, cycles	30,000	Typical

7.20.1 Air Ducts

A large use of epichlorohydrin is in automotive air duct tubing because of the excellent resistance to fuel and gas permeation, fuel, heat, and low-temperature resistance of the polymer. Table 7.19 illustrates recommended compounds for Ford WSE M2D470-A4 Air Ducts. In addition, three other Hydrin C2000L compounds,

TABLE 7.19
Recommended Formulas for Fuel and Oil Resistant Ford WSE M2D470-A4 Air Ducts

Hydrin C2000L	80	80	
Hydrin H1100	20	20	
N762	95	95	
Paraplex G-50	5		
TP95 Plasticizer	10	10	
Plasthall P-670	—	5	
Stearic acid	1.0	1.0	
Naugard 445	1.0	1.0	
Vanox MTI	0.5	0.5	
Calcium carbonate	5.0	—	
Magnesium oxide	2.0	3.0	
Zisnet F-PT	0.5	—	
Santogard PVI	0.3	—	
GND-75 (75% ETU in ECO)	—	0.8	
TOTAL	220.3	216.3	
Mooney viscosity			
ML 1 + 4 at 100°C	65	68	
Mooney scorch, ML at 125°C			
Minimum	53	59	
T_5, minutes	8.7	8.8	
T_{35}, minutes	14.5	19.0	
Rheometer, 20 min motor at 190°C			
MH, N·m	1.2	1	
ML, N·m	2	0.3	
Ts_2, minutes	1.0	1.7	
T_{90}, minutes	9.7	9.3	
			Ford WSE M2D470-A4
Originals, cured 20 min at 190°C			
Hardness, Shore A, pts	71	71	70–80
Modulus at 100%, MPa	3.7	3.6	3.3
Tensile strength, MPa	10.8	11.1	9.0
Elongation, %	370	370	250
Tear Die C, kNm	40.3	35	30

(*continued*)

TABLE 7.19   (Continued)
Recommended Formulas for Fuel and Oil Resistant Ford WSE M2D470-A4 Air Ducts

					Ford WSE M2D470-A4
Low temperature brittleness, ASTM D2137					
No cracks after 3 min at °C	−40		−43		−35
Compression set B, %, cured 30 min at 200°C					
70 h at 121°C, %					70
22 h at 135°C, %	45		57		
Low temperature retraction, TR10%					
In methanol, °C	−42		−42		−38
	Actual	Change, %	Actual	Change, %	
Aged in air oven 1000 h at 110°C					
Hardness, Shore A, pts	81	10	80	9	+15
Modulus at 100%, MPa	6.2	67	7.0	96	—
Tensile strength, MPa	10.7	−1	11.01	0	−50
Elongation, %	200	−46	180	−51	−70
Aged in ASTM #3 Oil 168 h at 130°C					
Hardness, Shore A, pts	79	8	81	10	+/−10
Modulus at 100%, MPa	6.62	78	7.52	110	—
Tensile strength, MPa	10.55	−2	9.58	−14	−30
Elongation, %	170	−54	140	−62	−70
Volume change, %		2		4	+10
Aged in Fuel C 70 h at 23°C					
Hardness, Shore A, pts	55	−16	58	−13	−25
Modulus at 100%, MPa	3.86	4	4.62	29	—
Tensile strength, MPa	9.58	−11	8.69	−22	−40
Elongation, %	230	−38	180	−51	−70
Volume change, %		25		30	+38
Aged in 85% Fuel B/15% methanol 70 h at 23°C					
Hardness, Shore A, pts	50	−21	53	−18	−35
Modulus at 100%, MPa	3.79	2	5.03	40	—
Tensile strength, MPa	7.65	−29	7.38	−34	55
Elongation, %	180	−51	140	−62	−70
Volume change, %		45		54	+80
Weight change, %		23		30	

with various cost structures, are presented in Table 7.20 with the requirements of specifications from Chrysler, Ford, and General Motors.

7.20.2 FUEL HOSE

Fuel Line hose is the largest application for epichlorohydrin, again due to the excellent balance of fuel, permeation, low temperature, ozone, and weathering

TABLE 7.20
Air Duct Formulation Recommendations

	Basic	Lower Cost	Value Enhanced	Chrysler MS-BZ120	GM-CPE ECO 026	Ford WSE-M2D470
Hydrin C2000L	100	100	100			
Bromobutyl X2	4	—	5			
Chlorobutyl HT1068	—	5	—			
N330	—	15	110			
N234	40	—	—			
N762	60	100	—			
Cabosil M5	15	—	—			
Paraplex G57	—	—	40			
Paraplex G50	15	5	—			
TP-95	15	—	—			
DIDP	—	10	10			
Stearic acid	1.5	1	1.5			
Maglite D	3	—	—			
Irganox MD-1024	1.5	—	—			
NBC	—	0.5	—			
Calcium carbonate	5	5	10			
Santogard PVI	—	0.4	—			
Vulkalent E/C	1	—	—			
Vulkanox MB-2/MG/C	—	1	—			
Zisnet F-PT	1.1	0.5	0.5			
DPG	0.3	0.7	1			
TOTAL	262.4	244.1	279.5			

(continued)

TABLE 7.20 (Continued)
Air Duct Formulation Recommendations

	Basic	Lower Cost	Value Enhanced	Chrysler MS-BZ120	GM-CPE ECO 026	Ford WSE-M2D470
Mooney viscosity						
ML 1 + 4 at 100°C	154	95	82			
Mooney scorch ML at 125°C						
Minimum	157	98	80			
T_5, min	1.6	3.1	4.4			
Monsanto rheometer at 200°C						
MH, N·m	10.7	8.9	5			
ML, N·m	3.4	2.2	1.6			
Ts_2, minutes	0.7	1.2	1.2			
T_{90}, minutes	4.1	10.1	10.7			
Originals cured at 200°C	12 min	20 min	20 min			
Hardness, Shore A, pts	79	77	74	75–85	70–80	70–80
Modulus at 100%, MPa	5	7.2	4.7	—	4.5	3.3
Tensile strength, MPa	10.2	8.3	7.4	8.3	—	9.0
Elongation, %	230	140	180	200	225	250
Tear, Die C, kN/m	31.5	29.8	24.9	25.9	21.9	30
Compression set B						
70 h at 121°C						70
22 h at 100°C						
70 h at 100°C			39			

						1000 h at 110°C
Aged in air oven 70 h at 125°C-change						
Hardness, pts	6	8	8	+/-10	0 to 15	+18
Modulus at 100%, %	23	19	29	—	25	—
Tensile strength, %	-14	4	-12	-20	—	-50
Elongation, %	-22	-21	-33	-50	-55	-72
Aged in oil 70 h at 125°C-change	ASTM #3	IRM 903	IRM 903	ASTM #3	ASTM #3	168 h at 130°C
Hardness, pts	4	0	4	-10	0-10	+/-10
Modulus at 100%, %	8	10	37	—	40	—
Tensile strength, %	-17	12	21	-10	—	-30
Elongation, %	-22	0	-17	-50	-45	-70
Volume change, %	1.1	6	0.1	+15	2-25	+10
Aged in Fuel C 48 h at 23°C-change						70 h
Hardness, pts	-20	-19	-20	-25	0 to -25	—
Modulus at 100%, %	-18	-26	-24	—	-30	—
Tensile strength, %	-21	-43	-29	-30	—	-40
Elongation, %	-22	-43	-22	-40	-50	-70
Volume change, %	23	35	23	+40	0-40	+38
Low temperature brittleness, ASTM D2137						
F17, 3 min at −40°C	Pass	Pass	No cracks	No cracks	—	—
F16, 3 min at −35°C			—	—	No cracks	—
TR 10°C					—	-35
Ozone resistance, 70 h at 50 pphm, 40°C						
Cracking	Pass	Pass	No cracks	No cracks	No cracks	No cracks
Aged in air 192 h at 150°C on mandrel						
Straighten	Pass	Pass	Fail	No cracks	No tears	—

resistance of this elastomer [7]. The very stringent EPA permeation requirements often dictate even better impermeability than either CO or ECO; hence an inner veneer of FKM is employed, with a tie layer and cover of ECO to offer additional protection to the veneer. Please note that if improved adhesion to the FKM compound is required, substitute the Dynamar or a peroxide cure system [8]. Recommendations for a Chrysler fuel hose are given in Table 7.21; a General Motors tubing in Tables 7.22 and 7.23; and a Ford hose in Table 7.24.

TABLE 7.21
Chrysler MSEA 235 Recommended Formula

Hydrin C2000	100		
N990	40		
SRF-LM N762	45		
N326	15		
DOP (or DIDP)	5		
Paraplex G-50	3		
Methyl Niclate	0.6		
NBC	0.9		
Stearic acid	0.5		
Calcium hydroxide	2.0		
Magnesium oxide	3.0		
Zisnet F-PT	1.0		
Santogard PVI	1.0		
TOTAL	217.0		
Mooney viscosity			
ML 1 + 4 at 100°C	115		
Mooney scorch, ML at 125°C			
Minimum	94		
T5, minutes	5.0		
Monsanto rheometer, 12 min motor at 200°C			
MH, $N \cdot m$	12		
ML, $N \cdot m$	2.9		
Ts_2, minutes	0.7		
T_{90}, minutes	5.6		
Monsanto rheometer, 30 min motor at 162°C			
MH, $M \cdot m$	11.4		
ML, $M \cdot m$	3.2		
Ts_2, minutes	1.8		
T_{90}, minutes	15.1		
Originals, cured 15 min at 162°C		Tube and Tie Layer	Cover
Hardness, Shore A, pts	78	70–80	70–80
Modulus at 100%, MPa	5		
Tensile strength, MPa	9.7	6.9	8.27
Elongation, %	230	200	200

TABLE 7.21 (Continued)
Chrysler MSEA 235 Recommended Formula

Aged in air oven 70 h at 150°C		Change %	Change %	Change %
Hardness, Shore A, pts	83	5	±10	±10
Modulus at 100%, MPa	7	40		
Tensile strength, MPa	9.5	−3	−10	−10
Elongation, %	150	−35	−50	−50
Aged in ASTM #3 Oil 70 h at 150°C				
Hardness, Shore A, pts	82	5		±10
Modulus at 100%, MPa	8.3	68		
Tensile strength, MPa	11.3	16		−10
Elongation, %	140	−39		−50
Volume change, %		5		+15
Aged in Fuel C 48 h at 23°C				
Hardness, Shore A, pts	60	−18	−25	−20
Modulus at 100%, MPa	4.8	−3		
Tensile strength, MPa	7.9	−19	−40	−40
Elongation, %	170	−26	−40	−40
Volume change, %		26	+40	+40
Weight change, %		14		

Ozone resistance, 70 h at 40°C in 50 pphm, 50% Elong., Precond. 24 h at 23°C

7X magnification	Pass

TABLE 7.22
Recommended Formula for GM 6148

Oil, and fuel, resistant tube

Hydrin C2000	100
N762	85
Hi-Sil 233	20
DIOP	8
Paraplex G-50	5
Irganox MD-1024	0.5
NBC	0.3
Methyl Niclate	0.7
Stearic acid	1.0
Calcium carbonate	5.0
Magnesium oxide	3.0
Zisnet F-PT	0.9
Santogard PVI	1.0

(*continued*)

TABLE 7.22   (Continued)
Recommended Formula for GM 6148

		Change, %	GM 6148 Specification	Change, %
Antimony oxide	5.0			
TOTAL	235.4			
Mooney scorch, ML at 200°C				
Minimum	104			
T_5, minutes	5.8			
Rheometer at 200°C				
MH, N·m	10.9			
ML, N·m	2.5			
Ts_2, minutes	1.0			
T_{90}, minutes	6.9			
Results				
Originals, cured 10 min at 200°C				
Hardness, Shore A, pts	80		65–80	
Modulus at 100%, MPa	5.5		4.5	
Tensile strength, MPa	9.3		8.3	
Elongation, %	200		200	
Compression set B, % 22 h at 100°C				
Set, %	22		29	
Aged 70 h at 150°C in test tube air				
Hardness, Shore A, pts	89	9		
Modulus at 100%, MPa	7.2	32	4.5	
Tensile strength, MPa	8.1	−13	6.9	
Elongation, %	140	−30	125	
Aged 70 h at 150°C in ASTM #3 Oil				
Hardness, Shore A, pts	84	4		
Modulus at 100%, MPa	6.9	27		
Tensile strength, MPa	8.4	−10	6.6	−30
Elongation, %	130	−35	100	−60
Volume change, %	5		+2 to +10	
Weight change, %	3			
Aged 70 h at 23°C in Fuel C				
Hardness, Shore A, pts	60	−20		
Modulus at 100%, MPa	4.5	−18		
Tensile strength, MPa	5.7	−39	5.4	−40
Elongation, %	130	−35	130	−40
Volume change, %	24		+20 to +35	
Weight change, %	13			
Gasoline extractables, GM9061P g/m^2			2.32	

TABLE 7.23

Fuel Hose Recommendation for GM 289M Hose

Hydrin C2000	75
Hydrin H	25
Chlorobutyl 1068	10
N330	80
DIDP	15
Paraplex G-50	15
Vulkanox MB-2/MG/C	1.0
Stearic acid	1.0
Calcium carbonate	5.0
Zisnet F-PT	0.6
DPG	0.7
TOTAL	228.3
Mooney viscosity at 100°C	
ML 1 + 4	72
Mooney scorch, ML at 125°C	
Minimum	68
T_5, minutes	3.0
T_{35}, minutes	6.2
Rheometer at 190°C	
MH, N · m	4.8
ML, N · m	1.5
Ts_2, minutes	1.5
T_{90}, minutes	24.6

(continued)

TABLE 7.23 (Continued)
Fuel Hose Recommendation for GM 289M Hose

			GM 6289 M SAE J30 (30R7)			
			Cover		Tube	
	Actual	Change	Actual	Change	Actual	Change
Originals, cured 25 min at 190°C						
Hardness, Shore A, pts	69					
Modulus at 100%, MPa	3.9					
Tensile strength, MPa	11.2		8.3		8.3	
Elongation, %	300		200		225	
Aged 70 h at 125°C in air						
Hardness, Shore A, pts	80	11				
Modulus at 100%, MPa	6.9	79				
Tensile strength, MPa	10.9	−3				−30
Elongation, %	170	−43				−40
180° Bend			Pass			Pass
GM4486P and ASTM D518, Ozone resistance, 168 h at 38°C, 20% elong., 100 pphm, 2× mag.			No cracks			No cracks
Aged 70 h at 125°C in IRM 903 Oil						
Hardness, Shore A, pts	68	−1				

Modulus at 100%, MPa	5.2	34	6.6	6.6	-30
Tensile strength, MPa	11.9	6	150		-40
Elongation, %	250		0 to +60	-5 to +15	
Volume change, %	5				
Weight change, %	3				
Aged 48 h at 23°C in Fuel C					
Modulus at 100%, MPa	3.5	-9			
Tensile strength, MPa	6.2	-45			
Elongation, %	170	-43			
Hardness, Shore A, pts	46				
Volume change, %	38	-23	0 to -60		
Low-temperature flex, 70 h at -40°C					
Bend within 4s between flat plates	Pass		No cracks		
Fuel, heat, and ozone resistance, adhesion test					
Aged in Fuel C for 48 h at 23°C, + air oven for 96 h at 100°C					
Repeat above steps using fresh Fuel C					
Ozone per above	Pass				
Adhesion, 180° peel, N/mm	1.0				

TABLE 7.24
Ford M2D241A Fuel and Oil Resistant Tubing Recommended Formula

Hydrin C2000	75
Hydrin H	25
N762	70
TP-95	3
Paraplex G-50	9
Vulkanox MB-2/MG/C	1.0
Span 60 (or stearic acid)	2.0
Calcium carbonate	5.0
Zisnet F-PT	1.0
Santogard PVI	1.0
DPG	0.3
TOTAL	192.3
Mooney viscosity	
ML 1 + 4 at 100°C	70
Mooney scorch, ML at 125°C	
Minimum	58
T5, minutes	11.9
T35, minutes	0
Monsanto rheometer, 15 min motor at 200°C	
MH, N·m	13.4
ML, N·m	2.7
Ts_2, minutes	2.1
T_{90}, minutes	10.4

	Actual	Change, %	FORD SPEC M2D241-A Actual	Change, %
Originals, cured 12 min at 200°C				
Hardness, Shore A, pts	66		65–75	
Modulus at 100%, MPa	4.5		8.3	
Tensile strength, MPa	12.2		200	
Elongation, %	300			
Tear strength, Die C, min kNm	33.3		26.3	
Compression set B				
22 h at 100°C, %	22		30	
Aged in air oven 168 h at 150°C				
Hardness, Shore A, pts	68	2		
Modulus at 100%, MPa	4.2	−6		
Tensile strength, MPa	4.7	−62		−70
Elongation, %	110	−63		−65
Aged in ASTM #3 Oil 70 h at 150°C				
Hardness, Shore A, pts	68	2	−10 to +5	
Modulus at 100%, MPa	6.7	49		
Tensile strength, MPa	11.9	−2		−15
Elongation, %	170	−43		−45
Volume change, %	3		−5 to +15	
Weight change, %	1			
Aged in Fuel C 70 h at 23°C				
Hardness, Shore A, pts	50	−16	−20 to 0	
Modulus at 100%, MPa	4	−11		
Tensile strength, MPa	8.3	−32		−40
Elongation, %	200	−33		−35
Volume change, %	30		0 to +35	

(continued)

TABLE 7.24 (Continued)
Ford M2D241A Fuel and Oil Resistant Tubing Recommended Formula

		FORD SPEC M2D241-A
Volume change after drying in air 24 h at 70°C, %		−15 to +5
Extractables, ASTM D297		
Toluene, max., %	Expected to pass	10
Waxy hydrocarbons, max., %		0.5
Cold flexibility [on parts]		
FLTM BP 6–3, 70 h at −34°C		No cracks
Ozone rating [on parts]		
FLTM BP 101–1, 20% stretch, 7× magnification 70 h at 50 pphm, 38°C (24 and 48 h inspections)		No cracks

7.20.3 CURB PUMP HOSE

The fuel, ozone, low temperature, and weathering resistance of epichlorohydrin are ideal properties for a curb pump hose. In addition, the static dissipative properties, especially when a conductive carbon black is used, add to the safety of the hose. The EPA low-fuel vapor transmission requirements may dictate an FKM veneer lining, to which GECO will adhere, as well as being a low fuel transmission elastomer itself. Hose components are shown in Table 7.25.

TABLE 7.25
Curb Pump Hose

Compound	Cover	Static Dissipative Layer	Tie Layer
Hydrin T3100	100	—	—
Hydrin T3102	—	100	—
Hydrin T65	—	—	100
N762	85	—	85
N472	—	40	—
Hi-Sil 233	20	—	15
DIOP	8	—	5
Merrol P6412	5	—	5
Stearic acid	1	1	—
Potassium stearate	—	—	3
Calcium carbonate	5	5	—
Calcium oxide, 85%	—	—	4
Magnesium oxide	0.9	—	—
Antimony oxide	5	—	—
Vulkanox MB2	—	1	—
NBC	0.3	—	1
Methyl niclate	0.7	—	1
Irganox MD1024	0.5	—	—
TMPTMA	—	—	0.5
DiCup 40C	—	—	0.8
DPG	—	0.5	0.2
CTP	1	0.2	—
Zisnet F-PT, 85%	3	0.7	—
TOTAL	235.4	148.4	220.5
Mooney scorch, ML at 125°C			
Minimum viscosity	104	—	73
T_5, min	5.8	—	6
Rheometer, 3° arc, 160°C, 60 min chart			
ML, N·m	19.5	—	14.1
MH, N·m	85	—	40.7
Ts_2, minutes	1	—	2.5
Tc_{90}, minutes	6.9	—	36

(*continued*)

TABLE 7.25   (Continued)
Curb Pump Hose

Compound	Cover	Static Dissipative Layer	Tie Layer
Physical properties, cured	12 min at 200°C	—	50 min at 160°C
Hardness, Shore A, pts	80	—	75
Modulus at 100%, MPa	5.4	—	3.9
Tensile, MPa	9.3	—	8.6
Elongation, %	200	—	320
Aged 70 h at 150°C-change			
Hardness, Shore A, pts	9	—	14
Tensile, %	−13	—	−1
Elongation, %	−30	—	−41
Aged in oil 70 h at 150°C-change	IRM 903	—	ASTM # 3
Hardness, pts	4	—	5
Tensile, %	−10	—	24
Elongation, %	−35	—	−44
Volume, %	5	—	4
Aged in Fuel C-change	70 h	—	48 h
Hardness, pts	−20	—	−22
Tensile, %	−39	—	−27
Elongation, %	−35	—	−31
Volume, %	24	—	25
Dynamic O_3 resistance after aging 48 h at 23°C in Fuel C tested in 50 pphm O_3, 200 h at 40°C, 25% elongation, 30 cpm			
Rating	No cracks	—	—
Volume resistivity at 50% relative humidity, 23°C			
Reading, ohm-cm	—	3000	—
Adhesion to FKM Veneer, 180° Peel			
Adhesion, kNm	—		10.3

7.20.4 AIR CONDITIONING

The balance of impermeability, aging resistance, and low-temperature properties also makes epichlorohydrin an excellent candidate for air-conditioning hose and seals. Table 7.26 shows a recommended compound for an air-conditioning hose.

7.20.5 DYNAMIC COMPONENTS

The very stable, low hysteresis of ECO and GECO compounds over a broad range of temperatures suits them to be ideal polymers for dynamic applications. Natural rubber components subject to higher temperatures than 70°C may be replaced by ECO or GECO with excellent results up to 125°C. Table 7.27 gives a comparison of the dynamic properties of ECO and HNBR, which may be employed for even higher temperatures in the 150°C range.

TABLE 7.26
Hydrin H for Air Conditioning

Hydrin H	100	
N762	90	
DIDP	5	
Vulkanox MB-2/MG/C	1.0	
Stearic acid	1.0	
Calcium carbonate	10.0	
Zisnet F-PT	0.5	
DPG	1.0	
Total	208.5	
	Actual	Change, %
Mooney viscosity		
ML1 + 4 at 100°C	137	
Mooney scorch at 125°C		
Minimum viscosity	130	
t_5, minutes	1.6	
t_{35}, minutes	3.1	
Rheometer, 3° arc at 170°C		
ML, N·m	2.83	
MH, N·m	11.41	
ts_2, minutes	1.9	
t_{90}, minutes	28.5	
Original properties, cured 60 min at 170°C		
Hardness, Shore A, pts	83	
Modulus at 50%, MPa	5.65	
Modulus at 100%, MPa	11.17	
Tensile, MPa	12.41	
Elongation, %	130	
Explosive decompression		
Freon 134a, microscope	No bubbles	
Compression set B, plied discs, 22 h at 125°C		
Set, %	27	
Aged in PAG 72 h at 150°C		
Hardness, Shore A, pts	88	5
Tensile, MPa	13.58	9
Elongation, %	80	−38
Volume change, %	−3.5	
Weight change, %	−3.7	
180° Bend	Pass	

7.20.6 STATIC DISSIPATION

A light-colored Hydrin compound is given in Table 7.28, illustrating the ability of ECO to be used in non-black static dissipative applications such as mats or pads for sensitive electronic equipment.

TABLE 7.27
Dampening Characteristics of Hydrin versus Zetpol

Hydrin T3000	100	100	—	—
Zetpol 2030L	—	—	100	100
N762	—	90	—	—
N990	50	—	—	—
N774	—	—	20	70
TP-95	2	2	—	—
TOTM	—	—	5	5
Naugard 445	1.0	1.0	1.0	1.0
Stearic acid	1.0	1.0	1.0	1.0
Calcium oxide	5.0	5.0	—	—
Zinc oxide	0.5	0.5	3.0	3.0
TMTM	1.0	1.0	0.4	0.4
Sulfur	0.1	0.1	1.5	1.5
MBTS	—	—	1.5	1.5
Total	160.6	200.6	133.4	183.4
Mooney viscosity				
ML1 + 4 at 100°C	72	137	49	88
Mooney scorch ML at 125°C				
Minimum viscosity,	59	23	110	49
t_5, minutes	—	—	15.5	11.9
T_{35}, minutes	—	—	—	16.0
Rheometer, 3° arc, 30 min motor at 180°C				
MH, N·m	46	59	72	115
ML, N·m	18	27	7	10
ts_2, minutes	3.3	3.1	1.3	1.3
T_{90}, minutes	25.1	24.7	7.5	4.2
Original properties, cured 30 min at 180°C				
Hardness, Shore A, pts	44	75	55	75
Modulus at 50%, MPa	0.6	1.9	1.0	2.7
Modulus at 100%, MPa	0.8	2.8	1.4	6.0
Modulus at 200%, MPa	1.2	5.4	3.0	16.2
Modulus at 300%, MPa	2.0	7.4	6.7	21.8
Tensile, MPa	7.4	8.8	16.0	21.8
Elongation, %	850	400	450	300
Compression set B, plied discs, 94 h at 70°C				
Set, %	43	48	18	16
Shrinkage				
ASTM, %	2.8	2.3	2.8	2.3
Resilience, buttons cured 45 min at 180°C				
Rebound, %	44	22	36	25

TABLE 7.28
Non-Black Static-Dissipative Formula

Hydrin C2000L	100
Micral 632, Hydral 710 (Al trihydroxide)	35
Hi-Sil 532 EP	35
KP-140	5
Stearic acid	1.0
Vulkanox MB-2/MG/C	1.0
Carbowax 8000	2.0
Calcium carbonate	5.0
Zisnet F-PT	0.8
DPG	0.3
Santogard PVI	0.2
Total	185.3
Mooney scorch at 125°C	
Minimum value	60
T_5, minutes	6.4
T_{35}, minutes	9.2
Monsanto rheometer, 20 min at 180°C	
MH, N·m	9.27
ML, N·m	2.26
Ts_2, minutes	1.0
T_{90}, minutes	8.3
Original properties	
Hardness, points	61
Modulus at 100%, MPa	2.07
Tensile, MPa	4.96
Elongation, %	400
Electrical properties	
Resistance, ohm-cm	10^9

7.20.7 ELECTRICAL CONDUCTIVITY

Electro-assist printing rolls and copy machine feed rolls require very good electrical dissipation, hence epichlorohydrin becomes the polymer of choice for these parts. Table 7.29 shows the use of a conductive carbon black for these applications, but HAF, FEF, or even SRF may be used with ECO or GECO if the requirements are not too stringent.

7.20.8 PRIMER BULB

Outboard motors that use portable fuel tanks require a primer bulb to initiate flow of the fuel from the tank to the engine. The excellent fuel, ozone, weathering, low temperature, and fuel permeation resistance, plus the resilience of epichlorohydrin make it an ideal elastomer for this application. A suggested formulation is given in Table 7.30; however, it is suggested that the following replacements be made:

TABLE 7.29
Conductive Roll Formula

Hardness, Shore A, 90	
Hydrin C2000	100
N472	75
KP-140	5
Stearic acid	1
Vulkaonox MB-2/MG/C	1
Calcium carbonate	5
Zisnet7 F-PT	1
DPG	0.5
Santogard PVI	0.7
TOTAL	189.2
Resistivity	
Expected value, ohm-cm	7

TABLE 7.30
Primer Bulb for Outboard Engines

Hydrin C2000L	100
N550	40
TP-95	10
Stearic acid	1
Naugard 445	1
Magnesium oxide	3
ETU 75% dispersion	1
Sulfasan R	0.3
TOTAL	156.3
Mooney scorch, ML at 125°C	
Minimum viscosity	46
T_5, minutes	14.1
Rheometer, 3° arc, 15 min motor at 190°C	
MH, N · m	6.3
ML, N · m	1.4
Ts_2, minutes	1.5
Tc_{90}, minutes	7.3
Physical properties, cured 60 min at 160°C	
Hardness, Shore A, pts	59
Modulus at 50%, MPa	1.3
Modulus at 100%, MPa	2.3
Modulus at 200%, MPa	5.1
Modulus at 300%, MPa	7.6
Tensile, MPa	13.8
Elongation, %	700

magnesium oxide changed to calcium carbonate, the ETU switched to Zisnet F-PT, and Sulfasan R to DPG, all in the same amounts.

7.20.9 OTHER APPLICATIONS

There are many other applications for epichlorohydrin that use the unique combination of properties of this elastomer. Some of these are

1. Marine fuel hose and tubing
2. Oil and fuel transfer hose
3. Pump impellers and liners
4. Diaphragms for automotive as well as natural gas meters
5. Pressure accumulator bladders
6. O-Rings, seals, and gaskets
7. Special oil field components [4]
8. Sliding seals for floating tops of fuel and oil storage tanks

APPENDIX—LIST OF CHEMICALS

Names	Chemical	Source
Dynamar FC 5157	Proprietary (bis-phenol) cross linker	Dyneon
Dynamar FX 5166	Proprietary (phosphonium salt) accelerator	Dyneon
Dynamar PPA-790	Fluorochemical process aid	Dyneon
Dynamar RC-5251Q	Proprietary	Dyneon
Millathane Glob	Urethane + cleaning material	TSE Industries
Naugard 445	4,4′bis(a-dimethylbenzyl) diphenylamine	Uniroyal Chemical Co.
Santogard PVI—or CTP (PreVulcanization Inhibitor)	N-(cyclohexylthio) phthalimide	Flexsys America L.P.
DTDM, Sulfasan R or Vanax A	4,4′ Dithio-bis-morpholine	Elastochem, Inc., Flexsys, and R.T. Vanderbilt Co. Inc.
Echo MPS (thiadiazole)	Proprietary cross linker	Hercules, Inc.
TyPly BN	Rubber to metal bonding agent	Lord Elastomer Prod.
Vanax A—or DTDM	4,4′ Dithio-bis-morpholine	R.T. Vanderbilt Co. Inc.
Vanox MTI	2-Mercaptotolulimidazole	R.T. Vanderbilt Co. Inc.
Vanox ZMTI	Zinc 2-mercapto-toluimidazole	R.T. Vanderbilt Co. Inc.
Vulkanox MB-2/MG/C	4- and 5-Methylmercapto benzimidazole—oil treated	Lanxess Corporation
Vulkalent E/C	N-(trichloromethyl sulfenyl)- Benzene sulfone anilide, 5% whiting, and 1%–2% oil	Lanxess Corporation
Zisnet F-PT (triazine)	2,4,6-Trimercapto-s-triazine	Zeon Chemicals L.P.
DOP	Dioctyl phthalate	C.P. Hall and various suppliers
DIDP	Diisodecyl phthalate	C.P. Hall and various suppliers
DBEEA (TP 95)	Di(butoxy-ethoxy-ethyl) adipate	Morton Thiokol and various suppliers
TOTM	Trioctyl trimellitate	C.P. Hall and various suppliers

ACKNOWLEDGMENT

The very helpful comments by Clark Cable of Zeon Chemicals L.P. have been instrumental in the writing of this chapter, and we would also like to thank Zeon Chemicals L.P. for the sources of information on which this work was based.

REFERENCES

1. E.J. Vandenberg, *Journal of Polymer Science* 47, 486, 1960.
2. J.T. Oetzel, "The Long Term Heat Resistance of Acrylate and Epichlorohydrin Polymers," page 3, presented at The Rubber Chemistry Division of The Chemical Institute of Canada, Toronto meeting April 4–30, 1975 and Montreal meeting May 1, 1975.
3. Kinro Hashimoto, Masaaki Inagami, Akio Maeda, and Noboru Watanabe, "Epichlorohydrin Rubber Vulcanized with Triazine Thiols—An Advanced System" presented at The Rubber Division, American Chemical Society Meeting October 7–10, 1986, Atlanta, Georgia.
4. Robert C. Klingender, "Epichlorohydrin," presented at The Energy Rubber Group Educational Symposium, 1989.
5. Vernon J. Kyllingstad and Clark J. Cable, "Lead-ETU Free Curing Polyepichlorohydrin Rubber Compounds," Zeon Chemicals publication PH 500.1, presented at the Specialty and High Performance Rubber Conference, September 9–12, 1992, RAPRA Technology Limited, Shawbury, Shropshire, England.
6. Kathryn Owens and Vernon Kyllingstad, "Elastomers Synthetic Polyether," *Kirk-Othmer Encyclopedia of Chemical Technology*, Fourth edn., Volume 8, Copyright 1993 by John Wiley & Sons, Inc.
7. Clark Cable and Charles Smith, "Epichlorohydrin in Fuel Hose," presented at The Rubber Division, American Chemical Society Meeting, Louisville, KY, meeting October 8–11, 1996.
8. Anonymously disclosed 408104, "Multilayer Construction of Elastomers and Fluoropolymers," *Research Development*, page 415, April 1998, Kenneth Mason Publications Ltd.

8 Compounding with Chlorinated Polyethylene

Ray Laakso

CONTENTS

8.1 INTRODUCTION

Chlorinated polyethylene (CPE) is a synthetic elastomer produced by the controlled chlorination of polyethylene and has been in commercial production since the late 1960s [1–3]. The properties of the CPE product are influenced by a number of factors, which include the molecular weight and molecular weight distribution (MWD) of the starting polyethylene, the chlorine content, and the chlorine distribution. In addition, the chlorination process can be adjusted to produce amorphous, that is, low- to noncrystalline products, or products that contain higher levels of residual polyethylene crystallinity. A variety of different CPE grades can be produced in this manner and thereby offer the compounder numerous opportunities to achieve a wide range of properties for different applications.

The purpose of this overview is to provide the compounder with a basic understanding of CPE, how the molecular parameters affect the end-use performance in the compound, and typical formulating techniques that can be used to meet various end-use requirements.

8.2 INFLUENCE OF CPE CHEMICAL STRUCTURE

8.2.1 SATURATED BACKBONE

A generalized chemical structure for CPE is shown (Figure 8.1). The ASTM D1418 designation for CPE is "CM" where "C" denotes "chloro" and "M" denotes "a saturated chain of the polymethylene type." CPE and CM are often used interchangeably in the industry. In this chapter, the term CPE is used.

The saturated backbone of CPE imparts outstanding ozone-, oxidative-, and heat-resistance to a compound's performance [4]. The inherent nature of the polymer backbone allows compounds of CPE to be formulated that meet stringent high heat requirements, for example, up to 150°C for certain automotive applications and 105°C for various wire and cable applications using a peroxide cure system [5]. CPE typically provides better heat-aging resistance than polymers containing backbone unsaturation, for example, natural rubber and polychloroprene (CR) (Figure 8.2).

8.2.2 CHLORINE CONTENT

Typical commercial grades of CPE contain from 25 to 42 wt% chlorine (Table 8.1). The addition of chlorine to the backbone creates polarity in the polymer structure that imparts oil- and chemical-resistance to the polymer and subsequently to the compounded material. In addition, the chlorine on the backbone can help provide inherent flame retardance by providing a halogen source in a fire situation [6]. This is often advantageous for the compounder, for example, it may not be necessary to add a costly halogen-containing flame-retardant additive to the formulation if the recipe includes a CPE that contains a sufficiently high level of chlorine. However, as the compounder knows, to obtain one property, another is often sacrificed. In the case of chlorine content, it is necessary to balance the flame-retardant properties with the low-temperature performance (Figure 8.3).

8.2.3 VISCOSITY

The molecular weight of the polyethylene feedstock plays a key role in determining the viscosity of the CPE product. Higher molecular weight feedstocks, that is, those with low melt index (I_2) values, yield CPE resins with higher Mooney Viscosity values. For a given polyethylene I_2, other parameters directly affect viscosity, for example, MWD and comonomer type. The use of broader or narrower MWD polyethylene resins coupled perhaps with a copolymer of ethylene with, for example, butene, hexene, or octene, can provide a wide range of CPE viscosities. In addition, the chlorine content of the CPE resin can also affect the viscosity of the CPE. For a given polyethylene feedstock, the CPE viscosity increases with chlorine content.

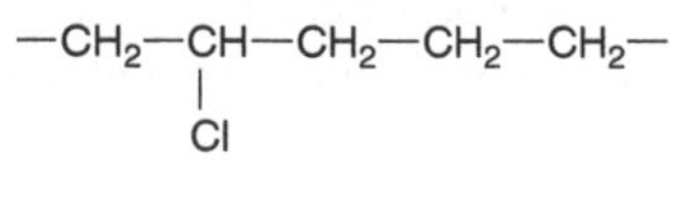

$$-CH_2-CH-CH_2-CH_2-CH_2-$$
$$|$$
$$Cl$$

FIGURE 8.1 Generalized chemical structure of CPE (~36 wt% chlorine).

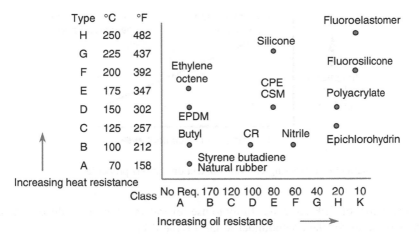

FIGURE 8.2 Heat and oil resistance of various elastomers (ASTM 2000/SAE J-200 Classification).

TABLE 8.1
Typical Commercial Grades of CPE for Elastomer Applications

Product	Chlorine Content,[a] wt%	Crystallinity H_f^a, J/g	Mooney Viscosity[a] (ML1 + 4 at 121°C), Mooney Units	Specific Gravity	Key Properties
TYRIN™ CM 0136	36	<2	80	1.16	General purpose elastomer grade
TYRIN™ CM 0836	36	<2	94	1.16	General purpose elastomer/high molecular weight
TYRIN™ CM 566	36	<2	80	1.16	Low electrolyte grade
TYRIN™ CM 0730	30	<2	65	1.14	Best balance of low and high temperature performance
TYRIN™ 3611P	36	<2	30	1.16	Good for viscosity modification
TYRIN™ 4211P	42	<2	42	1.22	Higher chlorine viscosity modification

Note: ™ is the Trademark of The Dow Chemical Company. Other product grades are available. Contact the polymer producers for additional product information. The Dow Chemical Company is the only North American producer of chlorinated polyethylene. Other producers of chlorinated polyethylene include Showa Denko (Japan), Osaka Soda (Japan), and Weifang (China).

[a] Reported values represent typical measurements.

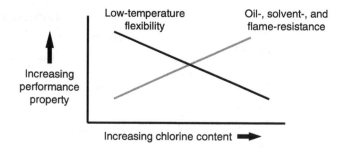

FIGURE 8.3 Effect of increasing chlorine content on various properties.

Manufacturers of CPE have the ability to use several variables to produce a range of CPE viscosities for the compounder to use (Table 8.1).

8.3 COMPOUNDING WITH CPE

8.3.1 INTRODUCTION TO COMPOUNDING

CPE, like most elastomers, needs to be properly "compounded" or "formulated" to meet the product performance requirements for any given application. The process of selecting types and levels of compounding ingredients can involve a complex combination of factors for the CPE compounder developing a formulation. Factors to consider include:

1. CPE Polymer type
   - Viscosity
   - Chlorine content
   - Residual crystallinity
   - Additives
2. Performance requirements
   - Original and aged physical properties
   - Processing
   - Electrical properties (where applicable)
   - Applicable standards, for example, automotive, wire and cable
3. Thermoset or thermoplastic application
4. Safety
   - Environmental issues
   - Industrial hygiene issues
5. Compounding ingredients
   - Black or colorable compound
   - Cure system
   - Proper selection to meet necessary performance requirements
6. Total cost of producing the final article

The consideration of these factors (and others) in the development of a suitable compound for a particular application can be a difficult task. The effects of various compounding ingredients on physical properties can usually be easily measured in the laboratory, but conducting trials at the production level is a critical step of the compound development process. The production trial reveals factors that may not have been predicted in the lab, for example, property differences due to scale-up in a larger extruder. The compounder can make modifications and refinements to the compound based on information from the factory experience.

General references on the subject of "compounding" elastomers can be found in the literature and usually cover a broad range of polymers [7–9]. In this section, the compounder is provided with information on factors to be considered in developing a CPE recipe for typical applications. Specific starting-point formulations for various applications are included later in this report and in the literature [10].

8.3.2 COMPOUNDING INGREDIENTS AND THEIR FUNCTION

The list of ingredients available to the compounder is quite impressive and can sometimes be overwhelming [11]. Most CPE formulations contain one or more ingredients from each of the following categories:

1. CPE
2. Fillers
3. Acid acceptors
4. Colorants
5. Plasticizers
6. Flame retardants
7. Antioxidants
8. Processing aids
9. Curatives
10. Coagents

A summary of several of the generic compounding ingredients and their function is included (Table 8.2). As noted, certain ingredients are specific to the cure system that is used. Undesirable chemical reactions can interfere with the curing mechanisms or even with the stability of the base polymer. For example, zinc-containing compounds, for instance, zinc oxide, should be avoided in all CPE compounds. Zinc oxide can cause dehydrochlorination of the CPE (regardless of the type of cure system) resulting in degradation of the polymer structure and a subsequent reduction in physical property performance. In addition, small quantities of zinc can adversely affect the cure mechanism in thiadiazole-cured compounds. The CPE manufacturer should be contacted for additional information on interactions/potential interactions of various compounding ingredients.

8.3.2.1 Choosing a CPE

The initial step in developing a CPE recipe typically involves the selection of the proper base polymer. Some of the most important variables to consider in a CPE product are viscosity, chlorine content, and degree of residual crystallinity.

TABLE 8.2
Compounding Ingredients and Their Function

Function	Ingredient[a]
Acid acceptor	Magnesium oxide, magnesium hydroxide, calcium hydroxide, epoxy compounds, lead compounds
Coagents or accelerators	Allylic, methacrylate
Colorant	Carbon black, titanium dioxide, organic pigments
Filler	Carbon black, mineral
Flame retardant	Antimony oxide, hydrated alumina, halogenated hydrocarbons
Plasticizers	Petroleum oils (aromatic and napthenic), esters, chlorinated paraffins, polymeric polyesters
Processing aids	Waxes, stearic acid, low molecular weight polyethylene, EPDM, EVA
Vulcanizing (curing) agents	Peroxide, thiadiazole, electron beam
Antioxidants	Variety of oxidative stabilizers are suitable

[a] Some ingredients are cure-system specific.

8.3.2.1.1 CPE Viscosity

The viscosity of CPE is typically influenced by a combination of the molecular weight of the starting polyethylene and the level of chlorine added to the polyethylene. The combination of these factors contributes to the viscosity of the CPE, as measured by such tests as Mooney viscosity or capillary rheology. In general, the viscosity of CPE is controlled by the combination of the polyethylene feedstock molecular weight and the amount of chlorine added to the polyethylene backbone. However, other factors can affect the viscosity such as the MWD of the polyethylene, additives in the final CPE product, and the presence of residual crystallinity.

When developing a new formulation, the compounder may be uncertain about which viscosity to choose in the initial formulations. A good starting-point for most general purpose, thermoset CPE compounds is an amorphous product, that is, essentially no residual crystallinity, with a Mooney viscosity (ML1 + 4 at 121°C) of around 75–80 and a chlorine content of 36 wt%. This starting-point CPE product generally provides a compound with suitable physicals and good processability. However, the compounder can evaluate the mechanical properties and processability characteristics to determine if a CPE product with higher or lower viscosity is needed to fulfill the desired application requirements. Blends of different CPE grades can be used to provide additional processability or physical property advantages. CPE is also compatible with a variety of other elastomers and thereby provides an additional method of tailoring the processability and physical property characteristics in polymer blends.

8.3.2.1.2 Chlorine Content

The chlorine content of commercial grades of CPE typically ranges from 25 to 42 wt% chlorine. The chlorine content of the resin is an important choice for the compounder. CPE resins with ~36% chlorine are most commonly used for thermoset elastomeric applications. This level of chlorine in the CPE typically provides a vulcanizate that possesses a good balance of chemical and oil resistance, flame retardance, low-temperature performance, processability, and good response to a peroxide cure system. If additional flame-, chemical-, or oil-resistance is desired, it may be necessary to use a CPE with higher chlorine content, for example, a 42% chlorine product. To achieve a balance between oil-, chemical-, and flame-resistance and good low- and high-temperature performance, 30% chlorine content amorphous CPE can provide a vulcanizate with suitable performance. CPE resins with even lower chlorine content (25% chlorine) are also available. Usually, the 25% chlorine grades contain residual polyethylene crystallinity that improves the compatibility when blended with other polymers such as polyethylene.

8.3.2.1.3   Residual Crystallinity
The parameters in the CPE production process allow the producer to tailor-make the desired degree of residual crystallinity in the final product. For thermoplastic applications of CPE or for applications requiring higher degrees of compatibility with another polymer such as polyethylene, it is often desirable to use a CPE that contains some portion of the original high-temperature crystallinity (melting point ~125°C–130°C) of the polyethylene. This crystallinity imparts stiffness to the CPE and thereby yields a higher modulus material. The resultant "polyethylene-type" characteristics in these semicrystalline CPE products can be used to good advantage when one blends the resin with high-ethylene content polymers to improve compatibilization.

Most CPE products are designed to be amorphous, that is, they contain essentially no residual crystallinity from the high-temperature portion of the polyethylene. This yields an elastomer with more rubbery qualities and the resultant products find use in a wider variety of applications than do the semi-crystalline CPE products.

8.3.2.2   Curatives

Cure agents for CPE compounds arc typically based on (1) peroxide cure systems with coagents; (2) thiadiazole-based chemistries; or (3) irradiation cross-linking techniques [12,13]. The choice of cure system depends upon a number of factors such as compound cost, processing equipment, and curing equipment.

Peroxide cures are preferred when extra scorch safety, shelf-life, or bin stability, low-permanent set, and high-temperature performance are desired. Thiadiazole cure provides the ability to cure over a wider range of temperature and pressure conditions while generating fewer volatile by-products than do peroxide cures. Irradiation-curable compounds are usually formulated in a similar manner to the peroxide-curable compounds except that no peroxide is necessary.

8.3.2.3   Fillers, Plasticizers, Other Ingredients

Fillers are used in CPE compounds for the same reasons they are used in most rubber compounds—fillers provide a means of obtaining a good balance between the necessary physical property characteristics and the economic requirements of the end-use article. The fillers used with CPE are common to the rubber industry: carbon black and mineral fillers (clay, whiting, talc, silica, etc.).

A variety of plasticizers are compatible with CPE. The most commonly used plasticizers include the ester-types. The type of plasticizer that can be used is often dependent upon the cure system. For example, aromatic plasticizers typically are not used in peroxide-cured recipes (due to hydrogen abstraction of the aromatic protons), whereas the aromatic plasticizers can be used effectively with the thiadiazole-cured compounds.

The final recipe usually includes a stabilizer, such as MgO or some similar acid acceptor. Other common rubber compounding ingredients are usually added to meet the physical property and processing requirements of the compound.

8.4   END-USE APPLICATIONS

Starting-point formulations for CPE compounds are available from polymer producers and the open literature [10]. End-use elastomeric applications for CPE are wide-ranging, for example, Wire and Cable, Automotive, Industrial, and General Rubber markets. CPE is also widely used in the Impact Modification market segment to improve the performance of vinyl siding and other vinyl-related products [14].

To aid the user in understanding typical CPE elastomer formulations, examples from different market segments are included (Tables 8.3 through 8.5). A wide range of performance can be achieved using CPE elastomers. The compounder is encouraged to explore the possibilities available with CPE: Oil Resistance, Ozone Resistance, Weatherability, Oil and Chemical Resistance, Heat-Aging Resistance, Low-Temperature Flexibility, Processability, Blend Additive, and more.

TABLE 8.3
Typical CPE "HPN Heater Cord" Jacket (90°C Rating)

Ingredient	phr
TYRIN™ CM 0136	100
Calcium carbonate	50
Diisononyl phthalate	25
Amino silane-functionalized hydrated aluminum silicate	60
Magnesium oxide	5
Polymerized 1,2-dihydro 2,2,4-trimethylquinoline	0.2
α, α'-Bis-(tert-butylperoxy)-diisopropylbenzene dispersed on clay (40% active)	5
85% Antimony oxide on CM binder	6
Trimethylolpropane trimethacrylate	5
Total phr	256.2
Specific gravity	1.55
Stock properties	
Mooney scorch MS + 1 at 121°C (ASTM D1646)	CPE Compound
Minimum viscosity, Mooney units	30.1
t_3 (time to 3-unit rise), minutes	>25
t_5 (time to 5-unit rise), minutes	>25

TABLE 8.3 (Continued)
Typical CPE "HPN Heater Cord" Jacket (90°C Rating)

Ingredient

ODR at 204°C, 0.051 rad, 1.66 Hz, microdie, 6 min (ASTM D2084)

Minimum torque, M_L, dN-m	12.2
Maximum torque, M_H, dN-m	64.1
Delta torque, dN-m	55.3
Time to 90% Cure (t_{90}), minutes	1.8

Vulcanizate properties
1.16 mm Insulation on 14 AWG aluminum wire
Cured 2 min in 1.72 MPa gauge steam

Original stress–strain properties	UL 62, 2.5 specification	CPE compound
Stress at 100% elongation, MPa		4.9
Stress at 200% elongation, MPa		9.0
Tensile strength at break, MPa	8.2 minimum	15.8
Elongation at break, %	200 minimum	402
Air-oven aged 10 days/110°C		
Tensile retention, %	50 minimum	105
Elongation retention, %	50 minimum	85
IRM 902 Oil immersion 18 h at 121°C		
Tensile retention, %	60 minimum	97
Elongation retention, %	60 minimum	78

Note: TM is the trademark of The Dow Chemical Company.

TABLE 8.4
Moisture, Flame-Resistant, Lead-Free, Heavy Duty Cable Jacket (90°C)

Ingredient	phr
TYRIN™ CM 566	100
Carbon black N550	5
Calcined and surface modified clay	60
Dioctyl adipate	15
Dow epoxy resin (D.E.R.™ 331)	5
Thiodiethylene bis-(3,5-di-tert-butyl-4-hydroxy)hydrocinnamate	1.5
α,α'-Bis-(tert-butylperoxy)-diisopropylbenzene dispersed on clay (40% active)	5
85% Antimony oxide on CM binder	7
80% Decabromodiphenyl oxide on CM binder	15
Trimethylolpropane trimethacrylate	5
Total phr/specific gravity	218.5/1.45

(*continued*)

TABLE 8.4   (Continued)
Moisture, Flame-Resistant, Lead-Free, Heavy Duty Cable Jacket (90°C)

Ingredient	phr
Stock properties	
Mooney scorch MS + 1 at 121°C (ASTM D1646)	CPE compound
Minimum viscosity, Mooney units	26.3
t_3 (time to 3-unit rise), minutes	>25
t_5 (time to 5-unit rise), minutes	>25

ODR at 204°C, 0.051 rad, 1.66 Hz, microdie, 6 min (ASTM D2084)	
Minimum torque, M_L, dN-m	11.1
Maximum torque, M_H, dN-m	57.3
Delta torque, dN-m	46.2
Time to 90% cure (t_{90}), minutes	1.9

Vulcanizate properties
1.16 mm Insulation on 14 AWG aluminum wire
Cured 2 min in 1.72 MPa gauge steam

Original stress–strain properties	ICEA S-68–516 4.4.11, ICEA S-19–81 4.13.11 specification	CPE compound
Stress at 100% elongation, MPa		4.1
Stress at 200% elongation, MPa	3.45 minimum	8.8
Tensile strength at break, MPa	12.4 minimum	17.4
Elongation at break, %	300 minimum	522
Air-oven aged 7 days/100°C		
Tensile retention, %	85 minimum	103
Elongation retention, %	55–65 minimum	83
IRM 902 Oil immersion 18 h at 121°C		
Tensile retention, %	60 minimum	70
Elongation retention, %	60 minimum	81

Note: ™ is the trademarks of The Dow Chemical Company.

TABLE 8.5
CPE Molded Goods Compound

Ingredient	phr
TYRIN™ CM 0136	100
Carbon black N550	50
Silica	10
Trioctyl trimellitate (TOTM)	35
Thiadiazole curative	3
Amine accelerator for thiadiazole cure	1
Magnesium hydroxide	5

TABLE 8.5   (Continued)
CPE Molded Goods Compound

Ingredient	phr
Styrenated diphenylamines antioxidant	1
Hindered phenolic antioxidant/metal deactivator	1
Total phr	206
Specific gravity	1.28

Compound properties
Mooney scorch MS + 1 at 121°C (ASTM D1646)

Minimum viscosity, Mooney units	38
t_3 (time to 3-unit rise), minutes	>25

ODR at 160°C, 0.051 rad, 1.66 Hz, microdie, 30 min (ASTM D2084)

Minimum torque, M_L, dN-m	12.5
Maximum torque, M_H, dN-m	52.0
Delta torque, dN-m	39.5
Time to 90% cure (t_{90}), minutes	13.0

Vulcanizate properties

Original stress–strain properties	CPE
(Cured 20 min at 160°C)	
Stress at 100% elongation, MPa	3.8
Stress at 200% elongation, MPa	7.6
Ultimate tensile, MPa	18.0
Elongation, %	530
Hardness, shore A	78
Die C Tear, ppi at 23°C	302
Die C Tear, ppi at 160°C	107
Low-temperature brittleness, °C	−34
Compression set 22 h/100°C, %	26
Ozone resistance, cured 5 min at 182°C	
72 h at 40°C, 100 pphm ozone, 55% humidity	Pass—no cracks
Aged in air oven 70 h at 150°C	
Tensile change, %	−9
Elongation change, %	−48

Note: ^{TM} is the trademark of The Dow Chemical Company.

REFERENCES

1. U.S. Patent 3,454,544, Process for the Chlorination of Polyolefins, Issued July 8, 1969 to Dow Chemical U.S.A.
2. U.S. Patent 3,429,865, Chlorinated Polyethylene Compositions, Issued February 25, 1969 to Dow Chemical U.S.A.
3. U.S. Patent 3,563,974, Linear Polyethylene Chlorination, Issued February 16, 1971 to Dow Chemical U.S.A.

4. W.H. Davis, Jr., R.L. Laakso, Jr., L.B. Hutchinson, and S.L. Watson, Peroxide-Cured Chlorinated Polyethylene Compounds Having Enhanced Resistance to Ozone-Induced Cracking, American Chemical Society Rubber Division, May 29–June 1, 1990, Paper No. 8.

5. R.R. Blanchard, Compounding Chlorinated Polyethylene Elastomers for High Temperature Service, *Advances in Synthetic Rubbers and Elastomers Science and Technology*, Technomic Publishing Co., Inc., 1973, pp. 1–13.

6. R.M. Aseeva and G.E. Zaikov, *Combustion of Polymer Materials*, Hanser Publishers, Munich Vienna, New York, 1985, pp. 214–219.

7. F.W. Barlow, *Rubber Compounding*, Second Edition, Marcel Dekker, Inc., New York, 1993.

8. J.E. Mark, B. Erman, and F.R. Eirich, *Science and Technology of Rubber*, Second Edition, Academic Press, Inc., San Diego, CA, 1994, Chapter 9.

9. M. Morton, *Rubber Technology*, Third Edition, Van Nostrand Reinhold, New York, 1987.

10. P.A. Ciullo and N. Hewitt, *The Rubber Formulary*, Noyes Publications, Norwich, New York, 1999, pp. 579–597.

11. *Rubber World Magazine's Blue Book 2002*, J.H. Lippincott, publisher, Lippincott & Peto Inc.

12. L.E. Sollberger and C.B. Carpenter, "Chlorinated Polyethylene Elastomers—A Comprehensive Characterization," presented at a meeting of the Rubber Division, American Chemical Society, Toronto, Canada, May 7–10, 1974.

13. J.H. Flynn and W.H. Davis, "Tyrin℗ Brand CPE Thiadiazole Cure System Studies— Chemistry and Dispersion," presented at a meeting of the Rubber Division, American Chemical Society, Los Angeles, California, April 23–26, 1985.

14. *Polymeric Materials Encyclopedia*, Volume 2/C, CRC Press, 1996, Editor-in-Chief J.C. Salamone, chapter on "Chlorinated Polyethylene," G.R. Marchand.

9 Chlorosulfonated Polyethylene and Alkylated Chlorosulfonated Polyethylene

Robert C. Klingender

CONTENTS

9.1 INTRODUCTION

The development of chlorosulfonated polyethylene (CSM) began during the early 1940s by Du Pont as part of the wartime effort using low cost ethylene and attempting to create a vulcanizable elastomer with the electrical and chemical properties of polyethylene [1].

The initial development was the creation of chlorinated polyethylene, however it was difficult to vulcanize with the limited peroxides available at the time. The process was altered to permit simultaneous chlorination and chlorosulfonation of polyethylene which allowed the elastomer to be cured with sulfur-bearing curatives. In 1951, Du Pont commercialized Hypalon© S-1, a chlorinated polyethylene, and Hypalon S-2, a chlorosulfonated polyethylene, later renamed Hypalon 20. S-1 was quickly withdrawn from the market due to the much improved Hypalon 20, which is still available today.

Chlorosulfonated polyethylene is produced by Du Pont with the name Hypalon in Beaumont, TX as well as Maydown, Northern Ireland, and by TOSOH Corporation as CSM in Japan. A cross-index of the various grades from these manufacturers is shown in Table 9.1.

In order to compete with HNBR and replace lost business for the CR previously used in timing belts, Du Pont-Dow developed a modified CSM called Ascium, an alkylated chlorosulfonated polyethylene (ACSM). This elastomer has the chemical and heat resistance of CSM and in addition has 20°C improved low-temperature properties compared with the base CSM. TOSOH also produces a competing grade called Extos. These products are also used in hose and other applications requiring a broader range of high and low-temperature properties. A listing of available grades of ACSM is given in Table 9.2.

The structure of CSM is shown in Figure 9.1 and of ACSM is given in Figure 9.2 [2]. The addition of alkyl groups to the polyethylene chain in ACSM interferes

---

TABLE 9.1
Cross-Index of CSM Grades

Du Pont Hypalon	Tosoh CSM	Chlorine Level %	Sulfur Level %	S. G.	ML 1+4 at 100°C	Remarks
H-45		24	1.0	1.07	37	Low viscosity and Cl₂, general purpose
	TS 320	25	1.0		37	Coatings, tile, magnets
HPG 6525		27	1.0	1.10	90	Industrial hose
H-20		29	1.4	1.12	28	Coatings, adhesives
	TS 530	30	1.0		56	Medium viscosity, general purpose
H-40		35	1.0	1.18	56	Seals, hose. General purpose
H-40S		35	1.0	1.18	46	Low viscosity. H-40, general purpose
	TS 430	35	1.0		46	Low viscosity. TS 530
H-4085		36	1.0	1.19	94	Magnets, hose, rolls, adhesives
	TS 930	36	1.0		97	High viscosity. TS 530
H-30		43	1.1	1.27	30	Coatings, adhesives
H-48		43	1.0	1.27	78	Coatings, hose
CPR 6140		40			100	Roofing

TABLE 9.2
Cross-Index of ACSM Grades

Grade		Chlorine Content (%)	Sulfur Content (%)	S. G.	ML 1+4 at 100°C	T_g, °C	Remarks
DuPont	Tosh Corp.						
6367S		27	1	1.11	34	−27	Low-viscosity grade
6367		27	1	1.11	43	−27	Standard grade, poly-V belts
6983		27	1	1.11	87.5	−27	High-viscosity, hose, etc
6932		30	1	1.14	55	−24	Higher oil resistance, timing and poly-V belts
	ET-8010	26	0.7		40		Standard grade, poly-V belts
	ET-8510	30	0.9		40		Higher oil resistance, timing belts

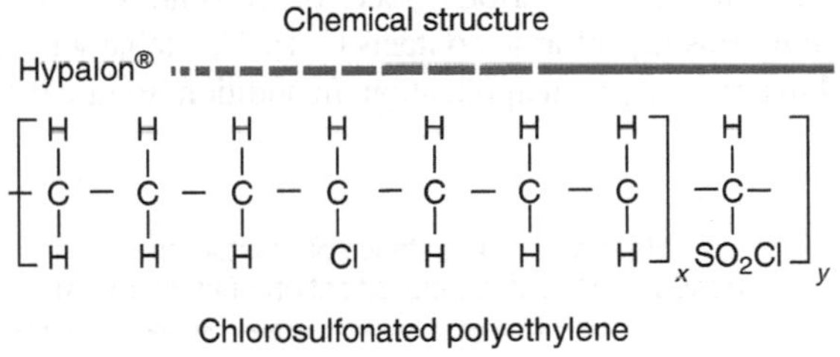

Chemical structure

FIGURE 9.1 CSM Structure.

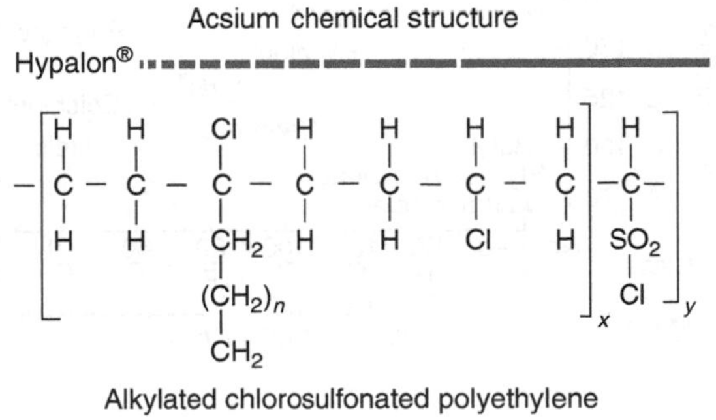

Acsium chemical structure

FIGURE 9.2 ACSM Structure.

with the crystalline properties of the polymer, thus improving the low-temperature properties.

9.2 CSM, CHLOROSULFONATED POLYETHYLENE

The saturated nature of CSM results in vulcanizates, which are highly resistant to heat, ozone, oxygen, weather, chemicals, and oil. This elastomer is also well suited for light-colored applications. Very good physical properties can be obtained with CSM vulcanizates, including abrasion resistance and high tensile strength [3]. A comparison of CSM with other polymers, as established in SAE J200, is given in Figure 9.3. CSM is classified as a DE material by SAE and ASTM indicating that it has an operating temperature of 150°C and moderate oil resistance. As with other chlorinated elastomers, CSM, with increase chlorine content, exhibits improved oil resistance and decreased low-temperature properties. The electrical properties allow it to be used in low-voltage insulation and in all wire and cable uses as a jacketing material. The low flammability also makes it an appropriate polymer to use for mining cable.

Applications for CSM include seals and gaskets, hose and cable jackets, low voltage insulation, roll covers, tank lining and coatings. The excellent green strength and stability of the unvulcanized elastomer allows it to be used in pond liners and roofing. Examples of some of these are given later in this chapter.

The general characteristics of various grades of CSM are shown in Figure 9.4 and Table 9.3, listed in increasing chlorine contents [3,4]. This table will assist in selecting the grade of CSM to use for a given application. In addition, Figure 9.5 summarizes the

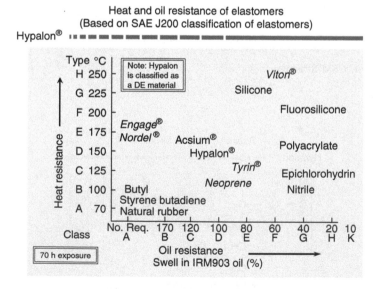

FIGURE 9.3 CSM (Hypalon) and ACSM (Ascium) comparison with other elastomers.

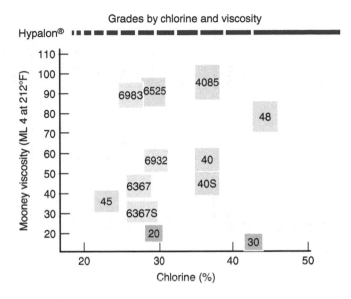

FIGURE 9.4 CSM Grades compared by chlorine and viscosity.

effect of the chlorine content on the properties of CSM. All grades of CSM exhibit excellent chemical, weathering, oxygen and ozone resistance, colorability, and color stability.

A visual relationship of the properties of CSM versus chlorine content is given in Figure 9.5.

9.3 COMPOUNDING CSM

9.3.1 SELECTING THE TYPE OF CSM

The two major properties to consider in selecting the grade of CSM to use in a given application are: chlorine content, which determines oil resistance and low-temperature flexibility, and viscosity to use for processing conditions that will be encountered. Figure 9.6 provides a good visual aid in determining the proper grade to use. Although this figure shows only Hypalon grades, the equivalent TOSOH elastomers would also follow these same precepts.

The chlorine content of these elastomers dictates the balance of swell in petroleum products and low-temperature flexibility. Increased chlorine up to 43% provides better resistance to flame, chemicals, oils, and aliphatic solvents, and poorer low-temperature properties. Excellent low-temperature properties are obtained if the chlorine levels are 24%–27%.

All grades have excellent weathering and ozone resistance and color stability. CSM compounds give excellent resistance to abrasion, mechanical abuse and may be formulated to provide very good electrical insulating properties.

TABLE 9.3
Basic Properties of CSM Grades

Grade	45	6525	20	40S	40	4085	30	48
	TS 320		TS 530	TS 430		TS 930		
Chlorine content %	24	27	29	35	35	36	43	43
Sulfur content %	1.0	1.0	1.4	1.0	1.0	1.0	1.1	1.0
ML 1 + 4 at 100°C	37	90	28	46	56	94	30	78
Specific gravity	1.07	1.10	1.12	1.18	1.18	1.19	1.27	1.27
Processing								
Extrusion	Good	Excellent	Fair	Excellent	Excellent	Excellent	Fair	Good
Molding	Excellent	Excellent	Good	Excellent	Excellent	Excellent	Fair	Good
Calendering	Excellent	Good	Fair	Excellent	Excellent	Excellent	Fair	Good
Vulcanizate properties								
Hardness, Shore A	65–98	40–90	45–95	40–95	40–95	40–95	60–95	60–95
Tensile (MPa)								
Black loaded	Up to 27	Up to 27	Up to 20	Up to 27	Up to 27	Up to 27	Up to 24	Up to 27
Gum stock	Up to 27	Up to 27	Up to 8	Up to 27	Up to 27	Up to 27	Up to 17	Up to 24
Low temperature	Excellent	Excellent	Good	Good	Good	Good	Poor	Poor
Tear strength	Good	Good	Fair	Good	Good	Good	Fair	Good
Resistance to								
Abrasion	Very good	Excellent	Very good	Excellent	Excellent	Excellent	Very good	Excellent
Chemicals	Good	Very good	Good	Excellent	Excellent	Excellent	Excellent	Excellent
Compression set	Good	Very good	Fair	Good	Good	Good	Poor	Fair-good
Flame	Fair	Fair	Fair	Good	Good	Good	Very good	Very good
Heat	Very good	Very good	Very good	Very good	Very good	Very good	Fair	Good
Mineral oils	Fair	Fair	Fair	Good	Good	Good	Excellent	Excellent
Features	High green strength, good heat aging and low temperature	High viscosity, good processing. Good aging and low temperature	Good low temperature	Low viscosity for highly loaded stiff stocks	Medium viscosity for general purposes	High viscosity, good green strength for high loadings	Easily soluble for hard coatings	High viscosity and green strength, excellent in fluids

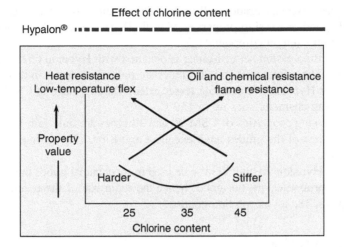

FIGURE 9.5 Properties of CSM versus chlorine content.

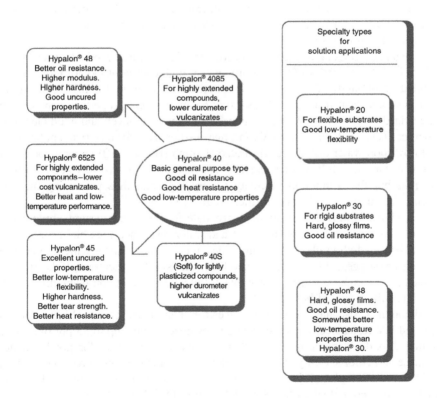

FIGURE 9.6 Selection guide for CSM.

The higher viscosity grades will accept high loadings of fillers and plasticizers and provide good processing properties as well. These also may be compounded to give high green strength compounds.

The best compression set resistance is obtained with Hypalon 6525 followed by the 40 series. The original 20 and 30 grades are only fair to poor in this regard.

Except for Hypalon 30, and to a lesser extent 48, all grades of CSM have very good heat aging characteristics up to 150°C.

The processing properties of CSM 20 and 30 types are only fair, Hypalon 48 is good and the rest of the grades have excellent extrusion, calendaring, and molding characteristics.

Although Hypalon 20 and 30 may be used in mechanical goods their processing and physical properties are not good, hence the main use of these, and to a lesser extent Hypalon 48, are as solution coatings.

9.3.2 FILLERS IN CSM

Carbon blacks are the preferred fillers for CSM compounds, although mineral fillers may be used in many applications [5]. Medium structure carbon blacks, N762 and N550 give the best processing properties with high strength, good resilience, and low-compression set, but higher or lower structure blacks can be selected for specific end properties. A comparison of various grades of carbon black at equal hardness loadings is given in Table 9.4.

Mineral fillers alone in CSM are needed for light-colored products or electrical applications. These are also used in combination with carbon blacks for more highly extended compounds. Whiting (calcium carbonate) is used for high elongation, lower cost, and heat resistance. Clay gives better physical properties than whiting and calcined clay, such as Whitetex, is used for low water swell and good electrical properties. Talc, such as Mistron Vapor, may also be used for electrical applications and fairly good water resistance. Hydrated alumina (Hydral 710) imparts good heat and flame resistance. Silicas such as Hi-Sil 233 or Cab-O-Sil give very good physical properties but the compound viscosity is very high and thus processing may be difficult. A comparison of mineral fillers is given in Table 9.5.

It is advisable to include a silane appropriate for the curing system employed in order to achieve lower compression set and water resistance plus higher modulus. Some fillers, including clays, talc, and silica, have a silane already incorporated in them and are suggested for electrical and low water swell applications. If the product is meant for acid resistant uses carbon black and acid resistant non-black fillers such as barium sulfate should be selected.

Although CSM has good basic weather resistance, carbon black imparts improved UV resistance. Light-colored articles need to have up to 35 phr of a rutile titanium dioxide, depending upon other non-black fillers used, for good weather resistance and color stability. Non-black fillers recommended are clays up to 50 phr, whitings up to 200 phr, all in combination with titanium dioxide. Low levels of talc may be employed and silica fillers are NOT good for UV resistance.

Flame resistance may be needed for some applications and the compound will need to have non-black fillers plus a hydrated alumina or antimony oxide.

TABLE 9.4

Comparison of Carbon Blacks at Equal Vulcanizate Hardness

Hypalon 40				100		
Aromatic process oil				25		
Low molecular weight polyethylene				2		
MBTS				0.5		
Tetrone A				2		
High activity MgO				5		
Pentaerthyritol (200 mesh)				3		

Carbon Black (Type as Shown)	MT	SRF	GPF	FEF	SAF	SPF
Type of carbon black						
ASTM Type designation	N990	N762	N650	N550	N330	N358
Carbon black, phr	110	60	50	45	40	35
Stock properties						
Mooney viscosity ML 1 + 4 at 100°C	52	48	55	49	46	46
Mooney scorch, MS at 125°C Minimum viscosity, μ	15	14	17	14	16	15
Time to 10-pt rise, min	14	14	13	14	13	13
MDR at 160°C, 0.5° arc, 30 min chart						
ML, dN·m	0.9	0.8	1.0	0.8	1.0	0.9
MH, dN·m	20	17	19	17	17	18
T_{s2}, min	1.5	1.4	1.2	1.4	1.1	1.2
T_{c90}, min	11.7	11.1	8.8	10.4	5.9	6.0
Vulcanizate properties						
Press cured at 160°C Cure time, min	5	18	23	18	13	13
100% Modulus, MPa	8.1	7.4	8.6	6.9	6.3	6.1

(continued)

TABLE 9.4 (Continued)
Comparison of Carbon Blacks at Equal Vulcanizate Hardness

	MT	SRF	GPF	FEF	SAF	SPF
Tensile strength, MPa	17.2	21.4	21.4	21.9	22.0	22.4
Elongation, %	255	275	238	315	254	283
Hardness, Durometer A, pts	75	70	72	70	71	70
Tear strength kN/m	39	38	36	41	39	42
Aged 70 h at 150°C-change						
100% Modulus, %	241	307	264	255	264	251
Tensile strength, %	−10	0	4	−7	−11	−10
Elongation, %	−68	−63	−56	−61	−59	−59
Hardness, pts	15	19	17	18	17	17
Compression set, method B, %						
After 22 h at 100°C	71	71	69	68	66	67
After 22 h at 125°C	85	85	85	86	81	82
DIN Abrasion Index, %	177	119	109	108	98	95
Resilience Zwick, %	21	23	22	24	23	23

TABLE 9.5
Comparison of Mineral Fillers at Equal Hardness

	Amount
Hypalon 40	100
High-activity MgO	4
Low molecular weight polyethylene	2
Mineral filler	As shown
Pentaerythritol (200 mesh)	3
Aromatic process oil	30
Tetrone A	2

Type of Filler	Atomite Whiting	Catalpo Clay	Whitetex Clay	Soft Clay	Hard Clay	Mistron Vapor Talc	Hi-Sil 233
Amount of filler, phr	150	140	110	90	80	60	40
Stock properties							
Mooney viscosity							
ML 1 + 4 at 100°C	47	50	42	34	34	27	58
Mooney scorch, MS at 121°C							
Minimum viscosity, units	14	17	13	10	10	9	20
Time to 10-unit rise, min	19	22	18	41	40	17	13
Vulcanizate properties							
Cured 30 min at 153°C							
100% Modulus, MPa	4.2	5.2	3.4	3.8	4.0	4.6	2.4
300% Modulus, MPa	5.2	7.0	5.8	6.2	8.2	6.8	8.8
Tensile strength, MPa	11.8	12.0	11.8	10.6	17.0	20.4	26.8

(continued)

TABLE 9.5 (Continued)
Comparison of Mineral Fillers at Equal Hardness

	Atomite Whiting	Catalpo Clay	Whitetex Clay	Soft Clay	Hard Clay	Mistron Vapor Talc	Hi-Sil 233
Elongation, %	510	430	530	490	560	560	550
Hardness, Durometer A, pts	71	71	69	70	68	69	70
Aged 7 days at 121°C-change							
Tensile strength, %	−12	20	−12	−8	−12	−37	−27
Elongation, %	−51	−44	−55	−69	−61	−30	−51
Hardness, pts	7	8	11	3	7	6	10
Abrasion resistance NBS							
Index, %	70	—	90	60	92	90	160
Resilience, Yerzley, %	60	—	58	60	69	63	58
Tear strength Die C, kN/m	34.1	45.5	36.8	45.5	50.8	47.2	57.8
Volume change, %							
Aged in IRM 903 oil, 70 h at 121°C	52	57	57	56	70	64	56
Aged in water, 7 days at 70°C	63	55	36	54	49	40	21
Compression set B[a], 22 h at 70°C							
Set, %	46	46	40	57	48	59	48

[a] Cure 35 min at 153°C.

Some other special applications require special fillers such as cork for good oil, heat and compression set resistance for gasketing applications or a magnetic ferrite filler for flexible magnetic uses. CSM is used as an unvulcanized sheet for roofing, pond and ditch liners, and as retention barriers around chemical storage tanks. The same principals are used in selecting fillers for these applications as for cured products based upon CSM.

9.3.3 Plasticizers and Process Aids for CSM

The selection of a plasticizer for CSM depends upon the end use of the part and the environment it will encounter in service [6].

Petroleum oils provide low cost, good processing characteristics and physical properties. Aromatic oils have the greatest compatibility but darken the compound and cause staining as well as interfering with a peroxide if that type of vulcanization agent is employed. Naphthenic oils may be used in limited amounts up to 15–20 phr due to lower compatibility. These reduce stickiness when combined with an aromatic oil. Paraffinic oils are not compatible with CSM and should be avoided.

Ester plasticizers provide much improved low temperature flexibility, are light colored and nonstaining and nondiscoloring but are not recommended for electrical uses. This class of plasticizer does not interfere with a peroxide cure system. Low volatility types such as DOS and TOTM should be used for higher heat applications with reasonable low-temperature properties.

Chlorinated paraffins may be selected for best flammability resistance, good physical properties, light colors, reasonable heat resistance up to 135°C and good low temperature flexibility. Commonly used products are Chlorowax LV, Chloroflo 42, or Unichlor 40. Plasticizers containing higher chlorine levels such as Chlorowax 50, Paroil 150-LV, or Unichlor 50 provide improved flame resistance at some sacrifice in low-temperature properties.

Polymeric esters such as Uniflex 330, or equivalent, give low volatility for good heat resistance but these are less effective in softening the vulcanizates, give poorer physical properties, and low temperature characteristics similar to an aromatic oil. Uniflex 300 gives better low-temperature properties but is slightly more volatile. TP-95 is similar to Uniflex 330 in physical properties but sacrifices some in processing. Other products are also effective, but as always there is a balance among volatility, low temperature, physical and processing properties.

Unvulcanized products based on Hypalon 48 have about a 10 point higher hardness, higher modulus, and lower resilience than one using Hypalon 40 series with the same aromatic oil levels. Aromatic oils are less compatible in Hypalon 48 than in the Hypalon 40 series hence the lowest aniline point grade should be selected to prevent bleeding, or blend with chlorinated paraffin or an ester plasticizer. Chlorinated paraffin or ester plasticizers do not exhibit any difference in properties in the Hypalon 48 compared with Hypalon 40 series compounds.

A comparison of plasticizers is given in Table 9.6, which will assist in selecting the type for a given application.

Common process aids for CSM are microcrystalline waxes such as Vanwax H and Multiwax 180M and low molecular weight polyethylene due to their compatibility and

TABLE 9.6
Comparison of Plasticizers in Black-Loaded CSM

	Hypalon 40	100
Low molecular weight polyethylene	N762 Carbon black	60
	Low molecular weight polyethylene	2
	High activity MgO	5
	Pentaerythriol (200 mesh)	3
	MBTS	0.5
	Tetrone A	2
	Plasticizer (As shown)	25

Plasticizer	Sundex 790	Shellflex 794	Chlorowax L.V.	DOP	DOS	Butyl Oleate	Plasthall 100	TOTM	Paraplex G62
Stock Properties									
ML 1 + 4 at 100°C	48.5	45.1	45.0	36.9	29.2	47.3	29.2	41.3	41.5
Mooney scorch, MS at 125°C									
Minimum viscosity, μ	13.9	13.6	12.4	9.2	7.4	17.7	7.9	10.8	11.4
Time to 10 pt rise, min.	14.0	13.8	26.9	20.0	17.4	7.6	17.7	22.4	20.0
MDR at 160°C, 0.5° arc, 30 min chart									
ML, dN·m	0.8	0.8	0.7	0.5	0.4	0.8	0.5	0.6	0.6
MH, dN·m	16.6	16.8	17.6	17.0	16.8	13.9	13.6	17.1	14.0
Ts_2, min	1.5	1.5	1.9	1.6	1.6	1.0	1.6	1.7	1.6
Tc_{90}, min	12.0	12.9	15.6	14.2	13.9	4.2	10.1	14.8	5.3
Cure rate, dNm/min	1.5	1.4	1.2	1.3	1.3	4.1	1.6	1.3	3.7

Vulcanizate properties								
Press cured at 160°C, min	30	30	30	30	30	12	20	30	12
100% Modulus, MPa	6.7	7.1	6.0	5.6	5.5	5.8	5.0	6.8	5.0
Tensile, MPa	21.7	21.1	20.2	19.8	19.0	17.2	18.1	20.5	19.6
Elongation, %	294	250	257	261	247	331	283	250	293
Hardness, Shore A, pts	69	67	68	67	67	70	65	69	65
Tear strength, Die C kN/m	40	35	41	36	31	42	33	34	38
Aged 168 h at 125°C-change								
100% Modulus, %	132	73	65	211	165	153	199	44	86
Tensile strength, %	3	0	-3	10	9	21	15	-11	-4
Elongation, %	-45	-31	-28	-47	-36	-49	-42	29	-30
Hardness, pts.	13	10	8	19	14	13	18	5	10
Aged 70 h at 150°C-change								
Tensile strength, %	-9	-8	-22	2	5	8	4	-28	-3
Elongation, %	-67	-57	-58	-69	-65	-76	-68	-57	-63
Hardness, pts	20	18	16	20	19	19	26	14	22
Weight, %	8	6	17	14	13	9	12	4	3
Volume change, %								
Aged 70 h/125°C/IRM903 Oil	55	53	57	52	49	62	52	52	54
Aged 168 h/70°C/Distilled Water	30	29	33	34	38	63	42	35	24
Abrasion resistance, DIN index, %	122	127	118	120	107	113	112	121	124
Compression set, method B 22 h at 100°C, %	71	68	64	66	67	89	71	68	56
Low-temperature properties, °C								
TR-10, unaged	-12	-15	-21	-25	-31	-26	-31	-23	-15
TR-10, aged 70 h at 150°C	-10	-13	-19	-10	-13	-13	-12	-19	-14
Brittle point	-45	-42	-50	-46	-58	-50	-56	-48	-34

good mill and Banbury release. Low-molecular weight polyethylene such as A-C Polyethylene 617A or Epolene N-14 are suggested for temperatures above 77°C and polyethylene glycol is very good below 77°C. A-C Polyethylene 1702 is good for lower temperatures encountered in mill mixing. Petrolatum combined with paraffin wax is very effective, but care needs to be taken to avoid bloom or bleeding in uncured compounds.

Stearic acid and stearates may also be used as release agents, but NOT with a litharge cure system as scorching occurs and a short induction time during curing. Zinc stearates should be avoided in heat-resistant compounds as zinc chloride forms and lead to dehydro-halogenation of the polymer upon aging, or exposure to weather.

Other internal lubricants such as Vanfre IL-2, a blend of fatty alcohols, also improves extrusion and mold flow. Fatty acid amides such as Kenamide S, Myverol 18–04, and Myvacet 9–40 are very good release and anti-blocking agents for calendering. Kenamide S at 0.5 phr is particularly effective in Hypalon 45 or 48 calendered compounds that may be rolled up on themselves without liners, after cooling. Both Myvacet and Myverol are very effective in lowering temperatures needed to calender Hypalon 40 types.

Many other proprietary process aids are available, but these should be tested first to determine any influence on scorch and/or cure rate. In addition, examine the uncured compound for decreased tack and any detrimental effects on adhesion to metal or other elastomers in a composite.

A small amount of another elastomer, such as EPDM or high *cis*-polybutadiene, blended with CSM increases compound green strength, especially in mineral-filled compounds, giving a stiffer, more easily handled compound. Better release of the stock is obtained with the polybutadiene.

Good building tack may be obtained by including 10–15 phr of a low-melting point coumarone indene resin such as Cumar P-10, avoiding waxes or polyethylene, and warming the stock to 50°C–60°C. Cumar P-10 imparts properties similar to that with an aromatic mineral oil, but exhibits poorer heat resistance. The effectiveness of a coumarone indene resin decreases with the age of the uncured compound. Freshening the surface of a CSM compound with a solvent such as cyclohexanone is the most effective method.

9.3.4 Selecting a Curing System for CSM

The sulfonyl chloride cure sites in CSM provides a wide range of choice in selecting a curing system, depending upon the end properties needed and factory process conditions [7]. A curing system combination requires an acid acceptor plus a cross-linking mechanism, with or without an accelerator.

Acid acceptors may be magnesium oxide, litharge, organic lead bases, and epoxy resins. NEVER use zinc oxide or zinc salts in a CSM compound as zinc chloride is formed during vulcanization, which causes degradation of the polymer as well as poor weathering and heat resistance. Zinc oxide may be required to effect a cure when CSM is blended with a diene elastomer, but the heat resistance of the blend is inferior to that of CSM alone.

Vulcanizing agents are sulfur, or sulfur-bearing accelerators, peroxides, or maleimides. Various activators or accelerators are used to achieve the desired cure rate and process safety, plus the specified physical properties. A selection chart in Table 9.7

TABLE 9.7
Summary of Typical Curing Systems for CSM

Desired Property or Characteristic	Suitable For		Curing System	
	Colored Cmpds.	Black Cmpds.	Components	Amounts, phr
General purpose	X	X	(1) Magnesia-PER-TMTD-Sulfur	4·3·2·1
	X	X	(2) Magnesia-PER-Tetrone A	4·3·2
		X	(3) Litharge-TMTD or TETD-Sulfur	25·2·1
		X	(4) Litharge-Tetrone A · MBTS	25·2·0.5
Maximum color stability	X		(5) Either of the above magnesia systems using 20 phr magnesia instead of 4 phr	See above
	X		(6) Magnesia-Varox Powder-TAC	10·6·4
Water and chemical resistance		X	(7) Litharge-TMTD or TETD-Sulfur	25·2·1
		X	(8) Litharge-Tetrone A · MBTS	25·2·0.5
	X		(9) Tribasic lead maleate · Staybelite Resin-MBT	40·2.5·1.5
	X		(10) Tribasic lead maleate-PER-MBT	40·3·1.5
	X		(11) Dibasic lead phthalate-Varox Powder-TAC	40·6·4
	X		(12) Magnesia-Epoxy Resin-Tetrone A	3·4·2
	X		(13) Epoxy Resin-Tetrone A-MBTS-DOTG	15·1.5·0.5·0.25
Heat resistance		X	(14) Litharge-Magnesia-TMTD-Sulfur-NBC	20·10·2·1·3
		X	(15) Litharge-Magnesia-Tetrone · A-MBTS-NBC	20·10·2·0.5·3
		X	(16) Litharge-Magnesia-Tetrone A · MBTS · HVA-2 · NBC	20·10·0.75·0.5·1·3
Compression set resistance	X		(17) Magnesia-Varox Powder-TAC	10·6·4
	X		(18) Dibasic lead phthalate-Varox Powder-TAC	40·6·4
		X	(19) Calcium hydroxide-HVA-2-Ionac GFI710	4·3·2
Processing safety and bin stability	X	X	(20) Magnesia-PER-Tetrone A · HVA-2	4·3·2·1

(continued)

TABLE 9.7 (Continued)
Summary of Typical Curing Systems for CSM

Desired Property or Characteristic	Suitable For		Curing System	
	Colored Cmpds.	Black Cmpds.	Components	Amounts, phr
Processing safety and bin stability	X	X	(21) Magnesia-Varox Powder-TAC	10·6·4
		X	(22) Calcium hydroxide-HVA-2- Ionac GFI710	4·3·2
		X	(23) Litharge-HVA-2-4,4 -dithio-morpholine	25·2·3
		X	(24) Polydispersion HC-11	30
Lead free	X	X	See Systems 1, 2, 5, 6, 12, 13, 20	
		X	See systems 1, 2, 5, 6, 12, 13, 19, 20	
LCM Curing	X	X	Most of the systems listed above, particularly those containing sulfur or sulfur donors	
Radiation curing	X	X	Acid acceptors and cure promoters only. Accelerators and curatives are not needed	
Lead press curing	X	X	Minimize or eliminate Pentaerythritol, sulfur and sulfur-bearing accelerators	
Steam and moisture curing	X	X	Systems-containing sulfur or sulfur donors. Peroxide and maleimide systems are satisfactory for high pressure steam cure, but have not been studied in other moisture cures	
Uncured applications			Acid acceptors only. Accelerators and curatives are not needed	

gives a quick reference to choices available, depending on properties needed. Keep in mind that this is general information and that certain compounding ingredients such as fillers, plasticizers, process aids, and other materials can influence cure activity and scorch properties.

Acid acceptors serve two purposes, first to neutralize any acid formed during vulcanization, and second to serve as cross-linking agents. The choice is determined by color of the vulcanizate, end-use properties, processing conditions, and health effects.

NOTE, the use of lead oxide or salts is NOT RECOMMENDED. Information provided on the use of lead in CSM compounds in this chapter is ONLY to serve as a basis for conversion to other acid acceptors and cross-linking agents. OSHA and EPA regulations dictate the discontinued use of lead complexes in CSM and ACSM compounds.

Magnesium oxide is low cost, light colored, and an effective cross-linking agent. It does increase water swell, hence should be avoided for low water or acid resistant applications. A high activity grade provides best results and paste forms improve dispersion and reduce water absorption, which reduces activity. The normal amounts used are 4 phr, but up to 20 phr is common for light-colored objects. If a CV system is used for vulcanization, 4–8 phr can be used. Increasing magnesium oxide from 1 to 10 phr has little effect on scorch time, but decreases compression set, tensile strength, and oil swell (due to tighter cross-linking) and increases modulus and resilience. There is little effect on elongation, except at higher amounts of Pentaerythritol.

Litharge provides very good resistance to water, chemicals, and compression sct. A drawback to litharge is the dark color imparted due to the formation of lead sulfide; hence it is restricted to dark or black applications. A high activity grades such as sublimed or fumed litharge is most effective and predispersed forms provide better dispersion and safer handling. The optimum state of cure and water resistance is obtained with 25–30 phr, but these levels reduce scorch time. *Note* that stearic acid or stearates should be avoided with litharge as these significantly reduce scorch safety. The water swell in a black compound with 10 phr litharge is about 65%, but decreases to about 3% at 30 phr. Similarly the oil resistance of CSM is improved with increased litharge levels due to the tighter state of cure, from 58% at 10 phr to 42% at 30 phr in a black-filled compound. As expected, modulus increases and elongation decreases with higher amounts of litharge. When using litharge, compliance with local and federal health and safety regulations needs to be addressed.

Combinations of MgO and PbO_2 are preferred for maximum heat resistance. Water resistance is also improved with the combination as compared with litharge alone. The best combination is 5 phr MgO with 5 phr PbO_2 for optimum retention of elongation and tensile strength. This also gives the better Mooney scorch time.

Organic lead salts such as tribasic lead maleate or dibasic lead phthalate are effective acid acceptors for both black and light-colored stocks and provide better water resistance than the epoxy/MgO combination. Colored compounds have fairly stable weather resistance but white compounds turn gray on exposure to sunlight.

Epoxy resins with an epoxide equivalent of 180–200 and epoxidized oils react with acidic byproducts to form water-resistant compounds. These are used in combination with MgO, hence are useful for light-colored products. Epoxy resins also serve as plasticizers and make uncured compounds tacky. The resultant vulcanizates have low modulus and high elongation.

Calcium hydroxide may be used as the acid acceptor with a maleimide/amine cure system to obtain maximum compression set resistance.

Cross-linking agents and activators are needed in combination with the magnesia, litharge, and others, to achieve a satisfactory rate and state of cure. The primary vulcanizing agents are sulfur or sulfur-bearing accelerators; peroxide, usually with a coagent; and maleimides such as N,N'-m-phenylene-dimaleimide (HVA-2). Various accelerators and coagents are used to activate the acid acceptor or accelerate the cross-linking activity of the main curing agent. The major combinations are given in Table 9.8, which provides a quick reference for selecting the best combination for a given application.

Sulfur and sulfur donors such as TMTD, TETD, and DPTT provide safe processing, low cost, and good general purpose curing agents for CSM, either black or light colored. Normally 2 phr of TMTD is used with 0.5–1.0 of sulfur, but TETD may be used for a

TABLE 9.8
Summary of Vulcanizing Agents and Activators for CSM

Sulfur and Sulfur-Donor Systems

TMTD/Sulfur or TETD/Sulfur	General purpose combination for non-black and black compounds lowest cost
Tetrone A	Primary general purpose curing agent for non-black and black compounds
	Higher cost, better electrical properties than TMTD/Sulfur
MBT, MBTS and DOTG	Secondary accelerators used with Tetrone A or TMTD/Sulfur (MBT can be used alone as a primary curative)
	MBT and MBTS improve scorch safety in some systems
HVA-2	Secondary accelerator and cure promoter to achieve a higher state of cure in heat-resistant compounds
	Improves insulation resistance of electrical compounds

Peroxide System

Organic Peroxide (Varox powder, Di-Cup 40C have been used satisfactorily; others should be suitable also)	Excellent scorch safety and compression set resistance in non-black or black compounds
	Special compounding is necessary (see text)
Coagent (TAC and HVA-2 have been used satisfactorily; others should be suitable also)	Increases state of cure of peroxide systems

Maleimide System

HVA-2	Primary curative
	Improves scorch safety in mineral-filled compounds. Excellent compression set resistance
Ionac GFI-710	Activators used with HVA-2 for maximum compression set
	Limited to black or dark-colored compounds because of discoloration and staining.

TABLE 9.8   (Continued)
Summary of Vulcanizing Agents and Activators for CSM

4,4′-Dithio Morpholine	Provides excellent bin stability in combination with HVA-2 and litharge in clay loaded compounds.
Hydrogenated Rosin	Cure activator with HVA-2 to improve electrical properties.
Other curatives and activators	
Pentaerythritol	Cure promoter in combination with magnesia
NBC	Activator used in heat-resistant compounds. Must be used in combination with Tetrone A and MBTS or TMTD and sulfur
Poly-dispersion HC-11	Complete curing system (activated lead oxides plus curing agent) in predispersed form
	Provides good processing safety and bin stability, moderate vulcanizate properties
PAPI	Acts as a retarder when scorchiness is caused by excess moisture

faster cure rate and better heat resistance, but at a sacrifice in process safety. Magnesia is normally used as the acid acceptor/cross-linking agent.

DPTT, dipentamethylene thiuram tetrasulfide, such as Tetrone A, as the primary curing agent for CSM results in short, flat cure rates, and a high state of cure. Effective activators for DPTT are thiazoles, thiurams, or guanidines and may be used with either magnesia for light-colored objects or litharge for black compounds. The usual amount of DPTT is 0.75–2.0 phr and as the levels are increased the state of cure is higher, resulting in higher tensile strength and modulus with lower elongation and compression set and scorch safety. Figure 9.7 illustrates the influence of DPTT on physical properties of a light-colored CSM compound.

Secondary accelerators for DPTT, TMTD, TETD, and sulfur are MBT, MBTS, DOTG, and MBM (HVA-2). The MBT and MBTS at 0.5–1.5 phr, act as retarder/ activators providing improved scorch safety and higher states of cure. The combination of DPTT and MBTS is the general purpose curing agent for black compounds. MBT is sometimes used as the primary curing agent for water resistant light-colored compounds with tribasic lead maleate as the acid acceptor. Another secondary accelerator occasionally used is DOTG at 0.25–0.5 phr with sulfur systems.

HVA-2 as a cure promoter results in improved state of cure, heat, and electrical insulation resistance. MBM is also used as the primary curative with the maleimide system and is effective as a coagent for peroxide cures.

A comparison of sulfur curing systems in dark colored ACM is given in Table 9.9. Compounds 3A, 3B, and 3C are general purpose systems from which the chemist may select, depending upon process safety, water and aging resistance and compression set. Additional general purpose compounds for dark compounds may be found in Table 9.10, compounds 4A, 4B, 4E, and 4F. Compounds 3B, 3C, 3D, and 4B provide the best water resistance. The best compression set resistance

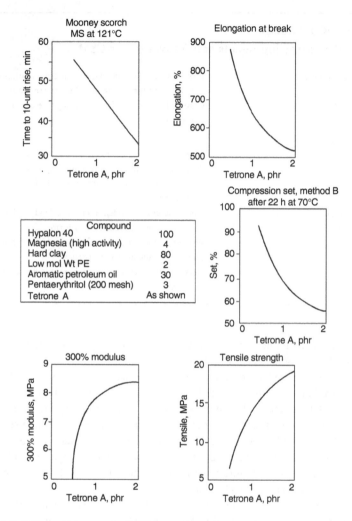

FIGURE 9.7 Effect of Tetrone A (DPTT) on physical properties.

was found in compounds 4G, 4H, 4C, and 4D. The better heat resistance was obtained in compounds 4G, 4F, 4C, 4D, 3F, 3H, 3G, and 3E. As with all elastomers, the mineral filled compounds exhibit the best heat resistance. The larger sized fillers are best for compression set.

9.3.4.1 Organic Peroxides

An organic peroxide vulcanizes CSM to give a compound with good processing safety and vulcanizates with good compression set. These may be used with either a magnesia or a lead oxide cross linker/acid acceptor. The selection of peroxide type depends upon curing temperature conditions and process safety desired. The most common types are dicumyl peroxide, such as DiCup 40C, or 2,5 dimethyl-2,5-bis[*t*-butyl proxy] hexane,

TABLE 9.9

Sulfur Curing Systems for Dark Colored CSM

Compound	3A	3B	3C	3D	3E	3F	3G	3H
Hypalon 40	100	100	100	100	100	100	100	100
N774 Carbon black	40	40	40	40	40	40	40	40
Acid acceptors and activators								
Magnesia (high activity)	4	—	—	—	10	10	10	10
Pentaerythritol (200 mesh)	3	—	—	—	—	—	—	—
Litharge (Sublimed)	—	25	25	—	20	20	20	20
Epoxy-resin	—	—	—	15	—	—	—	—
NBC	—	—	—	—	3	3	3	3
Sulfur	1	1	—	—	1	—	—	—
TMTD (or TETD)	2	2	2	—	2	—	—	—
Tetrone A	—	—	2	1.5	—	2	0.75	0.75
MBTS	—	—	0.5	0.5	—	0.5	0.5	0.5
DOTG	—	—	—	0.25	—	—	—	—
HVA-2	—	—	—	—	—	—	—	1
Compound properties								
Mooney scorch, MS at 121°C								
Minimum viscosity	34	33	34	19	47	42	44	40
Time to a 10-unit rise, min	23	26	20	19	13	9	9	10
Vulcanizate properties								
Cure: 30 min at 153°C								
100% Modulus, MPa	9.6	9.6	9.0	3.4	8.4	8.6	3.2	5.6
Tensile strength, MPa	25.6	26.0	25.8	24.4	26.8	25.4	15.8	20.0
Elongation, %	230	190	200	370	240	240	420	240
Hardness, Durometer A, Pts	75	82	81	69	81	82	67	67
Permanent set at break, %	2	1	1	—	4	4	13	4
Tear strength, Die B, kN/m	31.5	26.2	25.4	28.9	38.5	36.7	43.8	31.5

(continued)

TABLE 9.9 (Continued)

Sulfur Curing Systems for Dark Colored CSM

Compound	3A	3B	3C	3D	3E	3F	3G	3H
Aged 7 Days at 121°C								
Tensile strength, % change	—	-22	-16	—	-2	-2	-4	+7
Elongation, % change	—	-42	-40	—	-21	0	-7	0
Hardness, pts change	—	+5	+5	—	+1	+1	+6	+6
Aged 70 h at 100°C in ASTM No. 3 Oil								
100% Modulus, MPa	4.4	—	7.2	3.0	—	9.0	1.4	4.8
Tensile strength, MPa	13.8	—	17.4	15.4	—	12.4	8.2	10.4
Elongation, %	180	—	160	220	—	120	300	170
Hardness, Durometer A, Pts	54	—	62	47	—	65	38	49
Volume change, %	+76	—	+65	+67	+68	+65	+83	+66
Aged 28 days at 70°C in Water								
100% Modulus, MPa	8.6	—	8.8	2.8	—	8.4	4.2	6.2
Tensile strength, MPa	15.2	—	26.8	25.8	—	21.2	16.6	16.6
Elongation, %	140	—	150	300	—	180	310	200
Hardness, Durometer A, Pts	69	—	80	68	—	75	64	65
Volume change, %	+62	+6	+5	+4	+35	+29	+27	+26
Aged volume change, % after Immersion in								
Nitric Acid (70%) 14 Days at 24°C	+41	+13	+12	+23	+75	+70	+47	+41
Sodium hypochlorate (20%) 28 days at 70°C	No data	-2	-2	-5	No data	+2	+1	+1
Clash-Berg Torsional Stiffness T68.9 MPa, °C	-16	-15	—	-15	-14	—	-16	-19
Compression set, method B, %								
After 22 h at 70°C	31	14	14	23	20	23	38	20
After 70 h at 100°C	74	40	40	37	—	43	61	35

TABLE 9.10
Additional Curing Systems for Dark Colored CSM

Compound	4A	4C	4D	4E	4I	4J
Hypalon 40	100	100	100	100	—	—
Hypalon 4085	—	—	—	—	100	100
N774 Carbon black	50	50	50	3	50	50
Natural ground whiting	—	—	—	80	—	—
Whitetex clay	—	—	—	—	50	50
Chlorowax LV	30	30	30	20	30	30
AC617A	—	—	—	—	3	3
Paraffin wax	—	—	—	—	3	3
Maglite D	4	—	10	4	—	—
Pentaerythritol (200 mesh)	3	—	—	3	—	—
Calcium hydroxide	—	4	—	—	—	—
Kyowa DHT-4A (Hydrocalcite)	—	—	—	—	20	10
Vanox NBC	3	—	—	3	2	2
Tetrone A	2	—	—	2	1	1
MBTS	0.5	—	—	0.5	1	1
HVA-2	—	3	—	—	1	1
Ionac GFI 710	—	2	—	—	—	—
Varox powder	—	—	6	—	—	—
Triallyl Cyanurate (TAC)	—	—	3.75	—	—	—
Mooney scorch, MS at 121°C						
Minimum viscosity	10	10	13	17	43	31
Time to a 5-unit rise, min	14	23	16	15	15	22
ODR, 30 min at 160°C, ±3° arc						
t_g2, minutes	1.5	1.5	4	1.8	1.7	2.2
t_c90, minutes	11	5	24	12.5	14.8	19.2
Maximum torque, N·m	7.7	7.6	5.9	8.4	79	76
Unaged properties cured at 160°C (min)	10	5	25	15	15	20
Stress/strain original						
100% Modulus, MPa	2.8	3.8	2.4	1.4	6.4	4.9
Tensile strength, MPa	17.2	14.4	11.8	13.4	13.2	14.7
Elongation, %	340	210	290	560	260	335
Hardness, Durometer A, pts	60	62	64	56	76	71
Aged 14 days at 121°C-change					7 days	7 days
Tensile strength, %	+2	+5	+14	−6	4	−3
Elongation, %	−27	−5	+3	−4	−23	−34
Hardness, pts	+8	+5	+8	+6	5	8
Aged 5 days at 149°C-change					70 h	70 h
Tensile strength, %	−16	−2	−7	−43	1	−6
Elongation, %	−62	−29	−31	−46	−49	−58
Hardness, pts	+17	+8	+11	+16	10	14

(*continued*)

TABLE 9.10   (Continued)
Additional Curing Systems for Dark Colored CSM

Compound	4A	4C	4D	4E	4I	4J
Aged 70 h at 121°C in ASTM No. 3 Oil-change						
Tensile strength, %	−32	−26	−26	−18	−7	−11
Elongation at break, %	−27	−29	−28	+5	−9	−23
Hardness, pts	−25	−13	−22	−28	−23	−19
Volume, %	+53	+41	+48	+68	+42	+42
Aged in water 7 days at 100°C-change						
Volume, %	+98	+27	+85	+206	+19	+28
Compression set, method B, %						
After 22 h at 121°C	87	37	47	88	42	44

for example, Varox powder. When using a peroxide cure system it is important to cure at higher temperatures and for sufficient time to completely use up the peroxide otherwise poorer aging properties result. It is also important to avoid acidic materials, which interfere with peroxides as well as unsaturated and highly aromatic plasticizers. Chlorinated paraffins, esters, and polyesters normally do not interfere with peroxides. Magnesia at 10–20 phr is the most common acid acceptor, although dibasic lead phthalate may be used for better water resistance. Antioxidants such as NBC interferes with a peroxide cure and any alternates should be carefully selected.

Coagents should be used with the peroxide to obtain a satisfactory rate and state of cure. Commonly used coagents are TAC (triallyl cyanurate), TATM (triallyl trimellitate), HVA-2 (*N,N′-m*-phenylene-dimaleimide), di- or tri-methylacrylates, liquid high 1,2 polybutadiene, diallyl phthalates, and others. The allylic types provide better process safety and are more effective in CSM.

As the amount of either the peroxide or coagent, such as Triallyl Cyanurate, is increased, the modulus, or state of cure, increases, and the elongation, tensile strength, tear strength, and compression set decreases. Figure 9.8 illustrates this effect on the physical properties of a peroxide/coagent cured CSM.

Maleimides such as HVA-2 serve as either a primary curing agent for CSM, a cure promoter for sulfur cures, or as an activator for peroxide systems. It is particularly effective in mineral-filled compounds in giving good scorch safety. A combination of an amine activator with HVA-2 with calcium hydroxide as the acid acceptor gives compression set resistance equivalent to that of a peroxide cure.

Morfax (4-morpholinyl-2-benzothiazole disulfide) combined with HVA-2 and litharge acid acceptor provides excellent bin stability along with a good rate and state of cure, especially in highly loaded clay compounds.

The addition of a hydrogenated resin such as Stabelite Resin improves the electrical properties of a maleimide cure system in a CSM insulation compound and the vulcanizate exhibits little or no discoloration or adhesion to the copper.

Pentaerythritol (PER) functions as a cure promoter with magnesia, activating cure rate and increasing the state of cure. The level of magnesia may be reduced and long-term aging resistance is improved. These primary polyols are activators only

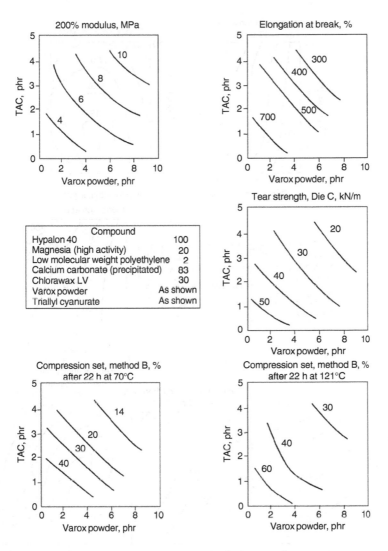

FIGURE 9.8 Effect of peroxide and coagent on physical properties.

and require the presence of sulfur, sulfur donors, or other curing agents. Although polyols may be used with litharge acid acceptors the resulting scorch times are extremely short. These also increase water absorption.

The processing safety with polyols is directly related to the melting point, the higher the better, and experience has shown that PER has the best balance of process safety, cost, and vulcanizate properties. The combination of 3 phr PER with 4 phr magnesia provides the best long-term aging resistance. The PER does not melt at process temperatures, hence a very fine particle size grade of 200 mesh, such as PER-200 from Hercules, is recommended. Figure 9.9 illustrates the influence of varying the amounts of both magnesia and Pentaerythritol on the physical properties

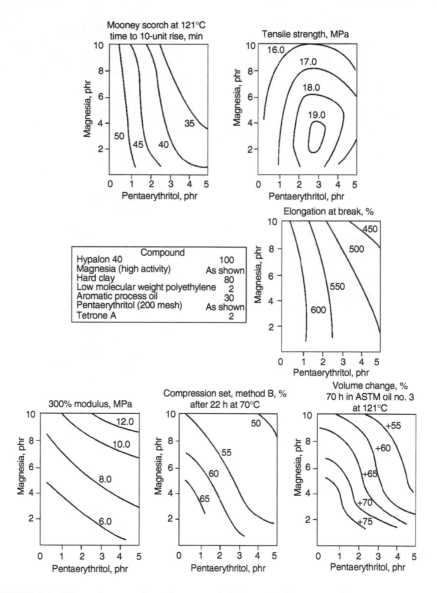

FIGURE 9.9 Effect of magnesia and pentaerythritol on physical properties.

of the vulcanizate. The recommended amounts of these cross-linking agents is magnesia 4 phr and Pentaerythritol 3 phr for the better aging resistance than with 10–20 phr magnesia without PER.

Miscellaneous Curatives include predispersed curative packages, PAPI (a polyisocyanate) and NBC (nickel dibutyldithiocarbamate). Poly-dispersion HC-11 is a curative package based on an activated lead oxide with undisclosed curatives in an elastomeric dispersing agent. This is fine for many applications as it provides simplicity, process safety, bin stability, and adequate rate and state of cure. PAPI serves as a moisture scavenger when added to curing systems sensitive to moisture, thus improves scorch resistance. Other isocyanates may also be effective. NBC functions as both an antioxidant as well as an accelerator. It imparts a green color to the compound, which needs to be recognized when selecting it. Moreover, as an accelerator the scorch time and cure rate is reduced, hence processing properties need to be considered.

9.3.4.2 Summary of Cross-Linking Systems

A Review of curatives and activators is provided in Table 9.8, which helps in selecting the best system for a given application and process conditions. The correct curative is important in achieving the desired properties for a given end use.

Color is often a factor to consider, hence magnesia and no NBC or amine activators should be used in light-colored products. Litharge produces a dark gray or black color.

Water resistance is very poor with a magnesia acid acceptor; hence litharge is commonly the material of choice. Compounds 3B and 3C in Table 9.9 show the best water resistance for a litharge cure. An epoxy acid acceptor also exhibits very good water resistance and may be used in either light or black colored rubber goods, as may be seen in Table 9.9 compound 3D. Compounds 4I and 4J in Table 9.10 using a hydrocalcite (Kyowa DHT-4A) as an acid acceptor also provides low water swell in light-colored products. An additional combination for a low water swell compound is with calcium hydroxide, HVA-2, and Ionac GFI 710 (N, N'-diphenyl propylenediamine from Sybron Chemicals) as illustrated in compound 4C in Table 9.10.

Compression set resistance is best obtained with a peroxide/coagent cured compound, as seen in Table 9.11 compound 2H plus 2I and 4D from Table 9.10. Another good cure system for good compression set is calcium hydroxide with HVA-2 and GFI 710 as illustrated by compound 4C in Table 9.10.

The best heat resistance may be obtained using a combination of magnesia and litharge, NBC antioxidant, and a sulfur-bearing curative as in Table 9.9 compounds 3E and 3F; however, this is only suitable for dark or black-colored compounds. Very good heat resistance is also available in light colored compounds with a magnesia acid acceptor and either a peroxide or a Tetrone A/HV-2 cure system as illustrated in Table 9.11 compounds 2I and 2J. Light colored fillers will enhance the heat resistance as well.

Processing safety and bin stability is improved by the addition of 1 phr of HVA-2 to a Tetrone A cure system as seen in a comparison of compounds 2B and 2J in Table 9.11. The peroxide cure as illustrated in compound 2I also gives safe, stable compounds, provided excess moisture or humidity and high-process temperatures are avoided.

Liquid curing medium (LCM) is a practical method provided that a rapid cure rate system and a high viscosity CSM are selected. A peroxide system may leave a tacky surface unless complete submersion or "rain" of the LCM on the part is provided.

Lead press curing is often used for long length hose, but the surface of a light-colored compound tends to darken in the presence of PER, water, and many sulfur and sulfur-bearing accelerators. The use of TMTD and sulfur or Tetrone A has been used successfully in a lead press cure.

TABLE 9.11
Cure Systems for Light-Colored Compounds

Compound	2A	2B	2E	2F	2G	2H	2I	2J
Hypalon 40	100	100	100	100	100	100	100	100
TiO₂ (rutile)	35	35	35	35	35	35	35	35
Whiting	50	50	—	—	—	—	50	50
Hydrated amorphous silica	—	—	20	—	—	—	—	—
Whitetex clay	—	—	—	50	50	50	—	—
Magnesia (high activity)	4	4	3	—	3	—	10	4
Pentaerythritol (200 mesh)	3	3	—	—	—	—	—	3
Epoxy resin	—	—	4	—	—	—	—	—
Tribasic lead maleate	—	—	—	40	40	—	—	—
Dibasic lead phthalate	—	—	—	—	—	40	—	—
Sulfur	1	—	—	—	—	—	—	—
Tetrone A	2	2	2	—	—	—	—	2
TMTD	—	—	—	—	1.5	—	—	—
MBT	—	—	—	1.5	—	—	—	—
Stabelite Resin	—	—	—	2.5	—	—	—	—
Varox powder	—	—	—	—	—	6	6	—
Triallyl Cyanurate	—	—	—	—	—	4	4	—
HVA-2	—	—	—	—	—	—	—	1
TOTAL	195	194	164	229	229.5	235	205	195
Mooney scorch, MS at 121°C								
Minimum viscosity, units	36	36	43	44	50	33	36	40
Time to 10 pt rise, min	26	20	15	18	23	>30	>30	>30

Original properties cured 30 min at 153°C								
Modulus at 100%, MPa	3.8	3.4	2.4	10.5	6.2	10.4	4.2	4.0
Modulus at 300%, MPa	7.4	6.8	7.0	12.4	—	—	11.0	7.0
Tensile, MPa	17.6	17.2	18.0	13.0	10.4	13.2	13.8	16.4
Elongation, %	440	460	480	330	240	140	360	500
Shore A, pts	77	75	78	84	71	80	71	76
Aged 7 days at 121°C in air-change								
Tensile, %	−12	−14	—	+43	+112	+41	0	−7
Elongation, %	−45	−41	—	−48	−33	−28	−28	−50
Shore A, pts	+6	+5	—	—	+7	+1	+3	+4
Aged 70 h at 100°C in ASTM #3 Oil-change								
Tensile, %	−30	−30	−21	−26	−13	+4.5	−18	−33
Elongation, %	+5	−4	−20	−6	−8	−14	−31	−26
Shore A, pts	−30	−27	−32	−41	−26	−11	−13	−29
Volume, %	+56	+63	+80	+89	—	+40	+49	+68
Aged in water at 70°C—volume change								
7 days, %	+53	+55	+9	+1	+15	+3	+33	+69
14 days, %	+85	+90	+6	+2	+22	+5	+51	—
28 days, %	+133	+135	+5	+2	+35	—	—	—
Clash–Berg low-temperature torsional stiffness								
T 68.9 MPa, °C	−16	−14	—	−15	−15	−18	−18	—
Compression set B								
Set after 22 h at 70°C, %	40	39	38	81	82	6	18	42
Set after 70 h at 100°C	90	82	91	93	—	15	22	87
NBS Abrasion resistance								
% of standard	90	88	130	—	—	—	—	108

Steam and moisture curing methods can be used successfully with CSM as the sulfonyl chloride groups are hydrolyzed by water making them more reactive with metal oxides and curatives. Warm or boiling water takes from weeks to hours to effect a cure but in steam at 202.9°C the time needed is 1 min. Sulfur and sulfur donor systems are satisfactory but maleimide systems are deactivated by moisture, unless rapid cures are used in high pressure steam. Peroxide cures require higher temperatures and pressures if used in steam and the autoclave needs to be vented at the beginning of the cure to remove any air. Magnesia at 8–20 phr is most effective in steam as lead oxides have lower water absorption resulting in slow cure rates. A combination of magnesia and litharge provide an excellent cure, but are slower than magnesia alone.

Unvulcanized compounds are often used for roofing and pond and ditch liners. Accelerators or curatives are excluded in these compounds to avoid scorching and short storage life before seaming [8]. The compounds should contain a metal oxide acid acceptor and stabilizer during storage and the service life of the product. Magnesia at 4–5 phr is recommended for good color stability and a lead free system, especially for potable water collection. If water or acid resistance is needed 25 phr of sublimed litharge should be used in black compounds or 20–40 phr dibasic lead phthalate for light-colored articles. The presence of a metal oxide and moisture does affect a cure over time during the service life of the sheeting.

Processing conditions of CSM compounds require some modification be made of the usual operating temperatures used with other elastomers to allow for its more thermoplastic nature [9]. The polymer does not break down with extended high shear conditions making it necessary to select the right viscosity grade of CSM for the operating conditions to be used. The presence of highly reactive sulfonyl chloride cure sites and bonded chlorine requires halogen-resistant stainless steel or hard chromed steel molds to prevent corrosion, mold fouling, and sticking. The polymer also contains low levels of residual carbon tetrachloride and chloroform, hence proper ventilation needs to be observed. In addition, the sensitivity to zinc-bearing materials dictates that these materials be avoided, including zinc stearates, or other zinc salts as process aids or partitioning material.

CSM elastomers may be processed on typical rubber processing equipment much like other rubbers, except that the more thermoplastic nature of the polymer as shown in Figure 9.10 requires some adjustment in operating temperatures. Figure 9.11 assists in selecting the appropriate grade of CSM for the process temperatures to be encountered.

Process aids as release agents assist in release from process equipment as well as from molds during the vulcanization step. It may be advisable to use a combination of small amounts of a few different materials that are effective at different temperatures and conditions. Microcrystalline or Paraffinic waxes such as Multiwax 180M and Vanwax H and petrolatum are effective but blooms if used in excess. Polyethylene glycol, such as Carbowax 3350, is very effective at low-processing temperatures. At higher temperatures low molecular weight polyethylene such as A-C Polyethylene 617A and A-C Polyethylene 1702, or equivalents, are effective, but these must be added at or near their fluxing temperatures to prevent mill sticking. Low molecular weight polyethylene is rarely used alone. Vanfre HYP, a proprietary process aid, is effective at improving flow and mold release.

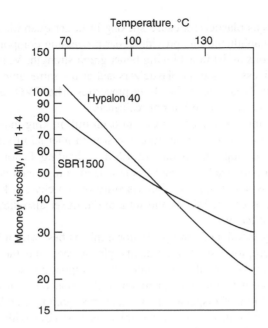

FIGURE 9.10 Viscosity/temperature relationship of CSM versus SBR.

A high *cis*-polybutadiene or EPDM at levels of 3–5 phr are useful at improving release and increasing green strength. Their presence has little effect on cure rates and there is no need to cure these elastomers.

Stearic acid, stearates (other than zinc), and many other process aids are safe and effective with magnesia cures. These should NOT be used with litharge as scorch safety is greatly reduced.

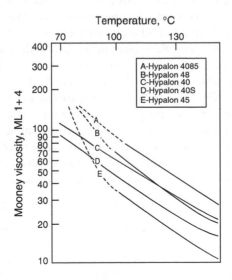

FIGURE 9.11 Viscosity/temperature relationship of various CSM polymers.

Aromatic oils as plasticizers cause sticking in larger quantities, so a blend with up to 15–20 phr naphthenic oils provide easier processing compounds. Reinforcing fillers with these oils assist in achieving better green strength. Vulcanized vegetable oil dries up stickiness caused by plasticizers and at the same time provides body to very soft compounds and better handling properties. The VVO reduces heat, light, and compression set resistance but not resilience.

Mixing in an internal mixer is best, but open mill mixing may also be satisfactory. An upside down mixing procedure is best with internal mixers.

Due to the thermoplastic nature of CSM, a loading factor of 70%–75% is most effective. Highly loaded or highly plasticized compounds need to be at 75% fill factor, whereas lower loaded, high viscosity or scorchy stocks require a lower load factor. The power capabilities and wear of the mixer also dictates some further adjustment in load factors.

As previously mentioned, an upside down mix is best with a loading sequence of fillers, acid acceptors, small ingredients, plasticizers, and the polymer on top. A single-pass mix is commonly used, unless the compound is scorchy in which case a two-pass mix, both in the internal mixer or the final pass on the mill with the curatives, in the form of dispersions for rapid incorporation, added and dispersed there. The sheet off mill of an internal mixer is best cold with the stock passed completely through the nip without banding to remove heat, then banding, quickly blending, and sheeting off and cooling to reduce heat history.

Mill mixing should be done with full cooling water and a tight nip at the beginning. High molecular weight grades, such as Hypalon 45 and 48, requires a hot roll at 60°C to obtain a smooth band. After incorporating the fillers and part of the plasticizer turn on full cooling water. Do not overload the mill as this will lead to poor dispersion and scorch. Once a band is formed begin adding the fillers. The magnesia and polyethylene should go in with the fillers as these will cause sticking if added alone. Stearic acid and process aids need to be incorporated after the magnesia is dispersed. Plasticizers may be added with nonreinforcing fillers but after reinforcing ones. If resins or polyethylene is used it may be necessary to shut off the water while fluxing them into the compound then return to full flow of the cooling water. Litharge and curatives, very preferably in the form of dispersions, are added at the end, unless the batch is too hot in which case cool the stock and add these in a second pass.

When extruding CSM compounds a cold feed extruder is best, but a hot feed one will also work, but uniform warming of the compound is essential for dimensional stability. Higher die temperatures are needed for a smooth surface and lower feed zone temperatures provide better back pressure for air removal. Operating temperatures of the equipment for both cold and hot feed extruders are

	Hot Feed	Cold Feed
Screw	Cool	55°C
Feed	50°C	50°C
Barrel	60°C	60°C
Head & die	95°C–105°C	93°C–107°C

Adjustments from these suggestions are needed to accommodate equipment characteristics. In addition, in the case of tougher CSM-based compounds, a higher temperature of 60°C in the feed zone and 120°C for the die. The extrudate needs to be cooled quickly to maintain dimensions due to the thermoplastic nature of CSM. A reminder, DO NOT use zinc stearates as an anti stick agent in the cooling tank.

Calendering of CSM is best with uneven roll speeds and the top roll about 10°C hotter than the middle roll. A temperature range of 60°C–95°C is usual depending upon the compound. Again tougher grades such as Hypalon 45 and 48 need a temperature of 95°C and even hotter, up to 150°C for unplasticized compounds. Cooler conditions help eliminate trapped air but higher temperatures provide a smoother surface. A balance is needed as too high a temperature may result in sagging of the sheet and poor dimensional control.

Molding of CSM requires some special care in the type of metal due to the acidic nature of CSM even with the presence of an acid acceptor. A hard surface (minimum 45 Rockwell C), abrasion and corrosion resistant and non-porous steel needs to be employed. A hard chrome plate at least 0.1 mm thick is recommended. Stainless steel is a more expensive but more effective alternative. Lower mold temperatures reduce mold fouling and corrosion if economics will allow. Molds need to be cleaned at least weekly with an alkali solution with ultrasonic vibration.

The compound and mold designs need to be coordinated to allow for the thermoplastic nature of CSM stocks.

Hand or machine building of articles based on CSM is best accomplished if the compound is pre-warmed to about 50°C–60°C to provide sufficient building tack. A tackifier such as Cumar P-10 at 10–15 phr will be of some help. Cyclohexanone provides good tack for extended times, provided that OSHA regulations for handling it are met. Rolls made from calendered sheet need the ends to be firmly blocked before wrapping tightly. If the stock is very stiff resting overnight may be advisable to allow any trapped air to escape. When vulcanizing the rolls start with a pre-warmed autoclave at full temperature to achieve fast curing before any excessive flow can occur. A layer of another CR or SBR compound over the CSM compound protects the wrapping tape from deterioration by any SO_4 generated. This layer can be removed during the grinding process. The inclusion of 4 phr of calcium hydroxide in the compound helps to absorb any SO_4 present.

Adhesion to metal is best accomplished with commercial adhesive such as Chemlok 250 or Thixon P-6-3 for solvent acceptable processes, and Chemlok 855, or Robond 5266 for water-based requirements. Contact your adhesive supplier for severe or unusual applications. Epoxy cured CSM compounds provide the best adhesion. Litharge cures are not as effective but are often adequate.

9.4 ALKYLATED CHLOROSULFONATED POLYETHYLENE

ACSM elastomers were developed primarily for timing and multi V-belt applications as well as other dynamic or low-temperature applications [10]. These polymers exhibit improved low-temperature resistance, lower hysteresis, and mechanical properties (due to their linear back bone) compared with CSM (Figure 9.12).

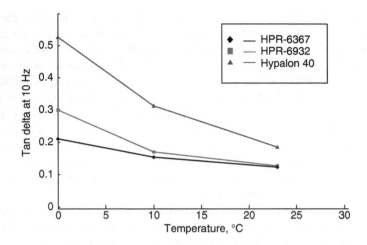

FIGURE 9.12 Dynamic properties of ACSM versus CSM.

The heat resistance of ACSM, CR, and HNBR, as measured by hardness change at 120°C, is given in Figure 9.13. As may be noted at the higher operating temperatures now encountered for timing belts, the CR based ones have very limited life at 120°C whereas the ACSM and HNBR compounds have excellent properties. When tested at 140°C there is an even greater advantage of ACSM and to a lesser extent, HNBR, versus CR as indicated in Figure 9.14. The compound properties of the three compounds used in this comparison, a typical sulfur modified CR, a sulfur-cured HNBR timing belt compound, and an ACSM of the same hardness as the CR one, are shown in Table 9.15.

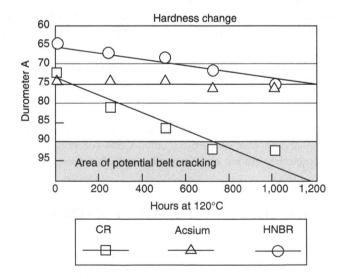

FIGURE 9.13 ACSM Comparative heat resistance at 120°C.

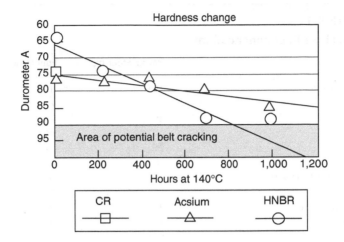

FIGURE 9.14 ACSM Comparative heat resistances at 140°C.

The dynamic properties, as measured by the Tan Delta at 10 Hz, of ACSM 6367, ACSM 6932, and CSM 40 are compared in Figure 9.12. The CSM compound at 25°C may be acceptable but at lower temperatures it loses the required low-temperature dynamic properties, and in the long term it would fail. The ACSM 6397 and especially ACSM 6932, at the lower temperatures needed for modern engines, would work well for extended periods of time.

The low-temperature resistance of ACSM is about 20°C lower than the base CSM rubber, the T_g being −24°C to −27°C.

Compounding of ACSM is the same as for CSM, taking into account the need to select fillers and cure systems that will provide good dynamic properties.

9.5 APPLICATIONS

9.5.1 CSM Hoses

Two different hose compounds are given in Table 9.12, one a Power Steering Hose and the other a Fuel Hose Jacket. The power steering hose is a peroxide cured compound for good compression set, heat aging, and flex resistance as well as resistance to Dexron III.

The fuel hose jacket compound is a HVA-2/Tetrone A cured compound for adhesion to the tie layer and fabric. It is a less expensive formulation as it does not have to meet the rigid dynamic properties of a power steering hose and is exposed to only occasional oil spills.

9.5.2 CSM Industrial Roll Cover

The formulation in Table 9.13 is for a general purpose, heat-resistant roll cover. If there are acids and water expected in service then a litharge cure would be needed for good performance.

TABLE 9.12

CSM Hose Recommendations

Hose	Power Steering	Fuel Hose Cover
Hypalon 4085	100	100
N762	75	60
N990	25	—
Atomite	—	90
TOTM	25	—
DOS	—	7.5
Kenflex A1 POD 93%	—	5
Sundex 790	—	22.5
Carbowax 3350	2	2
PE 617A	3	3
PE 200	5	3
Maglite D	10	5
Varox DCP 40KE	7	—
TAC DLC-A	4	—
HVA-2	—	1
MBTS	—	1
NBC	—	1
Tetrone A	—	1
Total	256	302
Mooney viscosity		
ML $1+4$ at 100°C	106.1	75.1
Mooney scorch MS at 125°C		
Time to 5 pt rise (min)	12.0	9.7
Time to 10 pt rise (min)	15.6	13.3
MDR at 160°C, 0.5° arc, 30 min chart		
ML (dN·m)	2.4	1.8
MH (dN·m)	29.5	15.5
Ts_2 (min)	0.7	1.2
Tc_{90} (min)	6.8	7.1
Physical properties, press cured	15 min at 175°C	15 min at 160°C
Hardness, Shore A	78	71
Modulus at 50%, MPa	4.1	2.6
Modulus at 100%, MPa	11.0	4.1
Tensile, MPa	19.6	11.8
Elongation, %	175	337
Tear Die C, kN/m	24	29.4
Low temperature		
DSC—T_g, °C	−31	−30.0
Aged in air, 70 h at 150°C-change		
Hardness, pts	7	19
Modulus at 50%, %	85	244
Tensile, %	−4.1	7.7
Elongation, %	−30.3	−60.1

TABLE 9.12   (Continued)
CSM Hose Recommendations

Hose	Power Steering	Fuel Hose Cover
Compression set B		
70 h at 125°C	35	79.6
Aged in Dexron III, 168 h at 150°C		
Hardness, pts	−9	NT
Modulus at 100%, %	10.0	NT
Tensile, %	−13.8	NT
Elongation, %	−29.1	NT
Volume swell, %	16.5	NT
Aged in IRM 903, 70 h at 125°C		
Hardness, pts	−18	−27
Modulus at 50%, %	−29.0	NT
Modulus at 100%, %	NT	−38.4
Tensile, %	−7.0	−21.0
Elongation, %	−23.4	−11.5
Volume swell, %	38	45

9.5.3   EXTRA HEAVY-DUTY MINING CABLE JACKETS

Formulations are given in Table 9.14, one of which uses treated silicates for higher modulus and lower elongation. These may be colored to meet customer requirements.

TABLE 9.13
CSM Industrial Roll Cover

Industrial Roll Cover	CSM
Hypalon 40	100
N 990	100
PE 617A	2
Carbowax 3350	1
Stearic acid	1
Light process oil	10
DOP	15
Pentaerythritol PE200	3
Maglite D	4
MBTS	0.5
Tetrone A	1.5
Total	238
Physical properties	Cured 20 min at 152°C
Hardness, Shore A	71
Tensile, MPa	15.2
Elongation, %	275

TABLE 9.14

Extra Heavy-Duty Colored Mining Cable

Compound	A	B	
Hypalon 4085	30	30	
Hypalon 40	70	70	
Magnesium oxide high activity	4	4	
Ti-Pure R-902	5	5	
Magnesium silicate	30	—	
Magnesium silicate (treated)	—	15	
Aluminum silicate (chemically modified)	—	15	
Precipitated hydrated silica	25	25	
Stearic acid	1	1	
Low mol. wt. polyethylene	1	1	
Microcrystalline wax	3	3	
Ester plasticizer	20	20	
TMTD	2	2	
Sulfur	0.5	0.5	
Total	191.5	191.5	
Mooney scorch, MS at 121°C			
Minimum viscosity, units	20	18.5	
Time to 5 pt rise, min	19	19	
Original properties, cured 90 min at 145°C			ICEA 7.2.18.1.6
Modulus at 200%, MPa	5.6	6.6	4.83 min
Tensile, MPa	23.4	20.6	16.55 min
Elongation, %	560	480	300 min
Specific gravity	1.37	1.36	
Permanent set, %	12	12	30 max
Tear, kN/m	11.1	9.98	7.01 min
Low-temperature properties			
Solenoid brittleness,°C	−45	−45	
Aged in air 7 days at 100°C	Retained	Retained	
Tensile, %	79	80	70 min
Elongation, %	78	77	60 min
Aged in air 7 days at 121°C	Retained	Retained	
Tensile, %	—	75	
Elongation, %	—	54	
Aged 18 h at 121°C in ASTM Oil #2	Retained	Retained	
Tensile, %	108	91	60 min
Elongation, %	87	100	60 min

9.5.4 Uncured CSM Roofing

The stability of high molecular weight CSM coupled with its good green strength make it an ideal candidate for roofing material. The black color further assists in providing good UV resistance as well as physical strength. When cool the calendered sheets may be rolled up on themselves without a liner (Table 9.15).

TABLE 9.15
CSM Uncured Roofing Compound

Compound #	1471
Hypalon 45	100
Magnesium oxide	6
Atomite whiting	80
TiPure R960	30
Irganox 1010	1
Kenamide S (stearamide)	1
Carbowax 3350	1.25
Tinuvin 622	0.8
Total	220.05
Uncured physical properties at RT	
Modulus at 100%, MPa	6.5
Modulus at 200%, MPa	7.7
Tensile, MPa	11.6
Elongation, %	720
Air aged 672 h at 70°C	Change
Modulus at 100%, %	−38.7
Tensile, %	−62.5
Elongation, %	−12.5
Aged 2000 h at 80°C in QUV Tester	Change
Modulus at 100%, %	59.2
Tensile, %	48.1
Elongation, %	6

9.5.5 ACSM TIMING BELTS

The properties of typical timing belts based on CR, ACSM, and HNBR are shown in Table 9.16. CR had been the traditional material used in timing belts; however, the higher engine operating temperatures have dictated better heat-resistant elastomers. ACSM is an ideal candidate for this application as it was designed to have the good

TABLE 9.16
Properties of Timing Belt Compounds

	CR	Acsium	HNBR
Vulcanizate properties Tensile strength, MPa	20.9	19.3	27.4
Elongation at break, %	435	297	539
Modulus at 100% extension, MPa	3.9	4.1	1.7
Hardness, A	73	75	64
Rebound resilience, %	46.0	43.4	52.2

heat resistance of CSM coupled with better low-temperature properties. The dynamic properties are also much better than CSM, which are essential for this application. More detail is given in the ACSM section of this chapter. The compound should use medium to large size black fillers, no plasticizer and a cure system to provide a high modulus and a low tan delta such as litharge 20 phr, magnesia 10 phr, Tetrone A 2 phr, MBTS 0.5 phr, and NBC 3 phr.

ACKNOWLEDGMENT

I thank DuPont Specialty Elastomers L.L.C. for supplying the technical information upon which this chapter is based and for the able assistance of Charles R. Baddorf.

REFERENCES

1. Maurice Morton (ed.), *Rubber Technology*, Third Edition, Chapter 12, Part II, Van Nostrand Reinhold, New York, 1987, 638 pp.
2. Charles R. Baddorf, Hypalon® Chlorosulfonated Polyethylene and Ascium®, Alkylated Chlorosulfonated Polyethylene lecture presented to The Chicago Rubber Group, Inc. February, 2003.
3. DuPont Hypalon® Technical Information HPE-H68577–00-C0701, "Types, Properties and Uses of Hypalon" Rev. 3 November, 2002.
4. TOSOH Synthetic Rubber Bulletin, extos, TOSO-CSM and SKYPRENE.
5. DuPont Hypalon® Technical Information HPE-H68570–00-E0401, "Selecting a Filler" Rev. 3 November, 2002.
6. DuPont Hypalon® Technical Information HPE-H73616–00-C1102, "Selecting a Plasticizer and Processing Aid" Rev. 2 January, 2001.
7. DuPont Hypalon® Technical Information HPE-H68578–00-C1102, "Selecting a Curing System" Rev. 2 November, 2002.
8. DuPont Hypalon® Technical Information HPE-H73616–00-C1102, "Compounding Hypalon® for Weather Resistance" Rev. 2 November, 2002.
9. DuPont Hypalon® Technical Information HPE-H68574–00-C1102, "Processing Hypalon®" Rev. 2 November, 2002.
10. DuPont Ascium® Technical Information H-79944, "Product and Properties Guide" June, 1998.

10 Ethylene Vinyl Acetate Elastomers (EVM) (ASTM Designation AEM)

Hermann Meisenheimer and Andrea Zens

CONTENTS

10.1　INTRODUCTION

Ethylene vinyl acetate elastomers belong to the class of specialty rubbers. The first patents on this copolymer were granted to ICI in 1938 [1]. But it took over 20 years until the polymerization process was optimized and the first products were commercially available [2]. Today, a variety of copolymers spanning a wide range of vinyl acetate contents are on the market as given in Table 10.1.

This chapter first discusses the polymerization conditions and production processes. An overview of the raw material properties and the influence of the comonomers ratio on the copolymer will follow. Some applications, compounding, and processing suggestions will be discussed in the last part.

10.2　POLYMERIZATION AND PRODUCTION PROCESS

The polymerization reaction is shown in Figure 10.1.

TABLE 10.1
EVM Commercial Products

VA-Content %	Trade Name	Company
28	Elvax 250	DuPont
	Elvax 260/265	DuPont
40	Levapren 400	Bayer
	Elvax 40L03/40L08	DuPont
	Elvax 40 W	DuPont
	Evaflux 40	Mitsui-DuPont
45	Levapren 450/450 HV	Bayer
	Elvax 46 L	DuPont
	Evaflux 45	Mitsui-DuPont
50	Levapren 500 HV	Bayer
60	Levapren 600 HV	Bayer
70	Levapren 700 HV	Bayer
80	Levapren 800 HV	Bayer

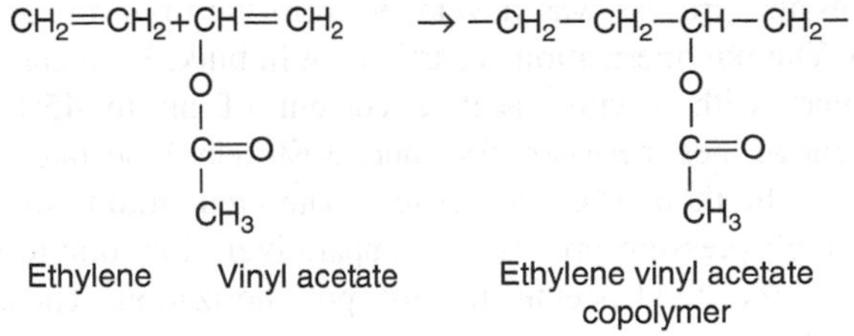

$$CH_2{=}CH_2 + CH{=}CH_2 \quad \rightarrow -CH_2-CH_2-CH-CH_2-$$

with:

CH₂=CH₂ + CH=CH₂ → −CH₂−CH₂−CH−CH₂−
|
O
|
C=O
|
CH₃

and right side:

−CH₂−CH₂−CH−CH₂−
|
O
|
C=O
|
CH₃

| Ethylene | Vinyl acetate | Ethylene vinyl acetate copolymer |

FIGURE 10.1 Polymerization reaction for EVM.

Ethylene and vinyl acetate are copolymerized to a poly (ethylene-co-vinyl acetate). The reaction can be carried out in every ratio of the comonomers as the reactivity ratios are both close to 1 [3,4]. Consequently, the distribution of the comonomers along the chain is random.

The versatility of the copolymer class is established by these copolymerization parameters. The mechanism of the reaction is a radical chain copolymerization, initiated by, for example, azo compounds.

Vinyl acetate acts as a chain-transfer agent giving rise to a broad molecular weight distribution and, in the case of a bulk polymerization, a limited molecular weight. This problem can be overcome by polymerizing in solution, for example, in toluene, benzene, heptane, or tertiary butanol [5].

Tertiary butanol is preferred because of its low chain-terminating properties. Three different methods for the production of ethylene vinyl acetate copolymers are known, these being: high-pressure process (0%–45% vinyl acetate); low-pressure emulsion process (55%–100% vinyl acetate); and medium-pressure process in solution (30%–100% vinyl acetate).

Figure 10.2 gives an overview of the production conditions and the vinyl acetate range covered.

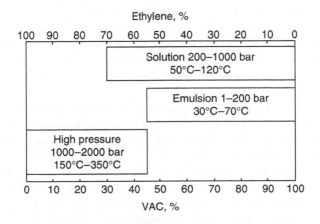

FIGURE 10.2 EVM production methods and products thereof.

The high-pressure process was developed from the high-pressure polyethylene process (LDPE). The polymerization is carried out in bulk. By means of this process mainly copolymers with a vinyl acetate content of up to 45% are produced. The important range lies between 5% and 30% vinyl acetate content, giving copolymers with thermoplastic properties. The maximum molecular weight achieved by the high-pressure process is comparatively low due to the high chain-transfer activity of the vinyl acetate in bulk polymerization. Therefore, the vinyl acetate content is limited.

The emulsion process yields material with vinyl acetate content between 45% and 100%. Emulsion-polymerized material is rarely used in the rubber industry, but rather in thermoplastic applications as an impact modifier.

The solution process is carried out in tertiary-butanol solution under medium pressures [6]. The maximum molecular weight reached is higher than that of the high-pressure process because the chain-transfer activity of the vinyl acetate molecule is reduced in solution. A more detailed description of the different processes is given by Gilby [2].

The main feature of poly(ethylene co-vinyl acetate) is the fully saturated backbone, which leads to some outstanding properties such as good resistance against heat, weathering, and ozone and, depending on the vinyl acetate content, oil. These elastomers also have excellent color stability, low compression set, and high capacity for filler loading and good processability. Since the polymers are halogen free, they are used for low smoke insulation and sheathing and can be compounded for excellent flame resistance.

10.3 STRUCTURE PROPERTY RELATIONSHIPS

10.3.1 CLASSIFICATION AND COMMERCIAL PRODUCTS

As already mentioned, ethylene vinyl acetate copolymers can be synthesized in every ratio of the comonomers. The resulting polymers exhibit the complete scale of properties from thermoplastic material to elastomers. Polyethylene is a well-known thermoplastic with a very low glass transition temperature at about −120°C and very high crystallinity of 40%–60%. Polyvinyl acetate is also a thermoplastic, but with a glass transition temperature of 28°C and completely amorphous structure.

If one adds vinyl acetate to the polyethylene chain, for example, up to about 33 by weight, the copolymers are still thermoplastic. Elastomers result by copolymerizing at least 40% by weight vinyl acetate [7].

As shown in Figure 10.3, the "rubber region" reaches up to 80% vinyl acetate. Above this point, the properties of the polyvinyl acetate become predominant.

As a result of the different properties of the material, the rubbery copolymers are indicated here as "EVM" according to the rubber nomenclature, with the "M" standing for the saturated backbone. We prefer the abbreviation "EVM" instead of "EAM," as prescribed in ASTM D 1418-85, to avoid confusion with ethylene-acrylate copolymers, coded as AEM. The thermoplastic material is denoted by "EVA," but this abbreviation is also frequently used for the whole class of ethylene vinyl acetate copolymers.

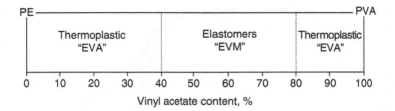

FIGURE 10.3 EVM content produces either elastic or plastic polymers.

A survey of the EVM grades available on the market is presented in the following.

Table 10.2 shows the different EVM grades from Bayer AG (Germany), which are known as "Levapren."

They differ in vinyl acetate content and Mooney viscosity or melt flow index (MFI).

The chart in Table 10.1 gives an outline of the major EVM elastomers on the market with respect to vinyl acetate content without laying claim to completeness.

The following discussion will primarily deal with copolymers of the 40–80 wt% vinyl acetate range, because they are the most interesting for the elastomer sector. Data for the thermoplastic ethylene vinyl acetate copolymers will be given as necessary for comparison with the elastomers.

10.3.2 THERMAL STABILITY AND HEAT AGING PROPERTIES

Thermo gravimetric measurements can give some information about the heat aging properties and the decomposition behavior of a polymer. Figure 10.4 shows typical

TABLE 10.2
Grades of Levapren EVM from Bayer

Trade Name	Weight% VAC	Viscosity ML (1 + 4) 100°C	Viscosity MFI (g/10 min)
Levapren 400	40 ± 1.5	20 ± 4	≤5
Levapren 450	45 ± 1.5	20 ± 4	≤5
Levapren 500 HV	50 ± 1.5	27 ± 4	≤5
Levapren 600 HV	60 ± 1.5	27 ± 4	≤5
Levapren 700 HV	70 ± 1.5	27 ± 4	≤6
Levpren 800 HV	80 ± 2.0	28 ± 6	
Levapren VP KA 8784	70 ± 1.5	≈60	
Levapren VP KA 8815	60 ± 1.5	≈55	
Levapren VP KA 8857	50 ± 1.5	≈55	
Levapren VP KA 8936	80 ± 2.0	≈55	
Levapren VP KA 8939	91 ± 2.0	≈38	
Levapren VP KA 8940	44 ± 1.5	≈30	
Levapren VP KA 8941	46 ± 1.5	≈30	

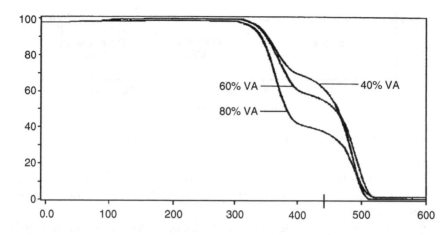

FIGURE 10.4 Thermo gravimetric measurements of 40, 60, and 80 wt% EVM copolymers. Heating rate 20°C/min.

thermo gravimetric measurements in a nitrogen atmosphere. The dynamic measurement was carried out with samples of 40, 60, and 80 wt% vinyl acetate. The decomposition of the copolymer occurs in two steps: the first varying with the level of vinyl acetate content of the sample; the higher the vinyl acetate content, the more pronounced is the first step. The initiation temperature of about 350°C is the same for all copolymers. This leads to the conclusion, that the first step is related to the cleavage of acetic acid from the copolymer. A plot of the height of the first step as a function of the vinyl acetate content, Figure 10.5 indicates a linear relationship, which reinforces the assumption.

The second step is attributed to the decomposition of the polymer backbone.

In air the same two-step mechanism is found, but the initiation temperature of the first step drops to 280°C–300°C. The heat of combustion (ΔH_c) of EVM polymers is given in Table 10.3.

The similarity of all copolymers as far as the heat-aging properties are concerned can also be demonstrated by investigating the heat aging of a compound.

A black-filled standard compound was aged for 14 days at 150°C. Figure 10.6 shows the test results, in terms of the tensile strength, Figure 10.7, shows elongation at break, and Figure 10.8 hardness. The results obtained after aging of the samples show a good retention of the properties.

10.3.3 GLASS TRANSITION TEMPERATURE AND CRYSTALLINITY

Both properties are very important for the cold flexibility of an elastomer. Polyethylene exhibits very high crystallinity, because of the regular structure of the polymer chain. Vinyl acetate molecules with their bulky acetate side-chains disturb the regularity and consequently reduce the crystallinity.

The crystalline part of a polymer is determined by measuring the heat of fusion in a differential scanning calorimeter (DSC). The glass transition temperature

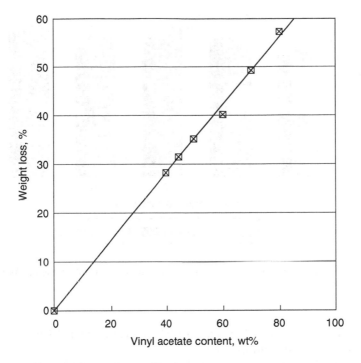

FIGURE 10.5 Weight losses during initial decomposition versus vinyl acetate content.

is measured in the same run. Two typical DSC-curves for a 40 and 60 wt% vinyl acetate copolymer are plotted in Figure 10.9.

The broad melting transition of the 40% EVM is found at −20°C to +60°C with a maximum at 35°C indicating a poor crystalline order. The thermogram of the 60 wt% EVM copolymer in Figure 10.10 exhibits only a glass transition.

The crystallinity can be calculated from the heat of fusion by dividing it by the value known for ideal crystallized polyethylene which is 293 J/g. The data of

TABLE 10.3

Heat of Combustion (ΔH_c)

of EVM Elastomers

Vinyl Acetate Content (wt%)	ΔH_c (MJ/kg)
40	36.7
45	35.9
50	34.4
60	32.0
70	29.1
80	27.2
10 (thermoplastic)	44.0

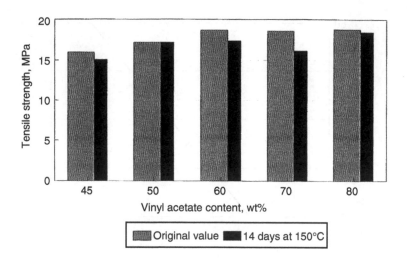

FIGURE 10.6 Original and aged tensile strength versus vinyl acetate content. (Aged 14 days at 150°C in air.)

the heat of fusion, melt transition, and crystallinity are listed in Table 10.4, together with the glass transition temperatures.

Melting endotherms are only found in samples with vinyl acetate contents up to 50 wt%. Copolymers with higher vinyl acetate content do not have the ability to crystallize because of the interruption of the regular chain structure. In Figure 10.11, the dependence of the heat of fusion on the vinyl acetate content is depicted. As the vinyl acetate content increases the heat of fusion decreases. As a result, the crystallinity shows the same trend, as seen in Figure 10.12.

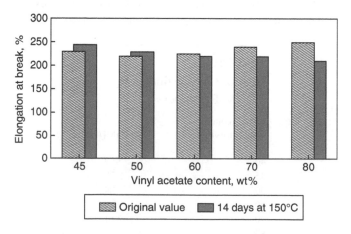

FIGURE 10.7 Original and aged elongation versus vinyl acetate content. (Aged 14 days at 150°C in air.)

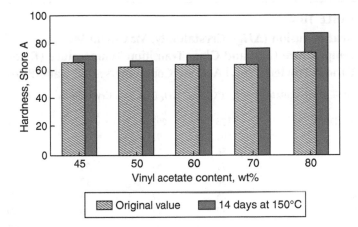

FIGURE 10.8 Original and aged hardness versus vinyl acetate content. (Aged 14 days at 150°C in air.)

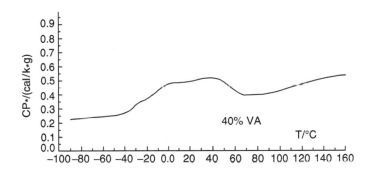

FIGURE 10.9 DSC Scan of 40 wt% vinyl acetate copolymer.

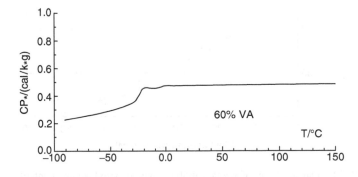

FIGURE 10.10 DSC Scan of 60 wt% vinyl acetate copolymer.

TABLE 10.4

Heat of Fusion (ΔH_ξ), Crystallinity, Maximum Melt Temperature (T_m), and Glass Transition Temperature (T_g) of Various Ethylene Vinyl Acetate Copolymers, HDPE and PVAC

Wt% Vinyl Acetate	T_m (°C)	ΔH_ξ (J/g)	Crystallinity	T_g (°C)
0 (HDPE)	133	293	—	−120
17	91	92	31	−35
22	86	75	26	−35
27	77	65	22	−34
40	36	36	12	−33
45	25	27	9	−32
50	12	17	6	−31
60	—	—	—	−25
70	—	—	—	−9
80	—	—	—	−4
100 (PVAC)	—	—	—	28

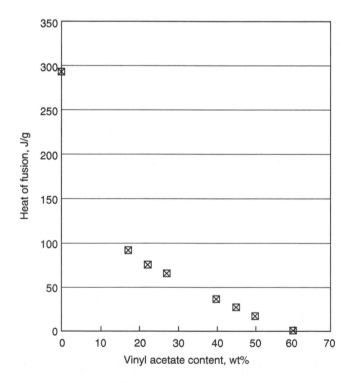

FIGURE 10.11 Heat of fusion of EVM copolymer versus vinyl acetate content.

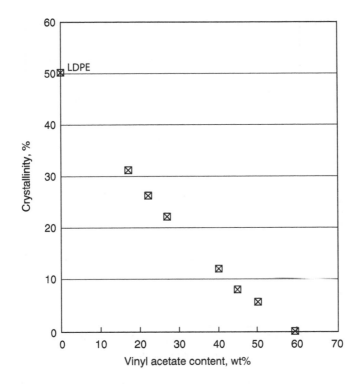

FIGURE 10.12 EVM Crystallinity versus vinyl acetate content.

The maximum temperature of the melting peaks decreases with increasing vinyl acetate content. This is due to a deterioration of the crystallite's perfection. The ethylene sequences, which are able to crystallize, become smaller and, thus, the resulting crystallites are smaller and less perfect. The biggest part of the melting endotherm of the 50 wt% EVM is found below room temperature. Therefore the contribution of crystallites to the tensile strength at room temperature is negligible. In the case of the 40 wt% EVM nearly 50 of the crystallites are already molten at room temperature.

The plot of the glass transition temperature versus the vinyl acetate content is shown in Figure 10.13.

Here the values are nearly constant up to an amount of 50 wt% vinyl acetate. Then the glass transition temperature increases steadily. By extrapolation of the last data to the value of polyvinyl acetate, the measured value is found.

Further studies carried out on an aluminum-trihydrate filled FRNC compound (compounding see Section 6.2) showed similar results. In Figure 10.14 the results of a torsion pendulum measurement are shown. The values of the first maximum are approximately equal to the 40, 45, and 50 wt% copolymer-based compound, whereas the higher vinyl acetate content leads to increasing maxima.

The glass transition temperature and the crystallinity of the FRNC compound, as detected by the DSC, remain nearly unchanged.

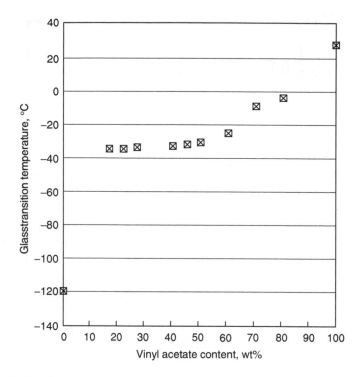

FIGURE 10.13 Glass transition temperature versus vinyl acetate content. (DSC measurements, heating rate 20°C/min.)

10.3.4 OIL RESISTANCE

Oil resistance is closely related to the polarity of an elastomer. The polarity of EVM elastomers is determined by the vinyl acetate content because of the polar nature

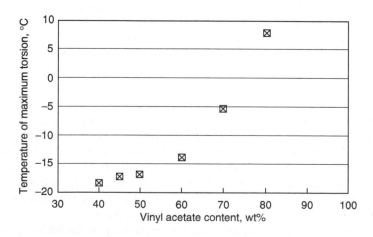

FIGURE 10.14 Maximum torsion temperature versus VA content.

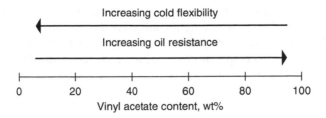

FIGURE 10.15 Relationship of cold and oil resistance with VA content.

of this comonomer. This behavior is monitored by studying the resistance of an FRNC compound to ASTM Oil No. 3. The results of the investigation are shown in Section 6.2.3.

The measurements confirm the assumption derived from the chemical structure of the copolymer. Good oil resistance is found for the high vinyl acetate content copolymers, whereas those with lower vinyl acetate content show poorer oil resistance, as schematically shown in Figure 10.15.

To summarize, EVM vulcanizates possess an excellent heat aging resistance but there is an inverse relationship between best cold flexibility and best oil resistance. The cold flexibility increases with decreasing vinyl acetate content and the oil resistance decreases. So it is up to the compounder to choose the appropriate copolymer in order to meet the requirements of the finished product.

10.4 COMPOUNDING

10.4.1 Curing System

EVM elastomers belong to the group of rubbers with fully saturated backbone. As there are no sulfur-curable sites in the side-chain, they have to be cured radically by means of peroxides or high-energy radiation.

The mechanism of the reaction is first an abstraction of a hydrogen atom from the polymer chain, leading to the formation of a reactive radical site [6]. Then two polymer radicals can combine, which results in a polymer network. The network formed by those reactions is very irregular. In order to improve the network the addition of compounds with multiple double bonds, for example, triallyl cyanurate (TAC), triallyl isocyanurate (TAIC), trimethylol propane trimethacrylate (TMPTA), or ethylene dimethacrylate (EDMA) is necessary [6]. They are often referred to as activators. The effect is described as an addition to the radical site at the polymer chain and transfer of the radical to the activator. The network is formed by reaction of the transferred radical with another chain.

The selection of the appropriate peroxide is determined by the mixing temperature and the possible need for odorless articles. Among the peroxides available on the market 1,4-bis(tertiary butyl peroxy isopropyl) benzene has found widespread use because of the low odor and process safety. 2,5-dimethyl-2,5-bis(tertiary butyl-peroxy) hexane also has a low odor but is slower curing. If the odor is acceptable, dicumyl peroxide is often used for curing at lower temperatures (150°C–160°C). The quantities recommended for typical light-colored compounds are 2–7 phr of peroxide (absorbed 40% on inactive fillers). The recommended amount of activator is 0.5–5 phr. Carbon black filled compounds may require 25% less peroxide.

Although zinc oxide is not necessary with peroxide cures, it does slightly improve aging properties.

10.4.2 ANTIOXIDANTS

Having no reactive double bonds EVM elastomers have an inherently excellent heat resistance. But for an optimum stability it is necessary to add approximately 1 phr of an antioxidant. Here it is necessary to select the right substance, because many well-known antioxidants cannot be used in peroxide cure. For an optimum heat resistance, the addition of 1 phr of styrenated or octylated diphenylamine (SDPA or ODPA) is recommended. Aralkylated phenols can be used as nonstaining antioxidants. To protect the vulcanizates from hydrolysis at elevated temperatures and higher humidity, the addition of a polycarbodiimide may be necessary.

10.4.3 FILLERS

The most suitable carbon blacks and typical ranges used are N 330, N 550 (up to 60 pphr), N 770 (up to 80 pphr), and N 990 (up to 100 pphr). N 220 can be used in a blend with other carbon blacks. Conductive compounds often use N 472 carbon black.

White fillers have to be selected more carefully because they can affect the degree of vulcanization. Good results have been achieved by using talcum (microtalcum), neutral clay (50–100 pphr), and silica (up to 60 pphr). Active silica provides higher tensile strengths but the degree of curing is reduced. A vinyl silane is recommended with white fillers.

For flame-resistant vulcanizates the use of 100–200 pphr aluminum trihydrate with a vinyl silane is recommended along with zinc borate at the 10–20 pphr level and 0–20 pphr of magnesium carbonate. Details of compounding principles will be discussed later.

10.4.4 PLASTICIZERS

The use of plasticizers is not mandatory in EVM grades with low Mooney viscosity. Plasticizers can be used especially in compounds based on high viscosity grades, or in highly filled mixtures.

The most appropriate extenders for EVM mixes are paraffinic mineral oils. Napthenic and aromatic oils are not recommended because they can affect the peroxide cure.

Synthetic plasticizers should be added to improve the cold flexibility of high vinyl acetate containing grades. Good results have been obtained by using adipate- and sebacate-type plasticizers.

10.4.5 MISCELLANEOUS

Stearic acid (1 phr) is frequently used to reduce sticking of mixes based on low-viscosity EVM. The dispersion of fillers is improved at the same time. Zinc stearate at 1–2 pphr is also an alternative.

Paraffin wax is not required as an ozone protector but 5 pphr is very helpful as a processing aid.

Coloring agents should be selected carefully as many organic pigments disturb the peroxide cure. Acid-free inorganic pigments are therefore recommended.

Unsaturated, low-pH, and sulfur-bearing materials such as resins, acidic light-colored fillers, and vulcanized vegetable oils should be avoided because of the presence of peroxides.

10.5 PROCESSING

10.5.1 MIXING

Normal methods of mixing may be employed with EVM such as either upside down or conventional mixing in an internal mixer. EVM with low viscosity bands quickly on a mill, whereas higher-viscosity grades may require a breakdown period to achieve a smooth band. Release agents such as stearic acid or zinc stearate should be added with the polymer to reduce mill sticking. Other release agents are vinyl silane, paraffin wax, or commercial internal lubricants.

Fillers should be added early without plasticizer for effective dispersion. The peroxide and coagent is best incorporated at the end of the mix.

Care should be taken to ensure that the equipment is clean to prevent contamination that will react with the peroxide curative.

10.5.2 EXTRUSION

Higher-filler loadings are advantageous to stiffen the low-viscosity EVM polymer. Higher-viscosity grades of EVM should be used if possible. If curing in open steam, add 3–5 pphr of a polycarbodiimide, such as Rhenogran P 50.

10.5.3 CALENDERING

As with extruding of EVM, a high compound viscosity through the use of higher-viscosity elastomers and higher-filler loadings is best. Increased amounts of stearic acid or zinc stearate are also recommended. Roll temperatures of 70°C are suggested.

10.5.4 FABRICATING

EVM has poorer tack at low temperatures, hence the components are best if prewarmed. Low-viscosity elastomers and the use of higher levels of plasticizer also help. Normal tackifiers will interfere with the peroxide cure.

10.5.5 MOLDING

Since EVM must be cured with peroxide, injection or transfer molding works well as air is excluded to a large extent. Mold design is important for intricate shapes since EVM has poor hot tear properties and demolding can be difficult.

10.5.6 STEAM CURING

As with any peroxide-cured compound, it is important to vent all the air from the autoclave to prevent a sticky surface on the vulcanizate. CV steam tube cures generally work well.

10.5.7 LCM SALT BATH CURING

EVM can be cured very effectively at 200°C to 240°C using this method as long as the extrusion is kept under the surface of the salt bath or a salt spray is used to exclude any air.

10.6 APPLICATIONS

The following section will identify some applications in which EVM is especially advantageous.

10.6.1 HIGH-TEMPERATURE INSULATION COMPOUNDS

This part presents the results of tests made with an insulation compound. Table 10.5 shows the formulation of a typical compound. EVM is used as an insulation material for special electrical conductors that have to withstand service temperatures up to 120°C. The required properties have been specified in VDE 0207/part 20 type 401.

10.6.1.1 Maximum Service Temperature

On the assumption that the elongation at break of an EVM vulcanizate changes at a constant rate in the course of aging, the maximum service temperature was

TABLE 10.5
Insulation Compound

EVM (40% VA)	100.0	Levapren 400, Bayer
Polycarbodiimide stabilizer	8.0	Rhenogran P 50, Rhein Chemie
Antioxidant SDPA	1.0	Vulkanox DDA, Bayer
Zinc stearate	2.0	
Micro talc	80.0	Mistron Vapor, Luzenac America
Paraffin wax (55°C)	5.0	
Peroxide	6.0	Perkadox 14/40, AKZO
TAC	4.0	Triallyl cyanurate, Cytec
Total	206.0	

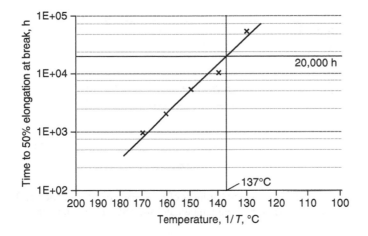

FIGURE 10.16 Arrhenius plot of EVM insulation compound (IEC 216).

determined according to IEC 216, taking the residual elongation at break as the relevant criterion. For each of the aging temperatures between 130°C to 170°C the time that elapsed until the elongation at break fell to 50% was plotted in hours (in logarithmic scale) versus the reciprocal test temperature (in K^{-1}). A straight line is formed (the Arrhenius plot), representing the service life of the insulation compound at the respective temperatures.

The point at which the straight line intersects the time limit (20,000 h) specified in IEC 216 gives a maximum service temperature of 137°C for this vulcanized EVM insulation compound, as shown in Figure 10.16.

10.6.1.2 Resistance to Ultraviolet Light, Weathering, and Ozone

Table 10.6 shows the changes in mechanical properties after UV exposure and weathering. Samples have been irradiated with an ultraviolet lamp of 300 W and 700 Lux.

Specimens exposed outdoor for 10 months at the Leverkusen-Engerfeld testing station, a nonindustrial area, showed minor changes in properties. No crazing was detected even after 2.5 years, as given in Table 10.6.

TABLE 10.6
Weathering Effects on EVM Properties
UV Light 300 W, 700 Lux

	Weathering 10 Months	Ultraviolet Light 7 Days at 47°C
Tensile strength, change %	−12	+3
Elongation, change %	−22	−7
Shore A, change pts	−2	+2

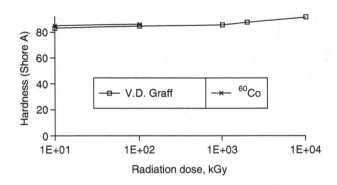

FIGURE 10.17 Effect of radiation on hardness of EVM compound. (Van de Graff accelerator to 100 kGy with a ^{60}Co source at room temperature and 65% relative humidity.)

In an ozone test, specimens were exposed at an elongation of 60% to a concentration of 250 ppm ozone at temperatures of 20°C and 50°C. No crack formation was observed during an exposure time of 14 days.

10.6.1.3 Radiation Resistance

Samples were exposed to two different sources of radiation:

1. A Van de Graff accelerator, current 10^{-4} A, voltage 1.5 MeV at 200 W; test at room temperature and a relative humidity of 65%; radiation rate 900 kGy/h.
2. A 5.55 TBq ^{60}Co source fort-irradiation at a rate of 9.2 Gy/h; test at room temperature and a relative humidity of 65%.

In Figures 10.17 and 10.18 the mechanical property data measured in the experiments are plotted versus the radiation dose. The influence of the difference in radiation is nearly equal.

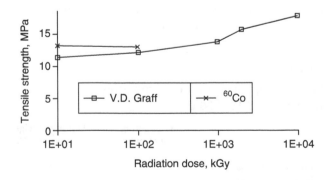

FIGURE 10.18 Effect of radiation on tensile of EVM compound. (Van de Graff accelerator to 10 kGy with a ^{60}Co source at room temperature and 65% relative humidity.)

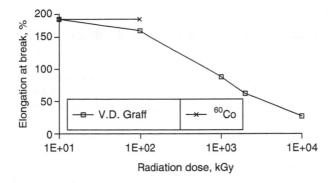

FIGURE 10.19 Effect of radiation on EVM compound. (Van de Graff accelerator to 100 kGy with a ^{60}Co source at room temperature and 65% relative humidity.)

Assuming a residual elongation of 50% as the lowest acceptable value, the EVM vulcanizates can be said to have a radiation resistance of 400 Mrad, as shown in Figure 10.19.

10.6.1.4 Electrical Properties

The electrical properties of the insulation compound were determined on 1 mm thick sheets that had been vulcanized in a press for 20 min at 160°C. The results are shown in Table 10.7.

10.6.2 EVM for Flame-Retardant Noncorrosive Applications

For certain safety reasons some rubber articles must meet stringent requirements, such as

1. Flame retardance
2. Low smoke formation
3. Noncorrosive combustion gases
4. Low toxicity of combustion gases

Table 10.8 shows the formulation of a typical flame-retardant noncorrosive (FRNC) compound.

The favorable performance of such a compound in case of fire is shown in Section 10.6.2.2. For mechanical and thermal properties see Section 10.3.2.

TABLE 10.7

Electrical Properties of an EVM Insulation Compound

Volume resistivity ξ_D	5.10^{13} $(\Omega \cdot cm)$	VDE 0303/Part 3, IEC 93
Dielectric constant E_r	3.9	VDE 0303/Part 2
Dissipation factor tan ς	0.011	IEC 250
Dielectric strength E_d	32 (kV/mm)	VDE 303/Part 2, IEC 243

TABLE 10.8

Flame-Resistant Noncorrosive Compound (FRNC)

EVM	100.0	Levapren, Bayer
Polycarbodiimide stabilizer	3.0	Rhenogran P 50, Rhein Chemie
ATH Aluminum trihydrate	190.0	Apyral B 120, Bayer
Silane	2.0	RC 1, Union Carbide
Zinc borate	10.0	
Antioxidant SDPA	1.0	Vulkanox DDA, Bayer
Dioctyl sebacate	6.0	
Trimethylolpropane trimethacrylate	0.5	
Peroxide	6.0	Perkadox 14/40, AKZO
Total	318.5	

10.6.2.1 Influence of Amount of Filler

In the case of polymers that are not "intrinsically flame retardant," good flame-retardant behavior is only possible if the compound is heavily loaded with flame-retardant fillers. The filler most widely used is aluminum trihydrate (ATH). Figure 10.20 shows the combustion behavior of a sample according to the limiting oxygen index (LOI) and critical oxygen index (COI). LOI is the minimum concentration of oxygen, expressed as a volume percent, in a mixture of oxygen and nitrogen that will just support flaming combustion of a material under test conditions (ASTM D 2863). COI is the temperature in degrees Celsius at which the LOI of a material becomes 21 (the approximate percentage of oxygen in air) under test conditions (NES 715).

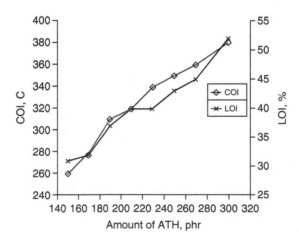

FIGURE 10.20 Influence of aluminum trihydrate amount on critical oxygen index in EVM FRNC compound. (NES 715 standard.)

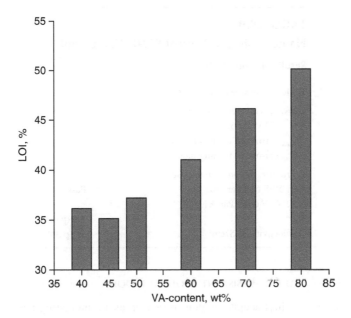

FIGURE 10.21 Influence of vinyl acetate content on limiting oxygen index. (EVM Compound with 190 pphr aluminum trihydrate.)

As ATH withdraws energy from the flames by decomposing into aluminum oxide and water it has been found that increasing levels of ATH steeply raise the LOI and COI.

Figure 10.21 shows how the LOI is related to another factor, the vinyl acetate content of the EVM, which is varied when the filler content of the compound is held constant at 190 phr ATH.

At approximately 50% vinyl acetate, the increase of the LOI is disproportionaly greater than that expected from linear reduction of the combustion enthalpy.

10.6.2.2 Toxicity, Smoke Density, and Corrosivity of Combustion Gases

All EVM polymers (VA-content 40%–80%) perform in a similar manner. Therefore, only the results of the EVM with 50% VA are exhibited in Table 10.9. The compound processes well with 190 pphr aluminum trihydrate and provides low smoke density and low corrosion combustion gases.

10.6.2.3 Influence of Vinyl Acetate Content on Oil Swelling

With increasing vinyl acetate content, the polarity of the EVM becomes higher. Figure 10.22 shows the volume swell after immersion in ASTM oil No. 3 for 70 h at 150°C and diesel fuel for 24 h at 70°C. The reduction in volume swell from 98% (40% VA) to 10% (80% VA) is remarkable. In addition, tensile strength remains nearly unchanged for an EVM containing 70–80% vinyl acetate, as may be seen in Figure 10.23.

TABLE 10.9

Flame Tests of a Typical FRNC Compound

Results of Flame Tests

Toxicity Index, acc. NES 713	1
Smoke density NBS chamber	
Acc. ASTM E 662-83	
(D_{max} corrosion) nonflaming mode	200
(D_{max} corrosion) flaming mode	160
Corrosivity of combustion gases	
Acc. NES 518 Part 403	Pass
Acc. VDE 0472 Part 813	
Ph	3.8 (pass)
Conductivity, μ S/cm	52 (pass)

10.6.3 HEAT- AND OIL-RESISTANT MOLDED GOODS

An EVM with 60% vinyl acetate content is chosen as an interesting material for oil- and heat-resistant gaskets, oil seals, and other molded goods for service temperatures up to 150°C–170°C. Variation of vinyl acetate content will change the properties (swelling, low temperature flexibility) in the way already described. In spite of peroxide cure, the material does not require post curing to fully develop optimal physical properties.

Considering that the compounds were peroxide cured, the demolding and mold-fouling characteristics in injection molding were surprisingly good. This has been proven by a plant trial using a Desma IM machine type 902 G using a two cavity molds for bellows. All tests were performed on samples press cured at 180°C for 10 min.

Table 10.10 shows the formulation of a typical compound for injection molding and accompanying properties.

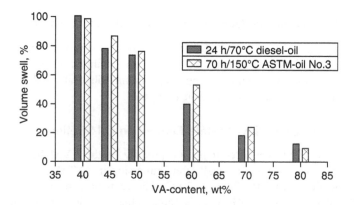

FIGURE 10.22 Influence of vinyl acetate content on swell in diesel and ASTM #3 oils. EVM FRNC compound.

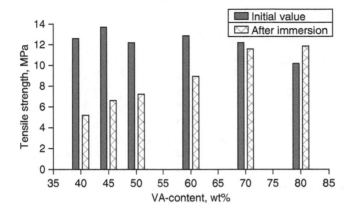

FIGURE 10.23 Influence of vinyl acetate content of EVM compound on tensile before and after aging in ASTM #3 Oil for 70 h at 150°C.

TABLE 10.10
EVM Injection Molding Compound

Injection Molding Compound

EVM (60% VA)	100.0	Levapren KA 8385, Bayer
N 550 carbon black	40.0	
Stearic acid	1.0	
Polycarbodiimide stabilizer	3.0	Rhenogran P 50, Rhein Chemie
Antioxidant SDPA	2.0	Vulkanox DDA, Bayer
Polyethylene wax	2.0	Hoechstwachs PE 520, Hoechst
Zinc salt of fatty acid	2.0	Aktiplast PP, Rhein Chemie
TAIC	1.5	Perkalink 301, AKZO
Peroxide (40%)	5.0	Perkadox 14/40, AKZO
Total	156.5	

Physical properties

Shore A at 23°C	62
Shore A at 70°C	50
Modulus at 100%, MPa	3.2
Tensile, MPa	16.0
Elongation, %	390
Rebound, %	29
Tear, ASTM D 624 B, N/mm	22
Tear, ASTM D 624 C, N/mm	27
DIN abrasion loss, mm^3	130
Compression set, 70 h at 150°C, %	25

(continued)

TABLE 10.10   (Continued)
EVM Injection Molding Compound

Low-temperature properties

Brittle point, °C	−46
TR 10, ASTM D 1329, °C	−23
TR 30, ASTM D 1329, °C	−15
TR 50, ASTM D 1329, °C	−13

Differential scanning calorimeter, 20°C/min, second heating cycle

T_E, °C	−34	Onset of glass transition process
T_G, °C	29	Glass transition temperature

Aged in air 14 days at 150°C-change

Shore A, pts	6
Tensile, %	−4
Elongation, %	−15

Aged in ASTM #3 oil 72 h at 150°C-change

Shore A, pts	−28
Tensile, %	−44
Elongation, %	−41
Volume, %	87

Aged in SAE 90 gear oil 72 h at 150°C-change

Shore A, pts	−17
Tensile, %	−7
Elongation, %	−3
Volume, %	20

Aged in SAE 20W40 engine oil 7 days at 150°C-change

Shore A, pts	22
Tensile, %	−19
Elongation, %	−26
Volume, %	37

10.6.4   EVM as a Blend Component

EVM is a soft copolymer and acts as a nonvolatile, nonextractable plasticizer, and processing aid when it is used as a blend component in compounds to improve processing by lowering the Mooney viscosity of the mix.

Two examples are given: first blend is based on chlorinated polyethylene (CM). By adding EVM to a light-colored CM compound the viscosity is reduced, as expected. The results are given in Table 10.11.

In addition, compression set and heat aging were improved without influencing tensile strength and elongation at break significantly.

The second example is a blend of EVM with a high vinyl acetate content and HNBR. As heat aging and oil resistance of both polymers are nearly identical both polymers are very suitable for blending. The two polymers show compatibility over a wide range of copolymer compositions. Again the viscosity of the compound is reduced with an increasing amount of EVM (Table 10.12).

TABLE 10.11
EVM/CM Blends in a Light-Colored Compound

EVM (50% VA)	—	25.0	50.0	Levapren 500, Bayer
CM	100.0	75	50	Tyrin CM 0136, Dow
Ground CaCO$_3$	60.0	60.0	60.0	Atomite, Polymer Valley Chemicals
Lead silicate	20.0	20.0	20.0	H(202) D 80, Rhein Chemie
Epoxidized soya oil	10.0	10.0	10.0	Drapex 6.8, Argus Chemical
Antioxidant SDPA	0.5	0.5	0.5	Vulkanox DDA, Bayer
Coagent	5.0	5.0	5.0	Saret 500, Sartomer
Peroxide	5.0	5.0	5.0	Perkodox 14/40, AKZO
Compound viscosity				
ML 1 + 4 at 100°C	74	64	41	
Physical properties, cured 15 min at 177°C				
Tensile, MPa	13.4	12.9	11.9	
Elongation, %	610	580	535	
Compression set B (ASTM D 395)				
70 h at 150°C, %	61	54	48	
Brittle point (ASTM D 2137)				
Temperature, °C	−25	−30	−32.5	
Aged in air 70 h at 150°C-change				
Tensile, %	−55	−42	−39	
Elongation, %	−54	−30	−23	

Compared with an unblended HNBR compound additional benefits can be seen in the EVM/HNBR blend:

1. Lower price
2. Better heat aging
3. Better compression set at high temperatures, whereas oil swelling and cold flexibility are better with unblended HNBR

10.6.5 OUTLOOK FOR NEW APPLICATIONS

EVM elastomers have traditionally been used in the cable industry. But now some new applications are developed and EVM has been recognized as a heat- and oil-resistant material, which can be used for automotive applications, such as gaskets, seals, and hoses.

The demand for more safety in public buildings, subway stations, computer centers, and so on, has led to acceptance of the FRNC concept for flooring materials. Here EVM elastomers are tested as a basic component. Noise reduction for example, in workshops gave rise to another development: noise-deadening sheets and curtains. They are based on EVM elastomers filled with barium sulfate.

Lastly, EVM is a good blend polymer. In blends with polar polymers, such as CM, HNBR, NBR, or TPU, it enhances, for example, the processing behavior and, depending on the blend component, vulcanizate properties and lowers compound price.

TABLE 10.12
EVM/HNBR Blends in Black Filled Compound

EVM (70% VA)	—	25.0	50.0	Levapren 700 HV, Bayer
HNBR (34% ACN)	100.0	75.0	50.0	Therban 1707, Bayer
ZnO	2.0	2.0	2.0	
MgO	2.0	2.0	2.0	
Antioxidant SDPA	1.0	1.0	1.0	Vulkanox DDA, Bayer
Antioxidant MMBI	0.4	0.4	0.4	Vulkanox ZMB-2/C, Bayer
N 550 carbon black	45.0	45.0	45.0	
TAIC	1.5	1.5	1.5	Perkalink 301, AKZO
Peroxide	7.0	7.0	7.0	Perkadox 14/40, AKZO
Total	158.9	158.9	158.9	
Compound viscosity				
ML 1 + 4 at 100°C	104	91	65	
Physical properties				
Shore A, pts	72	75	76	
Tensile, MPa	26.8	26.5	25.7	
Elongation, %	325	250	210	
Tear (DIN 53 515), N/mm	13.4	10.9	9.9	
Compression set (DIN 53 515)				
70 h at 23°C, %	12	10	11	
70 h at 150°C, %	27	24	22	
Low-temperature properties				
Ts, DIN 53 546, °C	<−70	−68	−65	
Tg, DIN 53 520, °C	−24	−25	−16	
Aged in air 500 h at 150°C-change				
Tensile, %	−12	−4	−1	
Elongation, %	−49	−34	−29	
Aged in ASTM #3 Oil 72 h at 150°C-change				
Volume	25	31	32	

As the need for specialty elastomers with good heat and oil resistance without halogen is increasing, expansion into other areas can be expected.

REFERENCES

1. Perrin et al., GB 497643 App.: ICI, 1938.
2. G.W. Gilby, In *Developments in Rubber Technology*, Vol. 3, Eds. A. Wheelan and K.S. Lee, Applied Science Publishers, New York, 1982, Chapter 4, "Ethylene-Vinyl Acetate Copolymers," pp. 101–144.
3. W.D. Kenneth, Low Density Polyethylene (High Pressure)—Ethylene Polymers, Vol. 6, Ed. H.F. Mark et al., John Wiley & Sons, New York, 1986, pp. 386–429.
4. R.D. Burkhardt, N.L. Zutty, *Journal of Polymer Science*. Part A 1, 1137, 1963.
5. H. Bartl, Kautschuk Gummi Kunststoffe, 452, 1972.
6. H. Bartl, J. Peter, Kautschuk Gummi Kunststoffe, 14, 23, 1961.
7. S. Koch, *Manual for the Rubber Industry*, 2nd Edition, Bayer AG, July, 1993.

11 Polysulfide Elastomers

Stephen K. Flanders* and Robert C. Klingender

CONTENTS

Polysulfides were first discovered and patented in about 1927 by J.C. Patrick by accident in attempting to develop antifreeze. The polymer was called Thiokol A, a condensation reaction of ethylene dichloride and sodium tetrasulfide, for which manufacturing began in 1929 [1]. This elastomer was quite difficult to process due to its variability and thermoplastic nature, leading to the discontinuation of manufacture after improved grades were developed.

The current types of polysulfides are Thiokol FA, the workhorse for specialty rollers, Thiokol ST designed for mechanical goods, and a series of liquid polymers, Thiokol LP, designed for sealants, coatings, and binders. Thiokol FA is an improved "A" type which uses approximately equal parts of ethylene dichloride and 1,1′ [methylene bis (oxy)] bis [2 chloroethane], or "formal." Polymerization takes place using a condensation reaction with sodium tetrasulfide for improved manufacturing control and processing, although the polymer is still very thermoplastic. Thiokol ST and LP are similar to each other in structure, except their molecular weights go from

---

* Stephen K. Flanders is deceased.

TABLE 11.1
Millable Polysulfide Types

Grade	FA	ST
Specific gravity	1.34	1.27
Moisture % maximum	0.5	—
Bin stability	Limited	Limited
Mooney viscosity		
ML 4 at 121°C	60–100	—
ML 3 at 100°C	—	28–55
Molecular weight	100,000+	80,000
Sulfur content, %	47	37
Cross-linking, %	0	2
Chain termination	-OH	-SH

the solid STs to the liquid LPs. These elastomers use primarily formal as the base monomer for the reaction with the tetrasulfide with some ethylene dichloride included to provide controlled cross-linking and thus better green strength and processing. The introduction of the "formal" groups, which allow splitting of the polymer to control molecular weight, was a very important discovery. The composition and properties of these elastomers is given in Tables 11.1 and 11.2.

TABLE 11.2
Liquid Polysulfide Types

Specification Limits	LP-31	LP-2	LP-32	LP-12	LP-3	LP-33
Color-MPQC-29-A maximum	150	100	100	70	50	30
Viscosity poises, 25°C	950–1550	410–525	410–525	410–525	9.4–14.4	15–20
Moisture content, %	0.12–0.22	0.15–0.25	0.15–0.25	0.12–0.22	<0.1	<0.1
Mercaptan content, %	1.0–1.5	1.5–2.0	1.5–2.0	1.5–2.0	5.9–7.7	5.0–6.5
General properties						
Average molecular weight	8,000	4,000	4,000	4,000	1,000	1,000
Refractive index	1.5728		1.5689		1.5649	
Pour point, °C	10	7	7	7	−26	−23
Flash point (PMCC), °C	225	208	212	208	174	186
Cross-linking agent, %	0.5	2.0	0.5	0.2	2.0	0.5
Specific gravity at 25°C	1.31	1.29	1.29	1.29	1.27	1.27
Average viscosity poises, 4°C	7400	3800	3800	3800	90	165
Average viscosity poises, 65°C	140	65	65	65	1.5	2.1
Low-temperature flex, °C	−54	−54	−54	−54	−54	−54

G 703 kg/cm^2 cured compound.

The LP polysulfides are used primarily as sealants and coatings for construction, transportation and chemical protection applications, and as a binder for solid rocket propellants. These uses will not be covered in this book, but there is a use for the LP series as reactive plasticizers for ST polysulfides to control viscosity but not seriously detract from final properties.

11.1 BASIC PROPERTIES OF SOLID POLYSULFIDES

Standard test formulations are shown in Table 11.3 for the two types of solid polysulfide elastomers. These use the recommended cure systems for each with a normal amount of N774 carbon black as reinforcement.

The resistance of these two types of polysulfide to highly aromatic solvents and ketones commonly encountered is outstanding. The resistance of Thiokol FA and ST to a few of these solvents is shown in Table 11.3. Additional chemical resistance information on polysulfides may be found in "Chemical Resistance Guide for

TABLE 11.3
Basic Physical Properties of Thiokol FA and ST

Formulations	Thiokol FA	Thiokol ST
Thiokol FA	100	—
Thiokol ST	—	100
SRF N774	60	60
Stearic acid	0.5	1
Zinc oxide	10	—
Zinc peroxide	—	5
Calcium hydroxide	—	1
MBTS	0.3	—
DPG	0.1	—
Total	170.9	167
Physical properties		
Shore A	65–70	65–70
Modulus at 100%, MPa	5.1	3.7
Tensile strength, MPa	8.3	8.3
Elongation, %	380	220
Compression set 22 h at 70°C	100	20
Gehman low temperature, °C	−45	−50
Aged in fluid 70 h at 23°C, volume change, %		
Toluene	+55	+70
Carbon tetrachloride	+50	+46
Ethyl acetate	+17	+35
Ethyl alcohol	+2	+2
SAE 30 Oil	0	0
ASTM #3 Oil	−1	0
Methyl ethyl ketone	+28	+35

TABLE 11.4

Permeability was Tested on a 0.165 cm Sheet of Thiokol FA Compound at 23°C

Solvent	Specific Permeability $(gm)(cm^2)/(24\ h)(cm)$
SR 10	0.0024
SR 6	0.03
Methyl alcohol	0.006
Carbon tetrachloride	0.06
Ethyl acetate	0.24
Benzene	0.84

Elastomers II" Compass Publications, La Jolla, CA 92038-1275 and in Rohm and Haas literature TD FA/10-80/1M, FA Polysulfide Rubber [2], and TD ST-1/83-1M, ST Polysulfide Rubber [3].

When selecting the grade of polysulfide to use for a given application the service conditions are the key criteria to consider. Thiokol FA, as previously indicated, is used mainly in covered rolls and specialty hose where its excellent solvent resistance may be needed. It is often blended with other elastomers such as polychloroprene if some sacrifice in solvent resistance can be tolerated, or with a high acrylonitrile NBR if a minimal loss in solvent resistance is dictated.

Thiokol ST will provide much better compression set resistance, hence is used in seals, gaskets, O-rings, and other mechanical goods. The solvent resistance is not as good as the FA type, but is often satisfactory for most applications.

The two grades of solid polysulfide provide vulcanizates with excellent low-temperature properties, with a Gehman t_{10} of $-45°C$ to $-50°C$ without plasticizer, which is not usually employed. If even better low-temperature properties are needed, an ester type plasticizer such as TP95 may be used.

The impermeability of polysulfide to both solvents and gases is excellent. The permeation of various solvents through a sheet of Thiokol FA is shown in Table 11.4 and resistance to hydrogen and helium is given in Table 11.5.

The weathering and ozone resistance of FA type polysulfide is excellent which makes it ideally suited for weather strip and sealants for buildings and vehicles. After

TABLE 11.5

Permeability was Tested on a 0.0254 cm Sheet of a Thiokol FA Compound at 23°C

Gas	Specific Permeability $(cc)/(cm^2)(min)$
Hydrogen	14.8×10^{-6}
Helium	9.4×10^{-6}

exposure for 1000 h cycling from a wet period of 18 min and a dry period of 102 min at 66°C, while elongated 20% in a Sunshine Arc Weatherometer, the tensile strength dropped 20%, the elongation decreased by 50%, and the hardness increased 4 points with no surface crazing or cracking. Testing of a Thiokol FA-cured slab at 20% elongation for ozone resistance for 300 h at 37°C, 50 parts per 100 million O^3 resulted in no cracking.

11.2 COMPOUNDING OF POLYSULFIDE

Basic compounds with the appropriate cure systems of Thiokol FA and ST are shown in Table 11.6.

11.2.1 VULCANIZATION AGENTS

Zinc oxide is the major vulcanization agent required for Thiokol FA; however, the particular grade is important as certain impurities may affect the cure rate and final properties. A French process zinc oxide, such as Midwest AZO 66L or equivalent, is recommended. Other oxides such as lead oxide or dioxide, cadmium oxide, and zinc hydroxide may also be used with varying rates of cure and physical properties. The amount of zinc oxide over or under 10 phr is not critical and will not affect the rate of vulcanization. Note that the MBTS and DPG are peptizing agents, not accelerators.

Thiokol ST is best cured using zinc peroxide, but organic peroxides and quinoid compounds with metal oxides can also be employed. Zinc peroxide is often difficult to disperse, hence it is strongly suggested that a dispersion in NBR be employed. Examples of combinations that may be used are: GMF 1.5 plus ZnO 0.5; GMF 1.0 plus $ZnCrO_4$ 10; however, these are very scorchy and have very short bin lives and in addition exhibit poor compression set properties. ST type polysulfides have controlled viscosities, hence do not need peptizing agents. In fact the presence of normal accelerators such as thiurams and dithiocarbamates should be avoided as they

TABLE 11.6
Basic Formulations

	Thiokol FA	Thiokol ST
Thiokol FA	100	0
Thiokol ST	0	100
SRF N762	60	60
Stearic acid	0.5	1
MBTS	0.3	0
DPG	0.1	0
Zinc oxide	10	0
Zinc peroxide	0	5
Calcium hydroxide	0	1
Total	170.9	167

will cause chain scission. Moreover, low pH materials such as clay and materials containing sulfur such as sulfur-modified chloroprene should not be used.

The best compression set is obtained with ST type of polysulfide and with a zinc peroxide-cure system. Optimum results are obtained when the vulcanization temperature is 155°C and a post-cure of 24 h at 100°C is used. Compression sets tested for 22 h at 70°C when cured for 30 min at the following temperatures are 52% at 142°C, 36% at 147.4°C, and 24% at 154°C. These results may be improved after post-curing for 24 h at 100°C to 23%, 15%, and 14%, respectively. The maximum recommended operating temperature for Thiokol ST in compression is 70°C and Thiokol FA is not suitable for applications requiring even fair compression set at elevated temperatures. The maximum recommended operating temperature for polysulfides in parts not under compression is 100°C for continuous service. The temperature of vulcanization should not exceed 163°C.

11.2.2 PLASTICIZING AND SOFTENING

ST polysulfides are produced in a molecular weight range that is suitable for normal factory processing, hence do not need to be peptized. Highly filled Thiokol ST compounds may require some softening, hence the replacement of 3–7.5 phr of the polymer with Thiokol LP-3 is often done.

FA types are much higher in viscosity as supplied, hence do need to be softened in order to be able to incorporate compounding ingredients. The only recommended type and amount of peptizer is 0.3–0.6 phr of MBTS plus 0.1 phr of DPG, which is to be added to the polymer either on a mill or better in an internal mixer, at 72°C for 3–5 min before adding other ingredients. The rubber will be rough at the beginning, hence it is important that all of the crumbs be swept up and blended back in early otherwise they will not be softened and will create polymer lumps that will not accept compounding ingredients and cause defects in the part.

If required for low-hardness compounds, ester type plasticizers or highly aromatic oils may be used as well as crumb-form (aerated) vulcanized vegetable oils. Solid forms of vulcanized vegetable oils will need to be passed through a tight mill to form a crumb and allowed to rest overnight to dissipate any trapped gases. Avoid sulfur containing VVO with Thiokol ST.

Thiokol FA parts that require good tack for building may need a slight increase in the level of MBTS to 0.5 or 0.6 phr if the compound is calendered, but this would probably be too high for extrusion. The addition of 5 phr of a coumarone indene resin such as Cumar P25 will also improve tack. Blends of Thiokol FA with neoprene are usually tacky enough for most applications.

11.2.3 INTERNAL LUBRICANTS

Stearic acid has a double function in Thiokol compounds. The first benefit is as in internal lubricant to improve mill release and also to aid in releasing from the mold after vulcanization. Note that 0.5 phr is normal for most compounds; however, if good tack and ply to ply adhesion is needed, such as for rollers, the level may be reduced to 0.25 phr. The second function of the stearic acid is to aid in the dispersion of fillers and to activate the cure system for a tighter state of cure.

11.3 FILLERS

Carbon black, as with other elastomers, serves as a good reinforcing agent as well as reducing nerve during factory processing and decreasing cost. The most common and effective carbon black is SRF followed by FEF, MAF, and MT. A study with various levels of SRF in Thiokol FA is given in Table 11.7. Thiokol ST is studied with various types of fillers in Table 11.8.

Light-colored fillers are not as effective as carbon black in reinforcing polysulfide elastomers. It is very important that low pH fillers, such as clay, be avoided as these will retard the rate of vulcanization, as may be seen in Table 11.8.

11.4 ELASTOMER BLENDS

Polysulfides often are blended with other elastomers such as nitrile rubber, NBR, or neoprene, CR, for improved physical properties and factory processing. Traditionally, Thiokol FA is blended with neoprene for improved strength and processing for rollers, at some sacrifice in solvent resistance. Table 11.9 has information on Thiokol FA blends. If minimal loss in chemical resistance is indicated, then a blend with a high ACN nitrile is employed. A cure system that is compatible with both rubbers needs to be used in all cases. The zinc peroxide-cure system is NBR specific, hence it is important that the recommended one be used or others be tested since many NBRs do not cure with zinc peroxide. Best results are obtained with blends if separate masterbatches are made with the individual polymers, which are then blended in the

TABLE 11.7
SRF Carbon Black Study in Thiokol FA

Base Formulation	phr		
Thiokol FA	100.0		
Zinc oxide	10.0		
Stearic acid	0.5		
MBTS	0.3		
DPG	0.1		
SRF N762	Variable		
Results			
SRF carbon black	Hardness, Shore A	Tensile strength, MPa	Elongation, %
0	40	1.07	450
10	45	2.07	550
20	49	4.19	700
40	58	6.9	600
60	68	8.28	380
80	78	8.28	230
100	82	7.59	210

TABLE 11.8
Various Fillers in Thiokol ST

Base Formulation	phr
Thiokol ST	100.0
Filler	As shown
Stearic acid	1.0
Zinc peroxide	5.0
Calcium hydroxide	1.0

Results

Filler	Amount phr	Hardness, Shore A	Modulus, MPa	Tensile Strength, MPa	Elongation, %
MT N990	60	50	2.07	5.52	600
SRF N762	60	62	7.80	8.80	390
Hydrated alumina	65	62	2.73	7.38	650
Calcium carbonate	70	45	1.07	1.07	340
Hard clay	70	18	—	0.48	640
Titanium dioxide	110	52	2.07	5.00	710

desired ratio. Blends of Thiokol ST with an NBR is shown in Table 11.9 and with CR may be found in Table 11.10.

11.5   PROCESSING

11.5.1   THIOKOL FA

Mill mixing of FA type polysulfide should be in two stages, the first being the peptizing step followed by the final stage in which the remaining ingredients are incorporated. A small batch size is best and mixing needs to be done on a very tight mill at 72°C. The rubber will not band; however, add the MBTS and DPG and mill with sweeping of crumbs of polymer and peptizer back into the batch for at least 4 min to ensure that there are no high viscosity polymer particles that will cause defects to the part. If vulcanized vegetable oil is used, it should also be added at the beginning. The next stage is to add the filler alternately to half of the band at a time as it will break up with each new addition until it is fully incorporated, hence the one side will hold the batch on the mill and pull the crumbled side back into the nip. Adjust the water flow to maintain the roll temperature at 72°C throughout the mix. The curatives should be added at the end and the batch cooled immediately. If the Thiokol is to be blended with another elastomer, make up separate masterbatches of each polymer, then blend the two together as a final step.

Mixing in an internal mixer needs to have some adjustment in batch size as a slightly larger volume is recommended with the starting temperature at 70°C. The polymers, if other elastomers are also included, should be added and mixed with the ram down for 1 min. The MBTS and DPG need to be put in next and mixed for 6 min, maintaining the temperature at 70°C throughout this time. The cooling water

TABLE 11.9
Blends of Thiokol ST with NIPOL 1001

Formulation	MB-1 phr			MB-2 phr		
Thiokol ST	100			—		
Nipol 1001	—			100		
Stearic acid	1			1		
SRF N762	60			55		
Zinc peroxide	5			4		
Calcium hydroxide	1			1		
Blend ratio ST/1001	100/0	80/20	60/40	40/60	20/80	0/100
MB-1	167	134	100	67	33	—
MB-2	—	32	65	97	130	161
Physical properties, cured 30 min at 160°C						
Hardness, Shore A	72	72	70	70	68	73
Modulus at 100%, MPa	3.07	2.86	2.73	2.48	2.14	4.35
Tensile strength, MPa	7.8	8.7	9.9	10.14	11.56	13.9
Elongation, %	265	330	420	520	590	555
Low temperature stiffness, °C	−51	−47	−23	−20	−17	−15
Compression set B, %						
22 h at 70°C	25	31	41	68	64	65
22 h at 100°C	92	89	88	91	83	77
Physical properties, cured 30 min at 160°C, post-cured 24 h at 100°C						
Hardness, Shore A	73	75	73	74	74	73
Modulus at 100%, MPa	3.62	3.59	3.38	3.11	2.62	3.14
Tensile strength, MPa	7.9	8.97	11.32	13.59	13.7	18.35
Elongation, %	225	270	365	460	530	500
Compression set B, %						
22 h at 70°C	16	20	29	45	43	42
22 h at 100°C	81	81	80	81	70	55
Aged in solvents 1 week at RT, volume swell, %						
Benzene	117	139	129	142	154	139
Toluene	81	95	92	106	113	111
Xylene	47	59	59	69	73	69
Fuel A	−1	−1	0	−2	0	−1
Fuel B	11	12	12	14	17	19
Fuel C	18	23	23	24	28	28
Methyl ethyl ketone	51	93	127	191	252	269
Acetone	34	75	104	164	242	238
Ethyl acetate	39	65	81	116	166	173
Water	0	0	0	0	0	0

TABLE 11.10
Blends of Thiokol ST with Neoprene W

Formulations	MB-1			MB-3		
Thiokol ST	100			—		
Neoprene W	—			100		
Stearic acid	1			1		
SRF N762	60			45		
Zinc peroxide	5			4		
Calcium hydroxide	1			1		
NA-22	—			0.67		
Ex. Lt. Calcined magnesia	—			4		
Blend ratio ST/W	100/0	80/20	60/40	40/60	20/80	0/100
MB-1	167	134	100	67	33	0
MB-3	0	31	62.5	94	125	155.17
Physical properties, cured 30 min at 160°C						
Hardness, Shore A	74	72	73	70	68	68
Modulus at 100%, MPa	4.14	3.14	4.83	4.21	3.97	3.52
Tensile strength, MPa	7.76	7.66	11.49	13.25	18.15	21.11
Elongation, %	185	240	220	250	330	350
Low temperature stiffness, °C	−51	−47	−42	−41	−39	−37
Compression set B, %						
22 h at 70°C	28	27	28	36	22	14
22 h at 100°C	93	86	70	73	60	33
Physical properties cured 30 min at 160°C, post-cured 24 h at 100°C						
Hardness, Shore A	76	75	76	73	69	69
Modulus at 100%, MPa	5.0	3.93	5.76	4.52	3.95	4.35
Tensile strength, MPa	8.73	7.94	11.35	13.66	16.28	22.91
Elongation, %	170	205	190	240	310	330
Compression set B, %						
22 h at 70°C	15	21	23	45	22	14
22 h at 100°C	85	86	75	73	52	26
Aged 1 week in solvent at RT, volume swell, %						
Benzene	119	128	145	162	190	199
Toluene	77	101	124	154	190	209
Xylene	47	75	104	138	176	196
Fuel A	−1	2	6	11	15	16
Fuel B	10	20	32	43	56	64
Fuel C	18	36	55	75	97	113
Methyl ethyl ketone	50	59	69	80	92	96
Acetone	37	37	37	37	37	34
Ethyl acetate	40	49	57	68	76	81
Water	0	2	2	2	2	2

needs to be turned on full now and all of the rest of the ingredients, except the zinc oxide, should be added and mixed until they are completely incorporated. The zinc oxide may be added if the stock temperature is not too high and mixed for 1 min, then the batch is dumped onto a cool tight mill. If scorch has been a problem, it may be better to add the zinc oxide as a separate stage on the mill. The batch should be cut down on the sheet of mill immediately to dissipate as much heat as possible, then rebanded to complete the blending followed by the usual sheet off and cooling.

Calendering of Thiokol FA compounds is best accomplished with the top roll at 42°C and the middle one at 39°C, and the maximum thickness set at 0.08 cm. The feedstock should be well warmed, fed in small strips, and a small rolling bank be maintained. If the compound is to be plied up to obtain a greater thickness, then the compound will need to have tackifiers included for satisfactory ply adhesion.

Extrusion of polysulfide compounds should be done in as large a tuber as available to obtain maximum head pressure. The screw and barrel needs to be at 37°C–48°C with the head temperature a bit higher and the die hot. The stock should be fed in steady, small, thin strips, taking care not to allow the compound to back up in the throat and choke the feed. The feedstock should not be warmed on a cool mill as sticking to the roll surface may occur.

Press curing of Thiokol compounds should be done in a press with cooling capabilities for best results. A minimum press temperature of 150°C is recommended for best physical properties for a time dictated by the rheometer curve at the curing temperature. At the end of the cure cycle the press should be cooled under pressure to approximately 40°C before opening the press to avoid pock marks and other surface defects. Typical cure times at 154°C for Thiokol FA is 40 min and for Thiokol ST is 30 min.

Open steam vulcanization should begin in an autoclave that has been preheated to the curing temperature of a minimum of 148°C. Thick article may need a step cure to allow any trapped air to escape, thus avoiding blisters. Products such as rollers or hose need a solid end plate or end clamp and a very tight cross wrap with wet nylon tape. Wrapping of rollers or hose both needs to start in the middle, winding towards one end first, then back to the other end and returning to the starting point in the middle. The parts do not need to be cooled in the autoclave, but do not unwind the tape until after the roll or hose has cooled to room temperature.

11.5.2 THIOKOL ST

The processing of ST polysulfide is similar to that of FA, except that it is not necessary to peptize or soften the polymer first.

Mill mixing should be on a cool, 42°C–50°C, tight mill with cooling water on full throughout the mix. Band the Thiokol ST and add LP-3, if being used, and blend thoroughly. Add the curative if powder zinc peroxide is employed, although it is highly recommended that a dispersion of the curative in NBR be employed, and when dispersed, add the stearic acid and fillers. The plasticizers or tackifiers should be incorporated with the last small amount of filler and blended well. If a peroxide dispersion is used it may be added just before the plasticizer incorporation.

The bin shelf life of polysulfide polymers is about 90 days, hence only order the amount needed at any one time. There is strict control of manufacturing only to produce what is required to keep inventory turned frequently and that the shelf life be kept within the 90 day limit.

The bin stability of mixed compounds is about 3 weeks for Thiokol FA and 6 days for Thiokol ST if the storage temperature is kept at or below 23°C; however, it is recommended that a lower storage temperature be adopted if possible.

11.6 APPLICATIONS

Polysulfide polymers, with their excellent solvent resistance, are well suited for roller applications requiring resistance to ketones, highly aromatic solvents, and some chlorinated solvents. Paint and lacquer coating rolls as well as printing rolls applying a wood grain appearance are typical applications. The inclusion of some mercaptan-modified CR elastomer in all, but the most stringent applications, helps in processing and physical properties. Typical compounds are given in Tables 11.11 and 11.12.

TABLE 11.11
Black Thiokol FA Printing Rollers

Formulations			Thiokol FA Masterbatch		
Thiokol FA			100		
Neophax A or 2L Brown			20		
MBTS			0.35		
DPG			0.15		
Compound	20	30	40	60	65
Thiokol FA masterbatch	120.5	120.5	120.5	120.5	120.5
Neoprene W	15	15	15	15	15
Zinc oxide	10	10	10	10	10
Magnesium oxide	0.6	0.6	0.6	0.6	0.6
ETU 75% dispersion	0.1	0.1	0.1	0.1	0.075
Stearic acid	0.5	0.5	0.5	0.5	0.5
SRF N774	40	40	40	40	55
Cumar P25	10	10	10	10	10
Highly aromatic oil	40	30	20	—	—
Physical properties, cured 40 min at 150°C					
Hardness, Shore A	20	30	40	60	68
Tensile strength, MPa	2.5	2.8	3.6	6.6	7.42
Elongation, %	350	700	530	230	225
Aged 1 week at 23°C in solvents, volume change, %					
Xylene	82	83	82	NT	NT
Toluene	120	112	112	NT	NT
Acetone	−2	−4	0	NT	NT
Methyl ethyl ketone	33	30	30	NT	NT
Methyl isobutyl ketone	30	27	26	NT	NT
Ethyl acetate	9	14	18	NT	NT
Solvesso 100	NT	NT	NT	NT	55

TABLE 11.12
White Thiokol FA Rolls

Formulations	WFA-MB1	WFA-MB2
Thiokol FA	100	100
MBTS	0.35	0.4
DPG	0.15	0.1
Neophax A	20	5
Compound	FA 43	FA 68
WFA-MB1	120.5	—
WFA-MB2	—	105.5
Neoprene W	20	15
Titanox AMO	90	150
Ex. Lt. Calcined magnesia	1.2	0.6
Stearic acid	1.0	0.5
Highly aromatic oil	10	—
Cumar P25	7	10
ETU 75% dispersion	0.4	0.1
Zinc oxide	10	10
Physical properties, cured 40 min at 150°C		
Hardness, Shore A	43	68
Modulus at 100%, MPa	0.3	0.8
Modulus at 200%, MPa	0.6	1.7
Tensile strength, MPa	3.2	4.8
Elongation, %	720	400

The use of titanium dioxide is quite expensive, hence other high pH light-colored fillers such as Zeolex 23, Wollastonite, Silene 732 D, Mistron, and so on is suggested.

The recommended cure time for these rollers with a thin wall is 2 h at 154°C. For rolls with wall thickness of 25 mm the cure time needs to be 3.5 h; for a thickness of 25–32 mm the time should be 4.5 h; and a thickness of 32–41 mm a cure time of 6 h is required. In addition, remember to preheat the autoclave and allow the rolls to cool to room temperature after removing from the vulcanization chamber before unwrapping.

11.6.1 PAINT CAN GASKET

Paint and lacquer spray can gaskets, as seen in Table 11.13, must resist highly aromatic solvents, ketones, acetone, and chlorinated solvents, which makes Thiokol FA a good choice for this application. Although Thiokol ST compounds exhibit better compression set properties, FA type is usually employed because of the superior swell resistance it imparts as the operating temperatures are invariably at room temperature. The room temperature compression set of Thiokol FA is 11%.

TABLE 11.13
Thiokol FA Paint Can Gasket

Formulation	Gasket
Thiokol FA	100
MBTS	0.4
DPG	0.1
Stearic acid	1
MT N990	130
Zinc oxide	10
Sulfur	1
Physical properties, cured 40 min at 149°C	
Hardness, Shore A	82
Tensile strength, MPa	5.18
Elongation, %	150

11.6.2 PAINT HOSE INNER LINER

Paint spray hose tubes shown in Table 11.14 have the same requirements as paint can gaskets, hence the selection of FA type polysulfide for these applications as well.

TABLE 11.14
Thiokol FA Hose Liner

Formulation	Liner	Masterbatch W
Thiokol FA	100	—
Neoprene W	—	100
Ex. Lt. Calcined magnesia	—	4
MBTS	0.4	—
DPG	0.1	—
SRF N762	60	55
Stearic acid	0.5	0.5
Zinc oxide	10	5
ETU 75% dispersion	0.133	—
Masterbatch W	29	—
Physical properties, cured 40 min at 150°C		
Hardness, Shore A	74	
Modulus at 300%, MPa	9.49	
Tensile strength, MPa	9.52	
Elongation, %	300	

TABLE 11.14 (Continued)
Thiokol FA Hose Liner

Formulation	Liner	Masterbatch W
Aged 30 days at 26°C in solvent, volume swell, %		
SR-6	20	
SR-10	4	
Ethyl acetate	24	
Acetone	18	
Methyl ethyl ketone	36	
Methyl isobutyl ketone	28	
Carbon tetrachloride	60	
Water	4	

11.7 ASTM D2000 M, TYPE A, CLASS K, WITH THIOKOL ST

Compounds to meet ASTM D2000 designations given in Table 11.15 are always based upon Thiokol ST in order to meet the compression set basic requirements.

There are many other applications for Thiokol FA and ST such as gas meter diaphragms and parts requiring very good low-temperature properties. The liquid LP grades are not covered in this book, but have very broad use in curtain wall and thermal pane sealants, protective coatings, and binders for rocket propellants. The odor of these polymers is a deterrent for some end applications and in factory processing.

TABLE 11.15
ASTM D2000 Compounds

Formulations	AK403	AK503	AK603	AK707	AK807	AK907
Thiokol ST	100	100	100	100	97	95
Thiokol LP-3	—	—	—	—	3	5
Stearic acid	1	1	1	1	1	1
SRF N774	30	30	35	—	—	—
HAF N231	—	—	—	50	70	90
Highly aromatic oil	10	5	—	—	—	—
Calcium hydroxide	1	1	1	1	1	1
Zinc peroxide	5	5	5	5	5	5
Physical properties, cured 30 min at 154°C						
Hardness, Shore A	42	50	60	70	80	90
Modulus at 300%, MPa	2.1	2.7	4.2	6.2	9.1	—
Tensile strength, MPa	3.8	4.3	5.4	11.2	12.1	9.8
Elongation, %	450	430	375	480	390	230
Compression set B, %						
22 h at 70°C	34	33	29	31	36	48

11.8 TROUBLE SHOOTING

In the event that problems arise either in processing or in product performance, a few suggested steps towards solving these are given here:

1. Variations in the viscosity of FA polysulfide compounds may be attributed to either inaccurate weighing of the MBTS and DPG or to a loss of these materials through inefficient sweeping during the mix. In addition, insufficient time or temperature on the mill may not have allowed for complete softening of the polymer.
2. Reversion of the vulcanizates could be caused by the substitution of other similar accelerators for MBTS or DPG. Other materials do not work as efficiently as these and could cause poor processing and physical properties.
3. Polymer lumps in the mixed compound are caused by not sweeping up the crumbs early and often throughout the mix. If these crumbs are added late they will not be softened by the MBTS and will float in the softened matrix.
4. A greasy surface on the uncured or cured compound may be caused by the use of nonaromatic process aids that are not compatible with Thiokol.
5. Porosity in the finished part may be from several sources. These elastomers are very thermoplastic and in molded goods it is very important that the preform is dense and shaped so that the air is pushed out of the cavity as the mold is closed. Little or no bumping is recommended. If vulcanized vegetable oil is used, it is important that it is in crumb-form or that solid material is passed through a very tight mill to convert it to crumb and be allowed to rest overnight to allow H_2S or other gases to escape. Loose wrapping of rolls, lack of an end plate, and incorrect wrapping procedure may also fail to squeeze the air out of the stock and result in porosity.
6. Poor or logy cures may be caused by contamination of the compound with foreign accelerators, acidic materials such as clay in a dip, curatives on the surface of wrapping tapes, and so on. Always be sure that processing equipment is clean and free of any previously handled compounds and new or only tapes reserved for Thiokol compounds be used. Avoid acidic compounding materials such as clays or other low pH fillers or process aids.
7. A poor state of cure will also result if a mercaptan-modified neoprene such as Neoprene W is replaced with a sulfur-modified grade, which will severely retard the vulcanization.
8. Pock-marked molded parts may be caused by a mold design that does not allow the air to escape readily, too small and incorrectly shaped performs, bumping or porous performs.
9. Removing the tape from the roll before it has cooled to room temperature may cause distortion or porosity.

REFERENCES

1. R.C. Klingender. Miscellaneous Elastomers, Chapter 17 in M. Morton, *Rubber Technology, Third Edition.* Van Nostrand Reinhold Company, 1987, pp. 482–488.

2. Polysulfide Rubber, FA Polysulfide Rubber, TD FA/10–80/1M. Morton Thiokol Inc., Specialty Chemicals Division (Rohm & Haas now own Morton Thiokol Inc. and Morton International).
3. Polysulfide Rubber, ST Polysulfide Rubber, TD-916, 2/77. Morton Thiokol Inc., Specialty Chemicals Division (Rohm & Haas now own Morton Thiokol Inc. and Morton International).

12 Plasticizers, Process Oils, Vulcanized Vegetable Oils

Peter C. Rand

CONTENTS

12.1 INTRODUCTION

In his preface to Volume 1 of *Plasticizer Technology*, Paul F. Bruins wrote in 1965 that "the technology relating to plasticizers has been discussed in many books and technical articles but, despite this, research people and others in the (polymer) industry find the selection of the optimum plasticizer system a difficult task." Forty years later, nothing much has changed. The problem may stem from the almost overwhelming variety of esters, process oils, waxes, and factices, which are available to the compounder.

The field of synthetic esters itself comprises literally thousands of products. Refined petroleum process oils for the rubber industry number in the hundreds because of the varieties of crude oil available around the world. Similarly, with the dozens of vegetable oils available for the production of vulcanized vegetable oils, we are baffled by the variety of seemingly similar but surprisingly different factices offered throughout the world rubber industry.

The concepts, opinions, and recommendations in this chapter are the distillation of the author's 40-plus years of experience in the rubber industry, especially in plasticizer manufacturing and use and in consulting. The aim is to guide the specialty elastomer compounder to the simplest and most practical formulations within a reasonable time frame. Practicality takes precedence over strict scientific method in most cases.

12.2 SCOPE

This chapter will attempt to simplify the process of choosing the best plasticizing system from within the family of synthetic esters and from the commonly available factice grades. The process oils present the more difficult task of selecting the optimum hydrocarbon process oil knowing that each product is unique to a specific crude feedstock and refinery process. After all, petroleum is natural and quite complex. Even vegetable oils are more consistent place to place. There will be very little organic chemistry discussed in this chapter beyond naming the alcohols, glycols, acids, and perhaps general carbon chain length components of petroleum oils. The focus will be on a few basic principles required to select the general class of plasticizer at first, then narrowing the choice to a specific grade from that larger group.

Over the years many have been interested in studying plasticizer compatibility using hydrogen bonding and solubility parameter concepts. However, for practical day-to-day compounding that requires quick selection of one or two likely candidates, certain shortcuts, practical rules of thumb (PROT) and occasional warnings, avoid this mistake (ATM), while not always academically acceptable, will get the formulation to the mixer in the shortest time. No attempt is made in this chapter to characterize certain hydrocarbon resins as plasticizers, although some fall close to the definition. More use is made of these as tackifying agents and compatibilizers.

12.3 DEFINITIONS

Plasticizer: Any material (primarily liquid) that is added to a compound to facilitate easier mixing, extruding, molding, and curing through modification of the basic physical properties of the uncured and cured compound (e.g., viscosity, hardness, modulus, low-temperature flex).

Monomeric types: It is common to characterize plasticizers generally as either polymeric or monomeric. For our purposes here, monomerics are all plasticizers (glycol esters, monoesters, diesters, and triesters), which are not polymeric. Viscosities range from 4 to 400 cps.

Polymeric types: This group is technically comprised of polyesters with viscosities from 400 to 200,000 cps. To create polyester, a glycol rather than an alcohol is reacted with a fatty acid. For purposes of this chapter we exclude the very high polymers used as polymeric plasticizers such as Nipol 1312, a low-molecular weight NBR elastomer.

Synthetic plasticizers: These are chemically reacted, man-made, from highly specified components to form a plasticizer with consistently reproducible properties. Esters comprise the major portion of this category.

Low-temperature improvement: This is achieved through the use of plasticizers primarily related to their molecular weight as measured by viscosity. A rating system has been applied to a number of plasticizers in Table 12.1. Compatibility is also a factor to be considered, hence the ratings in both polar and semipolar elastomers is given.

High-temperature performance: It is also related to the molecular weight of the plasticizer as measured by volatility, as may be seen in Table 12.2. The relative solubility of a number of plasticizers is also given, which relates to their extractability in various fluids in service. Both of these factors must be considered when designing a specialty elastomer compound.

Concept of plasticizer polarity: Plasticizers are ranked (Table 12.3) according to how compatible they are with well-known polar elastomers. As a very rough guide (PROT), polar constituents of an elastomer such as acrylates, acrylonitriles, glycols, aromatic/phenyl groups, and halogens are usually a sign of some degree of polarity. Olefins are nonpolar. For example, polyethylene is nonpolar but chlorinated polyethylene is deemed to be semipolar. Similarly, paraffinic process oil has little to no polar characteristics while process oil with significant aromatics content is very compatible with semipolar elastomers (e.g., SBR, CR).

TABLE 12.1
Viscosity versus Low Temperature

Product	Type	Viscosity-Centipoise at 25°C	Low-Temperature Efficiency Rating[a]	
			Polar	Semipolar
Octyl oleate (AO)	Monoester	10	NR	1
Butyl oleate (BO)	Monoester	9	NR	1
Acetyl methyl ricinoleate (P-4)	Monoester	18	NR	1–2
Tributoxyethyl phosphate (TBEP)	Glycol ether ester	12	1	2
Tributyl phosphate (TBP)	Triester	4	1	1
DBEEF (TP-90B, 4221)	Acetal	14	1	1
Diisobutyl adipate	Diester	15	1–2	1
Dioctyl adipate (DOA)	Diester	18	1–2	1
Dibutyl sebacate (DBS)	Diester	15	1	1
2-Ethylhexyl diphenyl phosphate	Alkylaryl ester	18	1–2	2
Dibutyl phthalate (DBP)	Diester	20	2	2–3
Isodecyl diphenyl phosphate	Alkylaryl ester	23	2	2
Epoxidized octyl tallate	Epoxidized monoester	20	NR	1
Dioctyl azelate (DOZ)	Diester	23	1–2	2
TP-759 (4426, RS-700)	Glycol ether ester	30	2	2
Dioctyl sebacate (DOS)	Diester	25	1–2	2
DBEEA (TP-95, 4226)	Glycol ether ester	25	2	NR
Triethylene glycol dioctoate	Glycol ester	24	1–2	2
Acetyl tributyl citrate	Triester	45	NR	3–4
Linear phthalate (810P)	Diester	35	2–3	2–3
Plasthall 7050 (glutarate)	Ether ester	50	2–3	NR
Dioctyl phthalate (DOP)	Diester	65	3	3
Triaryl phosphate (Reofos 65)	Triester	70	3–4	4
Diisononyl phthalate (DINP)	Diester	75	3–4	3–4
Butyl benzyl phthalate (BBP)	Diester	50	4	NR
Diisodecyl phthalate (DIDP)	Diester	90	4	4
Linear trimellitate (810 TM)	Triester	110	3–4	3–4
Propylene glycol dibenzoate	Glycol ester	140	4	5
Trioctyl trimellitate (TOTM)	Triester	220	4–5	4–5
Paraffinic process oil (total aromatics 16%)	Mineral oil	60	NR	NR
Naphthenic process oil (total aromatics 36%)	Mineral oil	40	NR	4–5
Aromatic process oil (total aromatics 75%)	Mineral oil	1100	4–5	4–5

TABLE 12.1 (Continued)
Viscosity versus Low Temperature

Product	Type	Viscosity-Centipoise at 25°C	Low-Temperature Efficiency Rating[a] Polar	Semipolar
Epoxidized soybean oil (ESO)	Epoxidized vegetable oil	400	NR	4–5
Polymeric adipate	Polyester	1,500	5[b]	NR
Polymeric adipate (G-50, P-6310)	Polyester	2,800	NR	NR
Polymeric adipate (G-59, P-6424)	Polyester	25,000	NR	NR
Polymeric sebacate (G-25, P-1030)	Polyester	160,000	NR	NR

NR—Not recommended.
1–5 Excellent-to-poor rating.
[a] Rating in (a) polar and (b) semipolar elastomer compounds. A general guide to how efficient a plasticizer is in improving low-temperature flex and torque.
[b] Best choice in the polyester/polymeric plasticizer group for low-temperature properties.

12.4 SYNTHETIC PLASTICIZERS

The great beauty of synthetic ester plasticizers is that there is an unlimited number of combinations of alcohols and acids from which to tailor the most suitable grade for a given application. Table 12.4 lists some of the most common precursors for rubber-oriented plasticizers. Unfortunately this great variety of available potential products has been a source of frustration for producers and compounders alike for decades. Table 12.5 illustrates some of the specialty synthetic plasticizers on the market along with many of the producers. Because the plasticizer loading ranks behind three or four other compounding ingredients deemed more important, (e.g., the reinforcing filler, the curing additives, antioxidants) it has always been hard to justify time spent selecting a perfect type. Most rubber compounders develop three or four favorites for low temperature, good oil resistance, and high-temperature permanence that they employ regularly. With a more studied approach and with the help of the plasticizer supplier, some of the selection frustration can be eliminated.

12.4.1 MONOESTERS—ONE ALCOHOL GROUP (OR RADICAL) WITH ONE ACID GROUP

Esters based on unsaturated oleic and tall oil fatty acids are the most common low-temperature plasticizers for CR, CR/SBR, and CSM.

They are very effective at imparting extremely low-temperature flex properties because of their low viscosity and long carbon chain lengths.

TABLE 12.2

Molecular Weight versus High-Temperature Permanence[a]

Product	Type	MW (Calc)	Solubility					Volatility
			Al	Ar	Et	Ac	H$_2$O	
Butyl oleate	Monoester	338	S	S	S	S	I	High
Methyl acetyl ricinoleate (P4)	Monoester	346	—	—	—	—	—	High
Tributyl phosphate (TBP)	Triester	266	—	S	—	S	—	High
Dibutyl phthalate (DBP)	Diester	278	S	S	S	S	I	High
DBEEF (TP 90B, 4221)	Acetal	200 (est.)	S	S	S	S	I	High
2-Ethylhexyl diphenyl phosphate	Alkylaryl ester	362	—	—	—	—	—	Med-High
Dioctyl adipate (DOA)	Diester	370	S	S	S	S	I	Med-High
Dibutyl sebacate (DBS)	Diester	314	S	S	S	S	I	High
Dioctyl azelate (DOZ)	Diester	412	S	S	S	S	I	Med
Dioctyl sebacate (DOS)	Diester	426	S	S	S	S	I	Med
DBEEA (TP 95, 4226)	Glycol ether ester	434	S	S	S	S	I	Med-High
Triethylene glycol dioctoate (3800)	Glycol ester	410	S	S	S	S	I	Med
TP-759 (4426)	Glycol ether ester	518	I	S	S	S	PS	Med

Plasticizer	Type	Mol. wt.	Al	Ar	Et	H$_2$O	Ac	Volatility
Trioctyl trimellitate	Triester	546	S	S	I	S	I	Low
Linear trimellitate	Triester	588	S	S	I	S	I	Low-Med
Paraffinic process oil (total aromatics 16%)	Mineral oil	410	S	S	—	—	I	High
Naphthenic process oil (total aromatics 36%)	Mineral oil	326	S	S	—	—	I	Med-High
Aromatic process oil (total aromatics 75%)	Mineral oil	486	S	S	—	—	I	Low-Med
Acetyl tributyl citrate (A-4)	Triester	402	—	—	—	—	—	Med
Adkcizer RS-735	Glycol ether ester	812	S	S	S	S	PS	Low-Med
Epoxidized soybean oil (G-62)	Epoxidized vegetable oil	1,000	PS	S	S	PS	I	Low
Polymeric adipate (P-6320)	Polyester	900	I	S	PS	S	I	Low-Med
Polymeric adipate (G-50,P-6310)	Polyester	2,200	PS	S	PS	S	I	Low
Polymeric sebacate (G-25)	Polyester	7,000	PS	S	PS	S	I	Very low

[a] A guide to the volatility of a plasticizer assuming a direct correlation with heat aging properties, for example, high volatility would result in high loss of plasticizer affecting heat-aged properties such as hardness, flex, modulus, and so on. Ratings in terms of % loss after 24 h at 150°C are High >25; Med-High 11–25; Med 6–10; Low-Med 2.5–5; Low 1–2.4; Very low <1.0. Solubility determinations are in Al—aliphatic; Ar—aromatic; Et—ethanol; H$_2$O—water; and Ac—acetone.

TABLE 12.3
Relative Polarity Ratings

Liquids	Rating[a]
Sulfonamides (e.g., *n*-BBSA)	10
Glycol ether esters (e.g., DBEP, DBES, TBEP, DBEEA, DBEEF)	9
Triaryl phosphates (e.g., TPP, isopropyl triphenyl)	8
Glycol esters (e.g., TEG-CC, TEG-2EH, Benzoates)	7
Mixed alkyl/aryl phosphates (e.g., Santicizer 148)	6
Trimellitates (e.g., TOTM, TINTM, TIOTM)	5
Phthalates diesters (e.g., DOP, DINP, DBP, DIDP)	5
Trialkyl phosphates (e.g., TBP, TOF, TEP)	5
Monoesters (e.g., butyl oleate, butyl tallate, other alkyl oleates, ricinoleates, stearates)	3–4
Rubber process oils—aromatic	3
Rubber process oils—naphthenic	1
Rubber process oils—paraffinic	0
Solid factice	Rating
Castor oil-based factice	3
Rapeseed oil-based factice	0
Soybean oil-based factice	0

[a] Rating indicates a scale of relative polarity: 10 (ten) is the highest polarity and 0 (zero) shows no significant polarity. This rating is subjective having been derived from the author's experience and observations of the compatibility of the plasticizers in elastomer formulations.

Alcohols (Monomeric): Methyl (C_1); Ethyl (C_2); *n*-Butyl (C_4); Isobutyl (C_4); *n*-Hexyl (C_6); 2-ethylhexyl (C_8); *n*-Octyl (C_8); Isooctyl (C_8); Isononyl (C_9); *n*-Decyl (C_{10}); Undecyl (C_{11}); Tridecyl (C_{13})

Glycols (Polyhydric Alcohols): Ethylene/Diethylene; Propylene/Dipropylene; 1,3 butylene; 1,4 butylene; Hexylene; Neopentyl; MP diol

Glycol Ethers: Ethylene; Glycol monobutyl ether; Diethylene glycol monobutyl ether; Triethylene glycol monobutyl ether

Esterification Examples: Alcohol + Acid → Ester + Water; Glycol + Acid → Polyester + Water; Glycol + Methyl ester → Glycol ester + Methanol

However, there are trade-offs as with all esters. Crystallization rate is increased, in CR especially, and permanence after oven aging is poor even at lower oven temperatures. Extraction resistance is also relatively poor.

ATM—never switch sources of monoester types without checking the Iodine numbers. Depending on the derivation of an unsaturated fatty acid, its Iodine number, which is a measure of unsaturation, can swing from 75 (animal) to 100 (vegetable) to 130 (tall oil). The impact on cure rate can be disastrous with a swing from high to low or vice versa. The unsaturation uses up curing ingredients and a switch to high unsaturation can slow the cure significantly depending on the plasticizer loading.

TABLE 12.4
Synthetic/Ester Plasticizer Precursors

	Acids	
Monofunctional	**Difunctional**	**Trifunctional**
Caproic (hexoic) (C_6)	Succinic (C_4)	Citric (C_6)
Caprylic (C_8)	Glutaric (C_5)	Trimellitic (C_9)[a]
Pelargonic (C_9)	Adipic (C_6)	
Capric (C_{10})	Azelaic (C_9)	
Lauric (C_{12})	Sebacic (C_{10})	
Stearic (C_{18})	Phthalic (C_8)[a]	
Oleic (C_{18} unsaturated)		
Benzoic (C_7)[a]		

[a] Aromatic.

TABLE 12.5
Synthetic Plasticizers: Equivalent Products and Producers

Product/Chemistry	Designation	Producers
Butyl oleate	BO	1, 3, 4, 6
Acetyl methyl ricinoleate	P-4	10
Epoxidized octyl tallate	4.4, E-45, EP-8	1, 3, 6
2-Ethylhexyl diphenyl phosphate	141, 362	12, 13, 17
Tributoxyethyl phosphate	TBEP, KP-140	6, 12, 14, 15, 20
Isodecyl diphenyl phosphate	148, 390	12, 13
Tributyl phosphate	TBP	6, 12, 14, 15, 20
Trioctyl phosphate	TOF	6, 15, 17, 20
Dibutoxyethoxyethyl formal	TP 90B, 4221	6, 7, 16
Dibutoxyethoxyethyl adipate	TP-95, 4226	3, 4, 5, 6, 7, 8, 16, 20
Glycol ether adipate	RS-735	8
Triethylene glycol butyl adipate	TP-759, 4426, RS-700	5, 6, 8, 16
Dibutyl sebacate	DBS	3, 4, 6
Diisobutyl adipate	DIBA	6
Dioctyl sebacate	DOS	3, 4, 6, 7, 10
Dioctyl adipate	DOA	2, 3, 4, 6, 7, 11
Dioctyl azelate	DOZ	3, 6, 19
Triethylene glycol dioctoate	SC-E, 3800, 803	3, 4, 6
Triethylene glycol caprate-caprylate	SC-B, 4141, 3810	3, 4, 6, 7
Dipropylene glycol dibenzoate	9-88	2, 9
Linear trimellitate	810TM	3, 6, 7
Linear phthalate	800P, 810P	6
Butyl benzyl phthalate	160, BBP	6, 13, 17
Acetyl tributyl citrate	ATBC	11, 18

(*continued*)

TABLE 12.5   (Continued)
Synthetic Plasticizers: Equivalent Products and Producers

Product/Chemistry	Designation	Producers
Triaryl phosphates	50/65, 519/521	6, 12, 14, 15
Polyester adipate	P-6320, P-670	2, 3, 6, 14
Polyester adipate	P-6310, G-50	3, 6, 17
Polyester adipate	P-6410, G-54, 330	2, 3, 6, 10, 14
Polyester glutarate	P-5521, P-7046	3, 6
Polyester sebacate	P-1030, G-25	3, 6
Polyester adipate	P-6303, G-40	2, 3, 6
Polyester adipate	P-6420, G-57, 9776	2, 3, 6, 19

1. Chemtura Corporation (Humco)—USA
2. Velsicol Corporation (Admex, Drapex, Nuoplaz)—USA
3. HallStar Co. (Plasthall, Paraplex)—USA
4. Harwick Standard Distribution (Polycizer, Merrol, SC)—USA
5. S.T.P.C./NYCO (Nycoflex)—Belgium
6. Merrand International Corp. (Merrox)—USA
7. Cognis Performance Chemicals (Bisoflex)—UK
8. Adeka Denka Co. (Adkcizer)—Japan
9. Noveon (K-Flex)—USA
10. Caschem/Rutherford Chemicals (Flexricin)—USA
11. Unitex Chemicals (Uniplex)—USA
12. Supresta (Phosflex)—USA
13. Ferro Corp. (Santicizer, Plaschek)—USA, UK
14. Chemtura (Reofos, Kronitex)—UK
15. Rhodia (Antiblaze, Amgard)—UK
16. Rohm & Haas (TP)—USA
17. Bayer AG (Disflamoll, Ultramoll)—Germany
18. Morflex Chemical (Citroflex)—USA
19. Cognis (Edenol)—USA
20. Daihachi Kagaku—Japan

12.4.2   DIESTERS—TWO ALCOHOL GROUPS REACTED WITH ONE ACID GROUP

Diester plasticizers represent the largest group of synthetic plasticizers. They are typically fully saturated and have no direct effect on curing. The two most common ways they affect cure indirectly are related to the polarity and mobility of the plasticizer in the compound. Lower compound viscosity is a function of the external lubricating effect and degree of polarity of the diester. The mobility of the synthetic plasticizer aids dispersion especially of ester-soluble ingredients but a tendency to migrate to the surface transporting some of the solubilized curatives and stabilizers calls for careful selection of the ester.

12.4.3 POLYESTERS—A POLYOL/GLYCOL GROUP REACTED WITH AN ACID GROUP

These polyesters are called the polymeric plasticizer family and are characterized by molecular weights of approximately 800–8000 (mono/diesters in the range MW 200–600) and viscosities from 350 to 200,000 cps. They are esterified to different molecular weights by controlling the linear chain extension using varying techniques including termination of the reaction at a calculated point in the process with an alcohol or monobasic fatty acid "cap." The majority of polymerics today are terminated to achieve consistent viscosity and residual hydroxyl properties. Older, higher MW grades such as Merrox P-1030/G-25 are unterminated and tend to have much wider viscosity and hydroxyl specifications. Unterminated polyesters produced with relatively high hydroxyl numbers (e.g., 35–80) are precursors in the polyester polyol polyurethane market.

ATM—never assume that higher the viscosity of a polymeric, the better it is for extraction resistance. Certain polyesters can be very poor for resistance to water but excellent against hydrocarbons. It is a matter of picking the correct glycol and acid combination with the help of the supplier.

12.4.4 MISCELLANEOUS

There will be considerable discussion of glycol ether ester plasticizers under the application section for polar elastomers. This group of plasticizers is the most polar of all and along with polymerics is probably the most important for purposes of this chapter.

The glycol esters have played an important part in the history of specialty elastomer compounding and several favorites comes from this group, largely for pure low temperature considerations due to their excellent compatibility and viscosity.

Esters made with inorganic acids also will come into the discussion. The main group is the alkyl, alkyl-aryl, and aryl phosphates that are flame retardants with added value as antifungal agents, surfactants, and low-temperature plasticizers.

12.5 HYDROCARBON PROCESS OILS/MINERAL OILS

Many compounders who gained their early experience in the tire industry are going to have a reasonably strong knowledge of naphthenic and aromatic process oils. Process oils are the major plasticizers for nonpolar NR, BR, IIR, and the slightly polar SBR. Similarly, those with experience formulating non-tire nonpolar EPDM as well as the more polar CR and CSM will have a feel for how paraffinic and naphthenic process oils contribute to the compounds. However, when the elastomer types become more polar and more specialized in their applications, process oils, especially those with high percentages of paraffinic carbon chains, no longer are suitable plasticizers.

As we discuss the plasticizer compounding techniques for the specialty elastomers later in this chapter, it will be seen that process oil choices are restricted to very specialized circumstances and even unconventional use. Classification of rubber process oils (Table 12.6) is based primarily on the carbon type analysis (%), aniline

TABLE 12.6

Hydrocarbon Process Oils Classification

Type	ASTM #	Saturates, %	Polar Compounds, %	Asphaltenes, %	Viscosity–Gravity Constant
Highly aromatic	101	20.0 max	25 max	0.75	0.940–0.999
Aromatic	102	20.1–35.0	12 max	0.50	0.900–0.939
Naphthenic	103	35.1–65.0	6 max	0.30	0.850–0.889
Naphthenic	104A	65.1 min	1 max	0.10	0.82 min
Paraffinic	104B	65.1 min	1 max	0.10	0.82 max

Note: Derived from ASTM D2226 with an additional rating by VGC.

point, viscosity–gravity constant, and viscosity. Color is also a factor for light-colored elastomer compounds.

12.5.1 PARAFFINIC PROCESS OIL

The industry defines this group with a minimum 55% paraffinic side chains (psc). Because of its high level of saturates, this group has excellent resistance to color development and the oxidation resistance increases with molecular weight. However, low aromatic content gives poor solvency and this group of process oils is not recommended for polar elastomer compounding.

12.5.2 NAPHTHENIC PROCESS OIL

Total aromatics (by clay-gel analysis) of 30% minimum are the general boundary although not hard and fast. Generally, naphthenic process oils are not recommended for polar elastomers but can be used to good advantage in semipolar elastomers. Naphthenic oils do not need to be labeled as carcinogenic if they have been adequately hydro-treated or extracted to meet industry and OSHA standards.

12.5.3 AROMATIC PROCESS OIL

Total aromatics content of 70% minimum describes this group. Aromatic content combined with viscosity considerations are the main factors when selecting suitable process oil for a semipolar to polar elastomer compound. In general, the high unsaturated content impacts color stability, oxidation resistance, and cure characteristics. All aromatic extract process oils must be labeled as carcinogenic under OSHA regulations.

12.5.4 PROCESS OIL PRODUCERS

At one time the principal influence on the American rubber process oil market was Sun Oil with familiar trademarks Circosol, Sundex, Sunpar, and Sunthene. In recent years many changes have taken place. Exxon Mobil increased their presence and then dropped out of aromatic types while Sun exited the naphthenic side of the business. Shell has also left the rubber process oils business and the popular Golden Bear refinery (California) closed several years ago. Below is a short list of some of the North American producers and the process oils they feature:

1. Cross Oil (naphthenic)
2. Exxon Mobil (paraffinic, naphthenic)
3. Sunoco (paraffinic, aromatic)
4. San Joaquin Refining (naphthenic, aromatic)
5. Ergon (paraffinic, naphthenic)
6. Valero (paraffinic, aromatic)
7. Calumet (paraffinic, aromatic)
8. Ashland (aromatic)

12.5.5 ASTM CLASSIFICATION AND COMMENTS

Table 12.6 has the ASTM D 2226 Classification and also shows the limits for classifying process oils by VGC. All aromatic process oils must be labeled as potential carcinogens. Paraffinic types and treated naphthenic do not need to be labeled. There is a growing availability problem with naphthenic process oils due to a shortage of naphthenic-rich crude and the exit of producers not able to justify the capital cost of solvent or hydro-treatment. This will result in increases in the cost of naphthenic oils for use in the rubber industry in the near future.

12.6 VULCANIZED VEGETABLE OIL (VVO)—FACTICE

12.6.1 TYPICAL VEGETABLE OILS IN FACTICE

Before the discovery of vulcanization of natural rubber using elemental sulfur in the mid-nineteenth Century, there had already been attempts to make a rubber substitute (the French term is caoutchouc factice meaning artificial rubber) using a variety of natural vegetable oils. There is an almost inexhaustible supply of vegetable oils available today with appropriate unsaturation for conversion to factice but cost, geography, and overall efficiency considerations have narrowed the field to a handful of oil candidates: soybean, rapeseed, sunflower seed, castor, safflower seed. The dominant oils today are rapeseed and soybean depending primarily on the cost of the oil and customer preference. The major types and their suppliers are listed in Table 12.7.

12.6.2 VULCANIZING AGENTS IN FACTICE PRODUCTION

The choice of vulcanizing agents has been much the same for the past 150 years with the exception of the use of an isocyanate cross-linker added during the last 50–60 years. Elemental sulfur (brown factice), sulfur chloride (white factice), hydrogen sulfide (light yellow/translucent factice), isocyanate, and peroxide (very white/hard/translucent factice) comprise the main types employed for vulcanizing vegetable oils.

TABLE 12.7
Vulcanized Vegetable Oil (Factice) Producers

Soybean oil—Brown—Sulfur
Similar grades—Akrofax A, Neophax A, 2L Brown, 5L Brown

Rapeseed oil—Brown—Sulfur
Similar grades—Vulcanol Brown BN, Rhenopren 14, VVO 350, Rhenopren C, VVO 20 Brown

Rapeseed oil—White—Sulfur monochloride
Similar grades—Vulcanol White XL, Rhenopren R, VVO 50B White, Hansa O
Similar grades—Vulcanol White Special 32, VVO French Special 32 (soybean), Akrofax 57 White

Castor oil—Brown—Sulfur or TDI
Similar grades—Vulcanol G-5, Neophax F, Vulcanol G-2
Similar grades—Vulcanol Nitrex, Rhenopren Asolvan
Similar grades (TDI) —Adaphax 758/759, Akrofax 758

Specialty—Light color—Hydrogen sulfide and Sulfur
Similar grades—Amberex BR (rapeseed), Akrofax BR (rapeseed), Amberex SR (soybean)

Each factice type identifies the base oil, resulting color, and cross-linking agent. Manufacturers:
1. D.O.G. (Hansa, Corona, NQ)—Germany
2. Lefrant-Rubco (Akrofax, Brown)—France
3. J. Alcott Manufacturing—UK
4. Polyone Corp. (Neophax, VVO)—USA (ceased operations in USA in 2004)
5. Puneet Polymers (Vulcanol)—India
6. Rhein-Chemie (Faktogel, Rhenopren)—Germany

12.6.3 MODIFIERS

During production of factice, other additives may be employed to lower the cost or buffer the pH, or adjust the processability. Calcium carbonate, process oils, and magnesium oxide buffer are some of the items that are added at relatively low levels to modify factice. Certain proprietary process aids may be added to improve processing characteristics.

12.6.4 ADDITIONAL COMMENTS

In the discussions on application formulations, most of the focus will be on the unique grades of castor oil factice. Both the sulfur cross-linked and the isocyanate cross-linked grades hold special interest for the compounder. The polarity of the castor grades allows them to be compounded into the more polar elastomers using both sulfur and peroxide cure systems. Cost of the castor/sulfur grades is only slightly higher than the rapeseed/soybean versions but the improved stress–strain properties, increased oil resistance, and thermoplastic-like processability make them a strong choice for polar elastomers. The isocyanate-crosslinked castor grades tend to be relatively expensive.

Castor factice is initially harder to warm up on open mills than soybean (easiest) or rapeseed oil grades but then exhibits a pseudo-thermoplastic behavior as it reaches temperatures over 115°C and typically produces 1–2 points lower Shore A hardness in the finished compound. Factice softness can cover a wide range depending on how tightly it is cross-linked. A sense for the degree of cross-link versus softness characteristic can be derived from the acetone (or hexane in the case of castor grades) extracts. Factice gets softer the higher the percentage extract.

12.7 APPLICATIONS AND COMPOUNDING

12.7.1 Nitrile Rubber (NBR, NBR/PVC, HNBR, XNBR)

12.7.1.1 NBR

The nitrile elastomers group is the largest consumer of the specialty synthetic plasticizers. It is primarily the acrylonitrile content of the NBR that needs to be managed from the processing, extraction/swell, and low-temperature flex points of view. Because NBR can be highly polar, care must be taken to compound with highly polar glycol ether esters for the maximum efficiency, compatibility, extraction requirements, and low-temperature properties.

NBR grades with 40% or higher ACN content can be compounded with high-levels (35–40 phr) of glycol ether esters such as DBEEA, DBES, and DBEEF. NBR with medium ACN content (30%–38%) can be formulated with lower loadings of glycol ether esters (25 phr) and the commodity diesters such as DIDP and DOA, as the ACN content goes down. Straight glycol esters TEG-2EH, TEG-CC are also popular.

NBR in the ACN range 18%–28% will require less glycol ether ester and the lower cost DOP, DINP grades tend to be used but at the 12–20 phr level.

One way to control swell/volume increase once the ACN level is picked is by addition of relatively low levels (5–12 phr) of medium-high to high viscosity polymeric plasticizer. High viscosity means high molecular weight and a very low tendency to migrate or be displaced; that is, it stays in the compound. More migratory fluids such as the fuels and other swell media in which the NBR compound is immersed will tend to be absorbed by the NBR polymer as well as the polymeric plasticizer. This creates a positive swell, which counterbalances the extraction of the more migratory low-viscosity compound ingredients including monomeric plasticizers, which create a negative swell.

The commodity diesters (e.g., DOP, TOTM, and DIDA) are relatively easy to extract. Moreover, when they have been added at a level beyond their compatibility limit, they will quickly exude and form an oily surface on the application part.

For low-temperature (LT) effects, several grades from the straight glycol esters family (TEG-CC, TEG-2EH, TEG-C9, etc.) are very efficient. Though not as compatible as the glycol ether types, they create their LT effect from their greater lubricity and not so much from their affinity for the polar constituents, although they are more polar than any of the commodity diester LT types DOA, DOZ, and so on. They are also more permanent (less extractable) than the commodity diesters.

Part for part, glycol esters will maintain higher modulus and tensile than the glycol ether types but have a lower threshold of incompatibility.

PROT—there are no effective low-temperature synthetic plasticizers with viscosities above 50 cps. See Table 12.3.

Compression set is largely controlled by the cure ingredients. If plasticizers have low acid numbers (<1.50) and are fully saturated, their influence on Set is minimal. However, the glycol ether and glycol thioether esters vary greatly in their effect on cure. For example, Vulcanol 81, 88, and OT grades speed the onset of cure (scorch) significantly due apparently to their thio content. Moreover, the high acid number of TP-95 compared to most other competitive DBEEA grades is known to slow cure onset in salt-bath cured hose.

Heat-aged properties of NBR compounds are directly influenced by the MW of the plasticizers used. High MW/high viscosity polymeric plasticizers have the best heat aging results while low MW grades used for improving low-temperature properties have the worst heat aging.

PROT—use the ACN content as the numerical guide to the maximum loading of high polarity plasticizers. For example, if working with a 38% ACN NBR, do not exceed 38 phr of DBEEA. When working with commodity diesters, the ACN content −8 should be maximum loading, for example, using an NBR with 28 ACN should signal a maximum addition of 20 phr of DOP, DOA, and so on.

ATM—it is possible to over plasticize an NBR compound with glycol ether esters without exudation appearing. This is detected by a precipitous falloff in physical properties. The plasticizer has literally solvated the polymer backbone. Variations of this condition can be found in all highly polar polymer systems.

ATM—do not try to manipulate modulus and tensile at break by varying the type and loading of plasticizer. There are too many possible combinations and the resultant changes too small to make the effort worthwhile. Adjust the plasticizer type and loading to optimize the specific characteristic desired and (e.g., low-temperature flex, hardness) then go back to the cure system, polymer grade, and fillers to correct the other physical properties.

12.7.1.2 NBR/PVC

With this fluxed polymer blend, try to keep the plasticizer choice as simple and as basic as possible. Use a commodity phthalate (DOP, DINP, and DIDP) to keep the cost down. If better low-temperature flex is required use a 100% linear phthalate (810P) or DOA/DIDA for maximum low-temperature performance. Rarely does an NBR/PVC application call for the expensive low-temperature types DOZ, DOS, TOF. To improve extraction resistance and to control swell use a medium viscosity (4,000–15,000 cps) adipic or glutaric polymeric plasticizer for 20%–40% of the total plasticizer content. Poor low-temperature flex and softening efficiency are the weak points of all polymerics. Monoesters (oleates, stearates, tallates) and process oils are not recommended for NBR/PVC.

For mandrel-built and pan-cured hose, the use of 10–20 phr of factice to improve green strength and reduce thin-out and sag is common. Castor-based factice has better extraction resistance and imparts higher green strength allowing reduced usage levels. For extruded hose, factice is used more as a processing aid and for surface effect. In fine, closed-cell NBR/PVC sponge applications, more highly cross-linked castor factice helps processing and uniformity of the cell structure.

In NBR and NBR/PVC-based low Durometer printing roll covers, castor factice is more effective than rapeseed or soybean factice at holding in, and in some cases replacing, liquid plasticizer thereby reducing the tendency for exudation. Factice addition of 10–20 phr improves the precision grinding characteristics and allows more control of roll surface finish.

PROT—replace 2 parts of DOP liquid with 1–2 parts of factice up to 30 phr depending on the factice type and softness.

ATM—never use a paraffinic or low naphthenic process oil above 5 phr when compounding with castor oil factice since they are incompatible and exudation occurs immediately.

12.7.1.3   HNBR

In a perfect world no rubber compounder should want to formulate HNBR using any kind of plasticizer. The very reason for using expensive HNBR is to reduce or eliminate the high temperature and extraction/swell deficiencies of regular NBR. Any volatile or extractable process aid or plasticizer is to be avoided. However, in spite of recent advances in the processability and LT characteristics of HNBR, there are times when use of a plasticizer is still indicated. The choice should hinge on the three properties of: softening efficiency for processing, volatility, and cost. The conventional choices are: 810TM for processing and reasonably low volatility; or a polymeric plasticizer in the 800–2000 cps viscosity range for its low volatility. But, ultimately, the plasticizer will volatilize and not survive the severe testing and application temperatures and the compounder needs to minimize the amount of polar plasticizer recognizing there may be a slight negative effect on dimensional stability and compression set. In most cases, the effect of synthetic plasticizers at 10–15 phr on the cured compound properties is negligible.

PROT—think of plasticizers for HNBR as temporary additions to aid processing and dispersion of fillers. Do not ascribe any permanent property effects to the use of synthetic plasticizers.

ATM—do not attempt to use factice of any kind or process oils in HNBR compound.

12.7.1.4   Carboxylated NBR—XNBR

The typical plasticizers used in regular NBR formulating are also suitable for XNBR. Usually the lower cost commodity diester types (TOTM, DOP, DIDA, etc.) are recommended since the typical applications for XNBR are less oriented toward low temperature uses than for abrasion resistance.

12.7.2   Styrene Butadiene Rubber—SBR and SBR/NBR Blends

Synthetic ester plasticizers are rarely recommended for SBR compounding. Process oils with medium to high aromatics content are quite compatible and much lower cost. Highly aromatic process oil improves green tack. Stress–strain properties are improved by the use of naphthenic process oil primarily since the aromatic oils are more compatible and tend to weaken the elastomer backbone thereby affecting stress–strain properties.

Various grades of factice are used in loadings of 10–150 phr with typical levels around 20 phr. Addition levels above this will lead directly to increasing compression set results and diminishing stress–stress properties. The addition of NBR to SBR to improve oil extraction resistance may dictate use of low levels of typical NBR plasticizers but it needs to be understood that the synthetic plasticizer has an affinity for NBR and the addition level should relate to NBR only. Some diester may migrate to the random styrene domains in the SBR but mainly to the ACN portions of the NBR.

12.7.3 STYRENE BLOCK COPOLYMERS

It is interesting to note that styrene-diene-styrene segmented/block polymers react exactly as expected when plasticized with polar and nonpolar plasticizers. Glycol ester with 2–3 phr (e.g., TEG-2EH) has an affinity for and softens the styrene end-blocks whereas the butadiene or isoprene center block is unaffected. The reverse is evident when a paraffinic/low naphthenic content process oil has little effect on the end-blocks. The proper selection of plasticizers in these S-B-S polymers leads to the optimum processing and end product flexibility.

Factice is used infrequently for special effects in S-B-S and similar thermoplastic elastomers. As an example, pencil erasers formulated with styrenic TPRs contain relatively high loadings of white factice.

12.7.4 ACRYLIC ELASTOMERS—ACM, AEM

There is little purpose in adding plasticizer to these high-temperature acrylic elastomers, other than to help with the initial processing. Polymeric plasticizers are ineffective. Low-temperature brittleness resistance can be a requirement for ACM and AEM gasket compounds and is usually accomplished with low loadings of glycol ether plasticizer. Compounding studies by Zeon and Unimatec have been done with TP-759 and ADK RS-735/RS-700, glycol ether diesters with an excellent combination of high-temperature permanence and LT flex in addition to good softening efficiency. As with all high-temperature application elastomers, not much is expected of the plasticizer other than improved processability since a significant portion of the plasticizer disappears early under the processing and application conditions especially in ACM. In AEM and ACM/AEM blends it is more common to use one of the high MW glycol ether esters (above). Recommended loading is 5–12 phr in ACM and 10–20 phr in AEM.

ATM—avoid thinking that the higher the molecular weight, the better. The best choice is a highly polar diester plasticizer with higher than normal MW for its class and low viscosity (preferably <40 cps).

12.7.5 EPICHLOROHYDRIN ELASTOMERS—ECO

Because of the superior properties in almost all categories and the availability of homopolymer, copolymer, and terpolymer grades allowing for flexibility in compounding for optimum performance, this class of elastomers requires relatively little plasticization. When plasticizer addition is indicated for fine-tuning a processing characteristic, 5–15 phr of DOP, DBEEA, or a low-viscosity polymeric in ascending order of extraction resistance and heat stability can be added. Factice and process oil are never recommended.

PROT—due to the tendency of ECO grades to exude, a maximum of 20 phr total plasticizer should be observed. If there are other incompatible ingredients such as wax or process aid, exudation may begin at an even lower level of plasticizer addition.

12.7.6 POLYCHLOROPRENE—CR

One of the most difficult elastomer families to plasticize with any confidence in the results is polychloroprene. The CR grades respond differently to plasticizers than most elastomers due to the ability of polar plasticizers and aromatic process oils to attack the molecular backbone of the CR. In addition, plain unsaturated vegetable oils (e.g., rapeseed, corn, and castor), which are relatively incompatible in most specialty elastomers, are seemingly quite compatible in CR up to about 30 phr.

Most types of CR are best plasticized with naphthenic process oils of varying viscosities and medium to high naphthenic/aromatic content. The trick in CR formulating after picking the appropriate grade is to balance the processability, heat resistance, UV resistance, crystallinity, and stress–strain properties through the selection of the plasticizer. This usually means a combination of vegetable oil, monoester, and process oil and becomes a trial and error exercise.

Highly aromatic process oils are very efficient CR plasticizers but have the negative effect of breaking down the polymer and creating a sticky compound with diminished stress–strain properties. Aromatic process oil tends to retard crystallization, whereas the monoesters (most commonly used to improve LT flex are butyl and octyl oleates) and other diesters have very low viscosity and are quite mobile in the compound. This makes them excellent LT-flex additives but has a side effect of allowing the CR molecules enough mobility to crystallize. A vegetable oil such as corn oil is useful as a low-cost plasticizer with good antioxidant properties due to its tendency to migrate to the surface and complex oxygen through its natural unsaturation. Compounding with pure vegetable oil promotes microbial and fungal growth and inclusion of antimicrobial additives may be necessary depending on the application.

One recommendation for a medium hardness compound with good LT flex would be 15–20 phr medium viscosity naphthenic oil, 8–10 phr low-viscosity phosphate ester (e.g., TOF) for some antifungal property as well as LT flex.

For a soft CR compound, soybean or rapeseed oil-based factice is indicated at loadings from 25 to 150 phr. If esters and aromatic oil are also used, castor-based factice is the better choice due to castor factice with greater polarity and plasticizer hold-in characteristics.

PROT—for enhancing the flame retardance of CR, triaryl phosphate ester and chlorinated paraffin can replace process oils part for part. The compatibility needs to be watched carefully.

ATM—too much factice will reduce stress–strain properties significantly with soybean grades having the greatest impact and castor grades the least on a part for part basis.

ATM—factice contains between 0.3% and 4.0% free sulfur. This needs to be recognized as a significant increase of curative in a sulfur compound. In a peroxide cure situation, only low levels (<10 phr) of sulfur factice can be tolerated unless a sulfurless grade is used (TDI cross-linked castor type).

ATM—paraffinic process oils are essentially incompatible with CR and are to be avoided.

12.7.7 CHLOROSULFONATED PE—CSM; CHLORINATED POLYETHYLENE—CPE

CSM and CPE can be treated similarly when it comes to choosing the appropriate plasticizers. They are considered semipolar polymers and the safest choice is one of the fully saturated commodity diesters for example, DOP, DINP, DOA, TOTM, 810TM (the latter two being triesters).

Many of the CPE compounds are peroxide-cured and need simple ester plasticizers which are efficient and do not interfere with the cure. In the case of CSM, sulfur cures are still common and all the commodity type plasticizers can be used. The use of factice in CSM is called for in roll compounds at 10–15 phr but otherwise rare.

ATM—this class of modified polyethylene has lower compatibility thresholds than other specialty elastomers and should not be over-plasticized, 10–20 phr is typical.

ATM—never use glycol ether esters in these grades. Try to avoid the use of unsaturated monoesters such as oleates, tallates, and ricinoleates unless you are trying to achieve a special property, recognizing that these monoesters impact the cure systems.

12.7.8 FLUOROELASTOMERS—FKM

Generally speaking, FKM compounders avoid the addition of any volatile plasticizers and typical process aids, which can have a negative impact on this family of expensive, precise, high-end applications. There is some use of DBS to act perhaps as a dispersion aid especially when reprocessed FKM is involved.

12.7.9 NONPOLAR ELASTOMERS—NR, EPDM, BR, IIR

This group of high volume nonpolar elastomers is typically plasticized with soybean and rapeseed oil factices and the full range of hydrocarbon process oils. Only when a special lubricating effect is wanted is a synthetic plasticizer used. To illustrate, the following two examples: (a) 10 phr DOA added to an EPDM Arctic CB radio coil cord where it acted as an extreme low temp (LT) flexibilizer; (b) a typical LT glycol ester (TEG-2EH) used at 8 phr in IIR to achieve both processing and LT enhancements. In both examples, the incompatible plasticizer acted as a lubricant.

Diesters are also used in proprietary high performance BR/SBR tire-tread formulations at approximately 10 phr to reduce hysteresis, improve hot traction in racing tires, increase flex life and other requirements of track and street car and motorcycle tires.

12.7.10 Environmental Considerations

When a compounder is trying to pick a plasticizer, what environmental situations impact that decision? First of all let us look at factice. Natural vegetable oils cross-linked with sulfur, hydrogen sulfide, sulfur chloride or peroxide with most grades approved for FDA food contact applications. None of these grades is regulated by any agency and all are environmentally friendly. (Note: they can be made non-FDA through addition of regulated process oils, process aids, and other non-FDA acceptable additives.)

In the area of hydrocarbon rubber process oils, straight paraffinic grades are essentially innocuous and safe mineral oils. The aromatics content and molecular configuration of the naphthenic and aromatic oils have been examined extensively and severe hydro-treating and solvent extraction has upgraded most of the naphthenic product used in the rubber industry today while the highly aromatic extracts need to be labeled as carcinogenic.

The field of the synthetic esters is diverse but only 40–50 ester and polyester types are used in the rubber industry. Of those only three are currently on the USA Hazardous Chemicals List: DOP, DBP, and BBP. These are used in high volume by the industry and when used over a certain quantity on an annual basis must be reported to local EPA authorities. It is interesting to note that Europe does not yet regulate these grades. In addition, several large volume triaryl phosphates are controlled/labeled when shipped in bulk where they might spill and contaminate fresh and saltwater fish habitats. However, other than taking the necessary precautions on these grades and following the MSDS recommendations on all these synthetic esters, this group of products is relatively environmentally friendly.

PROT—with the exception of the aromatic process oils, which are labeled carcinogenic (worldwide) and the three diesters, which are currently classified as potential carcinogens (North America), compounders have little to be concerned with the impact of the plasticizer products discussed in this chapter on the environment.

12.8 QUICK REFERENCE GUIDE

12.8.1 Synthetic Plasticizers

1. There are no effective low-temperature synthetic ester plasticizers with viscosities above 50 cps.
2. High-temperature volatility/aging is directly related to the MW of the plasticizer. Low viscosity/low MW equals poor aging and high viscosity/high MW equals good aging.
3. Extraction and swell characteristics are dependent on the chemistry of the extract and plasticizer and the MW. For example, high MW polymerics are relatively immobile and extraction fluids tend to be absorbed by the plasticizer (volume swell), whereas monomerics are much more mobile and easily exchange places with extracts and usually create more or less volume shrinkage depending also on the chemistry involved.

4. There is a *hold-in* characteristic of some plasticizers—especially high MW types—effectively increasing compatibility is related not only to size of the molecule (see 3 above) relative to a co-plasticizer or swell medium but also to the polarity of the plasticizers involved (see also castor factice below).

12.8.2 VULCANIZED VEGETABLE OIL/FACTICE

1. In highly polar elastomers use castor-based factice for highest compatibility, best stress–strain properties, and lowest volume swells.
2. Compression set is adversely affected by all sulfur-cured factice and should be monitored carefully.
3. Improve compatibility of highly plasticizer-loaded compounds by substituting 0.5–1.5 phr of factice for each phr of liquid plasticizer.
4. Depending on the formulation and loading, factice can have positive or negative effects on heat-aged properties in semipolar or polar compounds where use of factice is indicated.

12.8.3 RUBBER PROCESS OILS

1. Use hydrocarbon process oils sparingly in specialty elastomers.
2. Never use the paraffinic types in polar or semipolar elastomers.

ACKNOWLEDGMENTS

I thank Dr Vince P. Kuceski for his inspiration into the myriad possibilities of esters, John Ellis, my first mentor, who helped unlock the mysteries of the art of rubber compounding, Jack Eberly for input on Process Oils, and Arvind Kapoor for his guidance on the subject of vulcanized vegetable oil/factice.

13 Vulcanization Agents for Specialty Elastomers

Robert F. Ohm

CONTENTS

To prevent flow, elastomers must be cross-linked or vulcanized. This joining of adjacent polymer chains allows the rubber part to recover from large deformations quickly and forcibly.

A wide variety of vulcanization agents are used in the rubber industry. For specialty elastomers, peroxide cross-linking is probably the most widely used and fastest growing system. The various peroxides available for cross-linking and newer developments in this area will be reviewed. Other vulcanization agents for specialty elastomers will also be presented.

13.1 PEROXIDE CROSS-LINKING

When polymers are cross-linked with peroxides, the bond formed between individual polymer chains is a carbon-to-carbon bond. This C–C bond has the highest thermal stability of any vulcanization system, and optimizes the heat resistance and compression set of the particular elastomer.

The reaction sequence for peroxide vulcanization is shown in Figure 13.1. The first step is the homolytic cleavage of the oxygen–oxygen bond to provide two alkoxy free radicals (R–O$^{\bullet}$ and R$'$–O$^{\bullet}$). Most alkoxy radicals abstract a hydrogen atom from the polymer to generate a free radical on the polymer chain. If the polymer has double bonds, some alkoxy radicals can add to the polymer chain to generate slightly different free radicals.

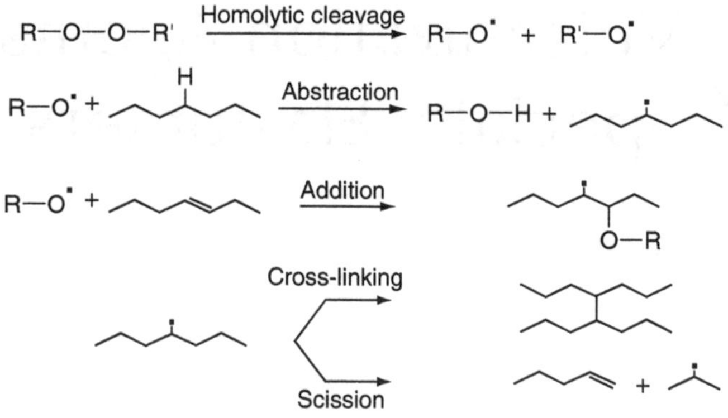

FIGURE 13.1 Peroxide reactions.

The final step in cross-linking with peroxides is the joining of two polymer free radicals to form the carbon–carbon cross-link.

A few polymers cannot be cross-linked with peroxides. These polymers have a high degree of branching or many side groups. In these polymers, such as butyl rubber, polyepichlorohydrin homo- and copolymers as well as polypropylene, the addition of peroxides causes chain scission and softening rather than cross-linking.

13.2 SELECTION OF THE PEROXIDE

The two major types of peroxides used for cross-linking elastomers are dialkyl peroxides and peroxyketals. A third class, diacyl peroxides as exemplified by benzoyl peroxide are used almost exclusively in cross-linking silicone rubbers.

The selection of the type of peroxide determines the speed of cross-linking. Generally, peroxyketals are faster curing than dialkyl peroxides, as indicated by the lower temperatures required for a 10 min cure in Table 13.1.

TABLE 13.1
Temperature for 10 min Cure in EPDM [1]

Trade Name	Chemical Name	10 min Cure, °C
Peroxyketals	$(R-O-O-\overset{\overset{R^1 \quad R^2}{\diagdown \diagup}}{C}-O-O-R^3)$	
VAROX 231	1,1-Bis(*t*-butylperoxy)-3,3,5-trimethylcyclohexane	152
VAROX 230	*n*-Butyl-4,4-bis(*t*-butylperoxy)valerate	169
Dialkyl peroxides	$(R-O-O-R^1)$	
VAROX DCP	Dicumyl peroxide	176
VAROX 802	α,α′-Bis(*t*-butylperoxy)-diisopropylbenzene	183
VAROX DBPH	2,5-Dimethyl-2,5-di(*t*-butylperoxy)hexane	185
VAROX 130	2,5-Dimethyl-2,5-di(*t*-butylperoxy)hexyne-3	193

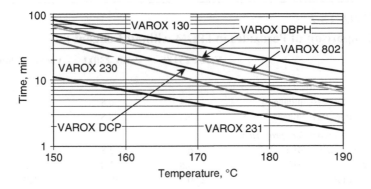

FIGURE 13.2 Cure time in EPDM [1].

The cure times necessary at other temperatures are shown in Figure 13.2. The cure times shown are only approximate because the particular polymer and compounding ingredients affect the cure time actually observed for a given formulation. Although most of the data on peroxides were generated with EPDM, the same relative results would be expected with other elastomers, except as noted.

Where shorter cure times are desired for maximum economics, higher cure temperatures can be employed. The maximum cure temperature for a given peroxide can be limited by its scorch time or the time before the onset of cross-linking. Plots of scorch times for the peroxides listed in Table 13.1 are shown in Figure 13.3.

Generally, scorch times follow the same relative trends as cure times. Nevertheless, a comparison of the two figures indicates that the peroxyketals have a more favorable ratio of scorch time to cure time. For example, VAROX 230 has a scorch time similar to that of VAROX DCP but the cure time at a given temperature is several minutes shorter for VAROX 230 as compared with VAROX DCP.

While peroxyketals have more attractive cure times, their cross-linking efficiency (i.e., the state of cure) is not always as good as that of the dialkyl peroxides. The two peroxyketals are compared with dicumyl peroxide in Table 13.2. In this EPDM compound, the peroxyketals show lower modulus and higher compression set.

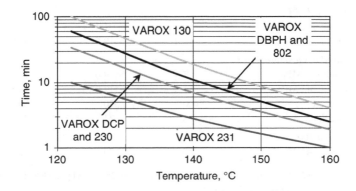

FIGURE 13.3 Scorch time in EPDM [1].

TABLE 13.2

Comparison of Peroxides in EPDM without Coagent [2]

	Varox 231XL	Varox 230XL	Varox DCP40C
Rheometer at 162°C			
Scorch time, ts_2, minutes	0.4	0.8	0.8
Cure time, tc_{90}, minutes	3.3	8.8	15.0
Press cured physical properties			
200% modulus, MPa	6.2	5.8	8.7
Tensile strength, MPa	14.8	13.4	15.0
Elongation, %	300	310	280
Hardness, Shore A	62	64	64
Aged 70 h at 100°C			
Compression set, %	14	16	6

Base Compound: 100 Vistalon 2504, 5 zinc oxide, 10 Sunpar 2280, 40 N990, 25 N550, 8 peroxide.

The difference in state of cure observed between dialkyl peroxides and peroxyketals depends on the particular polymer and the compounding ingredients. The use of a coagent is particularly important. With a coagent, the peroxide does not have to be a good abstractor of hydrogen atoms from the polymer, but rather the peroxide can add to the double bond of the coagent. The coagent can therefore increase the cross-linking efficiency of a peroxide. A comparison of dicumyl peroxide and two peroxyketals is shown in Table 13.3 for a chlorinated polyethylene (CPE) compound

TABLE 13.3

Comparison of Peroxides in CPE with a Coagent [3]

	Varox 231XL	Varox 230XL	Varox DCP40C
Rheometer at 149°C			
Scorch time, ts_2, minutes	0.5	1.4	2.0
Cure time, tc_{90}, minutes	8.4	19.6	45.0
Press cured physical properties			
200% modulus, MPa	7.4	9.0	11.3
Tensile strength, MPa	11.5	12.8	14.5
Elongation, %	400	330	300
Hardness, Shore A	68	72	71
Aged 70 h at 100°C			
Compression set, %	30	20	18

Base Compound: 100 Tyrin CM0136, 10 magnesium oxide, 25 dioctyl phthalate, 50 N990, 30 N660, 2 triallyl cyanurate, 8 peroxide.

with a coagent. The differences between VAROX 230XL and VAROX DCP40C peroxides in this CPE example are not as pronounced as in the EPDM compounds of Table 13.2.

13.2.1 COMPOUNDING WITH PEROXIDES

Peroxide cross-linking requires that special attention be given to the selection of compounding ingredients. Materials that compete with the polymer for free radicals should be avoided or minimized. This includes aromatic oils, acidic materials, and a limit on the amount or use of some phenolic and amine antioxidants. The antioxidants that show the least interference with peroxides are AGERITE RESIN D, its higher molecular weight analogue AGERITE MA, Naugard 445, and VANOX ZMTI. An example of the use of AGERITE RESIN D and VANOX ZMTI is shown in Table 13.4. *Para*-phenylenediamine-based antiozonants should be avoided in peroxide cures, although paraffin and microcrystalline waxes can be used for static ozone protection.

Highly acidic ingredients promote heterolytic decomposition of the peroxide without generating the free radicals necessary for cross-linking. Often this means avoiding silicas and air-floated kaolin clays that have a low pH. Another possible side effect is loss of cure during prolonged bin storage over several weeks or months. In this case, fresh peroxide can be added to the mixed compound to rejuvenate it. Different peroxides are sensitive to acidic materials in varying degrees. The peroxyketals are the

---

TABLE 13.4
Effect of Antioxidants in Peroxide Cure of EPDM [4]

VANOX ZMTI	—	3	3
AGERITE RESIN D	—	—	2
Rheometer at 171°C			
Scorch time, ts_2, minutes	0.4	0.5	0.6
Cure time, tc_{90}, minutes	8.0	7.8	7.3
Press cured physical properties			
100% modulus, MPa	5.5	5.3	5.0
Tensile strength, MPa	10.1	9.7	10.1
Elongation, %	180	180	210
Hardness, Shore A	75	76	77
Aged 168 h at 177°C			
Tensile change, %	−49	−32	−25
Elongation change, %	−56	−33	−29
Hardness change, points	+12	+14	+14
Aged 70 h at 150°C			
Compression set, %	20	24	28

Base Compound: 100 Nordel 1040, 15 zinc oxide, 5 magnesium oxide, 8 Neoprene W, 20 Sunpar 2280, 65 N550, 35 N990, 2 VANAX MBM, 10 VAROX DCP-40C.

TABLE 13.5
Effect of Coagent in Peroxide Cure of HNBR [5]

VANAX MBM, phr	0	4	8
Rheometer at 170°C			
Scorch time, ts_2, minutes	1.1	0.5	0.5
Cure time, tc_{90}, minutes	10.2	8.1	7.6
Press cured physical properties			
100% modulus, MPa	4.8	11.2	15.2
Tensile strength, MPa	28.9	28.3	26.3
Elongation, %	315	205	165
Hardness, Shore A	68	74	80
Aged 70 h at 150°C			
Compression set, %	20	16	14

Base Compound: 100 Zetpol 2010, 5 zinc oxide, 50 N774, 1.5 Naugard 445, 1 VANOX ZMTI, 8 VAROX 802-40KE.

most sensitive, followed by VAROX DCP and then VAROX DBPH. VAROX 130 is the least sensitive to acids.

Coagents containing unsaturation can be beneficial in increasing cross-link density. Coagents become part of the cross-link network and can also affect cure speed. Coagents containing electron donating groups, such as methacrylates and butadiene resins, may slow the cure. Coagents with electron withdrawing groups, such as triallyl cyanurate (TAC) and VANAX MBM can shorten cure time. An example is given in Table 13.5 for hydrogenated nitrile rubber (HNBR).

Examples of typical coagents are shown in Table 13.6. Some suppliers offer variants of the basic chemical with different alkyl groups, formulated products to control scorch or polymer bound dispersions for rapid mixing.

TABLE 13.6
Types of Coagents [6]

Trade Name(s)	Chemical Name	Features/Benefits
Ricon 150, Ricon 156	Liquid, hi vinyl polybutadiene	Low viscosity, high cure state
Saret 517, Saret 519	Diacrylate	Low compression set (C/S)
SR-706, SR-633	Zinc diacrylate	Rubber to metal adhesion
Saret 634	Zinc dimethacrylate	Hot tear strength
TAC	Triallyl cyanurate	Fast cure, low C/S
TAIC, Diak No. 7	Triallyl-iso-cyanurate	Fast cure, low C/S
TATM	Triallyltrimellitate	Low compression set
TMPTMA	Trimethylolpropane trimethacrylate	
VANAX MBM, HVA-2	*m*-Phenylenebismaleimide	Fast cure, low C/S

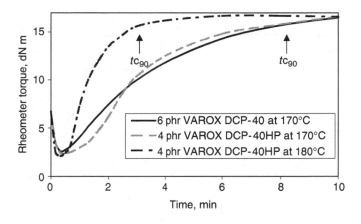

FIGURE 13.4 High performance peroxide in Engage EG 8180 EOM [7].

New peroxide types are being developed, and some have recently been introduced. For example, high performance (HP) grades provide a better balance of scorch and cure times. Figure 13.4 shows comparative rheometer cure curves in ethylene-octene copolymer (EOM). At the same curing temperature, the HP peroxide provides longer mold flow or scorch time than its traditional counterpart, with no sacrifice in cure time. In compounds in which the traditional peroxide gives acceptable mold flow times, the use of the HP peroxide allows a higher cure temperature, providing shorter cure time and greater productivity with no sacrifice in scorch time. Note also that the amount of HP peroxide may be decreased by one-third and still provide the same state of cure in EOM.

Both the HP and regular peroxides may exhibit different relative cure performance in other types of polymers. The HP peroxides typically provide higher states of cure in EOM, some grades of EPDM and Hypalon CSM. The regular grades often provide a higher state of cure in regular and hydrogenated nitrile rubber (NBR, HNBR) and Tyrin CPE. The HP peroxides may extend both scorch and cure times in peroxide curable fluoroelastomer (FKM) and silicone (VMQ).

When peroxide compounds are cured in contact with air, a tacky surface results because oxygen causes chain scission of the polymer free radical. This can create problems in deflashing molded parts, and prevent the use of peroxide cures in some extruded or autoclave processes. Peroxide formulations that resist air inhibition of the cure are under development.

13.3 SULFUR DONOR CURES

Some materials can donate one or more atoms of sulfur from their structure for cross-linking purposes. The resulting monosulfide cross-link of the sulfur donor cure systems has slightly poorer thermal stability than the C–C bond formed by peroxides but does have good aging and thermal stability. They are also relatively economical,

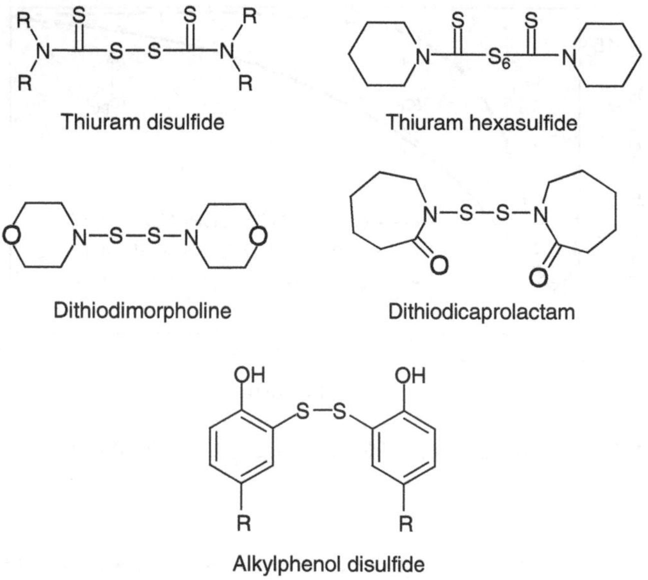

FIGURE 13.5 Sulfur donors.

and avoid some of the problems with peroxide cures, including tacky flash, relatively poor scorch to cure ratio, and limitations on the compounding ingredients and protective system. Some of the chemicals which function as sulfur donors are shown in Figure 13.5.

The alkyl group (R) of the thiuram disulfides is typically methyl, ethyl, or butyl. For low levels of airborne nitrosamines, the R group can be iso-butyl or benzyl. The thiuram hexasulfide based on piperidine is unique in that the molecule contains six atoms of sulfur linked together. The thiurams are potent accelerators as well as sulfur donors.

Dithiodimorpholine has relatively little acceleration activity and must be used with other, more traditional accelerators. Dithiodicaprolactam (DTDC) finds similar application in low nitrosamine cure systems.

The alkyl phenol disulfides are mainly used in tire inner liners for heat resistance, along with a conventional accelerator system. The R group is t-butyl or amyl. In addition to donating sulfur for cross-linking purposes, these materials can also function as bisphenolic curatives (see Section 13.6).

Comparisons of sulfur donor cures with peroxide cures in nitrile rubber are shown in Table 13.7. Each cure system is shown both with and without an antioxidant package. The peroxide cure is inherently more thermally stable than the sulfur donor cure, which provides less change in physical properties during long-term aging and better (lower) compression set. In each cure system, the inclusion of a good antioxidant package can further improve the heat and compression set performance.

TABLE 13.7

Comparison of Peroxide and Sulfur Donor Cures in NBR [8]

Compound	A	B	C	D
VAROX DCP-40KE	4	4	0	0
Sulfur	0	0	0.3	0.3
ETHYL CADMATE	0	0	2.5	2.5
MORFAX	0	0	2.5	2.5
AGERITE SUPERFLEX Solid	0	1	0	1
VANOX ZMTI	0	2	0	2
Press cured physical properties—cured 10 min at 171°C				
100% modulus, MPa	6.9	5.7	4.2	6.7
Tensile strength, MPa	18.9	19.2	15.8	15.6
Elongation, %	210	250	530	340
Hardness, Shore A	70	67	66	65
Aged 1000 h at 100°C in air				
Tensile change, %	−19	+4	+8	+5
Elongation change, %	−62	−20	−66	−39
Hardness change, points	+11	+8	+10	+7
Aged 70 h at 150°C				
Compression set, %	54	42	77	58

Base Compound: 100 Nipol 1032, 5 zinc oxide, 65 N774, 15 Paraplex G50.

13.4 SULFUR CURES

Elemental sulfur was the first material used by Charles Goodyear to cure rubber of its tendency to become sticky in summer and brittle in winter. It is still the most widely used cure system for general purpose rubbers because it provides articles with excellent toughness and resistance to flex fatigue, as well as being exceptionally economical.

For specialty elastomers, a significant drawback to elemental sulfur cure systems is that they provide relatively poor compression set at elevated temperature and lack good high temperature aging. Nevertheless, for specialty elastomer parts in dynamic service, sulfur cure systems can provide good performance and high resistance to flex fatigue.

Examples of a regular sulfur and an efficient vulcanization (EV) sulfurless cure in nitrile rubber are shown in Table 13.8. Compared to the normal sulfur control, the EV cure system provides comparable initial physical properties, lower (better) compression set, and less change in properties during high temperature aging.

A comparison of the three curing systems for NBR is illustrated in Table 13.9 in which a sulfur, sulfur donor, and peroxide cures in two NBRs are given.

13.5 METAL OXIDE CROSS-LINKING

Some polymers are cross-linked with metal oxides alone. Polychloroprenes that have been copolymerized with sulfur (the "G" type Neoprenes) are one example.

TABLE 13.8

Comparison of Regular and Sulfur Less Cure Systems in NBR [9]

Compound	Regular Sulfur	EV (Sulfur Less)
Sulfur	1.5	0
Zinc oxide	5.0	0
Magnesium oxide	0.10	0
ALTAX (MBTS)	1.5	0
UNADS (TMTM)	0.15	0
DURAX (CBS)	0	1.75
METHYL TUADS (TMTD)	0	1.4
ETHYL TUADS (TETD)	0	0.9
VANAX A (DTDM)	0	1.1
Rheometer at 171°C		
Scorch time, ts_2, minutes	1.7	1.7
Cure time, tc_{90}, minutes	3.7	9.0
Press cured physical properties		
100% modulus, MPa	5.4	5.5
Tensile strength, MPa	14.6	12.4
Elongation, %	530	400
Hardness, Shore A	64	65
Aged 70 h at 149°C in air-change		
Tensile, %	−42	−17
Elongation, %	−85	−48
Hardness, points	+13	+9
Aged 70 h at 100°C		
Compression set, %	30	17

Base Compound: 100 Nipol 1042, 2 VANFRE AP-2, 1.5 stearic acid, 2 AgeRite Stalite S, 5 Paraplex G25, 30 N550, 30 N990.

Typically 5 phr of zinc oxide and 4 phr of magnesium oxide are used. The speed of cure can be increased with accelerators such as thioureas.

In polychloroprenes that have been modified with a thiuram (the "W" types), an organic accelerator is essential, and a thiourea (e.g., ETU) or, less commonly DOTG and sulfur can be used.

Polymers that contain carboxylic acid functionality, such as carboxylated nitrile rubber (XNBR), can also be cross-linked with a divalent metal oxide, either zinc oxide, zinc peroxide, or magnesium oxide. The organo-metallic cross-link supplements the traditional sulfur or peroxide generated cross-links to provide superior toughness and abrasion resistance. In XNBR, the amount of ZnO depends on its surface area, with 5 phr at 9 m^2/g surface area being used compared with 9 phr at 3.5 m^2/g surface area. Standard XNBR grades are best cured with zinc peroxide or low

TABLE 13.9
Comparison of Sulfur, Sulfur Donor, and Peroxide Cure Systems in NBR

Nitrile Cure Systems	A	B	C	D	E	F
Paraclean 42L60	100.0	100.0	100.0	—	—	—
Paracril BJLT M50	—	—	—	100.0	100.0	100.0
IRB #6 Black	40.0	40.0	40.0	40.0	40.0	40.0
Kadox 911C	3.0	3.0	3.0	3.0	3.0	3.0
Stearic acid	1.0	1.0	1.0	1.0	1.0	1.0
Spider Sulfur	1.5	0.5	—	1.5	0.5	—
Delac NS	0.7	—	—	0.7	—	—
Delac S	—	1.0	—	—	1.0	—
Tuex	—	2.0	—	—	2.0	—
Di-Cup 40C	—	—	3.8	—	—	3.8
Naugard Q	—	—	1.0	—	—	1.0
Total	146.2	147.5	148.8	146.2	147.5	148.8
Rheometer at °C	160	160	177	—	—	—
ts_2, minutes	3.5	2.2	1.2	—	—	—
tc_{90}, minutes	13.5	5.6	4.3	—	—	—
Physical properties—cured 40 min at 150°C						
Shore A, points	75	73	70	72	68	73
300% modulus, MPa	14.5	17.4	12.8	10.5	12.0	14.8
Tensile, MPa	30.9	23.4	29.0	29.5	21.9	28.2
Elongation, %	550	380	590	630	470	490
Tear Die C, kN/m	3.9	3.2	3.6	7.1	4.9	5.5
Brittle point, ASTM D2137, °C	−26	−24	−26	−38	−40	−42
Compression set B, 22 h at °C	100	125	125	100	125	100
Set, %	43	12	23	54	16	16
Aged in air 70 h at °C-change	100	125	125	100	125	125
Shore A, points	1	3	7	5	8	8
Tensile, %	−3	−4	−4	−12	−14	6
Elongation, %	−19	−32	−37	−48	−38	−31
Aged in IRM Oil #903 70 h at °C-change	100	125	125	100	125	100
Shore A, points	2	4	4	−13	−9	−12
Tensile, %	10	0	−30	−9	0	−56
Elongation, %	−5	−4	−35	−13	−2	−48
Volume, %	4.6	4.6	5.6	13.7	15.5	16.0
Aged in water 70 h at 100°C-change						
Volume, %	7.2	4.6	4.9	7	6	5

activity zinc oxide for better scorch safety and bin stability. Inhibited grades of XNBR such as Nipol NX775 work well with standard zinc oxide. Coated grades are preferred because ZnO is typically incorporated late in the mix to prevent scorch.

HYPALON chlorosulfonated polyethylene (CSM) will react slowly with magnesium oxide in the presence of moisture. A cure system with 4 phr of MgO is used in applications that cure in place, such as membranes for roofing and pond and ditch liquid containment.

Linear polysulfide polymers (Thiokol FA types) are cross-linked with 5–10 phr of zinc oxide. Typical sulfur accelerators such as ALTAX (MBTS) and VANAX DPG are also incorporated in FA polymers, but their function is to lower the viscosity to a workable range.

13.6 MULTIFUNCTIONAL CROSS-LINKING AGENTS

Various multifunctional molecules can be used to react with rubber polymers that incorporate special cure site comonomers, usually labile halogen atoms. Some examples are shown in Figure 13.6.

Diamines were the first materials used to cross-link polyacrylates and fluorocarbon polymers. They provide excellent compression set and aging characteristics, but require an oven post-cure after molding to develop optimum properties. These curatives are also corrosive to mild steel molds. See Table 13.10 for a comparison of two diamines and a bisphenol curative in a fluoroelastomer.

Bisphenols were developed to overcome some of the limitations of the diamines. These curatives require an accelerator for fast cure rates, and a phosphonium salt is employed for this purpose, as shown in Table 13.10. Higher levels of the bisphenol, Curative 30, increase the state of cure (hardness and modulus) whereas higher levels of the onium salt, Curative 20, increase the rate of cure.

Di- and tri-functional thiol (or mercapto) molecules are used as cross-linkers for halogen-containing polymers. Like the bisphenols, they may require an accelerator, depending on the inherent reactivity of the halogen cure site and the particular

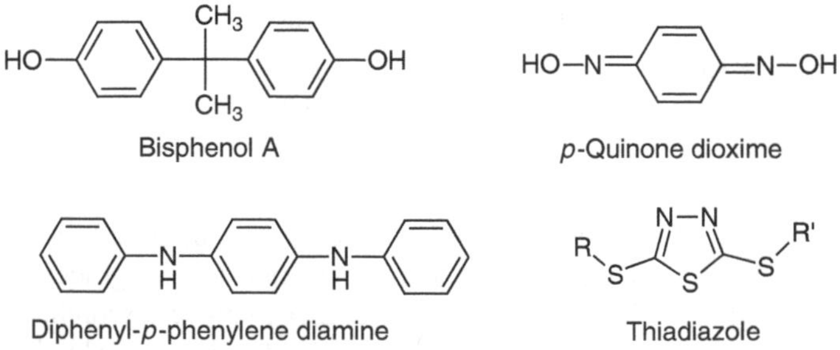

4,4'-Methylenebis
(cyclohexyamine)
carbamate

Trithiocyanuric acid

Bisphenol A

p-Quinone dioxime

Diphenyl-p-phenylene diamine

Thiadiazole

FIGURE 13.6 Multifunctional cross-linking agents.

TABLE 13.10

Comparison of Cross-linking Agents for Fluoroelastomer (FKM) [11]

	Diamine A	Diamine B	Bisphenol
Maglite Y, medium activity MgO	15	15	—
Maglite D, high activity MgO	—	—	3.0
Calcium hydroxide	—	—	6.0
Diak No. 1 diamine	1.5	—	—
Diak No. 3 diamine	—	3.0	—
Curative 20 onium sal	—	—	2.0
Curative 30 bisphenol	—	—	4.0
Rheometer at 177°C			
Scorch time, ts_2, minutes	1.5	3.3	4.2
Cure time, tc_{90}, minutes	7.5	17.3	8.5
Press cured physical properties			
100% modulus, MPa	6.8	6.6	5.8
Tensile strength, MPa	16.4	12.4	12.6
Elongation, %	210	210	180
Hardness, Shore A	76	78	78
Aged 70 h at 200°C			
Compression set, %	53	56	26

Base Compound: 100 Viton A, 30 N990.

multifunctional thiol. The accelerator is typically either an amine or a thiuram. Dithiocarbamates can also be used, provided that the chlorine content is not too high (<2%) because zinc tends to promote degradation by dehydrohalogenation.

A comparison of two thiadiazoles and two amine accelerators in CPE is shown in Table 13.11. Compound A shows that the first thiadiazole has a very fast cure that is useful in molded applications. The second thiadiazole has improved bin storage stability, particularly with the chemically stable amine, VANAX 882B. CPE compounds cured with the second thiadiazole are suitable for hose and tubing made by autoclave processes. Other thiadiazole derivatives for CPE are being investigated.

A comparison of multifunctional curatives in polyacrylate (ACM) is shown in Table 5.13, Chapter 5 on ACM elastomers. A dithiocarbamate, BUTYL ZIMATE, was used in this example because the chlorine content of the polymer is only ~1% (because of the cure site), and all the chlorine reacts during cure. The trifunctional trithiocyanuric acid gives excellent compression set but relatively low initial elongation. Various thiadiazoles are offered in which R and R′ may be the same or different, including alkyl groups and hydrogen. In VANCHEM DMTD, both R and R′ are hydrogen, and this provides the fastest reactivity. VANAX 829 thiadiazole derivative would more closely match the cure speed of the trithiocyanuric acid. It is

TABLE 13.11

Comparison of Thiadiazoles and Accelerators in CPE [10]

Compound	A	B	C
VANAX 808 amine	0.8	0.8	0
VANAX 882B amine	0	0	1.25
Echo thiadiazole	2.5	0	0
VANAX 829 thiadiazole	0	2.5	2.5
Mooney viscosity at 121°C-original			
Viscosity, ML	48	49.5	46.5
Scorch, t_5, minutes	22	39	11
Mooney at 121°C—after 2 weeks at 38°C			
Viscosity, ML	+150	69	56.5
Points change	+100	+19.5	+10.0
Rheometer at 171°C			
ML, N·m	0.8	0.9	0.8
MH, N·m	3.9	6.2	11.2
ts_2, minutes	1.2	2.5	1.6
tc_{90}, minutes	3.0	13.6	16.8
Physical properties			
Hardness, Shore A	78	82	75
200% modulus, MPa	8.2	9.1	8.8
Tensile strength, MPa	14.4	15.1	21.1
Elongation, %	450	390	420
Aged in air 14 days at 121°C-change			
Hardness change, points	+5	+5	+5
Tensile change, %	0	−4	−19
Elongation, %	−17	−20	−19
Aged 70 h at 100°C			
Compression set B, set, %	32	22	14

Base Formulation: 100 Tyrin CM 0136, 10 Maglite D, 50 N774, 30 Aromatic oil, Amine accelerator (as indicated), 2.5 Thiadiazole (as indicated).

possible that blends of these tri- and di-functional curatives could provide intermediate values of elongation and compression set.

A comparison of cure systems in epichlorohydrin-ethylene oxide copolymer (ECO) is shown in Table 13.12. ETU lead oxide cure system provides good technical properties, but is replaced because of health and environmental concerns. Vulcanizates cured with non-lead cure systems generally have poorer properties after immersion in water. This can be overcome in the Zisnet cure system by the substitution of a synthetic hydrotalcite (Hysafe 310) for calcium oxide, with some sacrifice in compression set. The substitution of the hydrotalcite for lead oxide in the ETU cure system does not typically provide good water resistance

TABLE 13.12

Comparison of Cure Systems in Epichlorohydrin [12]

Compound	A	B	C	D
GRD-75 (ETU)	1.5	0	0	0
GRD-90 (Lead)	5	0	0	0
Calcium carbonate	0	5	0	0
Hysafe 510	0	0	5	0
Zisnet F-PT	0	0.8	0.8	0
VANAX DPG	0	0.5	0.5	0
Santogard PVI	0	0.3	0.3	0
Magnesium oxide	0	0	0	5
Echo A	0	0	0	1
Echo MPS	0	0	0	3
Physical properties—cured 15 min at 200°C				
100% modulus, MPa	8.6	4.6	4.9	6.9
Tensile strength, MPa	18.7	13.7	15.4	14.4
Elongation, %	200	280	290	200
Hardness, Shore A	77	72	68	77
Aged 72 h at 90°C in distilled water-change				
Tensile, %	+3	−33	−7	−28
Elongation, %	0	−46	−13	−20
Hardness, points	−3	−8	−3	−8
Volume, %	+13	+33	+11	+25
Aged 22 h at 125°C				
Compression set B, set, %	14	19	28	13

Base Compound: 100 Hydrin C2000, 50 N330, 1 Stearic Acid, 1 Vulkanox MB-2.

and its substitution for the magnesium oxide in the Echo cure system leads to scorchy compounds.

The A and B cure systems can be used in epichlorohydrin homopolymer (CO). The B and D cure systems can be used in unsaturated terpolymers (GECO). In addition, GECO can be cured with sulfur and peroxide cure systems.

APPENDIX

Abbreviation	Composition	Trade Name
DPPD[a]	N,N'-Diphenyl-p-phenylene diamine	AgeRite[b] DPPD
TMQ	2,2,4-Trimethyl-1,2-dihydroquinoline polymer	AgeRite[b] MA
TMQ	2,2,4-Trimethyl-1,2-dihydroquinoline polymer	AgeRite[b] 2Resin D[b]
ODPA	Octylated diphenylamine	AgeRite[b] Stalite S[b]
ADPA	Acetone-diphenylamine reaction product	AgeRite[b] Superflex[a]
MBTS	Mercaptobenzothiazole disulfide	ALTAX[b]
ZDBC	Zinc di-n-butyl dithiocarbamate	BUTYL ZIMATE[b]
MBT	2-Mercaptobenzothiazole	CAPTAX[b]
HMDC	Hexamethylene diamine carbamate	Diak[c] No. 1
	N,N'-Dicinnamylidene-1,6-hexanediamine	Diak No. 3
TAIC	Triallyl-iso-cyanurate	Diak No. 7
CBS	N-Cyclohexyl-2-benzothiazole sulfonamide	DURAX[b]
	Monobenzoyl thiadiazole derivative	Echo[d]
EOM	Ethylene-octene copolymer	ENGAGE[e]
CdEC	Cadmium diethyl dithiocarbamate	ETHYL CADMATE[b]
TETD	Tetraethyl thiuram disulfide	ETHYL TUADS[b]
BIMS	Polyisobutylene/brominated-p-methylstyrene	EXXPRO[f]
ECO	Epichlorohydrin-ethylene oxide copolymer	Hydrin[g]
CSM	Chlorosulfonated polyethylene	HYPALON[c]
	Synthetic hydrotalcite	Hysafe[h]
ACM	Polyacrylate rubber	HyTemp[g]
MBM	Meta-phenylene bismaleimide	HVA[d]-2
TiBTD	Tetra-iso-butylthiuramdisulfide	ISOBUTYL TUADS
ZDiBC	Zinc di-iso-butyldithiocarbamate	ISOBUTYL ZIMATE
MgO	Magnesium oxide (high and medium activity)	Maglite[i] D and Y
TMTD	Tetramethyl thiuram disulfide	METHYL TUADS
MBS	2-Morpholino benzothiazole disulfide	MORFAX[b]
PIDPA	4,4'-Bis(α-dimethylbenzyl) diphenylamine	Naugard[j] 445
CR-W	Polychloroprene, thiuram modified	Neoprene[k] W
NBR	Nitrile-butadiene rubber	Nipol[g]
EPDM	Ethylene-propylene-diene terpolymer	NORDEL[e]
	Polymeric ester plasticizers	Paraplex[l] G25 and G50
	Liquid polybutadiene	Ricon[m] 150 and 156
	Zinc diacrylates	Saret[m] 517 and 519
MgO	Magnesium oxide dispersion	Scorchguard[n] O
	Paraffinic processing oil	Sunpar[o]
T	Polysulfide copolymer with ethylene dichloride	Thiokol[p]
CPE	Chlorinated polyethylene	Tyrin[e]
TMTM	Tetramethyl thiuram monosulfide	UNADS[b]
	2,5-Bis(alkoxyalkylthio)-1,3,4-thiadiazole	VANAX[b] 189
BA	Butyraldehyde-aniline reaction product	VANAX[b] 808
	Proprietary thiadiazole derivative	VANAX[b] 829
	Proprietary fatty amine	VANAX[b] 882B

(Continued)

Abbreviation	Composition	Trade Name
DTDM	Dithiodimorpholine	VANAX[b] A
DPG	Diphenyl guanidine	VANAX[b] DPG
MBM	*Meta*-phenylene bismaleimide	VANAX[b] MBM
DMTD	2,5-Dimercapto-1,3,4-thiadiazole	VANCHEM[b] DMTD
	Proprietary processing aid	VANFRE[b] AP-2
ADPA	Acetone-diphenylamine reaction product	VANOX[b] AM
ZMTI	Zinc 2-mercaptotoluimidazole	VANOX ZMTI
PVI	*N*-Cyclohexylthio maleimide	VANTARD[b] PVI
	1,1-Bis(*t*-butylperoxy)-3,3,5-trimethylcyclohexane	VAROX[b] 231
	n-Butyl-4,4-bis(*t*-butylperoxy)valerate	VAROX[b] 230
DCP	Dicumyl peroxide	VAROX DCP
	α,α'-Bis(*t*-butylperoxy)-diisopropylbenzene	VAROX[b] 802
DBPH	2,5-Dimethyl-2,5-di(*t*-butylperoxy)hexane	VAROX[b] DBPH
	2,5-Dimethyl-2,5-di(*t*-butylperoxy)hexyne-3	VAROX[b] 130
EPDM	Ethylene-propylene-diene terpolymer	Vistalon[f]
FKM	Fluorocarbon rubber	Viton[d]
MTI	2-Mercapto-4,5-methylbenzidazole	Vulkanox[q] MB-2
HNBR	Hydrogenated nitrile-butadiene rubber	Zetpol[g]
TCY	Trithiocyanuric acid	Zisnet[r] F-PT

[a] DPPD, RESIN D, STALITE, and SUPERFLEX are registered trademarks of The B.F. Goodrich Company, Akron, OH.

[b] AGERITE, ALTAX, CAPTAX, CADMATE, DURAX, ETHYL TUADS, MORFAX, TUADS, UNADS, VANAX, VANCHEM, VANFRE, VANOX, VANTARD, VAROX, and ZIMATE are registered trademarks of R. T. Vanderbilt Company, Inc., Norwalk, CT.

[c] DIAK, HYPALON, HVA, and Viton are registered trademarks of DuPont, Wilmington, DE.

[d] Echo is a registered trademark of Hercules Inc., Wilmington, DE.

[e] ENGAGE, NORDEL, and TYRIN are registered trademarks of Dow Chemical.

[f] EXXPRO and Vistalon are registered trademarks of Exxon Corp., Houston, TX.

[g] Nipol, Hydrin, Hytemp, and Zetpol are registered trademarks of Zeon Chemicals, LP, Louisville, KY.

[h] Hysafe is a registered trademark of Huber Corp., Havre-de-Grace, MD.

[i] Maglite is a registered trademark of Marine Magnesium Co., Coraopolis, PA.

[j] Naugard, Delac, and Tuex are registered trademarks of Chemtura Corp., Middlebury, CT.

[k] Neoprene is manufactured by DuPont, Wilmington, DE.

[l] Paraplex is a registered trademark of C.P. Hall Co., Chicago, IL.

[m] Ricon and Saret are registered trademarks of Sartomer Company Inc., Exton, PA.

[n] Scorchguard is a registered trademark of Rhein Chemie Corp., Trenton, NJ.

[o] Sunpar is a registered trademark of Sun Refining and Marketing Co., Philadelphia, PA.

[p] Thiokol is a registered trademark of Rohm and Haas Co., Philadelphia, PA.

[q] Vulkanox is a registered trademark of Bayer Corp., Pittsburgh, PA.

[r] Zisnet is a registered trademark of Kawaguchi Chemical, Tokyo, Japan.

Cross Reference List of Suppliers

ASTM Abbreviation	Trade Name	Supplier
DPPD	AgeRite DPPD	R.T. Vanderbilt
		Chemtura Corporation
TMQ (low MW)	AgeRite MA	R.T. Vanderbilt
	Naugard Q	Chemtura Corporation
	Flectol TMQ	Harwick/Flexsys
TMQ (high MW)	AgeRite Resin D	R.T. Vanderbilt
	Flectol TMQ Pastilles	Harwick/Flexsys
	Naugard	Chemtura Corporation
	Vulkanox HS	LanXess
ODPA or 8DPA	AgeRite Stalite S	R.T. Vanderbilt
	Octamine	Chemtura Corporation
	Vulkanox OCD	LanXess
	Stangard ODP	Harwick
ADPA (flake)	Vanox AM	R.T. Vanderbilt
	Aminox Flake	Chemtura Corporation
ADPA (liquid)	AgeRite Superflex	R.T. Vanderbilt
	BLE	Chemtura Corporation
ZMTI or ZnMMBI	Vanox ZMTI	R.T. Vanderbilt
	Vulkanox ZMB-2	LanXess
MMBI	Vanox MTI	R.T. Vanderbilt
	Vulkanox MB-2	LanXess
MBTS	Altax	R.T. Vanderbilt
	Naugex MBTS	Chemtura Corporation
	Vulkacit ZM	LanXess
	Perkacit MBTS	Flexsys/Harwick
MBT	Captax	R.T. Vanderbilt
	Naugex MBT	Chemtura Corporation
	Perkacit MBT	Flexsys/Harwick
	Vulkacit Merkapto	LanXess
ZDBC or ZnDBC	Butyl Zimate	R.T. Vanderbilt
	Butazate	Chemtura Corporation
	Perkacit ZDBC	Flexsys/Harwick
	Vulkacit LDB	LanXess
ZDiBC or ZnDBzC	Benzyl Zimate	R.T. Vanderbilt
	Arazate	Chemtura Corporation
	Perkacit ZBEC	Flexsys/Harwick
	Vulkacit ZBEC	LanXess
TETD	Ethyl Tuads	R.T. Vanderbilt
		Chemtura Corporation
	Perkacit TETD	Flexsys/Harwick
TMTD	Methyl Tuads	R.T. Vanderbilt
	Tuex	Chemtura Corporation
	Perkacit TMTD	Flexsys/Harwick
	Vulkacit Thiuram	LanXess

Cross Reference List of Suppliers (Continued)

ASTM Abbreviation	Trade Name	Supplier
TMTM	Unads	R.T. Vanderbilt
	Monex	Chemtura Corporation
	Perkacit TMTM	Flexsys/Harwick
	Vulkacit Thiuram MS	LanXess
MBM or m-PBM	Vanax MBM	R.T. Vanderbilt
	HVA #2	DuPont
DTDM	Vanax A	R.T. Vanderbilt
	Naugex SD-1	Chemtura Corporation
	Rhenocure M	Rhein Chemie
	Sulfasan DTDM	Flexsys/Harwick
DPG	Vanax DPG	R.T. Vanderbilt
	DPG	Various
	Perkacit DPG	Flexsys/Harwick
	Vulkacit D	LanXess
BAA	Vanax 833	R.T. Vanderbilt
	Vulkacit 576	LanXess
TAIC	Diak #7	DuPont
	TAIC	Mitsubishi
CBS	Durax	R.T. Vanderbilt
	Delac S	Chemtura Corporation
	Santocure CBS	Flexsys/Harwick
	Vulkacit CZ	LanXess
DCP	Varox DCP, DCP-40C, DCP-40KE	R.T. Vanderbilt
	Di-Cup, 40C, 40KE	Geo Specialty Chemicals
	Perkadox BC, BC-40B-pd, BC-40K-pd	Akzo Nobel
	Peroximon DC 40 KE	Akrochem
BDBPC	Varox 231, 231-XL	R.T. Vanderbilt
	Trigonox 29-40B-pd	Akzo Nobel/Harwick
BPV	Varox 230, 230-XL	R.T. Vanderbilt
	Trigonox 17-40B	Harwick
DBPH or DMBPHa	Varox DBPH, DBPH-50	R.T. Vanderbilt
	Trigonox 101	Akzo Nobel/Harwick
DMBPHy	Varox 130, 130-XL	R.T. Vanderbilt
	Trigonox 1145-45B-pd	Harwick
DBDB or BBPIB	Varox 802, 802-KE	R.T. Vanderbilt
	Vul-Cup R, 40KE	Geo Specialty Chemicals
	Perkadox 14-40B-pd, 14-40K-pd	Akzo Nobel

REFERENCES

1. Anon., "Selecting a Peroxide for Optimum Curing Speed," R.T. Vanderbilt Company, Inc. technical literature (1992).
2. Anon., Data Sheet 889, R.T. Vanderbilt Company, Inc. technical literature.
3. Anon., Data Sheet 1139, R.T. Vanderbilt Company, Inc. technical literature.

4. Anon., *Vanderbilt News*, Vol. 41 (No. 1), page 13 (1988).
5. M.J. Recchio and W.G. Bradford, "Innovative Peroxide and Coagent Cure Systems for use with HNBR Elastomers," *Rubber World*, pages 29–36 and 45 (Nov. 1995), based on Paper No. 17 presented at a meeting of the Rubber Division ACS (Oct. 11–14, 1994).
6. A.J. Johansson, Ph.D., "Peroxide Dispersions and their Applications," paper presented to the Southern Rubber Group (March 7, 2000).
7. L.H. Palys, P.A. Callais, M.F. Novits, and M.G. Moskal, "New Peroxide Crosslinking Formulations for Metallocene Based Poly (ethylene octene) Copolymer," Paper No. 88 presented at a meeting of the Rubber Division ACS (Oct. 8–11, 1996).
8. Anon., *Vanderbilt News*, Vol. 39 (No. 1), page 15 (1983).
9. R.F. Ohm, ed., *The Vanderbilt Rubber Handbook*, Vol. 13, page 483 (1990).
10. R.F. Ohm, "New Developments in Crosslinking Halogen-containing Polymers," *Rubber World*, Vol. 209 (No. 1), pages 26–32 (Oct. 1998).
11. J.G. Bauerle, "The A Types of Viton Fluoroelastomer A-35, A, A-HV," DuPont technical literature, Publication E-23074.
12. C. Cable, Zeon Chemicals Inc., private communication.

14 Antioxidants for Specialty Elastomers

Russell A. Mazzeo

CONTENTS

14.1 INTRODUCTION

Since the discovery and usage of vulcanized rubber, rubber chemists noted that rubber, more so unvulcanized than vulcanized, degraded rapidly when exposed to the elements, namely, oxygen, ozone, heat, ultraviolet (UV) light, and so on. These early investigators later discovered that the degradation was related to reactions involving ozone and the double bonds or unsaturation in the polymeric molecular chains, which lead to a severed polymer chain. Still other reactions involved the formation and further reaction of radicals which resulted in the elastomer becoming either completely resinified (hard) or depolymerized (soft). It soon became evident that the prevention of these various forms of degradation was vital to the future of the rubber industry.

14.2 OZONE DEGRADATION

Ozone, a degradant of vulcanized rubber, occurs naturally and is formed in the earth's atmosphere by the action of the sun's ultraviolet light on atmospheric oxygen. The ozone is carried into the atmosphere by winds and, depending on the seasons and geographic locations, can be found on the earth in normal concentrations of 6 parts per hundred million to concentrations of as high as 25 parts per hundred million.

The effects of ozone on vulcanized diene rubber are best noted when the rubber is stressed or stretched in use. A series of cracks develop, in time, which are perpendicular to the applied stress. Further exposure to ozone of these cracked surfaces causes the cracks to become wider and deeper until the rubber fails.

The detrimental effects of ozone are noted also on nonstressed rubber surfaces and manifest themselves as frosting, which is the exposure of the non-black fillers resulting from the formation of minute cracks on the rubber surface formed by the severing of the molecular chains. This phenomenon is quite common in footwear compounds [1].

The mechanism proposed by Criegee best describes the degradation initiated by ozone called ozonolysis. Ozone, a very reactive material, reacts at the surface, across the double bond, in an unsaturated polymer to form a trioxolane structure. This structure undergoes decomposition to give a carbonyl compound and a zwitterion, resulting in a severed molecular chain. The zwitterion can recombine to form either an ozonide, diperoxide, or higher peroxide.

Many theories have been proposed to explain the mechanism of how antiozonants protect elastomeric articles from the effects of ozone. Most support the scavenger model, and indicate that an ideal antiozonant should be capable of migrating in a rubber matrix to the surface whenever the equilibrium of the antiozonant concentration in the compound is upset by the formation of ozonized antiozonant at the surface, but yet not freely migrate to the surface and volatilize out of the elastomer without first reacting with ozone at the surface. On the other hand, the antiozonant should not be so slow in migrating through the elastomer matrix that it arrives at the surface after the ozone has already reacted with the elastomeric polymer. It has been stated that ozone is 200 times more reactive with an antiozonant than with the double bonds in an elastomer. Since a vulcanized elastomer matrix is a dynamic chemical system, the model for the mechanism of the function of an antiozonant is merely the restating of basic chemical laws derived from measurements made upon chemical systems in equilibrium [2].

14.2.1 ANTIOZONANTS

The most important group of chemical antiozonants is the organic group of p-phenylenediamines. The p-phenylenediamines are divided into three classes, depending on the nature of the substituents, R_1 and R_2, which are attached to the nitrogen atoms of the p-phenylenediamine molecule, shown below. In one class, R_1 is a phenyl (aryl) substituent, and R_2 is an alkyl group. In a second class, both R_1 and R_2 are alkyl groups; and in a third class, both R_1 and R_2 are large aryl groups. Compounds in which R_1 and R_2 are aryl groups are weak antiozonants, but strong antioxidants.

The three classes of p-phenylenediamines differ in their ability to protect rubber articles against degradation from ozone based upon their molecular size, solubility in the rubber, and their rate of migration through a rubber matrix. The rate of migration is somewhat affected by the intended application of the rubber article, that is, dynamic or static as well as the molecular size of the antiozonant. The N-alkyl-N'-aryl-p-phenylenediamines offer the best ozone protection to rubber articles intended for dynamic and intermittent dynamic uses, whereas the N-N'-dialkyl-p-phenylenediamines offer the best ozone protection to rubber articles intended for slightly, intermittent dynamic or static uses. The N-N'-diaryl-p-phenylenediamines offer good ozone protection to rubber articles intended for dynamic uses and poor protection to those intended for static uses. However, it should be noted that all the three mentioned classes of antiozonants stain and discolor to some degree [3].

The most common antiozonant from the class, N-alkyl-N'-aryl-p-phenylenediamine, is N-(1,3-dimethylbutyl)-N'-phenyl-p-phenylenediamine (6PPD), and the ozonation mechanism reported in the literature for this class of antiozonant is shown on the following page in a simplified form. The antiozonant reacts with ozone, Step (1), and proceeds through the formation of an unstable intermediate which rearranges to the *N*-hydroxydiamino compound, Step (2). The release of water by this compound leads further to the formation of a quinine diimine, Step (2), which can further react with a second mole of ozone to form a nitrone and oxygen, Step (3). The nitrone can eliminate the alkyl group to form the less active 4-nitrosodiphenylamine, Step (4), which is a good antioxidant but poor antiozonant.

The oxygen molecules which are formed as singlet oxygens are quenched by additional antiozonant to the lesser active triplet state.

The ozonation mechanism for the *N*-*N'*-diaryl-*p*-phenylenediamine antiozonants is identical to the above described mechanism except that the reaction terminates with the formation of the nitrone and the release of oxygen after reacting with the ozone, Step (3) [4].

The Ozonation Mechanism for an *N*-Alkyl-*N'*-Aryl-*p*-Phenylenediamine [(*N*-(1,3-Dimethylbutyl)-*N'*-Phenyl-*p*-Phenylenediamine] (6PPD)

The most common antiozonant from the class, *N*-*N'*-dialkyl-*p*-phenylenediamine, is *N,N'*-bis-(1,4-dimethylpentyl)-*p*-phenylenediamine (77PD), and the ozonation mechanism reported in the literature for this class of antiozonant is shown on the following page in a simplified form, and is similar to the one reported for the *N*-alkyl-*N'*-aryl-*p*-phenylenediamine up to the point where the second mole of ozone reacts with the reaction product previously formed and yields the nitrone, Step (3), which can react further with more ozone to form the dinitrone compound, Step (4) [5].

The Ozonation Mechanism for an *N-N'*-Dialkyl-*p*-Phenylenediamine, [*N-N'*-Bis (1,4-Dimethylpentyl)-*p*-Phenylenediamine] (77PD)

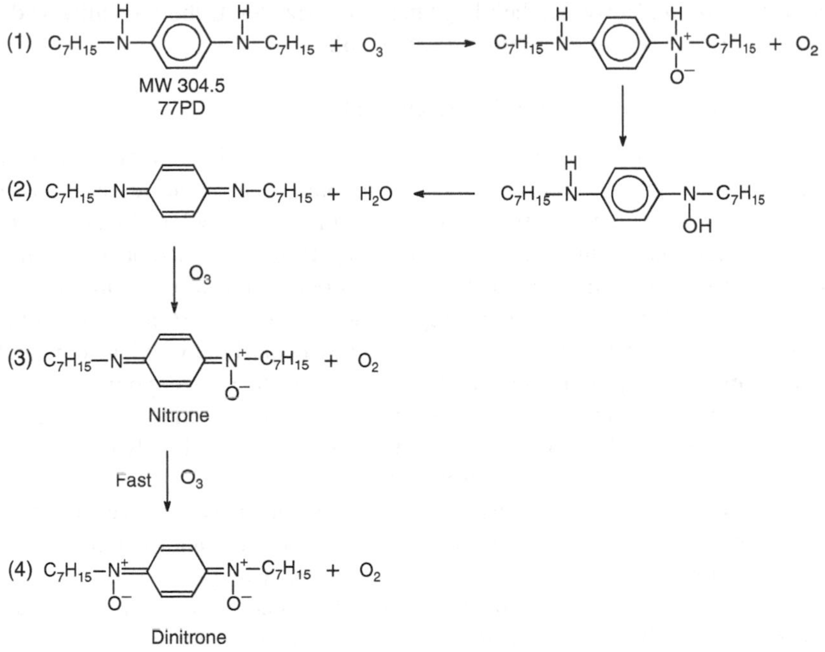

(1) reaction showing 77PD (MW 304.5) + O_3

(2) reaction yielding product + H_2O

(3) Nitrone + O_2

(4) Dinitrone + O_2

14.2.2 NONSTAINING, TRIAZINE BASED ANTIOZONANTS

Until recently, there were no effective, nonstaining, chemical antiozonants, since most chemical antiozonants were based upon *p*-phenylenediamine which severely stains rubber compounds. However, recent developments in the rubber chemicals industry have resulted in the production of a nonstaining, slightly discoloring antiozonant based upon triazine chemistry. This antiozonant is 2,4,6-tris-(*N*-1,4-dimethyl-pentyl-*p*-phenylenediamino)-1,3,5-triazine (TAPDT), and the molecular structure is shown below.

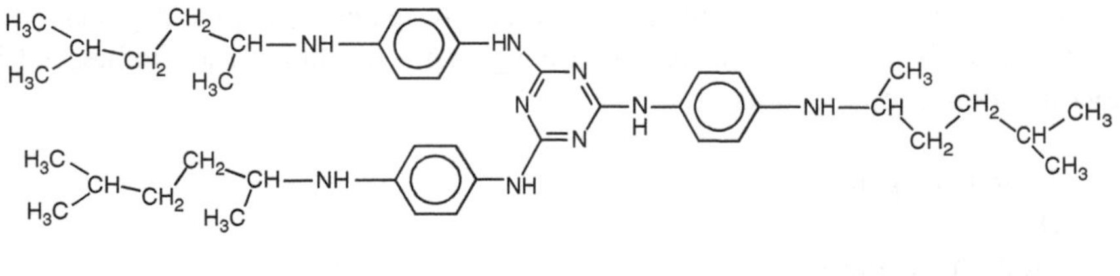

2,4,6-Tris-(*N*-1,4-Dimethylpentyl-*p*-Phenylenediamino)-1,3,5-Triazine (TAPDT)

TAPDT is unique because it offers excellent antiozonant protection without staining. It also offers excellent static ozone protection and improved flex life. The triazine based antiozonants behave similarly to the *N*-alkyl-*N'*-aryl-*p*-phenylenediamine class of antiozonants, and the ozonation mechanism of TAPDT is also identical to this class of antiozonant. A feature of the triazine based antiozonants is that they are extremely reactive with ozone. However, their large molecular size limits their mobility, and as a result, they offer excellent long-term protection [6].

14.2.3 WAXES AND POLYMER ANTIOZONANTS

Another way of rendering ozone resistance to an elastomeric compound is to either incorporate waxes into the compound or to form a blend of an ozone resistant polymer, such as EPDM, with other polymers in the compound which lack resistance to ozone.

Waxes function as antiozonants because they tend to be incompatible with the polymer or polymers with which they are combined, and as a result, migrate to the surface to form a barrier which is impermeable to ozone. Hence, the ozone is prevented from reacting with the compound polymers until the wax barrier is broken or cracked by some dynamic performance of the elastomeric compound. Therefore, waxes are generally employed in elastomeric compounds, intended for static applications, but when used in elastomeric compounds, intended for dynamic applications, they are employed for short-term protection only.

Polymeric antiozonants, when used in blends with other polymers that are not ozone resistant, are usually added in a quantity where they are the minor component of the blend. The major component of the blend, or the polymer lacking ozone resistance, forms the continuous phase of the blend matrix, and the ozone resistant polymer, ideally, forms domains which are uniformly dispersed in the continuous phase, and which inhibit the progression of any ozone cracking that may originate in the continuous phase [7].

One may conclude that the mechanisms involved in describing the ozone resistance derived from either waxes or polymers are more of a physical rather than a chemical nature.

14.3 DEGRADATION CAUSED BY OXIDATION

Elastomers, like most organic materials, are subject to atmospheric oxidation, even at moderate temperatures. The ease of susceptibility to degradation depends, to a large degree, upon structure and environmental exposure. For example, saturated polymers are inherently more stable than unsaturated polymers because of their stronger bonds, or lack of double bonds in their backbone. Therefore, it would hold that EPDM and butyl rubber would be more stable than SBR or NR against oxidative degradation.

Oxidation is a complex process involving many reactions, each influenced by prevailing conditions such as:

1. Singlet oxygen
2. Ozone
3. Mechanical shear
4. Heat
5. Light
6. Metals
7. Fatigue

14.3.1 Effects of Oxygen, Ozone, and Shear

Most elastomers are subject to oxidation and it is known that the addition of only 1%–2% combined oxygen will render a rubber article useless.

Oxidation proceeds by two mechanisms:

1. Chain scission: Results from the attack of the polymer backbone which causes softening and weakness. It is the primary mechanism observed for natural rubber and butyl oxidation.
2. Cross-linking: Brittle compounds result because of radical cross-linking reactions, resulting in the formation of new cross-links and a stiffer material. This reaction occurs predominantly with SBR, polychloroprene, NBR, and EPDM.

In most cases, both types of attack occur and the one which prevails determines the final compound properties. It has been found that loss of elongation is the most sensitive criterion for aging measurements regardless of the mechanism, and it is favored over the measurement of tensile loss.

An important factor worth mentioning is that the selection of the cure system for an elastomeric compound also plays an important role in determining aging resistance. The effect of types of sulfur cross-links and how they can be varied should be considered when imparting oxidation resistance to an elastomeric compound, but since the explanation of this effect has been covered in another chapter, it will not be discussed here.

14.3.2 Effects of Heat

As expected, heat accelerates oxidation. Therefore, the effects described previously are observed sooner, and are more severe as temperature is increased. In order to distinguish the effects of heat from those of oxidation, aging tests should be carried out in an inert atmosphere to determine the effects due solely to heat. When aging tests are done with natural rubber in an inert atmosphere, for example, the formation of more cross-links occurs initially—followed by reversion, as both cross-links and the main polymer chain backbone are broken down. For example, at 60°C, it requires 1.2% combined oxygen to reduce the tensile strength of conventionally cured natural rubber in half. However, at 110°C, only 0.65% oxygen is required. Had a similar aging been done at 110°C in the absence of oxygen, essentially no tensile loss would have been observed [8].

As noted, a small percent of oxygen, combined with an elastomer, can seriously degrade the physical properties of that elastomer. Heat can greatly increase the rate at which oxygen reacts with the polymer. The rate of reaction approximately doubles for each 10°C increase in temperature. About a 50-fold increase in reaction rate occurs between room temperature and 70°C. Another perspective would be that a change in physical properties, observed at a service temperature of 150°C, would require about 8000 times longer for the same change to occur at ambient temperature [9].

14.3.3 EFFECTS OF LIGHT AND WEATHERING

UV light promotes free radical oxidation at the rubber surface which produces a film of oxidized rubber. Heat and humidity then accelerate the formation of a crazing or alligatoring effect, and this oxidized layer can be rubbed off—giving a chalking appearance.

Black compounds are more resistant to UV light than light colored ones. In order to prevent oxidative surface degradation of lighter compounds, larger amounts of a nonstaining antioxidant are required to replenish that, which is used up at the surface, that is, an antioxidant reservoir is required. The issue is more severe with thin parts since product performance can suffer as well as being merely an issue of cosmetics.

14.3.4 EFFECTS OF METALS

Heavy metal (principally cobalt, copper, manganese, iron) ions are believed to catalyze oxidative reactions in elastomers by influencing the breakdown of peroxides in such a way as to accelerate further attack by oxygen. The first corrective approach is to eliminate all sources of harmful metals. Compounds of copper and manganese, such as stearates and oleates, which are directly soluble in rubber, are particularly active, since they provide a direct source of heavy metal ions. Even the less soluble forms such as the oxides can cause problems by reacting with fatty acids used in compounding to produce more soluble forms.

Although some antioxidants are active against catalyzed oxidation of rubber, in general, the standard antioxidants do not give protection against the heavy metal ions. Since the activity of the metal depends on its being in an ionic form, it is possible to protect compounds by incorporating substances which react with ionic metals to give stable coordination complexes.

14.3.5 EFFECT OF FATIGUE

One of the major causes of failure in rubber is the development of cracks at the surface. The growth of these cracks under repeated deformation, or fatigue, leads to catastrophic failure. This fatigue failure is initiated at minute flaws where stresses are high and mechanical rupture at such points can lead to the development of cracks. Similarly, attack by ozone can cause cracks to occur at the surface whose rate of growth is directly proportional to the ozone concentration.

Many factors, both chemical and physical, are involved and as would be anticipated, corrective action often consists of redesign and recompounding to minimize excessive stress concentration. Different classes of antioxidants can have different effects, and practical experience has led to the recognition that certain products—known as antiflexcracking agents, can, in addition to being antioxidants in a more normal sense, possess a particular ability to reduce the rate of crack growth.

Another factor to be considered with flexing is the effect of sulfur concentration. The rate of oxidation is proportional to the amount of combined sulfur, and with lower sulfur levels, better aging is obtained. It would appear, therefore, that it is better to compound with low sulfur. However, these low sulfur cures yield poor fatigue resistance in natural rubber, but little fatigue loss upon aging is experienced. SBR, on the other hand, exhibits both good aging and good fatigue resistance with low sulfur cures [10].

14.4   POLYMER DEGRADATION

Polymeric degradation typically occurs via a free radical process. Chemical bonds, whether they are within the main chain of the polymer or in side groups, can be dissociated by energy resulting from energy sources such as heat, mechanical shearing, or radiation to create a free radical (R$^{•}$). The formation of (R$^{•}$) commences the initiation stage and can occur in any one of the various phases of a polymer's life cycle: polymerization, processing, and end use; and is illustrated, below, in Equations 14.1 through 14.3 [11].

Initiation reactions

$$ROOH \rightarrow RO^{•} + {^{•}}OH \qquad (14.1)$$

$$RH + O_3 \rightarrow RO^{•} + HOO^{•} \qquad (14.2)$$

$$RR \rightarrow 2R^{•} \qquad (14.3)$$

In addition, Figure 14.1 illustrates the cyclic nature of the initiation reactions. The carbon radicals (C$^{•}$) enter into a cyclic oxidation process. In the presence of oxygen, peroxy radicals (COO$^{•}$) are formed. The peroxy radicals can abstract a hydrogen atom from the backbone of the polymer chain to generate another carbon radical and a hydroperoxide group (COOH). The hydroperoxide group will decompose under heat to form an alkoxy radical (CO$^{•}$) and a hydroxyl radical (HO$^{•}$). Each of these radicals can abstract hydrogen from the polymer backbone to form yet more carbon radicals [12].

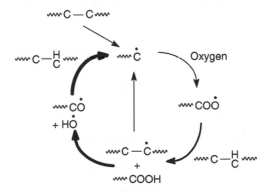

FIGURE 14.1 Cyclic oxidation process. (From Ohm, R.F., *Improving the Heat Resistance of HNBR, ACM, EAM by the Use of Antioxidants*, Crompton, presented at Connecticut Rubber Group Meeting, April 6, 2004. With permission.)

The propagation stage commences as the cycle, described earlier, is repeated, and additional free radicals, illustrated by the darker arrows, are generated. Two free radicals can combine to terminate each other, and result in a new cross-link. As more and more of these cross-links are formed, the polymer becomes stiff and embrittled. Radicals can also add to polymers containing double bonds (unsaturation), and generate cross-links without termination. One such radical can rapidly generate many cross-links, and this type of oxidation becomes much like peroxide cross-linking. Peroxides also generate free radicals that can cross-link the polymer [13].

Some polymers cannot be cross-linked by peroxides. These polymers tend to undergo reversion or chain scission during oxidative aging. Scission is favored in polymers that are branched or have many side groups. During aging, tertiary alkoxy or carbon radicals are generated, which undergo beta scission or unimolecular cleavage to lower molecular weight fragments.

The chemical reactions responsible for the propagation stage are illustrated in Equations 14.4 through 14.8. The propagation stage is usually quite rapid compared with initiation stage.

Propagation reactions

$$R^{\bullet} + O_2 \rightarrow ROO^{\bullet} \tag{14.4}$$

$$ROO^{\bullet} + RH \rightarrow ROOH + R^{\bullet} \tag{14.5}$$

$$ROOH \rightarrow RO^{\bullet} + {}^{\bullet}OH \tag{14.6}$$

$$RO^{\bullet} + RH \rightarrow ROH + R^{\bullet} \tag{14.7}$$

$$HO^{\bullet} + RH \rightarrow HOH + R^{\bullet} \tag{14.8}$$

The peroxy radical ($ROO^{\bullet}$) can further react with labile hydrogens of the polymer to yield unstable hydroperoxides. These hydroperoxides decompose into alkoxy and hydroxy radicals which, in turn, abstract more hydrogen atoms and generate more polymer radicals. The cycle becomes autocatalytic, as shown in Equations 14.5 through 14.8. These chemical equations illustrate some, but not all, of the reactions which can occur during the propagation stage [14].

Termination reactions

$$2R^{\bullet} \rightarrow R\text{-}R \tag{14.9}$$

$$\overset{\overset{\text{H}}{|}}{2R_2COO^{\bullet}} \rightarrow R_2C{=}O + R_2CHOH + O_2 \tag{14.10}$$

$$2R_3COO^{\bullet} \rightarrow R_3COOCR_3 + O_2 \tag{14.11}$$

$$2RO^{\bullet} \rightarrow ROOR \tag{14.12}$$

$$ROO^{\bullet} + {}^{\bullet}OH \rightarrow ROH + O_2 \tag{14.13}$$

Autoxidation will progress until termination results from the formation of stable products. Eventually, propagating radicals combine or disproportionate to form inert products and the process is terminated, as shown in Equations 14.9 through 14.13.

Equations 14.9, 14.11, and 14.12 represent cross-linking, and increase the molecular weight of the polymer. This type of degradation manifests itself as brittleness, gelation, and decreased elongation. Chain scission, Equations 14.10 and 14.13, results in a decrease in molecular weight leading to increased melt flow and reduced tensile strength.

The manner in which various common elastomers degrade, either by cross-linking or chain scission, is shown below.

Degradation of Elastomers

Natural rubber	Scission (softens)
Polyisoprene	Scission (softens)
Polychloroprene	Cross-linking and scission (hardens)
SBR	Cross-linking and scission (hardens)
NBR	Cross-linking (hardens)
BR	Cross-linking (hardens)
IIR	Scission (softens)
EPM	Cross-linking and scission (hardens)
EPDM	Cross-linking and scission (hardens)

14.5 INHIBITION OF DEGRADATION

Antioxidants do not completely eliminate oxidative degradation, but they markedly retard the rate of autoxidation by interfering with radical propagation. Depending on the types and combinations used antioxidants can provide suitable polymer protection during the phases of its life cycle [15].

Antioxidants are materials that can interfere with any of the steps in the oxidation process, as indicated in Figure 14.2. However, some steps are easier to circumvent. Initiation really depends on the strength of the carbon–carbon bond. If the polymer absorbs too much energy, the bond will break. Only the structure of the polymer determines how much energy can be tolerated.

Two general classifications can be used to categorize antioxidants—primary (chain terminating) and secondary (peroxide decomposing).

Most conventional antioxidants either trap the oxy radicals or decompose the hydroperoxides (COOH). Radical traps for peroxy and alkoxy radicals are the familiar amines and phenols. Examples of peroxide decomposing antioxidants are phosphate stabilizers such as tris(nonylphenyl) phosphite (Naugard P), thioester stabilizers such as dilauryl thiodiproprionate (DLTDT), and dithiocarbamate stabilizers such as nickel di-n-butyldithiocarbamate (Naugard NBC).

Generally, for maximum antioxidant effectiveness, inhibiting the oxidative process at more than one step can be very effective. Therefore, a combination of two

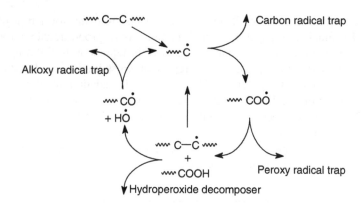

FIGURE 14.2 How antioxidants protect rubber. (From Ohm, R.F., *Improving the Heat Resistance of HNBR, ACM, EAM by the Use of Antioxidants*, Crompton, presented at Connecticut Rubber Group Meeting, April 6, 2004. With permission.)

types of antioxidants—a radical trap and a peroxide decomposer, can frequently provide better protection against oxidative aging than either type of antioxidant alone. However, it has been reported in the literature that in some high temperature applications, no apparent synergism is found to occur by the use of typical primary and secondary antioxidant combinations [16].

14.5.1 PRIMARY ANTIOXIDANTS

Hindered phenols and secondary aryl amines act as primary antioxidants by donating their reactive hydrogen (N–H, O–H) to free radicals, particularly peroxy radicals as shown here:

$$ROO^\bullet + AH \rightarrow ROOH + A^\bullet$$

Examples of alkoxy and peroxy radical trap antioxidants are hindered amines (i.e., octylated diphenylamine, Octamine) and phenols (i.e., butylated hydroxytoluene, Naugard BHT).

To sufficiently terminate the oxidative process, an antioxidant radical (A$^\bullet$) must be rendered stable so as not to continue propagation of new radicals. These radicals, in most cases, are stabilized via their electron delocalization or resonance.

14.5.2 PHENOLICS

Hindered phenolics, because of their nonstaining qualities, are the most preferred type of primary antioxidant for light colored applications such as footwear. This group can be further categorized into: simple phenolics (1), bisphenolics (2), polyphenolics (3), and thiobisphenolics (4), as shown on the following page [17].

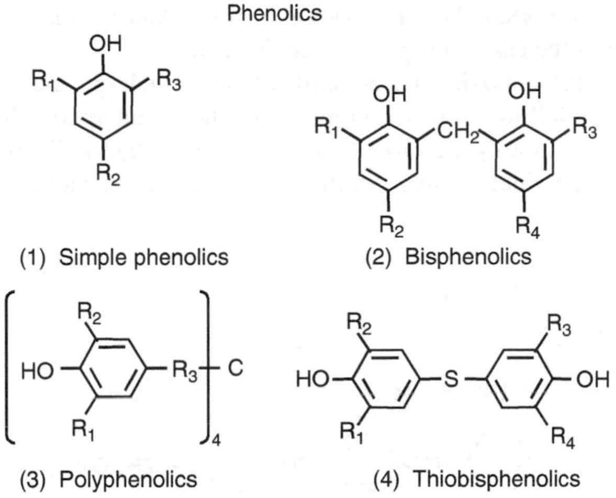

Phenolics

(1) Simple phenolics

(2) Bisphenolics

(3) Polyphenolics

(4) Thiobisphenolics

The mechanism by which BHT, a simple phenolic, functions is shown in Figure 14.3. A hindered phenol, when used alone, will react with two radicals. It then becomes spent and it will no longer protect the polymer. These reactions are only simplified overviews. More complex and undesirable side reactions can occur. For example, quinone formation by the reaction of an alkoxy radical at the para position of the phenoxy radical leads to color development because of the conjugated diene structure [18].

14.5.3 AMINES

Secondary aryl amines function by hydrogen donation, similar to the phenols; however, at higher temperatures they are also capable of decomposing peroxides. This single feature, which will be discussed later in this chapter, eliminates, most

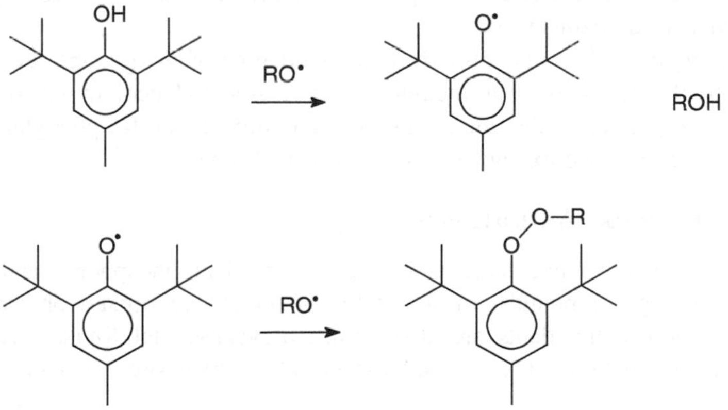

FIGURE 14.3 How alkoxy radical traps function. (From Ohm, R.F., *Improving the Heat Resistance of HNBR, ACM, EAM by the Use of Antioxidants*, Crompton, presented at Connecticut Rubber Group Meeting, April 6, 2004. With permission.)

times, the need for a secondary antioxidant in combination with an amine antioxidant, which is not the case with phenolic antioxidants.

Hindered amine antioxidants, as shown below, undergo the same two basic reactions as their fellow radical trapping phenolic counterparts. In addition, the amines can enter into a cyclic process to continuously trap radicals without being consumed. This can yield a longer lifetime at elevated temperature.

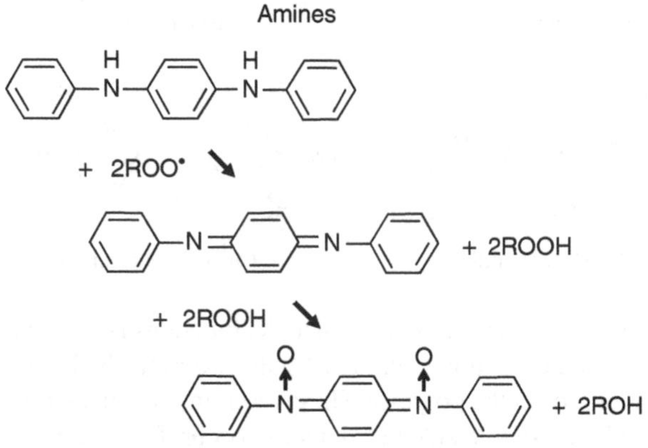

Amines

Although the amine class of primary antioxidants is usually more effective than the phenols, because of their ability to act both as chain terminators and peroxide decomposers, their use is generally limited to those applications where their discoloring characteristic can be tolerated or masked. The amines are perhaps most used in unsaturated polymers containing carbon black [19].

Most notable amine antioxidants are those derived from diphenylamine and p-phenylenediamines. Some of the alkylated diphenylamines are less discoloring than the phenylenediamines and find application in plastics. For example, 4,4'-bis (α-dimethylbenzyl)diphenylamine is widely used in flexible polyurethane foams and polyamide hot melt adhesives. The p-phenylenediamines are more recognized for their activity as antiozonants.

Additionally, carbon radicals exhibit a high degree of energy, and few materials can effectively trap these intermediates. Diaryl-p-phenylenediamines such as N, N'-diphenyl-p-phenylenediamine (Naugard J) and mixed diaryl-p-phenylenediamine (Novazone AS) are two examples of carbon radical traps.

14.5.4 SECONDARY ANTIOXIDANTS

This class of antioxidants consists of various trivalent phosphorus and divalent sulfur containing compounds, most notable of which are organo phosphites and thioesters. These antioxidants are also termed preventive stabilizers, because they prevent the proliferation of alkoxy and hydroxy radicals by destroying hydroperoxides.

14.5.5 PHOSPHITES

Phosphites function by reducing hydroperoxides to alcohols, and thus, converting themselves to phosphates.

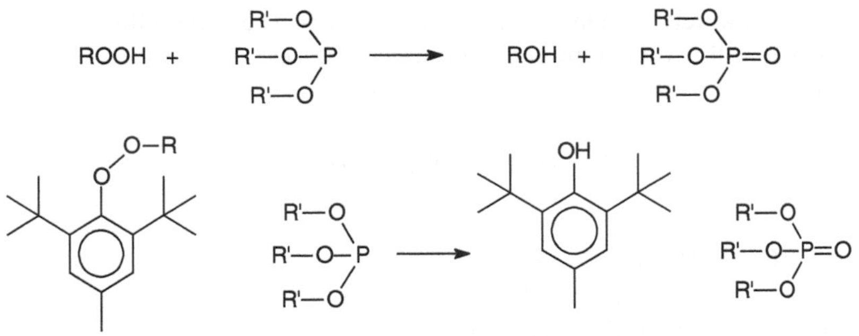

FIGURE 14.4 How peroxide decomposers function. (From Ohm, R.F., *Improving the Heat Resistance of HNBR, ACM, EAM by the Use of Antioxidants*, Crompton, presented at Connecticut Rubber Group Meeting, April 6, 2004. With permission.)

Phosphites

$$ROOH + (RO)_3\text{-}P \rightarrow ROH + (RO)_3\text{-}P = O$$

Not only are phosphites nondiscoloring, but they are color stabilizing, in that, they inhibit the formation of the discoloring quinoidal structures of phenolic antioxidants.

The most popular phosphite stabilizer is tris(nonylphenyl) phosphite (TNPP). The introduction of TNPP revolutionized the area of nondiscoloring rubber stabilization.

A serious drawback of phosphites is their sensitivity to hydrolysis. The commercially available phosphites vary significantly in their resistance to hydrolysis. Several phosphites are available, containing additives, to decrease susceptibility to hydrolysis. Hydrolysis of phosphites can ultimately lead to the formation of phosphorous acid which can accelerate degradation and also cause corrosion of processing equipment.

Phosphite stabilizers behave synergistically with hindered phenolics providing good processing protection; and in some cases, they enhance stability during ultraviolet exposure. They normally are the recommended secondary antioxidants for use in combination with primary phenolic antioxidants.

Peroxide decomposing antioxidants can react directly with the polymer hydroperoxide as shown for Naugard P in Figure 14.4. The phosphite antioxidant becomes oxidized to a phosphate whereas the peroxide is reduced to a harmless alcohol. In addition, peroxide decomposing antioxidants are believed to be capable of reacting with the spent, phenolic antioxidant to regenerate its activity. This technique is a manner by which a combination of two antioxidants can be used to provide longer service than either one alone. Moreover, the inclusion of a phosphite as a secondary antioxidant generally provides improved color compared with the use of a phenol alone.

14.5.6 THIOESTERS

Aliphatic esters of β-thiodipropionic acid are highly effective peroxide decomposers for long-term heat exposure applications when used in combination with phenolics.

They are more widely used in thermoplastic polymers, where the sulfur, contained in the thioester, will not interfere in a vulcanization process.

Thioesters

$$S - (CH_2 - CH_2 - CO_2 - R)_2$$

where

R = $C_{18}H_{37}$ for DSTDP (stearyl)
   $C_{12}H_{25}$ for DLTDP (lauryl)
   $C_{13}H_{27}$ for DTDTDP (tridecyl)
   $C_{14}H_{29}$ for DMTDP (myrisityl)

The major thioester stabilizers are dilauryl thiodiproprionate DLTDT and distearyl thiodiproprionate (DSTDP) [20].

14.6   CONSIDERATIONS FOR SELECTING THE PROPER ANTIOXIDANT(S)

The selection of a proper stabilizer system can sometimes be a difficult task. There are, as mentioned earlier, several types of antioxidants to consider and different combinations and concentrations to choose. A list of criteria to assist one in narrowing down his choice is as follows:

1. Color
2. Toxicity
3. Performance
4. Extractability
5. Cost
6. Physical form
7. Volatility
8. Compatibility
9. Odor

If an antidegradant is naturally highly colored, it will most likely discolor the polymer to which it is added. This phenomenon may be likened to pouring ink into water. The ink tints or discolors the entire container of water.

Many antidegradants which are not naturally colored and do not discolor a polymer when first added may still discolor the polymer because of chemical reactions which may occur during processing, curing, or while the polymer is in service.

Discoloration can also be caused by degradation at the surface. Ultraviolet discoloration is a degradative discoloration. It is very important to determine the type of discoloration since the remedy will require different solutions based upon the cause.

Staining is another color related issue. Basically, staining can be thought of as color in motion. If a chemical or color body is colored and has mobility, some of it can transfer from one compound to an adjacent one in contact with it. Contact

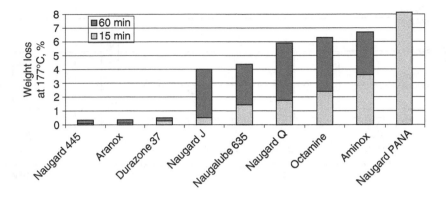

FIGURE 14.5 Volatility of antioxidants at 177°C. (From Ohm, R.F., *Improving the Heat Resistance of HNBR, ACM, EAM by the Use of Antioxidants*, Crompton, presented at Connecticut Rubber Group Meeting, April 6, 2004. With permission.)

staining is when the compound, from which the stain migrated, is removed, and the stain remains where the contact was made [21].

Antioxidants can also be lost by physical means. If they are volatile, they may evaporate from a substrate during high temperature service. A comparison of the volatility of several amine antioxidants at 177°C is shown in Figure 14.5. The alkylated phenyl-α-naphthylamine (Naugard PANA) is the most volatile of the amines in this comparison. The three least volatile antioxidants are the N-phenyl-N′-(p-toluenesulfonyl)-p-phenylenediamine (Aranox), 2,4,6-tris-(N-1, 4-dimethylpentyl-p-phenylenediamino)-1,3,5-triazine (Durazone 37), and α-methyl-styrenated diphenylamine (Naugard 445). Naugard 445 is far superior in resistance to volatilization when compared with its analogues, octylated (Octamine) and styrenated (Naugalube 635) diphenylamine.

In an application where an elastomeric article is immersed in a service fluid, the service fluid may extract the antioxidant from the elastomeric article. Figure 14.6 shows the relative solubility of several amine antioxidants in heptane, which is a good model for modern-day hydrocarbon motor oils. The least soluble antioxidants (Naugard 445, Naugard J, and Naugard Q) are more likely to remain in the rubber article and will provide protection for a longer service time when used in an application where the elastomeric part will be immersed in a hydrocarbon fluid [22].

14.7 EVALUATIONS OF ANTIOXIDANTS IN SPECIALTY ELASTOMER COMPOUNDS

In an initial study, various antioxidants were evaluated in hydrogenated nitrile-buta-diene rubber (HNBR), polyacrylic elastomer (ACM), and ethylene-acrylic elastomer (EAM), polymers which are frequently used in applications requiring high heat and oil resistance. Compounds of these elastomers were first immersed in ASTM 901 Oil to remove any soluble antioxidants that could be extracted in service, and later heat aged up to 21 days at 177°C in air. Similar agings were performed with samples placed in

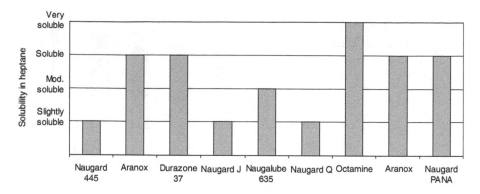

FIGURE 14.6 Solubility of antioxidants in heptane. (From Ohm, R.F., *Improving the Heat Resistance of HNBR, ACM, EAM by the Use of Antioxidants*, Crompton, presented at Connecticut Rubber Group Meeting, April 6, 2004. With permission.)

individual test tubes or cells. Primary or radical trap antioxidants were compared, optionally with a secondary or synergist antioxidant [23].

14.7.1 HYDROGENATED NITRILE-BUTADIENE RUBBER

HNBR suppliers recommend a combination of Naugard 445 and a toluimidazole in HNBR for the best resistance to high temperature. However, the optimum dose of each antioxidant has never been mentioned nor finalized. A central composite design experiment was undertaken to determine the optimum level of each antioxidant.

The HNBR base formulation and initial properties are shown in Table 14.1. The initial properties indicate that Naugard 445 significantly lowers maximum torque, as determined by a rheometer, and as would be expected for a primary, radical trap antioxidant in a peroxide-cured compound (see Figure 14.7). Since zinc 2-mercap-totoluimidazole (ZMTI) does not act as a free radical trap, it is surprising that the total level of antioxidant is a better single predictor of the maximum rheometer torque than is the level of Naugard 445, see Table 14.2.

Initial physical properties are relatively unaffected by either one of the antioxidants. Hardness and modulus values decrease modestly with total antioxidant dose; elongation increases with total antioxidant dose. Samples D and E appear to be outliers, giving relatively low tensile strength and elongation values. For this reason, the analysis of the data, after heat aging, looks only at the properties of heat aged specimens and not the change in properties from the original physical properties. Each antioxidant increases compression set by about 1%/phr.

Heat aging of the HNBR compounds at 177°C for 3 and 7 days provides a continuing loss in elongation and an increase in hardness and modulus, see Table 14.3. The change is comparable for all combinations of antioxidants. After 21 days at 177°C, Sample F, containing no Naugard 445 and 1 phr ZMTI, is nearly embrittled. All other samples continue to decrease in elongation and an increase in hardness and modulus by approximately the same amount.

TABLE 14.1
HNBR Formulations and Properties

	A	B	C	D	E	F	G	H	I
	MB								
Zetpol 2010[b]	100.0								
Zinc oxide	5.0								
N-762 Black	50.0	For all							
MBM	3.0								
Processing aid—1	2.0								
Total	160.0								
Naugard 445	2.0	0.5	0.5	1.5	1.0	0.0	1.0	1.5	1.0
ZMTI	1.0	0.5	1.5	0.5	0.0	1.0	2.0	1.5	1.0
Bis peroxide	8.0	8.0	8.0	8.0	8.0	8.0	8.0	8.0	8.0
Rheometer at 177°C									
ML, dN m	6.4	8.8	9.3	9.6	10.2	9.7	9.4	9.4	9.6
in-lb	5.7	7.8	8.2	8.5	9.0	8.6	8.3	8.3	8.5
MH, dN m	72.3	81.5	80.2	77.6	80.2	82.2	76.7	75.0	78.9
in-lb	64.0	72.1	71.0	68.7	71.0	72.7	67.9	66.4	69.8

(continued)

TABLE 14.1 (Continued)
HNBR Formulations and Properties

	A	B	C	D	E	F	G	H	I
ts_2, minutes	0.8	0.7	0.8	0.7	0.7	0.7	0.8	0.8	0.7
tc_{90}, minutes	5.7	5.4	5.5	5.3	5.5	5.3	5.5	5.5	5.4
Physical properties cured 14 min at 177°C									
Hardness, Shore A	70	72	71	71	71	71	71	71	70
100% modulus, MPa	8.2	10.7	10.8	9.4	10.5	11.7	9.7	10.0	11.2
psi	1190	1550	1570	1370	1530	1690	1410	1450	1620
Tensile strength, MPa	27.4	27.9	27.6	23.9	24.3	27.1	27.4	26.5	25.4
psi	3980	4040	4000	3460	3520	3930	3980	3840	3680
Elongation, %	250	210	220	190	180	190	230	220	190
Aged 70 h at 175°C									
Compression set, %	26	23	25	23	25	23	28	25	25

Source: From Ohm, R.F., *Improving the Heat Resistance of HNBR, ACM, EAM by the Use of Antioxidants,* Crompton, presented at Connecticut Rubber Group Meeting, April 6, 2004. With permission.

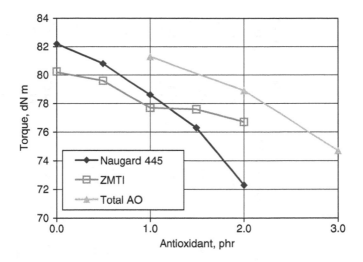

FIGURE 14.7 Rheometer maximum torque of peroxide-cured HNBR compounds. (From Ohm, R.F., *Improving the Heat Resistance of HNBR, ACM, EAM by the Use of Antioxidants*, Crompton, presented at Connecticut Rubber Group Meeting, April 6, 2004. With permission.)

The properties of most heat aged HNBR samples do not exhibit good correlation to the level of the antioxidants tested. Directionally, the antioxidants behave similarly. No significant synergism is evident. As noted in Figure 14.8, the only property that has a strong correlation with the antioxidant variables is the 100% modulus, after heat aging 7 days at 177°C. For this one condition, Naugard 445 is the main factor for preventing an increase in modulus.

Overall, the effects of the antioxidants on retained properties are only marginally above experimental error. Therefore, it is only possible to determine the main effects of the antioxidants. Nevertheless, there appears to be no significant synergism in using the toluimidazole with Naugard 445.

The HNBR compounds were immersed in ASTM 901 Oil for 3 days at 150°C, patted dry, and then heat aged for 3, 7, and 21 days at 177°C in an air oven. The purpose of this testing was to extract soluble antioxidants before air oven aging. The results are shown in Table 14.4.

The HNBR compounds that were pre-extracted in oil gave significantly poorer properties than those that were air oven aged only. Table 14.4 shows that, after oil immersion and heat aged 3 days at 177°C in air, Sample F, containing ZMTI alone, has <100% elongation. After oil immersion and heat aged 7 days at 177°C, this sample, as well as several others, are nearly embrittled. After heat aged 21 days, all samples are embrittled.

After an oil extraction and subsequent air oven aging, the amount of total antioxidant is the best single predictor of performance after heat aging 3 and 7 days. There is no evidence of synergism when using both antioxidants together.

TABLE 14.2
Regression Analysis of Antioxidant Effects in HNBR

	Constant	Naugard 445	ZMTI	X_1X_2	3-Term Model R^2	1-Term Models, R^2	
						N-445	Total AO
ODR MH	83.8	−3.4	~~−0.8~~	−1.4	**94.5**	**73.8**	**81.8**
Hardness	73.0	−1.5	−1.6	1.0	36.7		32.5
M-100	12.6	−1.9	−1.0	~~0.4~~	**65.0**	53.1	54.2
Tensile strength	28.7	−3.6	−1.6	2.9	54.1	0.9	0.6
Elongation	183	~~−3~~	~~12~~	~~20~~	**64.6**	52.4[a]	58.0
Compression set	22.3	~~1.0~~	~~1.3~~	~~0.0~~	46.0	32.1[a]	45.1
3 Days at 177°C							
Hardness	74	1.2	~~0.2~~	−1.0	58.7	1.9	20.1
M-100	14.1	−1.6	−1.4	0.9	41.9	22.2	37.7
Tensile	26.3	~~−0.1~~	~~0.6~~	~~0.0~~	12.8	22.8	37.9
Elongation	145	45	41	−30	46.2		
7 Days at 177°C							
Hardness	77	~~−0.5~~	−1.2	1.0	28.4	4.0[a]	1.7
M-100	17.3	−3.0	−1.5	1.7	**79.9**	**69.7**	20.1
Tensile	27.0	~~−0.4~~	~~−0.4~~	~~0.5~~	1.7	4.9	35.8
Elongation	152	23	7	~~0~~	47.8		
21 Days at 177°C							
Hardness	83	~~−2.0~~	~~−0.3~~	~~0.0~~	32.3	0.7	18.7

M-100 (n=8)	19.5	~~0.3~~	~~2.0~~	~~1.0~~	18.4	6.6	0.7
Tensile	10.8	9.7	~~3.2~~	~~2.2~~	34.3	0.8	20.1
Elongation	48	58	~~26~~	~~30~~	30.9	0.9	2.7
Pre-extracted in ASTM 901 Oil, then 3 days at 177°C							
Hardness	80.5	−1.8	−1.4	~~1.0~~	32.0	21.7	26.6
M-100 (n=9)	23.7	−5.7	−4.9	3.7	60.7	26.8	44.4
Tensile	18.6	1.4	1.3	~~0.4~~	**79.8**	35.1	**79.7**
Elongation	79	23	22	~~0~~	**74.2**	35.2	**74.1**
Pre-extracted in ASTM 901 Oil, then 7 days at 177°C							
Hardness	88.4	−2.5	−2.5	~~1.0~~	51.9	22.4	51.5
Tensile	8.3	2.4	~~1.5~~	~~0.7~~	**74.9**	45.3	**72.5**
Elongation	−10	28	25	~~1.0~~	**64.1**	33.1	**62.8**
Pre-extracted in ASTM 901 Oil, then 21 days at 177°C							
Hardness	104.6	−10.7	−8.8	10.0	32.0	5.7[a]	1.1
Tensile	2.2	5.6	5.0	−5.0	30.6	5.1	2.3
Elongation	4.0	−4.2	−4.3	5.0	46.8	9.6[a]	10.1
Aged 21 days at 177°C in cell oven							
Hardness	91.4	~~0.5~~	~~0.3~~	~~1.0~~	28.0	24.2[a]	23.1
Tensile	6.6	4.6	1.5	~~0.0~~	**89.5**	**79.6**	**68.1**
Elongation	1.1	~~0.5~~	~~0.8~~	5.0	75.9	41.7	**71.6**

Source: From Ohm, R.F., *Improving the Heat Resistance of HNBR, ACM, EAM by the Use of Antioxidants*, Crompton, presented at Connecticut Rubber Group Meeting, April 6, 2004. With permission.

[a] ZMTI used instead of αMSDPA.

Strikethrough = t-stat < 0.5.

TABLE 14.3
Air Oven Aging of HNBR

	A	B	C	D	E	F	G	H	I
Naugard 445	2.0	0.5	0.5	1.5	1.0	0.0	1.0	1.5	1.0
ZMTI	1.0	0.5	1.5	0.5	0.0	1.0	2.0	1.5	1.0
Aged 3 days at 177°C									
Hardness, Shore A	75	75	75	75	75	74	74	74	75
100% modulus, MPa	10.9	12.4	12.3	11.7	12.5	12.6	11.8	12.5	13.0
psi	1580	1800	1780	1700	1810	1830	1710	1810	1890
Tensile strength, MPa	26.9	28.1	27.9	27.4	24.0	26.6	26.9	27.2	27.1
psi	3900	4070	4040	3970	3480	3860	3900	3940	3930
Elongation, %	220	200	210	220	180	180	200	200	180
Tensile, % change	−2	1	1	15	−1	−2	−2	3	7
Elongation, % change	−12	−5	−5	16	0	−5	−13	−9	−5
100% modulus, % change	33	16	13	24	18	8	21	25	17
Hardness, points change	5	3	4	4	4	3	3	3	5
Aged 7 days at 177°C									
Hardness, Shore A	77	76	76	76	77	76	77	77	76
100% modulus, MPa	12.8	14.7	14.8	13.0	14.8	16.1	14.8	14.8	14.8
psi	1860	2130	2150	1880	2140	2330	2150	2140	2150
Tensile strength, MPa	26.8	27.8	27.2	26.8	25.5	25.9	26.9	26.8	26.3
psi	3880	4030	3950	3890	3700	3760	3900	3880	3810
Elongation, %	210	190	180	200	160	150	170	190	170
Tensile, % change	−3	0	−1	12	5	−4	−2	1	4
Elongation, % change	−16	−10	−18	5	−11	−21	−26	−14	−11
100% modulus, % change	56	37	37	37	40	38	52	48	33
Hardness, points change	7	4	5	5	6	5	6	6	6
Aged 21 days at 177°C									
Hardness, Shore A	81	81	81	80	80	86	82	80	81
100% modulus, MPa	20.4	20.9	21.9	20.5	17.7	—	—	20.6	24.0
psi	2960	3030	3180	2980	2560	—	—	2990	3480
Tensile strength, MPa	22.1	22.5	24.8	24.4	24.3	0.4	25.2	24.4	24.2
psi	3210	3270	3590	3540	3530	60	3660	3540	3510
Elongation, %	100	100	120	120	130	20	90	110	110
Tensile, % change	−19	−19	−10	2	0	−98	−8	−8	−5
Elongation, % change	−60	−52	−45	−37	−28	−89	−61	−50	−42
100% modulus, % change	149	95	103	118	67	−100	−100	106	115
Hardness, points change	11	9	10	9	9	15	11	9	11

Source: From Ohm, R.F., *Improving the Heat Resistance of HNBR, ACM, EAM by the Use of Antioxidants*, Crompton, presented at Connecticut Rubber Group Meeting, April 6, 2004. With permission.

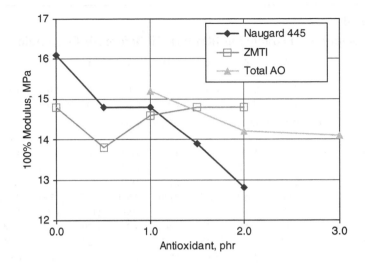

FIGURE 14.8 Air oven aging of HNBR compounds 7 days at 177°C. (From Ohm, R.F., *Improving the Heat Resistance of HNBR, ACM, EAM by the Use of Antioxidants*, Crompton, presented at Connecticut Rubber Group Meeting, April 6, 2004. With permission.)

Surprisingly, the HNBR samples, heat aged 21 days in individual test tubes (or cell ovens), are more severely deteriorated than the same compounds heat-aged communally in an air oven for the same time, despite the lower turnover of air and consequent more limited access to oxygen in the test tubes. While Naugard 445 appears to be somewhat more effective than ZMTI in protecting the samples in the cell oven test, Figure 14.9 shows that the total level of antioxidant is a reasonably good predictor of retained tensile strength and elongation properties after heat aging [24].

14.7.2 POLYACRYLIC ELASTOMER

Naugard 445 is also the recommended antioxidant for imparting the best high temperature resistance in ACM. Various antioxidant synergists were compared to supplement the performance of Naugard 445. They included a phosphite (Naugard P), a toluimidazole (ZMTI), a thioester, and sodium ethylhexyl sulfate in either a trithiocyanurate or a soap/sulfur cure system.

Table 14.5 shows the ACM compounds, cured with the trithiocyanurate cure system, had lower elongation and compression set values compared with the soap/sulfur-cured compounds. The ZMTI synergist interfered with both cure systems as both compounds (Compounds L and P) containing the ZMTI had high elongation, low tensile strength, and poor compression set values. Although not shown, the use of 2-mercaptotoluimidazole imparts a green corrosion of the stainless steel mold.

TABLE 14.4
HNBR Compounds Extracted in IRM 901 Oil Before Air Oven Aging

	A	B	C	D	E	F	G	H	I
Naugard 445	2.0	0.5	0.5	1.5	1.0	0.0	1.0	1.5	1.0
Vanox ZMTI	1.0	0.5	1.5	0.5	0.0	1.0	2.0	1.5	1.0
Aged 3 days at 150°C in IRM 901 Oil, then 3 days at 177°C in air									
Hardness, Shore A	78	79	79	77	79	79	79	78	78
100% modulus, MPa	15.0	18.8	16.9	14.7	18.2	—	17.2	16.5	17.0
psi	2170	2730	2450	2130	2640	—	2490	2390	2460
Tensile strength, MPa	23.7	21.2	21.9	21.7	19.2	19.0	23.1	22.7	22.8
psi	3440	3080	3170	3140	2780	2760	3350	3290	3310
Elongation, %	140	110	120	130	100	90	130	140	130
Tensile, % change	−14	−24	−21	−9	−21	−30	−16	−14	−10
Elongation, % change	−44	−48	−45	−32	−44	−53	−43	−36	−32
100% modulus, % change	82	76	56	55	73	—	77	65	52
Hardness, points change	8	7	8	6	8	8	8	7	8
Aged 3 days at 150°C in IRM 901 Oil, then 7 days at 177°C in air									
Hardness, Shore A	83	85	84	84	85	87	83	84	87
Tensile strength, MPa	15.4	11.4	10.9	14.9	10.8	9.9	17.2	15.0	10.8
psi	2230	1660	1580	2160	1560	1430	2500	2180	1560
Elongation, %	50	20	30	40	30	10	60	40	10
Tensile, % change	−44	−59	−60	−38	−56	−64	−37	−43	−58
Elongation, % change	−80	−90	−86	−79	−83	−95	−74	−82	−95
Hardness, points change	13	13	13	13	14	16	12	13	17
Aged 3 days at 150°C in IRM 901 Oil, then 21 days at 177°C in air									
Hardness, Shore A	92	98	91	92	95	93	97	95	100
Tensile strength, MPa	7.7	6.5	7.2	10.7	6.4	7.5	9.8	6.4	9.7
psi	1110	940	1040	1555	930	1085	1415	935	1405
Elongation, %	0	0	0	0	0	0	0	5	0
Tensile, % change	−72	−77	−74	−55	−74	−72	−64	−76	−62
Elongation, % change	−100	−100	−100	−100	−100	−100	−100	−98	−100
Hardness, points change	22	26	20	21	24	22	26	24	30

Source: From Ohm, R.F., *Improving the Heat Resistance of HNBR, ACM, EAM by the Use of Antioxidants*, Crompton, presented at Connecticut Rubber Group Meeting, April 6, 2004. With permission.

Table 14.6 illustrates the ACM compounds that were heat aged for 3, 7, and 21 days at 177°C in an air oven. All compounds increased in hardness and modulus, and decreased in tensile strength and elongation. All compounds containing antioxidants had better aged elongation than the unprotected control. While some of the compounds

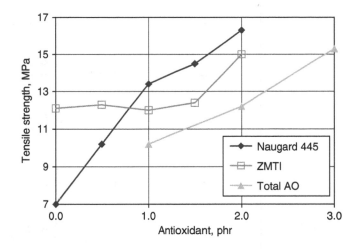

FIGURE 14.9 Cell oven aging of HNBR compounds 21 days at 177°C. (From Ohm, R.F., *Improving the Heat Resistance of HNBR, ACM, EAM by the Use of Antioxidants*, Crompton, presented at Connecticut Rubber Group Meeting, April 6, 2004. With permission.)

containing synergists gave better elongation after heat aging than the compound containing 2 phr Naugard 445, alone (Compound K), none was significantly better than the compound containing 4 phr Naugard 445, alone (Compound N), after heat aging 7 days at 177°C (see Figures 14.10 and 14.11). Therefore, there appears to be no benefit from the combination of an antioxidant synergist in this polymer.

After immersion, only, of all samples in IRM 901 Oil all of the ACM compounds showed a decrease in hardness, modulus, and tensile strength values and an increase in elongation, see Table 14.7. Subsequent air oven aging at 177°C led to an increase in hardness and a decrease in tensile strength and elongation values. The compounds containing 4 phr Naugard 445 alone showed values after heat aging that were equivalent to any of the synergistic combinations containing 2 phr Naugard 445.

Since volatile antioxidants can transfer between samples when communally exposed in an air oven, the ACM compounds were heat aged for 21 days in individual test tubes at 177°C, including both those that were and were not immersed in IRM 901 Oil. The data in Table 14.8 illustrate that all samples were severely degraded after these exposures, but not quite as badly as observed after air oven aging. The lower turnover of oxygen in the cell oven is likely responsible for this marginal improvement in cell oven aging [25].

14.7.3 ETHYLENE-ACRYLIC ELASTOMER

All of the evaluated EAM compounds are illustrated in Table 14.9. The initial physical properties of the uncured and cured compounds are also listed in this table. All of the listed compounds appear to possess generally similar properties.

TABLE 14.5

ACM Formulations and Properties

	J	K	L	M	N	O	P	Q	R	S	T
Hytemp AR72LF[b]	100.00	For all									
N550 Carbon black	65.00										
Stearic acid	2.00										
Processing aid—2	2.00										
TCY	1.00	1.00	1.00	1.00	1.00	—	—	—	—		
ZDBC	1.50	1.50	1.50	1.50	1.50	—	—	—	—		
CTP Retarder	0.20	0.20	0.20	0.20	0.20	—	—	—	—		
Sodium stearate	—	—	—	—	—	3.00	3.00	3.00	3.00	3.00	3.00
Potassium stearate	—	—	—	—	—	0.50	0.50	0.50	0.50	0.50	0.50
Sulfur	—	—	—	—	—	0.30	0.30	0.30	0.30	0.30	0.30
Naugard 445	—	2.00	2.00	2.00	4.00	2.00	2.00	2.00	2.00	2.00	4.00
ZMTI	—	—	2.00	—	—	—	2.00	—	—	—	—
Thioester	—	—	—	2.00	—	—	—	2.00	—	—	—
Phosphite	—	—	—	—	—	—	—	—	2.00	—	—
Sodium ethylhexyl sulfate	—	—	—	—	—	—	—	—	—	2.00	—

Rheometer at 177°C

ML, dN · m	5.7	5.6	5.7	5.1	5.1	5.9	8.1	5.3	5.1	5.4	5.5
in-lb	5.0	4.9	5.0	4.5	4.5	5.2	7.2	4.7	4.5	4.8	4.8
MH, dN · m	43.0	42.6	22.8	40.4	41.0	26.1	13.5	16.9	24.7	26.7	27.3
in-lb	38.1	37.7	20.2	35.8	36.3	23.1	12.0	15.0	21.9	23.6	24.2
ts_2, minutes	2.2	2.3	3.5	2.3	2.5	2.4	5.4	1.9	2.3	2.8	2.5
tc_{90}, minutes	9.7	10.2	14.2	9.9	11.3	17.3	19.1	15.0	17.5	18.3	16.9

Physical properties cured 30 min at 177°C post-cured 4 h at 175°C

Hardness, Shore A	72	67	68	70	71	65	62	65	62	63	62
100% modulus, MPa	9.3	8.7	3.9	8.7	9.0	6.0	1.6	5.1	4.6	5.2	5.4
psi	1350	1260	560	1260	1310	870	230	740	660	750	780
Tensile strength, MPa	11.7	9.4	9.2	11.0	11.3	10.5	6.4	10.7	9.0	9.8	10.1
psi	1690	1360	1330	1590	1640	1520	930	1550	1300	1420	1460
Elongation, %	130	110	280	150	150	190	500	220	260	200	210

Aged 70 h at 175°C

Compression set, %	49	44	60	44	41	75	100	69	73	76	73

Source: From Ohm, R.F., *Improving the Heat Resistance of HNBR, ACM, EAM by the Use of Antioxidants*, Crompton, presented at Connecticut Rubber Group Meeting, April 6, 2004. With permission.

TABLE 14.6
Air Oven Aging of ACM Formulations

	J	K	L	M	N	O	P	Q	R	S	T
TCY cure	2.70	2.70	2.70	2.70	2.70	—	—	—	—	—	—
Soap/sulfur cure	—	—	—	—	—	2.80	2.80	2.80	2.80	2.80	2.80
Naugard 445	—	2.00	2.00	2.00	4.00	2.00	2.00	2.00	2.00	2.00	4.00
ZMTI	—	—	2.00	—	—	—	2.00	—	—	—	—
Thioester	—	—	—	2.00	—	—	—	2.00	—	—	—
Phosphite	—	—	—	—	—	—	—	—	2.00	—	—
Sodium ethylhexyl sulfate	—	—	—	—	—	—	—	—	—	2.00	—
Aged 3 days at 177°C											
Hardness, Shore A	77	77	70	78	74	68	71	70	63	71	70
100% modulus, MPa	10.8	7.8	4.1	8.9	7.7	4.1	2.5	4.1	3.6	3.2	3.6
psi	1560	1130	600	1290	1110	590	360	600	520	470	520
Tensile strength, MPa	11.4	11.1	8.2	11.0	10.7	9.1	5.4	9.4	9.1	8.3	8.1
psi	1650	1610	1190	1590	1550	1320	780	1370	1320	1200	1180
Elongation, %	110	150	250	130	150	220	440	230	250	250	240
Tensile, % change	−2.4	18.4	−10.5	0.0	−5.5	−13.2	−16.1	−11.6	1.5	−15.5	−19.2
Elongation, % change	−15.4	36.4	−10.7	−13.3	0.0	15.8	−12.0	4.5	−3.8	25.0	14.3
Hardness, points change	5.0	10.0	2.0	8.0	3.0	3.0	9.0	5.0	1.0	8.0	8.0

Aged 7 days at 177°C

Hardness, Shore A	90	84	85	86	80	74	80	80	74	73	78
100% modulus, MPa	—	—	6.3	—	8.2	4.4	5.3	4.8	3.4	4.3	3.8
psi	—	—	920	—	1190	640	770	700	500	630	550
Tensile strength, MPa	6.1	8.0	6.6	7.8	8.0	6.5	6.3	7.1	6.4	6.6	5.9
psi	890	1160	960	1130	1160	940	920	1030	930	960	860
Elongation, %	10	90	110	70	100	160	150	150	190	160	180
Tensile, % change	−47.3	−14.7	−27.8	−28.9	−29.3	−38.2	−1.1	−33.5	−28.5	−32.4	−41.1
Elongation, % change	−92.3	−18.2	−60.7	−53.3	−33.3	−15.8	−70.0	−31.8	−26.9	−20.0	−14.3
Hardness, points change	18.0	17.0	17.0	16.0	9.0	9.0	18.0	15.0	12.0	10.0	16.0

Aged 21 days at 177°C

Hardness, Shore A	93	90	90	88	94	88	88	87	82	84	84
Tensile strength, MPa	0.6	7.5	7.3	1.9	7.7	8.5	8.5	9.0	8.5	8.6	7.9
psi	80	1090	1060	280	1120	1230	1230	1310	1240	1250	1150
Elongation, %	20	30	20	30	30	40	20	30	70	40	70
Tensile, % change	−95.3	−19.9	−20.3	−82.4	−31.7	−19.1	32.3	−15.5	−4.6	−12.0	−21.2
Elongation, % change	−84.6	−72.7	−92.9	−80.0	−80.0	−78.9	−96.0	−86.4	−73.1	−80.0	−66.7
Hardness, points change	21.0	23.0	22.0	18.0	23.0	23.0	26.0	22.0	20.0	21.0	22.0

Source: From Ohm, R.F., *Improving the Heat Resistance of HNBR, ACM, EAM by the Use of Antioxidants*, Crompton, presented at Connecticut Rubber Group Meeting, April 6, 2004. With permission.

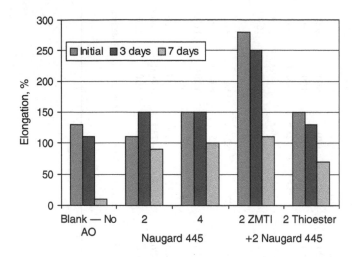

FIGURE 14.10 Air oven aging of TCY-cured ACM compounds at 177°C. (From Ohm, R.F., *Improving the Heat Resistance of HNBR, ACM, EAM by the Use of Antioxidants*, Crompton, presented at Connecticut Rubber Group Meeting, April 6, 2004. With permission.)

As illustrated in Table 14.10 all of the EAM compounds exhibit an increase in hardness and modulus values and a concurrent decrease in tensile strength and elongation values after air oven aging. The compounds containing Naugard 445

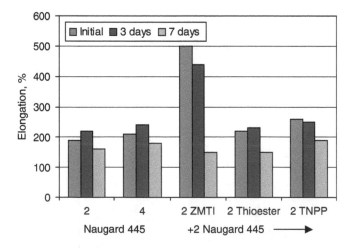

FIGURE 14.11 Air oven aging of soap/sulfur-cured ACM compounds at 177°C. (From Ohm, R.F., *Improving the Heat Resistance of HNBR, ACM, EAM by the Use of Antioxidants*, Crompton, presented at Connecticut Rubber Group Meeting, April 6, 2004. With permission.)

TABLE 14.7

ACM Compounds Extracted in IRM 901 Oil Before Air Oven Aging

	J	K	L	M	N	O	P	Q	R	S	T
TCY cure	2.70	2.70	2.70	2.70	2.70	—	—	—	—	—	—
Soap/sulfur cure	—	—	—	—	—	2.80	2.80	2.80	2.80	2.80	2.80
Naugard 445	—	2.00	2.00	2.00	4.00	2.00	2.00	2.00	2.00	2.00	4.00
ZMTI	—	—	2.00	—	—	—	2.00	—	—	—	—
Thioester	—	—	—	2.00	—	—	—	2.00	—	—	—
Phosphite	—	—	—	—	—	—	—	—	2.00	—	—
Sodium ethylhexyl sulfate	—	—	—	—	—	—	—	—	—	2.00	—
Aged 93 h at 150°C in ASTM 901 Oil											
Hardness, Shore A	69	71	68	69	70	65	65	66	65	68	65
100% modulus, MPa	7.9	7.7	3.4	7.0	8.1	5.1	1.6	4.5	5.2	4.4	5.2
psi	1150	1115	495	1020	1175	740	225	650	760	640	755
Tensile strength, MPa	10.3	11.6	8.7	10.1	11.3	10.4	5.0	10.7	10.3	10.3	10.3
psi	1495	1680	1265	1465	1640	1510	725	1545	1495	1490	1495
Elongation, %	130	175	295	150	170	230	590	220	225	220	215
Tensile, % change	-11.5	23.5	-4.9	-7.9	0.0	-0.7	-22.0	-0.3	15.0	4.9	2.4
Elongation, % change	0.0	59.1	5.4	0.0	13.3	21.1	18.0	0.0	-13.5	10.0	2.4
Hardness, points change	-3.0	4.0	0.0	-1.0	-1.0	0.0	3.0	1.0	3.0	5.0	3.0

(continued)

TABLE 14.7 (Continued)
ACM Compounds Extracted in IRM 901 Oil Before Air Oven Aging

	J	K	L	M	N	O	P	Q	R	S	T
Aged 3 days at 150°C in IRM 901 Oil, then aged 3 days at 177°C											
Hardness, Shore A	83	81	81	83	80	72	72	77	76	73	76
Tensile strength, MPa	3.9	5.7	9.2	7.1	12.1	9.7	8.9	11.0	10.2	10.3	11.0
psi	560	830	1340	1030	1760	1410	1290	1600	1480	1500	1600
Elongation, %	30	40	160	50	100	130	250	130	160	160	150
Tensile, % change	−66.9	−39.0	0.8	−35.2	7.3	−7.2	38.7	3.2	13.8	5.6	9.6
Elongation, % change	−76.9	−63.6	−42.9	−66.7	−33.3	−31.6	−50.0	−40.9	−38.5	−20.0	−28.6
Hardness, points change	11.0	14.0	13.0	13.0	9.0	7.0	10.0	12.0	14.0	10.0	14.0
Aged 3 days at 150°C in IRM 901 Oil, then aged 7 days at 177°C											
Hardness, Shore A	81	86	87	89	84	80	83	81	79	81	78
Tensile strength, MPa	5.0	4.1	8.1	3.8	4.1	9.7	9.6	9.4	9.4	8.1	9.0
psi	720	590	1180	550	590	1400	1390	1370	1370	1170	1310
Elongation, %	10	20	50	20	10	90	110	90	90	90	120
Tensile, % change	−57.4	−56.6	−11.3	−65.4	−64.0	−7.9	49.5	−11.6	5.4	−17.6	−10.3
Elongation, % change	−92.3	−81.8	−82.1	−86.7	−93.3	−52.6	−78.0	−59.1	−65.4	−55.0	−42.9
Hardness, points change	9.0	19.0	19.0	19.0	13.0	15.0	21.0	16.0	17.0	18.0	16.0

Source: From Ohm, R.F., *Improving the Heat Resistance of HNBR, ACM, EAM by the Use of Antioxidants*, Crompton, presented at Connecticut Rubber Group Meeting, April 6, 2004. With permission.

TABLE 14.8

ACM Compounds Extracted in IRM 901 Oil Before Air Oven Aging

	J	K	L	M	N	O	P	Q	R	S	T
TCY cure	2.70	2.70	2.70	2.70	2.70	—	—	—	—	—	—
Soap/sulfur cure	—	—	—	—	—	2.80	2.80	2.80	2.80	2.80	2.80
Naugard 445	—	2.00	2.00	2.00	4.00	2.00	2.00	2.00	2.00	2.00	4.00
ZMTI	—	—	—	—	—	—	2.00	—	—	—	—
Thioester	—	—	—	2.00	—	—	—	2.00	—	—	—
Phosphite	—	—	—	—	—	—	—	—	2.00	—	—
Sodium ethylhexyl sulfate	—	—	—	—	—	—	—	—	—	2.00	—
Aged 3 days at 150°C in IRM 901 Oil, then aged 21 days at 177°C											
Tensile strength, MPa		9.1	10.5	10.3	11.0	7.4	10.1	7.8	6.6	4.3	6.1
psi		1320	1530	1490	1600	1080	1460	1130	950	620	880
Elongation, %		10	10	10	10	10	—	—	—	—	—
Tensile, % change		−2.9	15.0	−6.3	−2.4	−28.9	57.0	−27.1	−26.9	−56.3	−39.7
Elongation, % change		−90.9	−96.4	−93.3	−93.3	−94.7	—	—	—	—	—
Hardness, points change		−67.0	−68.0	−70.0	−71.0	−65.0	—	—	—	—	—
Aged 21 days at 177°C in test tubes											
Hardness, Shore A	91	87	89	90	87	81	82	86	75	79	79
Tensile strength, MPa	8.5	7.8	7.9	8.3	8.8	4.8	3.8	6.9	3.1	4.2	2.5
psi	1230	1135	1150	1210	1275	700	555	995	445	615	365

(continued)

TABLE 14.8 (Continued)
ACM Compounds Extracted in IRM 901 Oil Before Air Oven Aging

	J	K	L	M	N	O	P	Q	R	S	T
Elongation, %	7	35	60	30	50	60	15	65	50	50	25
Tensile, % change	−27.2	−16.5	−13.5	−23.9	−22.3	−53.9	−40.3	−35.8	−65.8	−56.7	−75.0
Elongation, % change	−94.6	−68.2	−78.6	−80.0	−66.7	−68.4	−97.0	−70.5	−80.8	−75.0	−88.1
Hardness, points change	19.0	20.0	21.0	20.0	16.0	16.0	20.0	21.0	13.0	16.0	17.0
Aged 3 days at 150°C in IRM 901 Oil, then aged 21 days at 177°C in test tubes											
Hardness, Shore A	95	98	95	95	95	90	90	90	94	91	92
Tensile strength, MPa	—	—	12.0	12.0	13.8	9.0	9.4	7.5	7.0	5.3	6.1
psi	—	—	1740	1740	2000	1300	1360	1090	1020	770	890
Elongation, %	—	—	10	10	10	10	10	10	10	10	10
Tensile, % change	—	—	30.8	9.4	22.0	−14.5	46.2	−29.7	−21.5	−45.8	−39.0
Elongation, % change	—	—	−96.4	−93.3	−93.3	−94.7	−98.0	−95.5	−96.2	−95.0	−95.2
Hardness, points change	23.0	31.0	27.0	25.0	24.0	25.0	28.0	25.0	32.0	28.0	30.0

Source: From Ohm, R.F., *Improving the Heat Resistance of HNBR, ACM, EAM by the Use of Antioxidants,* Crompton, presented at Connecticut Rubber Group Meeting, April 6, 2004. With permission.

TABLE 14.9
EAM Formulations and Properties

	U	V	W	X	Y	Z	AA	AB	AC
Vamac G[c]	100.0								
Processing aid—3	0.5								
Stearic acid	1.5	For all							
N-762 Black	60.0								
HMDC	1.5								
DOTG	4.0								
Naugard 445	—	2.0	—	—	—	—	—	—	4.0
Naugard PANA	—	—	2.0	—	—	—	—	—	—
Aranox	—	—	—	2.0	—	—	—	—	—
Naugard 10	—	—	—	—	2.0	—	—	—	—
Durazone 37	—	—	—	—	—	2.0	—	—	—
Naugalube 635	—	—	—	—	—	—	2.0	—	—
Naugard 495	—	—	—	—	—	—	—	2.0	—
Rheometer at 177°C									
ML, dN · m	1.7	1.6	1.6	1.6	1.6	1.4	1.2	1.5	1.5
in-lb	1.5	1.4	1.4	1.4	1.4	1.2	1.1	1.3	1.3
MH, dN · m	37.7	37.1	37.6	37.2	35.9	36.7	35.9	37.1	36.2
in-lb	33.4	32.8	33.3	32.9	31.8	32.5	31.8	32.8	32.1
ts_2, minutes	1.1	1.1	1.1	1.2	1.2	1.2	1.2	1.2	1.3
tc_{90}, minutes	8.1	6.8	6.7	7.0	6.8	6.9	7.0	7.1	7.7

(continued)

TABLE 14.9 (Continued)
EAM Formulations and Properties

	U	V	W	X	Y	Z	AA	AB	AC
Physical properties cured 10 min at 177°C									
Post-cured 4 h at 175°C									
Hardness, Shore A	67	65	67	66	67	68	63	62	66
200% modulus, MPa	14.8	13.7	12.6	13.2	13.5	15.2	13.9	13.3	12.1
psi	2140	1990	1830	1920	1960	2210	2020	1930	1760
Tensile strength, MPa	18.1	17.9	17.5	18.2	18.1	19.0	18.5	17.4	17.2
psi	2630	2590	2540	2640	2620	2760	2690	2520	2500
Elongation, %	270	300	330	300	300	270	340	320	320
Aged 70 h at 175°C									
Compression set, %	11	12	14	14	11	11	11	12	19

Source: From Ohm, R.F., *Improving the Heat Resistance of HNBR, ACM, EAM by the Use of Antioxidants*, Crompton, presented at Connecticut Rubber Group Meeting, April 6, 2004. With permission.

TABLE 14.10
Air Oven Aging of EAM Formulations

	U	V	W	X	Y	Z	AA	AB	AC
Naugard 445	—	2.0	—	—	—	—	—	—	4.0
Naugard PANA	—	—	2.0	—	—	—	—	—	—
Aranox	—	—	—	2.0	—	—	—	—	—
Naugard 10	—	—	—	—	2.0	—	—	—	—
Durazone 37	—	—	—	—	—	2.0	—	—	—
Naugard 635	—	—	—	—	—	—	2.0	—	—
Naugard 495	—	—	—	—	—	—	—	2.0	—
Aged 3 days at 177°C									
Hardness, Shore A	69	65	70	72	70	75	65	64	—
200% modulus, MPa	17.2	14.0	15.3	15.5	18.0	—	17.6	16.9	—
psi	2490	2025	2225	2255	2605	—	2550	2450	—
Tensile strength, MPa	13.8	16.5	15.1	17.2	19.4	16.4	19.1	18.9	—
psi	2000	2390	2195	2500	2810	2380	2765	2740	—
Elongation, %	165	245	200	225	220	160	220	225	—
Tensile, % change	−24	−8	−14	−5	7	−14	3	9	—
Elongation, % change	−39	−18	−39	−25	−27	−41	−35	−30	—
Hardness, points change	2	0	3	6	3	7	2	2	—
Aged 7 days at 177°C									
Hardness, Shore A	71	71	69	75	73	75	73	71	69
100% modulus, MPa	7.9	5.7	7.1	8.0	8.6	10.0	6.3	7.7	12.8
psi	1150	830	1030	1160	1250	1450	920	1110	1860
Tensile strength, MPa	10.8	17.3	8.3	16.5	15.9	12.1	18.4	15.0	17.1
psi	1560	2510	1210	2390	2310	1750	2670	2180	2480
Elongation, %	120	260	120	180	160	110	210	160	290
Tensile, % change	−41	−3	−52	−9	−12	−37	−1	−13	−1
Elongation, % change	−56	−13	−64	−40	−47	−59	−38	−50	−9
Hardness, points change	4	6	2	9	6	7	10	9	3
Aged 21 days at 177°C in air oven									
Hardness, Shore A	83	73	78	79	79	85	71	80	73
Tensile strength, MPa	6.8	7.2	6.6	8.4	9.6	6.4	8.3	7.9	13.2
psi	990	1050	960	1220	1390	930	1200	1150	1920
Elongation, %	20	80	30	60	50	30	60	40	160
Tensile, % change	−62	−59	−62	−54	−47	−66	−55	−54	−24
Elongation, % change	−93	−73	−91	−80	−83	−89	−82	−88	−50
Hardness, points change	16	8	11	13	12	17	8	18	7

Source: From Ohm, R.F., *Improving the Heat Resistance of HNBR, ACM, EAM by the Use of Antioxidants*, Crompton, presented at Connecticut Rubber Group Meeting, April 6, 2004. With permission.

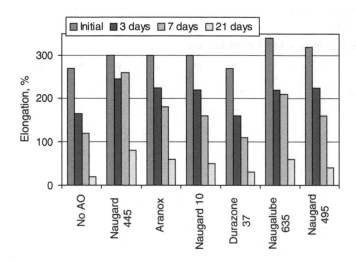

FIGURE 14.12 Air oven aging of EAM compounds at 177°C. (From Ohm, R.F., *Improving the Heat Resistance of HNBR, ACM, EAM by the Use of Antioxidants*, Crompton, presented at Connecticut Rubber Group Meeting, April 6, 2004. With permission.)

and Naugalube 635 (styrenated diphenylamine) exhibit the least change in physical properties after being subjected to the various heat aging conditions. The benefits of these antioxidants are particularly noticeable after heat aging 3 and 7 days at 177°C and noting the change in elongation. These data are illustrated graphically in Figure 14.12. The compound containing Durazone 37 (Compound Z) shows a greater increase in hardness and modulus during this aging regime than the control sample (Sample U) containing no antioxidant. It may be possible that this trifunctional material promotes cross-linking at elevated temperature.

Table 14.11 shows testing of EAM compounds after immersion in IRM 901 Oil. After oil immersion only, all compounds showed a decrease in hardness, modulus, and tensile strength values and an increase in elongation. Subsequent air oven aging at 177°C leads to an increase in hardness and a decrease in tensile and elongation values. The compound containing Naugard 445 (Compound V) showed the least change in hardness and elongation values after heat aging 3 days at 177°C. The compound containing Aranox (Compound X) showed the best retention of tensile strength and elongation after heat aging 7 and 21 days at 177°C. On the down side, the compound containing Durazone 37 (Compound Z) again exhibits higher hardness, lower tensile strength, and elongation values after heat aging 3 days at 177°C. These detrimental results are still evident after heat aging 7 days.

EAM compounds, after cell oven aging, are shown in Table 14.12. All samples were severely degraded after these exposures, but not quite as badly as observed after air oven aging. The lower turnover of oxygen in the cell oven is likely responsible for this marginal improvement in cell oven aging [26].

TABLE 14.11

EAM Compounds Extracted in IRM 901 Oil Before Air Oven Aging

	U	V	W	X	Y	Z	AA	AB
Naugard 445	—	2.0	—	—	—	—	—	—
Naugard PANA	—	—	2.0	—	—	—	—	—
Aranox	—	—	—	2.0	—	—	—	—
Naugard 10	—	—	—	—	2.0	—	—	—
Durazone 37	—	—	—	—	—	2.0	—	—
Naugard 635	—	—	—	—	—	—	2.0	—
Naugard 495	—	—	—	—	—	—	—	2.0
Aged 3 days at 150°C in RIM 901 Oil								
Hardness, Shore A	61	59	56	61	62	65	58	61
200% modulus, MPa	10.8	10.5	10.2	10.1	9.3	12.4	10.5	9.3
psi	1570	1520	1480	1460	1350	1800	1530	1350
Tensile strength, MPa	16.4	16.2	15.9	15.8	16.4	16.7	15.7	14.9
psi	2380	2350	2301	2290	2380	2420	2280	2160
Elongation, %	310	320	340	320	340	280	320	330
Tensile, % change	−10	−9	−9	−13	−9	−12	−15	−14
Elongation, % change	15	7	3	7	13	4	−6	3
Hardness, points change	−6	−6	−11	−5	−5	−3	−5	−1
Aged 3 days at 150°C in RIM 901 Oil, then 3 days at 177°C in air oven								
Hardness, Shore A	63	64	60	64	66	71	59	62
200% modulus, MPa	16.1	14.9	—	—	15.5	—	16.3	—
psi	2340	2160	—	—	2250	—	2370	—
Tensile strength, MPa	17.0	18.5	16.5	16.8	16.5	12.1	18.9	16.8
psi	2470	2680	2400	2440	2400	1750	2740	2430

(continued)

TABLE 14.11 (Continued)
EAM Compounds Extracted in IRM 901 Oil Before Air Oven Aging

	U	V	W	X	Y	Z	AA	AB
Elongation, %	210	260	180	190	220	130	220	190
Tensile, % change	−6	3	−6	−8	−8	−37	2	−4
Elongation, % change	−22	−13	−45	−37	−27	−52	−35	−41
Hardness, points change	−4	−1	−7	−2	−1	3	−4	0
Aged 3 days at 150°C in RIM 901 Oil, then 7 days at 177°C in air oven								
Hardness, Shore A	70	69	69	73	71	73	67	70
Tensile strength, MPa	9.9	11.0	10.2	14.1	11.1	9.4	11.4	14.3
psi	1430	1600	1480	2050	1610	1360	1650	2080
Elongation, %	110	160	130	180	130	110	130	140
Tensile, % change	−46	−38	−42	−22	−39	−51	−39	−17
Elongation, % change	−59	−47	−61	−40	−57	−59	−62	−56
Hardness, points change	3	4	2	7	4	5	4	8
Aged 3 days at 150°C in RIM 901 Oil, then 21 days at 177°C in air oven								
Hardness, Shore A	84	81	82	82	78	80	80	82
Tensile strength, MPa	5.6	5.7	6.2	8.4	7.4	6.9	6.6	7.0
psi	810	820	900	1220	1070	1000	950	1020
Elongation, %	20	30	30	60	40	30	30	20
Tensile, % change	−69	−68	−65	−54	−59	−64	−65	−60
Elongation, % change	−93	−90	−91	−80	−87	−89	−91	−94
Hardness, points change	17	16	15	16	11	12	17	20

Source: From Ohm, R.F., *Improving the Heat Resistance of HNBR, ACM, EAM by the Use of Antioxidants*, Crompton, presented at Connecticut Rubber Group Meeting, April 6, 2004. With permission.

TABLE 14.12
Cell Oven Aging of EAM Compounds

	U	V	W	X	Y	Z	AA	AB
Naugard 445	—	2.0	—	—	—	—	—	—
Naugard PANA	—	—	2.0	—	—	—	—	—
Aranox	—	—	—	2.0	—	—	—	—
Naugard 10	—	—	—	—	2.0	—	—	—
Durazone 37	—	—	—	—	—	2.0	—	—
Naugard 635	—	—	—	—	—	—	2.0	—
Naugard 495	—	—	—	—	—	—	—	2.0
Aged 21 days at 177°C in cell oven								
Hardness, Shore A	80	69	75	85	79	76	78	79
Tensile strength, MPa	5.4	11.0	9.3	9.9	9.2	7.7	10.1	11.9
psi	790	1590	1350	1440	1330	1110	1470	1720
Elongation, %	40	150	70	40	70	50	100	80
Tensile, % change	−70	−39	−47	−45	−49	−60	−45	−32
Elongation, % change	−85	−50	−79	−87	−77	−81	−71	−75
Hardness, points change	13	4	8	19	12	8	15	17
Aged 3 days at 150°C in RIM 901 Oil, then 21 days at 177°C in cell oven								
Hardness, Shore A	62	68	75	77	75	76	66	76
Tensile strength, MPa	5.7	5.5	6.0	6.7	5.5	6.1	5.7	6.2
psi	820	800	870	970	800	890	830	900
Elongation, %	30	60	50	30	40	40	50	40
Tensile, % change	−69	−69	−66	−63	−69	−68	−69	−64
Elongation, % change	−89	−80	−85	−90	−87	−85	−85	−88
Hardness, points change	−5	3	8	11	8	8	3	14
Aged 21 days at 177°C in air oven								
Hardness, Shore A	83	73	78	79	79	85	71	80
Tensile strength, MPa	6.8	7.2	6.6	8.4	9.6	6.4	8.3	7.9
psi	990	1050	960	1220	1390	930	1200	1150
Elongation, %	20	80	30	60	50	30	60	40
Tensile, % change	−62	−59	−62	−54	−47	−66	−55	−54
Elongation, % change	−93	−73	−91	−80	−83	−89	−82	−88
Hardness, points change	16	8	11	13	12	17	8	18
Aged 3 days at 150°C in RIM 901 Oil, then 21 days at 177°C in air oven								
Hardness, Shore A	84	81	82	82	78	80	80	82
Tensile strength, MPa	5.6	5.7	6.2	8.4	7.4	6.9	6.6	7.0
psi	810	820	900	1220	1070	1000	950	1020
Elongation, %	20	30	30	60	40	30	30	20
Tensile, % change	−69	−68	−65	−54	−59	−64	−65	−60
Elongation, % change	−93	−90	−91	−80	−87	−89	−91	−94
Hardness, points change	17	16	15	16	11	12	17	20

Source: From Ohm, R.F., *Improving the Heat Resistance of HNBR, ACM, EAM by the Use of Antioxidants*, Crompton, presented at Connecticut Rubber Group Meeting, April 6, 2004. With permission.

TABLE 14.13
Materials Evaluated or Discussed

Abbreviations	Composition	Trade Name
αMSDPA	α-Methylstyrenated diphenylamine	Naugard[a] 445
ACM	Polyacrylic elastomer	Hytemp[b] AR72LF
ADPA	Acetone–diphenylamine reaction product	Aminox[a]
Bis peroxide	α,α′-Bis(t-butylperoxy diisopropyl) benzene	VulCup[c] 40-KE
DTP	N-(Cyclohexylthio) Phthalimide	Santogard[d] PVI
DCP 40-KE	Dicumyl peroxide	DiCup[c]
DPPD	N,N′-Diphenyl-p-phenylenediamine	Naugard J
DTDM	4,4′-Dithio dimorpholine	Naugex[a] SD-1
EAM	Ethylene-acrylic elastomer	Vamac[e] G
HMDC	Hexamethylene diamine carbonate	Diak[f] 1
HNBR	Hydrogenated acrylonitrile-butadiene copolymer	Zetpol[b] 2010
MBM	m-Phenylene-bis(maleimide)	HVA-2[f]
MTI	2-Mercaptotoluimidazole	Vanox[g] MTI
N-10	Tetra(3,5-di-t-butyl-4-hydroxy hydrocinnamate) methane	Naugard 10
N-495	AO blend of radical trap and hydroperoxide decomposer	Naugard 495
ODPA	Octylated diphenylamine	Octamine[a]
PANA	N-Phenyl-α-naphthylamine	Naugard PANA
	Processing aid—1	Struktol[h] WB-212
	Processing aid—2	Struktol[h] WB-222
	Processing aid—3	Armeen[i] 18D
SDPA	Styrenated diphenylamine	Naugalube[a] 635
	Sodium ethylhexyl sulfate (40% in water)	Niaproof[j] 08
TAPTD	2,4,6-Tris-(N-1,4-dimethylpentyl-p-phenylenediamino)-1,3,5-triazine	Durazone[a] 37
TCY	1,3,5-Trithiocyanuric acid	Zisnet[b] F-PT
Thioester	1,11-(3,6,9-Trioxaundecyl)-bis-3-(dodecylthio) propionate	Wingstay[k] SN-1
TMQ	Polymerized 2,2,4-trimethyl-1,2-dihydroquinoline	Naugard Q
TNPP	Tris(mixed mono- and di-nonylphenyl) phosphite	Polygard[a]
TPPD	N-Phenyl-N′-(p-toluenesulfonyl)-p-phenylenediamine	Aranox[a]
ZMTI	Zinc 2-mercaptotoluimidazole	Vanox ZMTI
ZDBC	Zinc di-n-butyldithiocarbamate	Butazate[a]

[a] Registered trademark of Chemtura Corporation, Middlebury, CT.
[b] Registered trademark of Zeon Chemicals L.P., Louisville, KY.
[c] Registered trademark of Geo Specialty Chemicals, Wilmington, DE.
[d] Registered trademark of Flexsys America L.P., Akron, OH.
[e] Registered trademark of E. I. DuPont de Nemours, Wilmington, DE.
[f] Registered trademark of DuPont Dow Elastomers, L.L.C., Wilmington, DE.
[g] Registered trademark of R. T. Vanderbilt Company, Inc., Norwalk, CT.
[h] Registered trademark of Struktol Co. of America, Stow, OH.
[i] Registered trademark of Akzo Corporation, Chicago, IL.
[j] Registered trademark of Niacet Corporation, Niagara Falls, NY.
[k] Registered trademark of Eliokem, Akron, OH.

Source: From Ohm, R.F., *Improving the Heat Resistance of HNBR, ACM, EAM by the Use of Antioxidants*, Crompton, presented at Connecticut Rubber Group Meeting, April 6, 2004. With permission.

14.8 SUMMARY

The heat aging at a high temperature of HNBR, ACM, and AEM was examined under various aging conditions. Several primary antioxidants were compared, optionally, with a secondary antioxidant or "synergist." At 177°C, the best performance, in terms of heat aging resistance, in all three "under-the-hood" polymers was achieved with one antioxidant, Naugard 445. No significant, synergistic antioxidant combinations were found. In the few cases where a secondary antioxidant provided some additional benefit, the same improvement could be obtained by an incremental increase in the dose of the Naugard 445. The beneficial performance of Naugard 445 was likely attributed to its low volatility and low solubility in hydrocarbons, both factors leading to maximum permanence of the antioxidant in service.

Table 14.13 lists the chemical composition, as well as the trade name and abbreviated name listed in the text, of the ingredients used in the compound formulations of the samples of HNBR, ACM, and EAM that were evaluated [27].

In conclusion, the selection of the proper antidegradants for elastomeric compounds should be based upon a careful assessment of the elastomers used, the normal service requirements expected from these compounds, and the environment in which they are to perform. Only then, one can choose the ideal combination of antioxidants and antiozonants. It must be emphasized that the degradation of elastomers proceeds via several mechanisms, and as a result, it is essential to choose and use a combination of antidegradants which will inhibit all of the described mechanisms of degradation that are expected to be encountered during a products service life [28].

ACKNOWLEDGMENTS

The author wishes to acknowledge Chemtura Corporation and its predecessor companies, such as Uniroyal Chemical Company, Inc., for their published, technical information, which contributed significantly to the writing of this chapter.

ENDNOTES

1. Barnhart, R.R., "*Antioxidants and Antiozonants*," Uniroyal Chemical Company, Inc., 1966.
2. Mazzeo, R.A., Boisseau, N.A., Hong, S.W., and Wheeler, E.L., "*Functions and Mechanisms of Antidegradants to Prevent Polymer Degradation*," Uniroyal Chemical Company, Inc., presented at the 145th Rubber Division, A.C.S. Meeting in Chicago, April 19–22, 1994.
3. Barnhart, R.R., Op. Cit.
4. Lattimer, R.P., et al., "*Mechanisms of Ozonation of N-(1-3-Dimethylbutyl)-,N'-Phenyl-p-Phenylenediamine*," The B.F. Goodrich Research and Development Center, presented at the Rubber Division, A.C.S. Meeting in Chicago, October 5–8, 1982.
5. Lattimer, R.P., et al., "*Mechanisms of Ozonation of N,N'-Di-(1-Methylheptyl)-p-Phenylenediamine*," The B.F. Goodrich Research and Development Center, presented at the Rubber Division, A.C.S. Meeting in Las Vegas, May 20–23, 1980.
6. Mazzeo, R.A., "*Non-staining, Non-discoloring Antidegradants*," Paper E, Uniroyal Chemical Company, Inc., presented at the 143rd Rubber Division, A.C.S. Meeting in Denver, May 18–29, 1993.

7. Cesare, Frank, C., *"Natsyn Polyisoprene Rubber/EPDM/Trilene Blends,"* Uniroyal Chemical Company, Inc., September 1, 1989.
8. Barnhart, R.R., Op. Cit.
9. Ohm, R.F., "Improving *the Heat Resistance of HNBR, ACM, EAM by the Use of Antioxidants,"* Crompton, presented at Connecticut Rubber Group Meeting, April 6, 2004.
10. Barnhart, R.R., Op. Cit.
11. Paolino, P.R., *"Antioxidants,"* Uniroyal Chemical Company, Inc., 1980.
12. Ohm, R.F., Op. Cit.
13. Paolino, P.R., Op. Cit.
14. Ohm, R.F., Op. Cit.
15. Paolino, P.R., Op. Cit.
16. Ohm, R.F., Op. Cit.
17. Paolino, P.R., Op. Cit.
18. Ohm, R.F., Op. Cit.
19. Paolino, P.R., Op. Cit.
20. Ohm, R.F., Op. Cit.
21. Mazzeo, R.A., *"Non-staining, Non-discoloring Antidegradants,"* Op. Cit.
22. Ohm, R.F., Op. Cit.
23. Ibid.
24. Ibid.
25. Ibid.
26. Ibid.
27. Ibid.
28. Mazzeo, R.A., et al., *"Functions and Mechanisms of Antidegradants to Prevent Polymer Degradation,"* Op. Cit.

BIBLIOGRAPHY

1. Barnhart, R.R., *"Antioxidants and Antiozonants,"* Uniroyal Chemical Company, Inc., 1966.
2. Cesare, Frank, C., *"Natsyn Polyisoprene Rubber/EPDM/Trilene Blends,"* Uniroyal Chemical Company, Inc., September 1, 1989.
3. Hong, S.W., *"Improved Tire Performance Through the Use of Antidegradants,"* presented at ITEC, Akron, OH, September 10–12, 1996.
4. Hong, S.W. and Lin, C.Y., *"Improved Flex Fatigue and Dynamic Ozone Crack Resistance through the Use of Antidegradants in Tire Compounds,"* presented at the 156th Rubber Division, A.C.S. Meeting, September 21–24, 1999.
5. Hunter, B.A., *"Chemical Protection Against Degradation of Hydrocarbon Polymers,"* Uniroyal Chemical Company, Inc., 1966.
6. Lattimer, R.P., et al., *"Mechanisms of Ozonation of N,N'-Di-(1-Methylheptyl)-p-Phenylenediamine,"* The B.F. Goodrich Research and Development Center, presented at the Rubber Division, A.C.S. Meeting in Las Vegas, May 20–23, 1980.
7. Lattimer, R.P., et al., *"Mechanisms of Ozonation of N-(1-3-Dimethylbutyl)-,N'-Phenyl-p-Phenylenediamine,"* The B.F. Goodrich Research and Development Center, presented at the Rubber Division, A.C.S. Meeting in Chicago, October 5–8, 1982.
8. Lattimer, R.P., et al., *"Mechanisms of Antiozonant Protection: Antiozonant – Rubber Reactions During Ozone Exposure,"* The B.F. Goodrich Research and Development Center, presented at the Rubber Division, A.C.S. Meeting in Indianapolis, May 8–11, 1984.

9. Mazzeo, R.A., *"Non-staining, Non-discoloring Antidegradants,"* Paper E, Uniroyal Chemical Company, Inc., presented at the 143rd Rubber Division, A.C.S. Meeting in Denver, May 18–29, 1993.
10. Mazzeo, R.A., Boisseau, N.A., Hong, S.W., and Wheeler, E.L., *"Functions and Mechanisms of Antidegradants to Prevent Polymer Degradation,"* Uniroyal Chemical Company, Inc., presented at the 145th Rubber Division, A.C.S. Meeting in Chicago, April 19–22, 1994.
11. Mazzeo, R.A., Boisseau, N.A., Hong, S.W., and Wheeler, E.L., *"Functions and Mechanisms of Antidegradants to Prevent Polymer Degradation,"* reprint, Tire Technology International 1994, p. 36, London, England.
12. Ohm, R.F., *"The Selection and Use of Antioxidants,"* Uniroyal Chemical, a Crompton Business, presented at the Fort Wayne Rubber Group Meeting, November 7, 2002.
13. Ohm, R.F., *"Improving the Heat Resistance of HNBR, ACM, EAM by the Use of Antioxidants,"* Crompton, presented at Connecticut Rubber Group Meeting, April 6, 2004.
14. Paolino, P.R., *"Antioxidants,"* Uniroyal Chemical Company, Inc., 1980.
15. Paolino, P.R., *"Antidegradants,"* Uniroyal Chemical Company, Inc., presented at the 26th Annual Akron Rubber Group Lecture Series, April 24, 1989.
16. Pospisil, J.P., *"Developments in Polymer Stabilization,"* Vol. 1, pp. 19–20, Scott, G., Editor, Applied Science Publications, London, 1979.
17. Pospisil, Jan and Klemchuk, Peter P., Editors, *"Oxidation Inhibition in Organic Materials,"* Vol. 1, CRC Press, Inc., Boca Raton, Florida, 1990.

15 Processing Aids for Specialty Elastomers

Jerry M. Sherritt

CONTENTS

Chemists sometimes find it convenient or necessary to use process aids as a temporary fix to an existing factory formulation or as a permanent addition to a new formula. This is especially true when compounding specialty elastomers. There are many of these polymers available, but often only in a limited number of grades on hand in the factory, so the rubber chemist must make do with those grades which are readily available.

Obviously, the compounder's first challenge is to meet the requirements of all applicable physical property specifications. Once this is accomplished, the chemist may find that the resultant formula will not process well in his factory. At this point, either the processing equipment must be altered to handle the new formula, which really is impractical, or the compound formula must be adjusted to meet both the factory requirements and the material specification requirements. The simplest solution for that type of problem is to make use of an additive that will overcome the compound's processing deficiencies while retaining the physical properties within specification limits. Sometimes this can be accomplished by adding 1.00 pphr (parts per hundred parts of rubber hydrocarbon) of a process aid or a level equal to 1% of the compound's total batch weight. If the processing problem is extreme, a level of over 1% processing additive may be required.

These higher levels of additives will alter the physical properties and it may be necessary to further adjust the formula to meet specification requirements. If the need for processing aids is foreseen, then it would be wise to include those in the original formulation, thus eliminating a later need for an additional compound adjustment.

15.1 PROCESSING AIDS

Processing aids or additives are a class of chemicals which can be added to an elastomeric compound for the purpose of resolving processing problems. Those problems include poor release from the calender; mill and mixer; slow extrusion rates and high die swell; poor mold flow, mold release, and mold fouling.

15.2 PROPRIETARY PROCESSING AIDS

Process aids in this group are usually compounds or mixtures of chemicals that serve as either general purpose or specialized materials. Examples are chemical combinations of various stearates or a combination of a stearate and a zinc or other metal soap with petrolatum or a resin. The chemist may choose to combine simpler products to eliminate processing problems, such as low molecular weight polyethylene (LMWPE), paraffin wax, and petrolatum, but usually the simplest, and sometimes less costly approach, is to select a specialized proprietary additive to achieve the same or even better results. There are hundreds of proprietary processing aids on the market from which the chemist may choose. The emphasis in this section has been placed on processing aids manufactured by Struktol Company and its parent, Schill & Seilacher (GmbH & Co.) of Hamburg, Germany, since most of these products are available throughout the world. The processing aids recommended in this chapter are briefly described in Table 15.1. There are substitutions available for many of these products offered by manufacturers such as Blachford Ltd of Canada; Akrochem, Harwick Standard, R.T. Vanderbilt of the USA; Kettlitz and Rhein Chemie of Germany, to name a few. Table 15.2 lists a few alternatives for some of the processing aids recommended in this chapter.

15.3 MELTING AND SOFTENING TEMPERATURES
OF PROCESS AIDS

Some of the products recommended here may not be suitable for use in all mill mixing operations because mixing temperatures could be too low to sufficiently soften some process aids. Examples are mill mixing and mixing of low-viscosity

TABLE 15.1
Proprietary Processing Aids

Product	SG	Color/Form	d.p./ m.p.,°C	Description	Reference
Armeen 18D	0.79	White, waxy solid	55[a]	Stearylamine	[9]
Cumar P-10	1.04	Amber, viscous liquid	10[b]	Coumarone-indene resin	[10]
Pluoriol E4000	1.15	White powder	55[a]	Polyethylene glycol	[11]
Strukrez 160	1.10	Amber pastilles	85	Hydrocarbon resin	[12]
Struktol 60 NS	0.97	Amber pastilles	100	Mixture of aliphatic hydrocarbon resins	[12]
Struktol HPS 11	0.98	Light tan beads	85	Blend of fatty acid derivatives	[12]
Struktol HM 97	0.91	White pastilles	104	Blend of LMWPE waxes	[12]
Struktol PE H-100	0.91	White powder	98	Homopolymer, nonoxidized PE wax	[12]
Struktol WA 48A	1.01	Amber pastilles	87	Mixture of esters and zinc soaps of natural fatty acids	[12]
Struktol WB 42	0.93	Yellowish pastilles	90	Blend of fatty acid derivatives	[12]
Struktol WB 222	0.95	White beads	56	Highly concentrated water free blend of high molecular weight aliphatic fatty acid esters and condensation products	[12]
Struktol WS 180	1.00	Light brown pastilles	91	Condensation product of fatty acid derivatives and silicones	[12]
Struktol WS 280	0.98	Paste	—	Organosilicone product	[12]
Struktol TR 121	0.94	Light cream pellets	73	Wide specification, unsaturated fatty primary amide derived from oleic acid	[12,13]
Struktol TR 131	0.93	Off-white pellets	82	Unsaturated primary amide derived from erucic acid	[12]
Struktol TR 141	0.96	Cream color beads	103	A saturated primary amide with no double bond. Stearamide	[12]

[a] Approximate.
[b] Softening point (ASTM E-28).

compounds in an internal mixer where mix times are very short. In these situations, process aids that have melting or softening points over 100°C may not disperse well. The form of a process aid also influences whether or not it will disperse well. For example, consider a process aid that has a melting point of 100°C. If the form of this

TABLE 15.2
Substitute Proprietary Processing Aids

Process Aids	Substitute	Source	Equivalency
Armeen 18D	Kenamine P990D	Witco/Humko	Same
Pluoriol E4000	Carbowax 3350	Harwick Standard	Similar
Strukrez 160	Nevchem 100	Neville Chemical	Similar
Struktol 60 NS	Rhenogran 260	Rhein-Chemie	Similar
	Promix 300HA	Flow Polymers, Inc.	Similar
Struktol PE 1-1-100	Epolene N34	Eastman Chemical	Similar
Struktol WB 42	Aflux 42	Rhein-Chemie	Same function
	Deoflow S	D.O.G. (Germany)	Same function
Struktol WA 48	Seriac WA 48	Seriac (Brazil)	Same
Struktol WB 222	Aflux 54	Rhein-Chemie	Similar
	Vanfre AP-2	R.T. Vanderbilt	Same function
	Proaid AC-22	Akrochem	Similar
Struktol TR 121	Kenamide 0	Witco/Humko	Same
Struktol TR 131	Kenamide E	Witco/Humko	Same
Struktol TR 141	Kenamide S	Witco/Humko	Similar

material is a fine powder, then it may incorporate well into a mill mixed NBR with a Mooney viscosity of 50 (ML $1+4$ at 100°C). If that same process aid is in a large flake or pastille form, it may not reach a high enough temperature during mixing to soften or melt the entire particle in this same compound.

15.4   BLOOM, MIGRATION

Many processing aids will migrate to the surface of uncured and cured compounds and form a film. This film may be transparent, translucent, or even opaque. This migration, or bloom, can have either positive effects, such as preventing slabs in a stack from sticking together, or negative effects, such as interfering with a rubber to metal bond. To avoid adhesion problems, the compounder must take care in the selection of a processing aid as well as the level incorporated in a formulation. A low-level addition of a particular process aid may have no effect on a bond, but the addition of a higher level of that same processing aid may be detrimental to adhesion. Some processing aids cannot be tolerated at any level where a critical bonding situation exists.

15.5   SURFACE TACK

Some processing aids can either improve or eliminate the surface tackiness of uncured compounds. Tackifying resins, especially phenolic types, are excellent for increasing surface tack, while materials such as primary amides can be used to reduce or eliminate surface tack. The primary amides may not be desirable for use in compounds meant for compression molded products because the resultant surface bloom on the preforms may cause knitting problems during the molding operation.

15.6 INCORPORATION OF PROCESS AIDS

When an internal mixer is employed, processing additives are usually incorporated during the first stage of the mixing process. This is the most effective method where improved release from the mixer and mills is desired. If the melting or softening point of the processing aid is 100°C or lower and the form of the material is powder or small grains, then it may be added during the final mixing stage. Additions during the final mixing stage may be desirable if increased extrusion rates and improved mold flow are sought. Some processing aids, such as Struktols WB 222 and HPS 11, have been reported to function better in these areas when added during the final mixing stage. It is best to add the processing aid with the filler when mill mixing. If the process aid is one that may be difficult to incorporate, such as a hard hydrocarbon resin with a 105°C softening point, then it is advisable to mix it with part of the filler, and add this mixture in the early stage of the procedure.

15.6.1 GENERAL NOTES

Unless stated otherwise, the recommended additions for process aids will be expressed as a PERCENTAGE OF THE TOTAL WEIGHT OF THE FORMULA-TION. The reason for this is variations in formula weight due to the amount of fillers and plasticizers used will change the amount of process aid required. Although 1.00 pphr of a given process aid may function well in a formula totaling 150 pphr, that same amount may be ineffective in a compound totaling 300 pphr.

15.6.2 NBR, XNBR, AND HNBR

15.6.2.1 Milling, Calendering, and Extrusion

NBR, HNBR, and XNBR compounds, especially those based on polymers with lower ACN contents, can have a strong tendency to bag on mill and calender rolls. This bagging may be worsened if the compound is lightly plasticized, unplasticized, or has a high Mooney viscosity. The bagging can be reduced or eliminated by adding 3%–10% of a coumarone-indene resin, such as Cumar P-10. Struktol 60 NS is recommended at levels of 2%–3% to improve the extrusion rate and appearance of extrudates. Struktol 60 NS is not recommended for peroxide-cure systems.

15.6.2.2 Molding

Struktol WB 222 is recommended at 0.5%–2% as a general purpose processing aid for acrylonitrile-based elastomers. It is especially effective for improving mold flow, mold release, and to reduce or eliminate mold fouling. See Figure 15.1 for a comparison. In some compounds, Struktols HPS 11 or WA 48, at the same levels, may perform better than Struktol WB 222 [1]. Struktol WA 48 is not recommended with peroxide-cure systems or for XNBR compounds.

A study was conducted of Struktol WB 222 in a 50 Mooney, medium ACN compound, as shown in Table 15.3. The results shown in Figure 15.1 indicate that there was a negligible effect on physical properties but that there was a significant improvement in lowering viscosity and increasing flow of the compound.

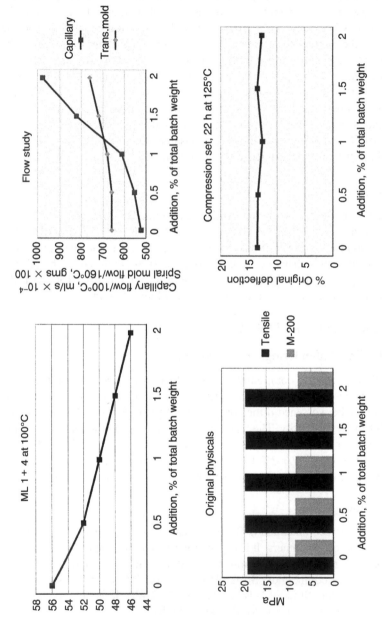

FIGURE 15.1 A study of Struktol WB 222 in a 50 Mooney, Med% CAN reinforced with 70 pphr N660 black.

TABLE 15.3
Struktol WB 222 at Various Levels in NBR Formulation

Additive (%)	0.0	0.5	1.0	1.5	2.0
Chemigum N615 B	100.00	100.00	100.00	100.00	100.00
N660 carbon black	70.00	70.00	70.00	70.00	70.00
Zinc oxide	5.00	5.00	5.00	5.00	5.00
Stearic acid	0.50	0.50	0.50	0.50	0.50
TMQ	2.00	2.00	2.00	2.00	2.00
DOP	10.00	10.00	10.00	10.00	10.00
Spider Sulfur	0.40	0.40	0.40	0.40	0.40
Benzyl Tuex	3.50	3.50	3.50	3.50	3.50
TBBS	1.50	1.50	1.50	1.50	1.50
Struktol WB 222	0.00	1.00	2.00	3.00	4.00
TOTAL	192.90	193.90	194.90	195.90	196.90
Physical properties, cured 12 min at 160°C					
Hardness, Shore A	71	72	72	72	72
Modulus at 200%, MPa	8.6	8.5	8.4	8.3	7.9
Tensile strength, MPa	19.3	19.8	19.8	19.6	19.7
Elongation, %	440	450	460	460	470
Compression set, 22 h at 125°C					
Set, %	13.5	13.4	12.6	13.5	12.7

An additional experiment was conducted with the medium Mooney and ACN NBR compound given in Table 15.4 in which a comparison was made of Struktols WB 222, WA 48, and HPS 11. The additives had a similar reduction in compound viscosity and spiral flow characteristics, but Struktol HPS 11 provided greater capillary rheometer flow at 125°C as seen in Figure 15.2. There was a very significant improvement in Mooney scorch times when Struktols WB 222 or WA 48 were included in the compound and to a lesser, but still significant extent, with Struktol HPS 11.

15.6.3 FKM

Attention must be given to the volatility of the processing aid when selecting one for FKM fluorinated rubber. A good example is dioctyl phthalate (DOP) since the inclusion of 2.00–3.00 pphr of DOP will result in volatilization of it during demolding and oven post-cure. The finished parts will be undersized, hence the mold cavities must be cut to allow for the increased shrinkage.

Carnauba wax is recommended at levels of 1%–2% to improve mill and mold release as well as extrusion rate, appearance and mold flow. Carnauba wax should not be used in peroxide-cured compounds. Beeswax or Struktol WS 280 paste may be incorporated at levels of 1%–2%, but are not as effective as carnauba wax. Struktol WS 280 paste is especially effective in reducing mold fouling of mineral filled FKM compounds. Peroxide-cured FKM compounds benefit from the inclusion of

TABLE 15.4

Comparing Various Process Aids in NBR Formulation

Additives (%)	None	Struktol WB 222	Struktol WA 48	Struktol HPS 11
Chemigum N615 B	100.00	100.00	100.00	100.00
N550 carbon black	60.00	60.00	60.00	60.0
Zinc oxide	5.00	5.00	5.00	5.00
Stearic acid	0.50	0.50	0.50	0.50
TMQ	2.00	2.00	2.00	2.00
Spider Sulfur	0.40	0.40	0.40	0.40
TMTD	1.00	1.00	1.00	1.00
TBBS	2.00	2.00	2.00	2.00
Struktol WB 222	0	2.50	0	0
Struktol WA 48	0	0	2.50	0
Struktol HPS 11	0	0	0	2.50
TOTAL	170.90	173.40	173.40	173.40
Physical properties—cured 10 min at 160°C				
Hardness, Shore A	78	77	77	77
Modulus at 300%, MPa	19.17	17.10	17.31	17.31
Tensile strength, MPa	24.17	22.68	21.79	22.79
Elongation, %	420	470	420	460
Compression set B, 22 h at 100°C				
Set, %	13.8	16.5	15.7	14.8

0.5%–1.0% stearamine such as Armeen 18D to inhibit mold fouling. Calcium stearate is a mildly effective processing aid when employed at a level of 1.5%. Calcium stearate has the added benefit that it tends not to interfere with bonding to metal.

15.6.4 ACM

Stearic acid is added at levels of 0.5%–1.0% in order to facilitate acceptable mill release, and to improve processing properties in general. In addition, Struktol WB 222 can be added at levels of 1.0%–1.5% to increase mold flow and reduce mold fouling. Care must be taken when using Struktol WB 222 combined with another processing aid such as stearic acid, as excessive levels will migrate to the surface of the ACM vulcanizate creating a light colored bloom. Even a 1% addition may result in migration of the process aid, which may not appear until months after vulcanization. Such a bloom was noted where stearic acid was used in the test formula as a level of 1.6%. An alternate material, although not effective as Struktol WB 222, is Struktol WS 280 as seen in Figure 15.3. Any migration to the surface of WB 280 will not appear as a discoloring bloom. Zinc containing chemicals, such as zinc stearate, should not be used either internally or as a dusting agent, as they can act as a pro-degradant [2].

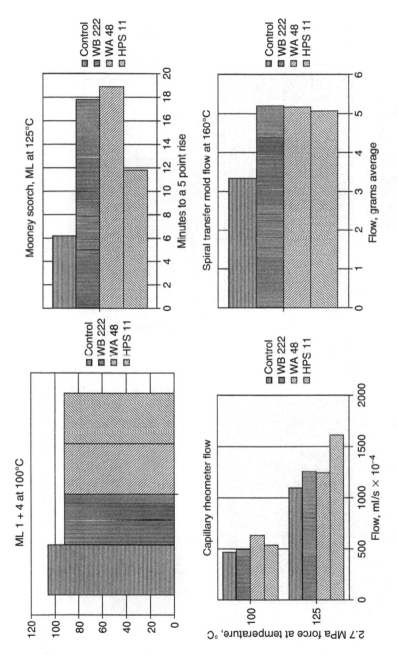

FIGURE 15.2 1.4% addition of various additives in a N550 reinforced, Med% CAN NBR.

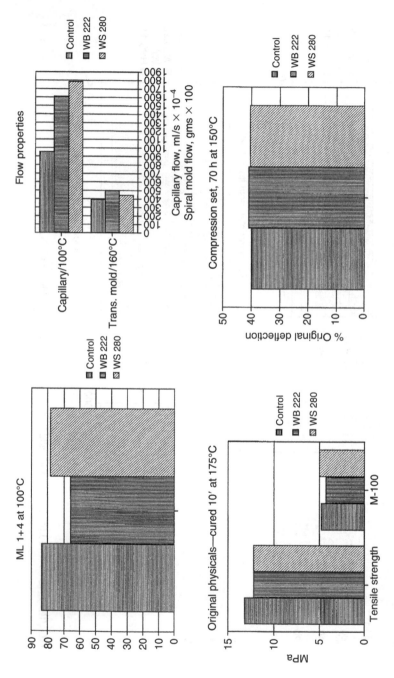

FIGURE 15.3 Struktol materials WB 222 and WS 280 at a level of 1.6%, in an NPC polyacrylate reinforced with N550.

TABLE 15.5
Comparing Various Process Aids in ACM Formulation

Additives	None	Struktol WB 222	Struktol WS 280
HyTemp 4051 EP	100.00	100.00	100.00
Stearic acid	3.00	3.00	3.00
N550 carbon black	70.00	70.00	70.00
AgeRite Stalite S	2.00	2.00	2.00
Sodium stearate	4.00	4.00	4.00
HyTemp NPC 50	2.00	2.00	2.00
Struktol WB 222	0	3.00	0
Struktol WS 280 powder (75% active)	0	0	4.00
TOTAL	181.00	184.00	185.00
Physical properties, cured 10 min at 176.6°C			
Hardness, Shore A	74	76	75
Modulus at 100%, MPa	4.76	4.34	4.96
Tensile strength, MPa	13.2	12.2	12.2
Elongation, %	220	230	210
Compression set B, 70 h at 148.9°C			
Set, %	39.9	40.9	40.4

A comparison of Struktols WB 222 and WS 280 in an ACM compound using an NPC system was made in the formula in Table 15.5. The physical properties of the vulcanizate were essentially unchanged as seen in Figure 15.3. The Mooney viscosity reduction and transfer mold flow with Struktol WB 222 was better than that of the Struktol WS 280; however, the capillary flow was better with the Struktol WS 280.

15.6.5 AEM

Struktol WB 42 at a 1% level, combined with 1% stearic acid, is a good combination for general purpose use in peroxide-cured AEM ethylene acrylic elastomer compounds. Vamac AEM stocks using hexamethylene diamine carbamate (Diak No. 1) as the curative, requires 1% stearic acid plus 0.3% of a stearamide such as Armeen 18D from Akzo and 0.8% of Struktol WS 180. These chemicals act as coagents for mill release. Struktol WB 42, Struktol WS 180, and Armeen 18D also serve as mold release agents.

15.6.6 ECO AND CO

Struktol WA 48, used at levels of 1%–2%, will improve release from the mill, mixer, and mold while improving mold flow. If the vulcanizate is to be exposed to long-term service temperature in excess of 135°C, care should be taken to use a good antidegradant system with Struktol WA 48. In lieu of an effective antioxidant system, such as 0.75 pphr nickel dibutyl dithiocarbamate, (Vanox NBC from R.T. Vanderbilt), plus 1.25 pphr nickel di-isobutyldithiocarbamate, (Isobutyl Niclate from R.T. Vanderbilt) [3], use Struktol WB 222 as an effective alternate to Struktol WA 48. WB 222 will not cause surface hardening, but it is not as effective for releasing sticky AEM compounds from the mill rolls. Struktol WB 222 can be used at levels up to 2%, and is very effective the reduction of mold fouling and mold flow improvement.

15.6.7 CM

Struktol WB 222, used at levels of 0.75%–2.0% in CM compounds, can eliminate sticking in the mixer and on mill rolls, increase extrusion rates, reduce extruder die swell, and improve the appearance of the extrudate. This process aid will also improve mold flow and mold release of chlorinated polyethylene. Chlorinated polyethylene is designated CM by ASTM, but may be commonly referred to as CPE.

15.6.8 CR

Stearic acid is generally added to aid in the release of CR compounds from mill and calender rolls. No more than 0.50 pphr of stearic acid is recommended with neoprene because of the retarding effect it has on cure rates [4]. Struktol HPS 11 is an excellent general purpose processing aid for neoprene. It should be used at a level of 1.5%–2% to prevent sticking to mill and calender rolls, and to improve mold flow and release. Whenever long-term storage life is critical, use the same levels of Struktol WB 222 in place of Struktol HPS 11. Struktol WB 222 is not as effective as HPS 11 for mold flow, but performs better as a mill release.

Struktol HPS 11 tends to slightly accelerate the cure rate of CR compounds, whereas Struktol WB 222 may slightly retard [5]. The blowing characteristics and surface appearance of neoprene sponge products can be improved by the addition of 1%–2% of Struktol WB 222. Splitting of neoprene compounds on calender rolls can be eliminated by adding 2% of Struktol HM 97 to the formula. Struktol HM 97 will function well only if the roll temperatures are maintained above 65°C. Combinations of LMWPE (0.5%–2.0%), 60°C–65°C MP (A-C Polyethylene 1702 from Allied Signal), paraffin wax (0.5%–1%), and petrolatum (0.7%–1%) usually can be used in various combinations to obtain good processing characteristics of CR compounds, however, the compounder may find it more convenient to use one of the proprietary materials such as Struktol HPS 11.

The test results seen in Figure 15.4 compare the effectiveness of Struktol WB 222 and HPS 11 when used in the compound described in Table 15.6. It should be noted that Struktol WB 222 improves scorch time slightly whereas Struktol HPS 11 speeds the cure rate and reduces scorch time. The faster cure obtained with Struktol HPS 11 also results in improved compression set properties of the vulcanizate.

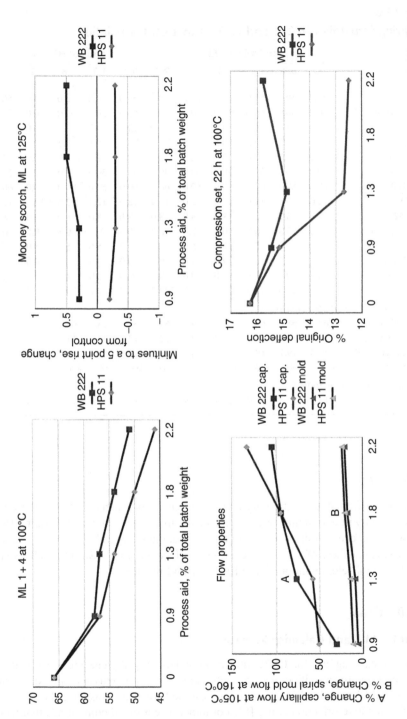

FIGURE 15.4 Struktol materials WB 222 and HPS 11 in a neoprene compound, reinforced with N330 and treated clay and cured with ETU.

TABLE 15.6

Comparing Struktols WB 222 and HPS 11 in a CR Formulation

Additive Level, %	Struktol WB 222						Struktol HPS 11				
	0.0	0.9	1.3	1.8	2.2	2.7	0.9	1.3	1.8	2.2	2.7
Neoprene W	50.0	50.0	50.0	50.0	50.0	50.0	50.0	50.0	50.0	50.0	50.0
Neoprene WHV 100	50.0	50.0	50.0	50.0	50.0	50.0	50.0	50.0	50.0	50.0	50.0
MgO	4.0	4.0	4.0	4.0	4.0	4.0	4.0	4.0	4.0	4.0	4.0
Stearic acid	0.5	0.5	0.5	0.5	0.5	0.5	0.5	0.5	0.5	0.5	0.5
Wingstay 100 AZ	2.0	2.0	2.0	2.0	2.0	2.0	2.0	2.0	2.0	2.0	2.0
N330 carbon black	30.0	30.0	30.0	30.0	30.0	30.0	30.0	30.0	30.0	30.0	30.0
Nu Cap 100	60.0	60.0	60.0	60.0	60.0	60.0	60.0	60.0	60.0	60.0	60.0
Sundex 790	12.0	12.0	12.0	12.0	12.0	12.0	12.0	12.0	12.0	12.0	12.0
Circosol 4240	8.0	8.0	8.0	8.0	8.0	8.0	8.0	8.0	8.0	8.0	8.0
ETU-75 dispersion	0.8	0.8	0.8	0.8	0.8	0.8	0.8	0.8	0.8	0.8	0.8
ZnO-90 dispersion	5.0	5.0	5.0	5.0	5.0	5.0	5.0	5.0	5.0	5.0	5.0
Struktol WB 222	0.0	2.0	3.0	4.0	5.0	6.0	0.0	0.0	0.0	0.0	0.0
Struktol HPS 11	0.0	0.0	0.0	0.0	0.0	0.0	2.0	3.0	4.0	5.0	6.0
TOTAL	222.3	224.3	225.3	226.3	227.3	228.3	224.3	225.3	226.3	227.3	228.3
Physical properties, cured 20 min at 160°C											
Hardness, Shore A	68	68	68	68	69	69	68	68	68	67	67
Modulus at 300%, MPa	14.8	14.2	13.8	13.0	12.6	12.6	12.8	12.4	11.6	11.2	10.6
Tensile strength, MPa	19.8	19.7	19.7	19.7	18.3	18.3	20.1	19.7	19.4	19.0	18.2
Elongation, %	460	490	480	480	470	480	490	500	490	480	470
Compression set B, 22 h at 100°C											
Set, %	15.9	15.5	14.9	15.8	15.8	14.9	15.2	12.7	11.4	12.5	11.2

15.6.9   EVM

The thermoplastic nature of the ethylene/vinyl acetate polymers leads to problems such as difficulty in demolding, sticking inside the mixer, and poor mill handling [6]. These problems can be eliminated or reduced by adding Struktol WS 180 at a level of 1%. Levels in excess of 1.5% can result in a surface bloom [7]. Struktol WS 180 is an excellent mill release and a fairly effective mold release.

15.6.10   CSM

15.6.10.1   Mill and Calender Release

Common release agents for CSM chlorosulfonated polyethylene are stearic acid, paraffin and microcrystalline waxes, petrolatum, and polyethylene glycol. Zinc stearate should be avoided in compounds designed for heat resistance, because of a potential negative effect on aging [8]. For mineral-reinforced compounds that tend to stick and split on the calender and mill rolls, use a combination of 1% Struktol WB

222, 1.5% Struktol HM 97 plus 2.5% of a hydrocarbon resin with a softening point of 80°C–95°C, such as Strukrez 160. Primary amides, such as Struktol TR 121 (oleamide), Struktol TR 131 (erucamide), and Struktol TR 141 (stearamide) used at a level of 0.25%–0.50%, are used to block the tendency of calendered sheets to coalesce.

15.6.10.2   Extrusion

Polyethylene glycol, such as Pluoriol E4000 from BASF or a LMWPE such as Struktol PE H-100, is used to improve the extrudability of Hypalon compounds. Struktol WB 222 is recommended at levels up to 2% to further improve the appearance of the extrudate.

15.6.10.3   General Purpose and Molding

Struktol WB 222 is recommended at levels of 1.5%–2% for improved mold release and as general purpose process aid.

REFERENCES

1. Struktol report no. 90048, STRUKTOL HPS I I in NBR, July 25, 1990, Jerry M. Sherritt and Struktol Product Bulletin, NBR PROCESSING IMPROVEMENT, March 13, 1989, extracted from report no. 89015, Jerry M. Sherritt.
2. HyTemp Bulletin PA 0600, Zeon Chemicals HyTemp product binder.
3. Struktol internal report no. 8524 (R24, R24A) January 24, 1985, Jerry M. Sherritt.
4. *"The Neoprenes,"* R.M. Murray and D.C. Thompson, E.I. Du Pont de Nemours & Co. Inc.
5. Struktol report no. 91064, STRUKTOL PROCESS AIDS IN VARIOUS ELASTO-MERS, Study IV, CR. 2/3/93, Jerry M. Sherritt.
6. Rhein Chemie Technical Report No. 38.
7. "Compatibility of Process Aids of the Lubricant and Release Agent Type at Low Filler Concentration," M. Krambeer, Schill & Seilacher. 8/12/91.
8. DuPont Technical Information Bulletin HP-340.1.
9. Akzo MSDS, Revision date 3/20/87.
10. Neville Technical Data Sheet dated 5/96.
11. BASF Technical Bulletin TI/P 257 1, May, 1985.
12. Struktol Company Technical Data Sheets.
13. *Industrial Chemical Thesaurus, Second edition,* Vol. 1, VCH Publishers.

16 Considerations in the Design of a Rubber Formulation

Robert C. Klingender

The previous chapters have provided an insight into various specialty elastomers, defined herein as heat- and oil-resistant polymers. In addition, other key ingredients used with these rubbers are described and their usage in these polymers is covered.

Now comes the important part, using this and other information along with the rubber chemist's experience, to design a satisfactory formulation to meet the customer's requirements, real or imagined. The MOST IMPORTANT THING IS COMMUNICATION with the customer! This is often extremely difficult as the customer often does not really understand the service needs of the part as he or she have not been given the complete requirements. "Black like a tire tread" is not an acceptable definition, so it is necessary to train not only the design engineers and sales staff, but also the most important cog in the wheel, the customer. It is highly recommended that a complete detailed general checklist be designed for the customer, sales personnel and design engineer. In addition, a second list needs to be prepared for the mold and die maker, process engineer, design engineer, and rubber chemist; so these requirements can be included in the compound requirements in addition to the service needs.

This chapter is divided into three parts to provide only some of the considerations to be accounted for in the creation of a satisfactory rubber compound.

Chapter 16A "Oil Field Elastomeric Products" by Robert C. Klingender is an insight into the energy development field and most important the service conditions that cannot be duplicated in the laboratory but must be tested in actual service. Here it is even more important to obtain as much detailed information about the temperatures, pressures, fluids, gases, other chemicals in drilling muds, work-over conditions and environments, fracturing of producing strata to free-up the crude oil or gas, and so on. In this section, it is very apparent that experience is the most important requirement for the proper design of drilling and production components of oil and gas wells.

Chapter 16B "Life Prediction" by John Vicic will give the rubber chemist an insight into considerations regarding the service life needs of specialty rubber parts. Although the customer may not state initially how long the part is expected to last, upon closer questioning the expectations of the customer may be revealed. It is very important that the life expectancy of the particular rubber component in the environment, as described by the customer, be spelled out to avoid any misunderstandings and potential law suits.

Chapter 16C "Compression, Transfer and Injection Molding of Specialty Elastomers" by Dr. Robert W. Keller will give the rubber compounder an excellent view of the considerations necessary in the molding operation in the factory. This is only one aspect of rubber processing, but similar approaches are needed for other processes such as extrusion, calendering, machine, and hand building of hose, belts, rolls, tank, and pipe linings, and so on.

It is not an easy task to design specialty elastomeric components, but perhaps this book will give the reader a better understanding of the many aspects and challenges facing him or her in this industry.

16 Part A: Oil Field Elastomeric Products

Robert C. Klingender

CONTENTS

The term elastomer is used in the energy industry to refer to rubber compounds and parts used in oil tools for the exploration, drilling, and production of oil, gas, and geothermal energy [1]. Elastomers used herein may either be natural rubber or synthetic polymers. Although both types are used extensively in the oil field, this discussion will be limited to those typically considered to be "oil and heat resistant."

16.1 MATERIALS

Though a library of information details material properties of elastomers used in the oil field, the following gives a synopsis of commonly used elastomers based on three key aspects: price, function, and chemical and temperature compatibility.

Nitrile or NBR. Nitrile is the king of the oil field. It is low priced and easy to process so it is readily available in standard and custom forms. It has very high strength so it can be used in about every type of application. It has good general chemical compatibility but does not perform well in sour crude containing hydrogen sulfide or strong acids. In closed environments and static conditions it will be satisfactory up to 125°C.

Nitriles are copolymers of butadiene and acrylonitrile, and it is the amount of acrylonitrile that determines the degree of oil resistance. Polymers are commercially available with acrylonitrile levels ranging from 18% to 50%, the most commonly used being 41%. Thus it becomes obvious that there are many variations of polymers within the Nitrile group. The higher its acrylonitrile level, the greater is the oil resistance of the compound. Increasing acrylonitrile to improve oil resistance, however, comes at the expense of low-temperature properties. As the acrylonitrile level of NBR increases, the low-temperature properties get worse.

HNBR—Highly Saturated Nitrile is moderately priced and is available in acrylonitrile levels similar to NBR. It has excellent strength and is generally used with a mechanical restraint or reduced extrusion gap. It has good general chemical and very good temperature compatibility, up to 175°C or 200°C. HNBR is unusual in that it does not lose hardness and modulus as the temperature increases to nearly the same extent as other elastomers.

These polymers are produced by hydrogenation of the butadiene units of Nitrile polymers and are available with different amounts of unsaturation. As the number of double bonds is reduced, the elastomer becomes more highly saturated and the polymers have better heat resistance. They require peroxide cures to develop optimum properties.

Fluorocarbon or FKM. Fluorocarbons have wide usage in the oil field. It is moderate to high priced depending on compound. It can require some special processing but is generally available in standard and custom forms. It has moderate strength and is generally used with a mechanical restraint or reduced extrusion gap. It has excellent acid, oil and fuel resistance, and temperature compatibility up to 200°C–250°C, depending upon the fluid encountered and type of FKM. TFE/P or Aflas does have good resistance to caustic, wet sour gas and amine containing drilling mud, which FKM does not resist.

Fluorocarbons are also copolymers or terpolymers and there are also many variations within this group.

16.2 APPLICATION OF ELASTOMERS IN THE ENERGY INDUSTRY

Elastomers are used in the energy industry in a variety of applications but are primarily used as dynamic or static seals. Other applications such as electrical insulation and dampening exist. These applications represent a very small segment of the overall elastomer usage in the oil field and they will not be discussed in this text. There is a great deal of information available on these applications. The aspect of elastomer usage in the energy industry that really sets it apart from other industries is the extremes of pressure, temperature, and chemical environment. Though there are extremes found in other industries, they generally exist only as a single parameter.

In the energy industry elastomers are often subjected to an environment where a differential pressure of 70 MPa exists at a temperature above 200°C in an extremely caustic or acidic fluid in very severe applications. The lack of oxygen in downhole applications will allow elastomers to perform at higher temperatures than would be normally expected of them, especially in static conditions. This environment offers a tremendous challenge to any elastomer and its performance is affected by both physical and chemical changes, which are very difficult to predict.

Most elastomers used in the energy industry that are used as dynamic or static seals fall into two general categories, compliant and energized seals. Compliant seals are designed to fit to a predetermined sealing configuration such as O-rings, V-packing, and swab cups. There is interference between the seal and its surroundings so that an initial seal is made. Energized seals are seals that are formed by applying an outside force to cause a prescribed distortion such as pipe rams and packer elements. This distortion supplies interference between the seal and its surroundings so that an initial seal is made. Both seal types are unique in their material and design selection.

Compliant seals are easier to design because there is a more controlled sealing environment, but they are more subject to seal loss due to physical and chemical changes.

Energized seals are more difficult to design to ensure that the distortion load forms a proper seal, but they tend to be less subject to physical and chemical changes once they are in place. Energized seals are sensitive to the initial physical properties of the elastomer, whereas compliant seals are sensitive to the prolonged physical properties. An example of this would be if we ran and set a packer by setting 10,000 kg of pipe weight on the packer. This causes the packer element to seal by forcing the weight to distort the rubber element to the casing wall. As time passes the weight continues to distort the element so that physical changes in the rubber are compensated for by continued distortion. All the time there is compensation in the element, the O-rings in the packer are undergoing physical changes, for which compensation does not occur. The packer begins to leak at the O-rings and so it is pulled from the well. Though all the elastomers in the packer have seen similar changes in physical properties, it was only critical in the compliant seals. This is just one of the challenges of using elastomers in the energy industry.

Packers are really known as down-hole packers that function as seals between layers of the oil- or gas-rich strata. It is intuitive that they must have the appropriate properties to permit deformation/compliance to provide a seal, while at the same time have adequate extrusion resistance to prevent loss of volume and sealing force. These requirements tend to be mutually exclusive and only an optimum balance of each will be successful. Extrusion resistance at the operating temperature is the most critical requirement for these packers. This property is related to modulus, generally the higher the modulus the better the extrusion resistance. This modulus is the combined result of both the elastic and viscous properties of the elastomer. High modulus may be achieved by using a low Mooney viscosity polymer with a high filler loading, or by using a higher Mooney viscosity elastomer with less reinforcing fillers.

Some of the largest rubber parts for drilling and production are annular packers for blow-out-prevention (BOP), with some of the larger ones using in excess of 500 kg of rubber for a single part. The rubber is bonded to a few hundred more kilos of large metal inserts, which provide support and extrusion resistance. This rubber must provide performance in two critical areas when closing on an open hole. It must have sufficient elongation and tear resistance to seal without rupturing when closed on the drill stem or on itself. The whole assembly then must recover to the original inside diameter to permit the drill stem and casing to be inserted without damaging the rubber, thus resilience and hysteresis are key properties. We are well aware that the physical properties of an elastomer change with temperature; however, a BOP must perform successfully over a wide range of conditions from those in the Arctic on the low end to the maximum imparted by the produce fluids in the well bore, including geothermal applications.

Inflatable packers are another type of packer manufactured to pass through the smallest restrictions while being sent downhole, and then inflated hydraulically to the casing wall or open-hole diameter to seal off the well. The inflatable packer is like a long tube with longitudinal metal strips within the rubber for reinforcement, which opens like an umbrella when inflated. This requires a high elongation of the rubber to 500% or more, coupled with excellent tear resistance at the operating temperature and in the chemical environment, making for a major challenge in compound design.

These examples of critical requirements of key rubber parts represent just a few uses for the various oil-field elastomers. Formulations used in these parts are highly proprietary and their development is strictly confidential. There are, however, several sources available to provide assistance with improvements for existing compounds or initial development when access to formulations is not readily available. Some of these sources include custom mixers and polymer and other compounds ingredient suppliers who are eager to assist, based on their own products. Other sources are educational seminars, text books and published technical papers on these subjects.

16.3 PHYSICAL CHANGES

Effects of temperature change and swelling are two types of physical changes of primary concern with oil-field elastomers. Normal laboratory tests used to indicate serviceability are not adequate in these applications. High-pressure exposure to combinations of gases and fluids unique to well-bore environments limit their serviceability. These are the combinations that are very difficult to duplicate in laboratory testing and impossible without a pressure vessel. And even with a pressure vessel, exposure of test pieces and subsequent testing at ambient laboratory conditions after removal is less than adequate. The test pieces are no longer at high temperature and pressure.

The effects of temperature alone on physical properties are well known. High-temperature test results for several elastomers are shown in Table 16A.1 [2]. Note the dramatic loss of tensile strength and elongation for temperatures above the standard laboratory test temperature.

Physical changes related to temperature alone occur as soon as the temperature of the elastomer falls closer and closer to its glass transition point but these changes tend to be completely reversible. The main exception is low-temperature crystallization, which is time dependent and associated primarily with natural and polychloroprene elastomers.

Oil-field elastomers are exposed to a broad range of ambient and service temperatures. A wellhead elastomer in the Arctic may have to function not only at

TABLE 16A.1
Comparison of Oil-Field Elastomers at Elevated Temperatures

Elastomer	Zetpol 2010	Zetpol 1020	Nipol 1051	Aflas 150P	Viton GF	Viton E-430
Zetpol 2010	100.0	—	—	—	—	—
Zetpol 1020	—	100.0	—	—	—	—
Nipol 1051	—	—	100.0	—	—	—
Aflas 150P	—	—	—	100.0	—	—
Viton GF	—	—	—	—	100.0	—
Viton E-430	—	—	—	—	—	100.0
Vanox ZMTI	1.5	1.5	1.5	—	—	—
Naugard 445	1.5	1.5	1.5	—	—	—
Carnauba wax	—	—	—	1.0	1.0	—
Stearic acid	0.5	0.5	0.5	—	—	—
Sodium stearate	—	—	—	2.0	—	—
Zinc oxide	5.0	5.0	5.0	—	—	—
Magnesium oxide	10.0	—	—	—	—	3.0
Lead monoxide	—	—	—	—	3.0	—
Calcium hydroxide	—	—	—	—	—	6.0
N110 carbon black	—	—	—	25.0	—	—
N330 carbon black	50.0	50.0	60.0	—	—	—
N990 carbon black	—	—	—	—	30.0	30.0
Spider sulfur	—	0.5	1.5	—	—	—
TMTD	—	1.5	—	—	—	—
MBT	—	0.5	—	—	—	—
TMTM	—	—	0.2	—	—	—
TETD	—	1.0	—	—	—	—
MBTS	—	—	1.5	—	—	—
VulCup 40KE	10.0	—	—	2.5	—	—
Perhexyn 2.5B	—		—	—	3.0	—
Sartomer SR-350	20.0	—	—	—	—	—
TAIC	—	—	—	5.0	3.0	—
Total	198.5	162.0	171.7	135.5	140.0	139.0
Cured at 160°C, min	—	20	20	—	—	—
Cured at 170°C, min	15	—	—	15	15	15
P. C. at 150°C, h	4	4	—	—	—	—
P. C. at 200°, h	—	—	—	4	24	24
Original properties at 25°C						
Shore A, pts	92	84	84	88	81	80
Modulus at 100%, MPa	26.3	7.7	9.3	11.6	9.5	5.8
Tensile, MPa	29.4	28.3	29.1	20.7	22.0	14.5
Elongation, %	110	310	270	200	190	250

(*continued*)

TABLE 16A.1 (Continued)
Comparison of Oil-Field Elastomers at Elevated Temperatures

Elastomer	Zetpol 2010	Zetpol 1020	Nipol 1051	Aflas 150P	Viton GF	Viton E-430
Properties tested at 100°C						
Shore A, pts	90	77	74	71	71	71
Modulus at 100%, MPa	—	4.5	6.6	4.9	6.8	6.8
Tensile, MPa	13.0	10.3	13.8	6.3	8.1	8.1
Elongation, %	110	190	170	150	110	160
Properties tested at 150°C						
Shore A, pts	89	75	75	67	69	70
Modulus at 100%, MPa	11.0	2.9	4.4	3.0	3.7	3.3
Tensile, MPa	11.2	7.5	11.3	4.7	5.7	5.6
Elongation, %	100	170	150	100	100	110

subzero temperatures during nonproduction periods, but also at temperatures above 150°C during production. It is important to understand that elastomer properties are different at different temperatures. This change of properties is due to their viscoelastic nature.

Elastomers are very stiff and brittle below their glass transition temperatures, T_g. This stiffening due to decrease in temperature does not occur abruptly at a specific temperature, but over a rather broad range. When a material has become brittle, it is said to have gone through its T_g.

The brittle point of a specific elastomer is very reproducible to within a degree or two at most when tested under the same conditions. It is, however, influenced by the frequency or speed of testing and may vary by several degrees if the speed of the test is increased by a factor of 10 or 100 or more. The important thing to remember is that low-temperature stiffening is a physical change that is completely reversible.

Generally, elastomers lose strength as the temperature increases above their glass transition temperature, T_g, which is near their brittle point. Although the high temperatures of downhole environments and produced fluids make high-temperature service capability a requirement more often than not for oil-field elastomers, low-temperature capability is often a requirement for these same elastomers.

The glass transition for each individual polymer is very reproducible, but different compounds of the same polymer can have very different glass transitions and brittle points. Although the polymer in a compound determines low-temperature characteristics for the most part, the brittle point can be affected by different compounding materials, and especially by plasticizers and blending with other polymers.

Most oil-field elastomers must also be oil resistant, and oil-resistant elastomers typically have poor low-temperature properties. When a polymer type is selected for its oil resistance and does not have acceptable low-temperature properties, plasticizers can be added to provide better low-temperature properties for the compound. Addition of plasticizers reduces modulus and changes other properties. Changing other compound ingredients, however, will help to compensate for these changes.

Most elastomers remain flexible as temperatures decrease to within 10°C, above their glass transition; then they stiffen exponentially. The most publicized failure due to low temperature stiffening was the fatal Challenger accident, January 28, 1986. The elastomeric O-ring seals designed to contain the hot gases produced during launch were too cold to conform to their glands, and thus allowed the hot gases to escape.

Oil-field elastomers also fail to perform satisfactorily when too near or below their glass transition temperatures. Although the consequences are not as dramatic or well publicized, if blow-out-preventer elastomers are too stiff to function properly, a blow out can occur. O-rings and other seals can fail when they lose resilience causing leaks of contained fluids and gases.

Beyond developing elastomers for acceptable service performance in low-temperature environments, consideration must be given to low temperatures during storage. In some locations, storage temperatures are below required service temperatures, and the elastomers must be warmed to permit acceptable installation and handling. This is frequently the case for downhole applications where low-temperature service is not a consideration. If the storage temperature is too low, that is, within 10°C of the elastomer's T_g, it might be cracked due to rough handling, that is, dropping or other sharp impact, or it may just be too stiff to install properly. Elastomers are very poor thermal conductors and the surface temperature of thick parts will change long before the entire cross section; therefore, adequate time must be allowed for warming the entire part. As a rough rule of thumb, it requires 15 min for each centimeter of thickness of the part for the elastomer to change temperature by 1°C.

When seals must perform over a wide range of temperatures, other failures can occur because of the thermal expansion and shrinkage characteristics of elastomers, which can be as much as 10 times as great as steel. Glands must be designed to permit thermal expansion at high temperatures without extrusion of the elastomer and without losing sealing force due to shrinkage of the elastomer at low temperatures.

In addition to exposure to extremes of temperature, oil-field elastomers must perform satisfactorily when exposed to high pressure. The primary problems related to high-pressure service are extrusion and explosive decompression. Parts are rated for differential service pressures in 34.48 MPa increments up to 137.9 MPa. A comparison of the extrusion resistance of various elastomers is given in Table 16A.2.

When exposed to high differential pressures, unconstrained elastomers tend to relieve the applied stress by changing shape, that is, extrude and flow in the direction of least stress. When an elastomer is forced through a tight gap, pieces are torn off, i.e., nibbled away, which reduces the volume of the sealing element and results in seal failure due to loss of volume. Reduction of extrusion gaps and increasing an elastomer's modulus are effective ways to reduce this type of failure. The disadvantage of increasing an elastomer's modulus is that it takes more force for it to seal. This should not be a major problem in most cases, but can be for large parts, for example, downhole packers.

Explosive decompression occurs due to expansion of trapped gases deep inside a part's cross-section. It is also related to an elastomer's strength, that is, shear modulus. Oil-field elastomers will absorb highly pressurized methane and carbon dioxide. When the pressure is released, these gasses want to expand. And when the

TABLE 16A.2

API Extrusion Test on Various Elastomer Compounds from Table 16A.1

Polymers	Zetpol 2010	Zetpol 1020	Nipol 1051	Aflas 100H	Viton GF	Viton E-430
Original properties						
Shore A, pts	90	90	93	92	90	92
Tensile, MPa	37.7	30.3	28.3	25.0	21.1	15.4
Elongation, %	140	170	150	140	100	150
Travel of Ram at 150°C, 34.5 MPa						
Extrusion resistance, in. $\times 10^{-3}$	31	69	78	168	83	134

pressure of the trapped gas exceeds the elastomer's shear strength, it creates splits and blisters within the rubber matrix. This same problem with internal blisters and splits occurs when pressure is released too soon during vulcanization and the parts are porous inside where the temperature was not adequate for a long enough period of time to provide optimum cure. It can also occur when parts are post-cured at high temperatures. Problems with explosive decompression can be alleviated by slowly reducing the pressure, which permits the trapped gases to permeate before expanding. When post-curing, using a step post cure alleviates the problem in the same way.

In addition, rapid reduction of pressure is often of greater concern. It can cause blistering and fracturing commonly referred to as explosive decompression. Polar elastomers like nitrile and most other oil-field elastomers absorb large quantities of carbon dioxide and methane under high pressure well-bore conditions. They swell, get soft, and lose strength depending on time of exposure, temperature, and pressure. When the pressure is released, these gasses want to expand. When the pressure of the trapped gas exceeds the shear strength of the elastomer, it creates splits and blisters within the rubber matrix. An example of explosive decompression is given in Table 16A.3 in which Zetpol 1020 and 3120 are exposed to Freon 134A. It was concluded that high modulus and impermeability, (from high-acrylonitrile content), provide good explosive decompression resistance. Although Freon 134A is not encountered in the drilling industry, CO_2 will react in a similar way.

This problem not only has the obvious affect on physical properties, but also permits the escape of dissolved gases, which also affects physical properties. When the pressure is released, the gas evolves and the elastomers may regain essentially all of their original mechanical properties lost during high-pressure exposure while the gases were still in solution within the elastomer [2].

Swelling is another physical change, which is time dependent but can be completely or at least partially reversed depending on the volatility of the absorbed fluid or gas. Polar elastomers like Nitrile readily absorb large quantities of CO_2 under high pressure well-bore conditions and the elastomers swell accordingly, get soft, and lose strength. When the pressure is released, however, the gas evolves and the elastomers regain essentially all of their original properties [3]. This is also true for volatile liquids. Even less volatile liquids can be extracted with recovery of nearly all of the original properties.

TABLE 16A.3
Explosive Decompression in Freon 134A

Zetpol 1020	100.0	100.0	100.0	100.0	—	—	—	—
Zetpol 3120	—	—	—	—	100.0	100.0	100.0	100.0
N220 Carbon black	60.0	60.0	—	—	60.0	60.0	—	—
N990 Carbon black	—	—	60.0	60.0	—	—	60.0	60.0
Plasthall TOTM	8.0	8.0	8.0	8.0	8.0	8.0	8.0	8.0
Zinc oxide	5.0	5.0	5.0	5.0	5.0	5.0	5.0	5.0
Maglite D	5.0	5.0	5.0	5.0	5.0	5.0	5.0	5.0
Naugard 445	1.5	1.5	1.5	1.5	1.5	1.5	1.5	1.5
Vanox ZMTI	1.0	1.0	1.0	1.0	1.0	1.0	1.0	1.0
HVA #2	4.0	4.0	4.0	4.0	4.0	4.0	4.0	4.0
VulCup 40KE	5.0	9.0	5.0	9.0	5.0	9.0	5.0	9.0
Total	189.5	193.5	189.5	193.5	189.5	193.5	189.5	193.5
Mooney scorch MS at 125°C								
Minimum viscosity	74.0	70.1	42.1	39.0	80.4	80.5	51.5	49.1
t_5, minutes	20.1	13.6	18.3	13.3	22.4	15.0	17.7	11.1
ODR Rheometer, 3° arc, 100 cpm, 170°C								
ML, dN·m	18.2	18.0	13.8	13.8	19.0	19.6	16.6	16.8
MH, dN·m	142.9	166.5	113.9	138.8	118.1	143.5	93.5	116.1
Ts_2, minutes	1.4	1.2	1.4	1.2	1.7	1.3	1.3	1.8
t'_{90}, minutes	12.9	11.6	14.3	13.4	12.8	12.0	14.5	14.4
Explosive decompression, 24 h at 23°C in Freon 134A in pressure vessel, +1 h in Air at 100°C								
Surface condition	Excel.	Excel.	Excel.	Excel.	Fair	Excel.	Fair	Fair
Physical properties—cured at 170°C								
Cure time, minutes	16	16	18	17	15	15	18	18
Shore A, pts	89	90	74	76	83	86	67	67
Modulus at 10%, MPa	1.9	2.1	0.7	0.9	1.5	1.5	0.6	0.6
Modulus at 20%, MPa	3.0	3.6	1.3	1.5	2.2	2.6	0.9	1.1
Modulus at 30%, MPa	7.2	10.4	2.9	4.5	5.4	7.7	2.00	3.1
Tensile, MPa	31.2	31.5	16.6	19.0	25.6	8.0	16.2	16.4
Elongation, %	155	115	427	164	162	49	231	149
Compression set B, 70 h at 50°C								
Set, %	18.2	16.3	9.8	9.5	22.2	17.8	12.5	11.0

With all the discussions about physical changes that occur in the elastomer, there is one thing that should always be considered and that is the change may make a significant difference in overall performance; however, there are ways to overcome these changes. Some swelling of static compressive seals may actually be beneficial. An elastomer may not form a seal and leak horribly at cold surface conditions, but work perfectly when at the elevated temperatures generally associated with downhole conditions. All this must be considered before determining the suitability of an elastomer. Nitriles are routinely used at temperatures over 200°C and in H_2S environments. How can this happen? First, these are short-term applications, generally less than 72 h. Second, nitrile is only used in static seal applications. Third, a seal is established as quickly as possible and left in its original sealing position. If the seal must be disturbed, then the assumption has been previously made that the seal cannot be set again and the equipment must be pulled out of service and replaced. These types of applications require a good working knowledge of the equipment and the environment before attempting, but they are possible and practical. What is a Nitrile seal like after being used for 60 h at 220°C? At the surface, it is so hard and brittle you can break it with a soft hammer blow but at the point of application it retained enough of its physical nature, given the special steps taken, to maintain a seal.

16.4 CHEMICAL CHANGES

Chemical changes are both time and temperature dependent and are not reversible. Thermal and oxidative degradation are the most common examples. Loss of elastomer strength after extended periods at elevated temperatures, as mentioned above, is immediate and irreversible. Loss of strength due to thermal degradation is chemical, time dependent, and irreversible. It is important to understand the significance between these two high-temperature effects. Too often, there are misapplications of elastomers because an elastomer with resistance to thermal degradation is specified and fails because it has poor high-temperature strength. Characterizing an elastomer as being heat resistant only implies that it can withstand long term, high temperature exposure with little or no change of properties when again tested at room temperature. This in no way implies that it will provide good physical results or performance if tested at elevated temperature.

The decrease in modulus at high temperatures noted previously for Nitrile compounds shown in Table 16A.1 is to be expected. As exposure at elevated temperatures of NBR elastomers progresses with time, the modulus then increases and the part will become stiff and brittle. Dumbell test pieces for this test were preheated 10 min inside the environmental chamber before testing. The HNBR compound on the other hand is much more heat resistant and did not show as great an effect of high-temperature exposure.

Standard oven aging involves circulating air heated to anticipated service temperatures. Oxygen is frequently absent in oil-field environments; consequently, the effects of oxidative aging as experienced in oven aging are not a factor. Although many elastomers can survive much higher temperatures when oxygen is eliminated, it is the combination of other gases and fluids under high pressure, which limits their serviceability. These are the combinations, which are very difficult to duplicate in laboratory testing and impossible without a pressure vessel. And even with a pressure vessel, exposure of test pieces and subsequent testing at ambient laboratory conditions after removal is less than adequate. The test pieces are no longer at high temperature and pressure.

While thermal and oxidative effects on elastomer properties are quite detrimental, the effect of high temperature, high-pressure exposure to H_2S is usually catastrophic. Trace amounts of H_2S are found in most wells, and even 5% can cause serious problems for most elastomers within a few weeks when temperatures exceed 120°C. In this environment, Nitrile elastomers become brittle and lose essentially all elongation and resilience. This effect is shown in Table 16A.4 in which several

TABLE 16A.4

Elastomer Comparison after Sour Environment Exposure, Compounds from Table 16A.1, Test Conditions in Table 16A.5

Polymer	Zetpol 2010	Zetpol 1020	Nipol 1051	Aflas 150P	Viton GF	Viton E-430
Original physicals (Table 16A.1)						
Shore A, pts	92	84	84	88	81	80
Tensile, MPa	29.4	28.3	29.1	20.0	22.0	14.5
Elongation, %	110	310	270	200	190	250
Gas phase 24 h-change						
Shore A, pts	−5	−8	−16	−15	−6	−1
Tensile, %	−10	−12	−72	−77	−28	−28
Elongation, %	0	3	−93	−50	0	−36
180° bend	Pass	Pass	Fail	Pass	Pass	Pass
Appearance	Smooth	Smooth	Smooth	Blister	Smooth	Smooth
Gas phase 72 h-change						
Shore A, pts	−4	−4	+6	−14	−4	−1
Tensile, %	−9	2	−73	−26	−40	−29
Elongation, %	−9	−52	−98	+5	−21	−32
180° bend	Pass	Pass	Fail	Pass	Pass	Pass
Appearance	Smooth	Smooth	Smooth	Soften	Smooth	Smooth
Gas phase 168 h-change						
Shore A, pts	−5	−7	−1	−20	−5	+3
Tensile, %	−20	−7	−84	−56	−42	−40
Elongation, %	−9	−55	−95	−25	−21	−52
180° bend	Pass	Pass	Fail	Pass	Fail	Fail
Appearance	Smooth	Smooth	Smooth	Soften	Smooth	Smooth
Liquid phase 24 h-change						
Shore A, pts	−5	−13	0	−17	−5	0
Tensile, %	−24	−18	−85	−70	−50	−45
Elongation, %	−9	−16	−81	−40	−32	−55
180° bend	Pass	Pass	Fail	Pass	Pass	Pass
Appearance	Smooth	Smooth	Smooth	Soften & blister	Blister	Smooth
Liquid phase 72 h-change						
Shore A, pts	−6	−9	−4	−16	−5	+1
Tensile, %	−21	−19	−90	−43	−74	−63
Elongation, %	−9	−44	−85	−25	−58	−68
180° bend	Pass	Pass	Fail	Pass	Fail	Fail
Appearance	Smooth	Smooth	Blister	Soften & blister	Smooth	Smooth
Liquid phase 168 h-change						
Shore A, pts	−7	−9	−2	−21	−3	+2
Tensile, %	−22	−29	−87	−79	−81	−70
Elongation, %	−9	−55	−92	−55	−63	−68
180° bend	Pass	Pass	Fail	Pass	Fail	Fail
Appearance	Smooth	Smooth	Blister	Soften & blister	Smooth	Smooth

TABLE 16A.5
Sour Environment (20% H₂S) Test Conditions

Vessel size	41 L
Gas phase composition	20% H_2S by volume
	5% CO_2 by volume
	75% CH_4 by volume
Liquid phase composition	95% Diesel by volume
	4% H_2O by volume
	1% NACE Amine B by volume
Temperature	150°C
Pressure	6.9 MPa
Exposure period	24, 72 & 168 h
Gas volume	20.5 L
Liquid volume	20.5 L
Gas phase/sample volume ratio	37/L
Liquid phase/sample volume ratio	44/L

elastomers commonly used in Ram Packers [2,4] are compared in a simulated sour environment containing 20% H_2S, as given in Table 16A.5. Under these severe conditions the HNBR was the only elastomer to perform well.

H&H Rubber Company, a Division of Stewart & Stevenson undertook to have comparisons made of commercial Ram Packers in a simulated sour environment and reported the results at an Energy Rubber Group meeting [4]. The initial screening of five commercial Ram Packers is shown in Table 16A.6 indicated that the Hi Temp parts performed much better than the Regular Packer elements in 5% H_2S up to 121.1°C. Further testing was carried out on the Hi Temp Ram Packers at temperatures ranging from 24.9°C in steps to 176.6°C and with both 5% and 20% H_2S, as shown in Table 16A.7. It was apparent that the presence of 20% H_2S was much more severe and could be tolerated up to 65.6°C, but was marginal at 121.1°C and not satisfactory at 148.9°C or 176.6°C. The lower concentration of H_2S was tolerated for up to 14 days at 121.1°C, marginal for 7 days at 148.9 and not acceptable for 14 days at 148.9°C and at any time at 176.6°C.

Although these tests somewhat simulate downhole conditions, there is such a variety of actual environmental conditions and types of drilling fluids that field testing is still the only sure method to check the performance of various elastomers.

16.5 WELL-BORE FLUIDS

There are three primary environments when considering well-bore fluids; drilling, completion, and work over and all three are very unique to themselves. The drilling environment is generally the most benign of the three because of the overall sensitivity of the wellbore. There is direct contact between all the well formations so extremes in the drilling fluid chemistry are usually avoided. There are, however,

TABLE 16A.6
Sour Environment Tests on Commercial Ram BOP Packers

Sample	MS710 Regular Slab	MS710 Regular Part	MS715 Hi Temp Slab	MS715 Hi Temp Part	X Regular Part	X Hi Temp Part	Y Regular Part	Z Regular Part
Original properties								
Shore A, pts	81	82	80	81	77	78	83	85
Tensile, MPa	20.6	20.5	21.5	18.5	19.0	17.9	16.5	19.8
Elongation, %	361	345	185	175	357	316	395	189
Aged 7 days in gas phase at 24.9°C-change								
Shore A, pts	−21	−21	−4	−4	−7	−9	−12	−16
Tensile, %	−31	−31	−18	−11	−14	−10	−33	−27
Elongation, %	36	23	−6	−6	−6	4	−12	58
Weight, %	3	3	2	2	3	2	2	4
Aged 7 days in liquid phase at 24.9°C-change								
Shore A, pts	−24	−20	−9	−8	−14	−15	−14	−22
Tensile, %	−36	−30	−29	−14	−19	−27	−11	−25
Elongation, %	28	26	−21	−11	−9	4	−12	75
Weight, %	5	4	7	6	8	11	7	11
Aged 14 days in gas phase at 24.9°C-change								
Shore A, pts	−19	−19	−4	−3	−6	−8	−10	−14
Tensile, %	−39	−35	−29	−24	−24	−29	−22	−35
Elongation, %	22	24	−13	−12	−8	8	−18	44
Weight, %	5	5	4	4	4	3	4	4
Aged 14 days in liquid phase at 24.9°C-change								
Shore A, pts	−19	−20	−7	−4	−9	−12	−12	−18
Tensile, %	−38	−38	−37	−24	−28	−21	−17	−38
Elongation, %	21	18	−19	−23	−9	−4	−21	32
Weight, %	4	4	6	6	7	11	7	10
Aged 7 days in gas phase at 65.6°C-change								
Shore A, pts	−19	−16	6	−5	−4	−9	8	−15
Tensile, %	−33	−39	−26	−20	−19	−26	12	−38
Elongation, %	−4	−9	−12	−20	−20	4	−88	6
Weight, %	8	8	6	6	8	10	11	12
Aged 7 days in liquid phase at 65.6°C-change								
Shore A, pts	−24	−20	−9	−8	−14	−15	−14	−22
Tensile, %	−30	−36	−27	−25	−12	−16	32	−36
Elongation, %	−13	−14	−11	−19	−24	−70	−99	−1
Weight, %	4	4	6	5	6	10	11	11
Aged 14 days in gas phase at 65.6°C-change								
Shore A, pts	2	9	−3	−2	6	−3	11	2
Tensile, %	−31	−27	−10	−22	−22	−2	145	−25
Elongation, %	−68	−71	−10	−16	−71	−15	−99	−60
Weight, %	9	9	7	6	11	11	16	15

(continued)

TABLE 16A.6 (Continued)
Sour Environment Tests on Commercial Ram BOP Packers

Sample	MS710 Regular Slab	MS710 Regular Part	MS715 Hi Temp Slab	MS715 Hi Temp Part	X Regular Part	X Hi Temp Part	Y Regular Part	Z Regular Part
Aged 14 days in liquid phase at 65.6°C-change								
Shore A, pts	2	9	−3	−2	6	−3	11	2
Tensile, %	−34	−34	−24	−24	−10	−14	74	−16
Elongation, %	−68	−65	−15	−17	−66	−20	−99	−53
Weight, %	6	6	6	6	8	10	16	14
Aged 7 days in gas phase at 121.1°C-change								
Shore A, pts	7	8	14	13	10	15	15	5
Tensile, %	−66	−68	68	89	−82	90	−4	−67
Elongation, %	−99	−99	−60	−61	−100	−61	−99	−98
Weight, %	8	8	9	9	12	16	10	15
Aged 7 days in liquid phase at 121.1°C-change								
Shore A, pts	9	15	14	12	12	13	11	7
Tensile, %	−62	−57	62	71	−76	100	−28	−54
Elongation, %	−99	−99	−66	−68	−100	−70	−100	−98
Weight, %	8	8	9	9	12	14	11	14
Aged 14 days in gas phase at 121.1°C-change								
Shore A, pts	16	14	14	12	15	14	9	8
Tensile, %	−33	−38	65	88	−64	133	−20	−28
Elongation, %	−99	−99	−78	−78	−99	−82	−99	−98
Weight, %	10	9	10	9	14	16	13	14
Aged 14 days in liquid phase at 121.1°C-change								
Shore A, pts	15	13	12	10	15	11	7	9
Tensile, %	−30	−34	46	67	−53	80	NA[a]	−26
Elongation, %	−99	−99	−74	−72	−100	−80	NA[a]	−99
Weight, %	10	9	10	9	13	14	10	15

Note: Gas phase 5/20/75 $H_2S/CO_2/CH_4$ at 6.9 MPa Liquid phase Jet Fuel A.

[a] NA indicates that the dumbell would not hold in the grips.

ester based drilling fluids that have compatibility problems with some elastomers. General compatibility is available from several sources for most drilling fluids.

The completion environment offers a mixed bag of chemical environments that varies from extremely acidic to extremely basic, from man made fluids to naturally produced fluids, from simple salt solutions to complex polymers, and from heavy immature hydrocarbons to solvents. This environmental diversity makes for a difficult design condition without taking into account pressure and temperature.

The overall completion process can also cause design difficulties since the mixed bag of fluid types above may occur at different times in the life of a producing well. The chemistry of production enhancement techniques is extremely complex and varies wildly with the downhole conditions. To add to this, extremes of pressure

TABLE 16A.7
Commercial Ram Packers Tested in 5% and 20% H₂S

Sample	MS715 Hi Temp Part		X Hi Temp Part	
H₂S, %	5	20	5	20
Original properties				
Shore A, pts	81	81	78	78
Tensile, MPa	18.5	18.5	17.9	17.9
Elongation, %	175	175	316	316
Aged 7 days at 24.9°C vapor phase-change				
Shore A, pts	−4	−12	−9	−19
Tensile, %	−11	−32	−10	−23
Elongation, %	−6	−11	4	0
Weight, %	2	8	2	12
Aged 7 days at 24.9°C liquid phase-change				
Shore A, pts	−8	−12	−15	−21
Tensile, %	−14	−31	−27	29
Elongation, %	−11	−25	4	−13
Weight, %	6	9	11	16
Aged 14 days at 24.9°C vapor phase-change				
Shore A, pts	−3	−7	−8	−11
Tensile, %	−24	−39	−29	−41
Elongation, %	−12	−24	8	−14
Weight, %	4	6	3	10
Aged 14 days at 24.9°C liquid phase-change				
Shore A, pts	−4	−7	−12	−13
Tensile, %	−24	−41	−21	−36
Elongation, %	−23	−29	−4	−9
Weight, %	6	6	11	9
Aged 7 days at 65.6°C vapor phase-change				
Shore A, pts	−5	−7	−9	−9
Tensile, %	−20	−17	−26	−5
Elongation, %	−20	2	4	−26
Weight, %	6	6	10	12
Aged 7 days at 65.6°C liquid phase-change				
Shore A, pts	−4	−8	−9	−15
Tensile, %	−25	−21	−16	−18
Elongation, %	−19	−5	−12	6
Weight, %	5	6	10	12
Aged 14 days at 65.6°C vapor phase-change				
Shore A, pts	−2	−1	−3	−5
Tensile, %	−22	−11	−2	1
Elongation, %	−16	−17	−15	−22
Weight, %	6	6	11	13

(*continued*)

TABLE 16A.7 (Continued)
Commercial Ram Packers Tested in 5% and 20% H_2S

Sample	MS715 Hi Temp Part		X Hi Temp Part	
H_2S, %	5	20	5	20
Aged 14 days at 65.6°C liquid phase-change				
Shore A, pts	−3	−5	−8	−10
Tensile, %	−24	−32	−14	−14
Elongation, %	−17	−25	−20	−24
Weight, %	6	6	10	11
Aged 7 days at 121.1°C vapor phase-change				
Shore A, pts	13	13	15	14
Tensile, %	89	−6	90	−8
Elongation, %	−61	−80	−61	−90
Weight, %	9	14	16	22
Aged 7 days at 121.1°C liquid phase-change				
Shore A, pts	12	12	13	14
Tensile, %	71	−19	100	−14
Elongation, %	−98	−85	−70	−93
Weight, %	9	14	14	20
Aged 14 days at 121.1°C vapor phase-change				
Shore A, pts	8	10	11	12
Tensile, %	88	−57	133	NA[a]
Elongation, %	−78	−94	−82	NA[a]
Weight, %	9	18	16	25
Aged 14 days at 121.1°C liquid phase-change				
Shore A, pts	10	11	11	12
Tensile, %	67	−51	80	−45
Elongation, %	−72	−95	−80	−96
Weight, %	9	17	14	24
Aged 7 days at 148.9°C vapor phase-change				
Shore A, pts	8	11	10	12
Tensile, %	−27	NA[a]	−4	NA[a]
Elongation, %	−81	NA[a]	−88	NA[a]
Weight, %	10	23	16	25
Aged 7 days at 148.9°C liquid phase-change				
Shore A, pts	8	12	9	13
Tensile, %	−37	NA[a]	−30	−65
Elongation, %	−84	NA[a]	−90	−97
Weight, %	9	23	18	24
Aged 14 days at 148.9°C vapor phase-change				
Shore A, pts	8	10	11	12
Tensile, %	−40	NA[a]	37	105
Elongation, %	−92	NA[a]	−94	−93
Weight, %	15	20	21	24

TABLE 16A.7   (Continued)
Commercial Ram Packers Tested in 5% and 20% H₂S

Sample	MS715 Hi Temp Part		X Hi Temp Part	
H₂S, %	5	20	5	20
Aged 14 days at 148.9°C liquid phase-change				
Shore A, pts	8	10	9	11
Tensile, %	−36	NA[a]	NA[a]	NA[a]
Elongation, %	−91	NA[a]	NA[a]	NA[a]
Weight, %	15	20	20	27
Aged 7 days at 176.6°C vapor phase-change				
Shore A, pts	7	11	6	14
Tensile, %	−48	NA[a]	−20	NA[a]
Elongation, %	−87	NA[a]	−89	NA[a]
Weight, %	11	16	17	23
Aged 7 days at 176.6°C liquid phase-change				
Shore A, pts	6	10	2	13
Tensile, %	−52	−82	−47	NA[a]
Elongation, %	−84	−97	−90	NA[a]
Weight, %	11	16	16	23
Aged 14 days at 176.6°C vapor phase-change				
Shore A, pts	8	9	9	12
Tensile, %	−49	NA[a]	−34	NA[a]
Elongation, %	−88	NA[a]	−91	NA[a]
Weight, %	13	15	20	26
Aged 14 days at 176.6°C liquid phase-change				
Shore A, pts	6	12	6	11
Tensile, %	−51	NA[a]	−59	−72
Elongation, %	−87	NA[a]	−93	−95
Weight, %	14	19	20	27

Note: Tested in Steps from 24.9 to 176.6°C. Gas Phases 5/20/75 and 20/20/60 H₂S/CO₂/CH₄ at 6.9 MPa. Liquid Phase Jet Fuel A.

[a]   NA indicates that the samples shattered before testing.

often are experienced during the completion process. Though a completed well may never see differential pressures that exceed 6.9 MPa, hydraulic fracturing to enhance production may place several times that differential across the completion equipment. This must be considered in the overall scope of the design or the consequences of this oversight could be devastating.

The work-over environment is much like a cross between the drilling and completion environment with one additional constraint, it has further cost consideration that the other two do not. This cost consideration can complicate the environment worse than some of the chemicals.

16.6 QUALIFICATION TESTS

Qualifying a compound involves a number of different tests to sufficiently characterize the compound's performance up to and including full-blown part testing when exposed to the extremes of service environments. These tests can be quite expensive and time consuming. When compounds for large parts used in blowout preventers are qualified, it can take several hundred pounds of rubber to manufacture just one part. Obviously, then, qualification tests cannot be a requirement for each batch of compound. Once the compound is qualified, short term, easy to run QC tests are specified for controlling the quality of production batches.

Probably the hardest aspect of elastomer design is determining the requirements. Since elastomer performance is determined by multiple factors, and their relationship to each other, it gets very difficult to develop a strict list of hard and fast physical properties that will relate to the prescribed performance. An excellent performance in one physical aspect may mask a perceived weakness of another. One of the best ways to answer the above dilemma is to develop requirements, or specifications, around application-based testing. There are two distinct ways to do this, based on physical characteristics and performance. One is to do comparative testing of potential rubber compounds with variations in the mixture. This will help establish the proper blend by determining the proper relationship of the physical properties. This makes for a lot of testing but will best blue print the compound. A second way is to develop performance-based specifications. Performance-based specifications are the result of an elastomer performing successfully at simulated worst conditions. These tests are usually pass/fail tests. Performance based specifications are usually much easier to develop and require less testing, but often do not supply a clear cut understanding of why a compound works, only that it does.

16.7 LIFE PREDICTION

In addition to required performance when exposed for short terms to severe environments, oil-field elastomers must provide extended lifetime service of up to 20 years in those environments. As noted previously, elastomer properties change with time and before 1980, estimates of life prediction were difficult to assess. At that time, a life prediction technique based on testing under worst case conditions at temperatures in excess of service requirements was introduced to our industry [5]. This technique has been used in our industry successfully since then and is covered in Chapter 16B as well as Rubber Division, ACS papers [6,7].

ACKNOWLEDGMENTS

This chapter is based upon information and suggestions of Charles Raines and Don Hushbeck, for which the author is very grateful.

REFERENCES

1. Raines, Charles, Review of Elastomeric Materials Commonly Used in the Oilfield. Presented to the Energy Rubber Group Educational Symposium, Houston, TX, September 17, 1996.
2. Hashimoto, K., Watanabe, N., and Todani, Y., Introduction of Zetpol, New Elastomers for Offshore Applications. Presented to Plastics & Rubber Institute on Offshore Engineering and the Institution of Mechanical Engineers "Polymers in Oil Exploration Seminar," London, June 12, 1986.
3. Raines, Charles, Effect of Severe H_2S Environments on Elastomers From Commercial Ram BOP Packers. Presented to the Spring Technical Meeting of the Rubber Division of ACS, Dallas, Texas, April 5, 2000.
4. Raines, C. and Callahan, M., Pitfalls of Severe Environment Testing. Presented to NACE Corrosion 90, Las Vegas, Nevada, April 22–27, 1990.
5. Vicic, John, Testing of Polymers for Oil and Gas Applications. Presented to the Energy Rubber Group, Houston, Texas, September 27, 1984.
6. Page, Nigel, Seal Life in Real Life. Presented to the Spring Technical Meeting of the Rubber Division of ACS, Dallas, Texas, April 5, 2000.
7. Vicic, John et al., A Review of Polymer Life Estimation Using Variable Temperature/ Stress Acceleration Methods. Presented to the Spring Technical Meeting of the Rubber Division of ACS, Dallas, Texas, April 5, 2000.

16 Part B: Life Prediction

John Vicic

CONTENTS

Elastomers are used in many critical service applications that require long, trouble-free life such as seals, gaskets, bridge-bearing pads, tires, and medical components. The ability to estimate long-term durability or the time to failure for elastomer materials and components presents a challenge to both scientists and technologists that work with these materials.

Accelerated aging methods wherein specific elastomer properties are measured at a number of elevated temperatures and stresses over different periods of time provide a means for estimating long-term performance. These same methods may be used, in some cases, to estimate the service performance of actual components. This type of methodology also enables an investigator to determine a rate expression that can be useful for predicting the behavior of an elastomer under other conditions of time, temperature, and stress.

Interest in this area is high and a number of international scientific meetings are dedicated to this subject. Some good general publications in the area include those by Vicic [1], Stevenson [2], Le Huy [3], and Gillen [4]. The American Petroleum Institute Committee 6 and the National Association of Corrosion Engineers Committee T1-G have task groups working on developing accelerated life estimation tests for elastomer seals. The API task group issued a recommended test practice API TR 6JI "Elastomer Life Estimation Test Procedures" [5]. The Norwegian Technology Centre also developed standard NORSOK M 710 Rev. 2 "Qualification of Non-metallic Sealing Materials and Manufacturers" that includes accelerated life estimation tests [6].

The test methodology is traceable to Arrhenius' theory of the thermal acceleration of reaction kinetics and the concept of the activation energy of a process. Arrhenius associated the variation in reaction velocity to temperature as shown in Equation 16B.1 [7]. From this relation, the natural logarithm of the reaction rate k is proportional to $1/T$, and the slope is equal to $-E_a/R$. Thus, if the reaction rates are known for several temperatures, E_a can be calculated or a plot can be constructed to determine E_a. This same plot can be used to extrapolate reaction rates at other temperatures.

$$Ln(k) = -E_a/RT + C \qquad (16B.1)$$

where
   k = reaction rate
   E_a = activation energy
   R = gas constant
   T = absolute temperature, K
   C = constant

In the simplest case, if viscoelastic behavior and degradation are taken as thermally activated molecular processes with constant activation energy, the Arrhenius equation can be used to estimate long-term changes in an elastomer [8]. Reaction rate data for an elastomer property are obtained at a series of elevated temperatures under otherwise constant conditions and plotted as described earlier. The plot can then be used to estimate changes at long periods of time at reduced temperatures.

One of the problems with using test schemes based on a simple Arrhenius relation is that it does not readily account for the influence of different imposed stresses on the reaction rate. This is important, since data measured at one stress level will not necessarily provide a good estimate of reaction rate under another condition. Zhurkov investigated the time to failure of polymeric, metallic, and nonmetallic crystalline materials in uniaxial tension at a number of elevated temperatures and different stresses [9]. His experimental results fit the empirically derived expression shown in Equation 16B.2 that he called the "thermofluctuational theory."

$$\tau = \tau_0 \exp\left[(U_o - \gamma\sigma)/kT\right] \tag{16B.2}$$

where
   τ = time to failure
   τ_0 = atomic vibration period
   U_o = activation energy
   γ = structure coefficient, $\gamma = \varphi \, V_a$
   σ = stress
   k = Boltzman's constant
   T = absolute temperature, K

The factor τ_0 was found to be 10^{-13} s, which coincides with the reciprocal of the atomic vibrational frequency. The parameter U_o is taken as the activation energy, and it decreases linearly with tensile stress. The coefficient γ is a structural factor that describes the orientation of the material. Zhurkov defined γ as a coefficient that relates the activation volume, V_a, and φ, the localized overstress on a bond, to the average stress in the specimen. The source of stress on the system can be from either mechanical- or thermo-chemical effects.

Thermal acceleration techniques were used by Kennelley et al. [10] to estimate the failure life of oil-field elastomer seals in sour gas/dimethyl disulfide environments at 135°C at constant pressure. The materials were tested in the form of O-rings, which were used to seal an autoclave containing the test environment at 14 MPa gas pressure. Tests were conducted to failure at three elevated temperatures for each type of elastomer seal. Failure was defined as the inability to maintain test pressure. The time to failure data for each material was correlated to Equation 16B.3.

$$\text{Log}(t) = \log(C) + E_a/[2.303R(1/T)] \tag{16B.3}$$

where

t = time to failure, hours
C = constant related to bond degradation
E_a = apparent activation energy for the failure process
R = gas constant
T = absolute temperature, K

Regression analysis of the failure life data, with replicate points, to the model equation gave correlation coefficients of 0.99–1.00. This demonstrates an excellent fit to the model for logarithm of time to failure in hours against $1/T$ for the various materials. Extrapolation of the results to service conditions revealed substantial differences in estimated seal life that facilitated selection of a seal material for the application. Additionally, the seal life prediction results were in reverse order from that observed in an earlier study that relied on single-point mechanical property determinations from autoclave immersion test.

Janoff, Vicic, and Cain [11] used accelerated life test methods to select the best material to use for lip seals used in water/glycol fluids contained in subsea oil-field hydraulic systems. The required seal life was 20 years at 93°C sustained temperature at 20.7 MPa pressure with 100 mechanical operations. Finite Element Analysis was used to determine the actual seal temperature in the installed component to establish the service temperature rating. Polyurethane and thermoplastic alloy seals were tested in water/glycol and hydrocarbon hydraulic fluids at 3–5 elevated temperatures at 20.7 MPa constant pressure. Failure was defined as seal leakage at pressure. Test results were analyzed using Least Squares regression in an equation like that used by Kennelley et al. The correlation coefficients exceeded 0.95 and show excellent fit to the model. Table 16B.1 shows the results of the study.

TABLE 16B.1
Life Estimation Comparison Table

TPA		Urethane	
Temperature, °C	Life Estimation Years	Temperature, °C	Life Estimation Years
na	na	16	42
na	na	27	12
38	7.7E + 08	38	3.9
66	3.9E + 05	66	.3
93	640	na	na
99	198	na	na
104	63	na	na
110	21	na	na

518 Handbook of Specialty Elastomers

Elastomer life prediction methods will continue to evolve and be an area of strong interest in the future. End-use applications, like seals used in ultra deep-water subsea production equipment, continue to increase in severity. Additionally, even more mundane uses of elastomers can benefit from life prediction through more cost-effective material selection and reduced warranty costs.

REFERENCES

1. J. Vicic et al., "A Review of Polymer Life Estimation Using Variable Stress/Temperature Acceleration Methods," Elastomer Service Life Prediction Symposium'97, August 1997, Akron, Ohio.
2. A. Stevenson, "An Overview of the Requirements for Seal Life Prediction," Paper 65, Corrosion 92, 1992.
3. M. Le Huy and G. Evrard, Methodologies for Lifetime Predictions of Rubber Using Arrhenius and WLF Models. *Die Angewandte Makromolekulare Chemie*, 261/262 (4624):135–142, 1998.
4. K.T. Gillen, R.L. Clough, and J. Wiese, Prediction of Elastomer Lifetimes from Accelerated Thermal Aging Experiments. In R.L. Clough, N.C. Billingham, and R.L. Clough, editors, *Advances in Chemistry Series*, number 249, pp. 557–575, American Chemical Society, Washington D.C., 1996.
5. API TR 6JI, "Elastomer Life Estimation Test Procedures," American Petroleum Institute, Washington D.C.
6. NORSOK M 710, "Qualification of Non-metallic Sealing Materials and Manufacturers," Rev. 2 October 2001, Norwegian Technical Centre, Oslo, Norway.
7. S. Maron and C. Prutton, *Principles of Physical Chemistry*, 4th Ed., Macmillan Company, New York, 1970, chapter 13, pp. 571–585.
8. I.M. Ward, *Mechanical Properties of Solid Polymers*, 2nd Ed., John Wiley & Sons, New York, 1983, chapter 7, pp. 142–144.
9. S.N. Zhurkov, *Intern. J. Fracture Mech.*, 1, 311 (1965).
10. K.J. Kennelley, P.J. Abrams, J.C. Vicic, and D. Cain, "Failure Life Determination of Oilfield Elastomer Seals in Sour Gas/Dimethyl Disulfide Environments," Paper 212 Presented at Corrosion 89, NACE, New Orleans, April 17–21, 1989.
11. D. Janoff, J. Vicic, and D. Cain, "Thermoplastic Elastomer Alloy, TPA, Sub sea Hydraulic Seal Development for Service Including Water-Based Fluids," Conference Papers, International Conference on Oilfield Engineering with Polymers, October 28–29, 1996, London, UK.

16 Part C: Compression, Transfer, and Injection Molding of Specialty Elastomers

Robert W. Keller

CONTENTS

16.1 INTRODUCTION

While the basic technology of compression, transfer, and injection molding of rubber has existed for many years, industry requirements for greater performance, improved quality, and improved value have increased. Along with these increasing requirements, the relentless pressure from Automotive, Consumer Products, Heavy Equipment, and Aerospace for reduced costs has forced rubber molders to rethink and reengineer existing processes as well as invest in new processes. During this time of change, frustrations arise when the promise of an easy conversion from, for example, compression to injection molding suddenly becomes a black hole of effort and money. Unfortunately, many of these changes are driven by business concerns and application of the existing scientific knowledge base and the use of good problem-solving techniques are often neglected in up-front design.

The purpose of this chapter is to review the basic technologies of rubber viscoelastic flow and the basic technologies of compression, transfer, and injection molding and to offer guidance in the use of existing scientific knowledge to assist in problem solving. It has been the experience of many that molding issues in production rubber manufacturing degenerate rapidly into emotional and political issues. In this chapter, much of the scientific background necessary for understanding rubber molding flow will be presented with applications in solving common problems encountered in rubber molding. With the tools presented here, solutions to many of the common problems facing production molding operations will be available. The reader can skip forward to the end of the chapter to some specific problem-solving examples, but reference to the early parts of the chapter will help in understanding cause and effect relationships in flow and molding. Hopefully, with a better understanding of rubber flow and viscoelasticity, the reader will be better able to resolve issues through the application of known scientific principles and disciplined problem solving.

16.2 BASICS OF RUBBER VISCOELASTIC FLOW

16.2.1 Viscosity

Rubber and plastic melts can be considered, to a first approximation, as extremely high-viscosity fluids. This is only an approximation and it must be remembered that polymers generally show viscoelastic properties—a combination of viscous flow and elastic recovery. Viscosity, in turn, is the quantitative measure of resistance to flow under a given set of circumstances. The Greek letter that usually designates viscosity is η. For an ideal, Newtonian fluid, viscosity is simply the ratio between Shear Stress (τ), the pressure placed on the fluid to create flow, and the Shear Rate (γ), the rate of flow over time as seen in Equation 16C.1:

$$\eta = \tau/\gamma \qquad\qquad (16C.1)$$

where
  η = viscosity expressed as Pa/s
  τ = shear stress in Pa
  γ = shear rate in 1/s

Thus, for an ideal fluid, the viscosity is constant over the entire range of shear rates at a given shear stress. Life would certainly be easy if viscosity were such a simple function. However, polymer melts such as rubber generally do not show a constant viscosity across a range of shear rates. Polymer melts generally show pseudoplastic behavior, which can be described as the reduction of viscosity with increasing shear rates. The best, everyday way to visualize pseudoplastic behavior is to visualize thick Tomato Ketchup: if you upend the bottle and hold, you will probably see the Ketchup stay in the bottle, but if you shake the bottle too vigorously up and down, copious amounts of Ketchup go everywhere including into your lap. At zero shear rate caused by simply upending the bottle, the viscosity (resistance to flow) is extremely high and nothing seems to happen. With vigorous up and down shaking, the shear rate goes up dramatically and the viscosity drops and too much of the product exits the bottle.

A typical viscosity curve as a function of shear rate is shown in Figure 16C.1 for a rubbery material. At low shear rates, the material shows Newtonian behavior and viscosity appears constant regardless of changes in shear rate. As shear rate increases, the material transforms from an apparent Newtonian fluid to a pseudo-plastic material—viscosity decreases with increasing shear rate. This is a curve that we will see many times throughout this chapter and we will explore ways of using this type of data in understanding flow in molding and how to solve flow-related problems. Please look at this curve and form a mental picture since it will help you solve problems and explain situations properly.

Now that we know that rubber materials are not ideal Newtonian fluids and we understand that viscosity decreases with increasing shear rate, we need a mathematical relationship to describe these observations. Fortunately, quite a lot of scientific analysis has been devoted to developing the mathematics needed to describe the actual behavior of rubber and other polymers in flow [1–11]. Two relationships developed to more accurately describe polymer flow are given in Equations 16C.2 and 16C.3:

$$\eta = K\gamma^{n-1} \tag{16C.2}$$

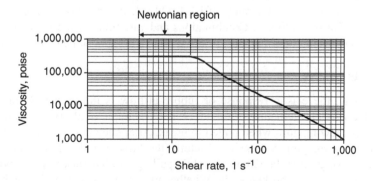

FIGURE 16C.1 Typical viscosity—shear rate behavior for rubber.

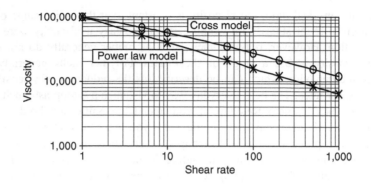

FIGURE 16C.2 Comparison of Power Law and Cross models for viscosity—shear rate behavior of rubber.

where

$\eta =$ viscosity

$\gamma =$ shear rate in 1/s

$K =$ curve-fitting parameter determined from experimental data

$n =$ curve-fitting parameter determined from experimental data

$$\eta = \eta_o/(1 + A\gamma^{1-n}) \tag{16C.3}$$

where

$\eta_o =$ zero shear rate viscosity

$A\ =$ curve-fitting parameter determined from experimental data

$n\ =$ curve-fitting parameter determined from experimental data

Equation 16C.2 is called the Power Law and Equation 16C.3 is called the Cross Model. Plots of these two models are shown in Figure 16C.2. In general, the Cross Model does a better job of describing the actual viscosity versus shear rate data observed on commercial polymers. Now we have a mathematical model that describes the actual viscosity versus shear rate profile of rubber materials.

In general, the shear rate of compression molding occurs in the range of $1–10\ s^{-1}$, transfer molding occurs in the range of $10–1,000\ s^{-1}$, and injection molding occurs in the range of $500–10,000\ s^{-1}$. Combinations and variations of these techniques can give some interesting intermediate shear rates and a number of new problems.

16.2.2 ELASTICITY

Unfortunately, the elastic response of rubber in flow is the least well-understood facet of the viscoelastic nature of rubber. A good example of elasticity in flow is the behavior of chewing gum when pulled to break. After chewing and softening, if one pulls the chewing gum to tension fracture, the broken ends of the gum retract. The flow of the gum is the extension while pulling and the elasticity is represented by the recovery of the broken ends. Whether in compression, transfer, or injection molding, the flow of the rubber compound is also accompanied by elastic rebound.

If we ignore the elastic component of rubber flow, the results can sometimes be quite expensive. One of the best ways to study this effect has been to study the extrudate swell, die swell, of polymers forced through various dies. As the polymer exits the die, it will swell in diameter and decrease in length. Unfortunately, at the high shear rates of injection molding or with lightly filled elastomers, the extrudate swell is anything but smooth and predictable. The knotted and twisted appearance of unfilled or lightly loaded elastomers when exiting a die at moderate to high shear rates is very difficult to quantify and study. This may also explain the knotted, twisted, or dimensionally noncompliant nature of molded rubber parts when the compound is lightly loaded or when shear rates are very high.

In one attempt to quantify and describe extrudate swell, Tanner [12] found that extrudate swell, B, could be estimated using Equation 16C.4:

$$B = C + \left[1 + \frac{1}{2}(\tau \gamma_w)^2\right]^{1/6} \tag{16C.4}$$

where
  C = constant approximately equal to 0.12
  τ = shear stress
  γ_w = shear rate at wall of die

Equation 16C.4 applies to molten polymers but does not describe fully filled and compounded elastomers well. However, it is useful for quantifying the changes in elastic rebound induced by increasing shear rates and increasing shear stresses. Using Equation 16C.4, we can set up the proportionality between elastic recovery and shear stress or shear rate. Figure 16C.3 shows the effects of percentage increases in shear rate on the function B or the elastic recovery. This type of relationship may be important in describing dimensional problems encountered in high shear rate molding such as transfer or injection molding. For example, a rubber part is originally established in injection molding with a 20 s injection time followed by a 60 s vulcanization time. Because of cost pressure on the manufacturer, reductions in cycle time are required. The manufacturer then raises the injection pressure on the machine to reduce injection time to 10 s to reduce overall cycle. Unfortunately, this increases the shear

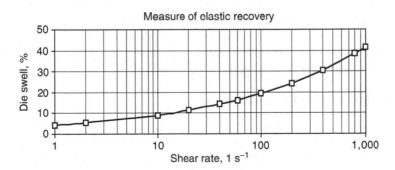

FIGURE 16C.3 Effects on Die Swell of increasing shear rate.

rate by a factor of 2, which increases the elastic rebound shown as B in Equation 16C.4. This rebound manifests itself as a sudden shift in the diameter of the finished part making the part undersized in diameter. Very quickly, the manufacturer has turned his injection press into a high-speed scrap machine. As mentioned previously, this is the least understood facet of rubber viscoelasticity in flow, but it is important in understanding some of the undesirable effects of process changes.

16.3 EFFECTS OF PROCESS VARIABLES

16.3.1 TEMPERATURE EFFECTS ON VISCOSITY

Like engine oil, polymer melts show reduced viscosity with increasing temperature. In the absence of experimental data defining these curves, which is mostly the case, this can be determined by using the Williams, Landel, and Ferry relationship for time-temperature superposition [13]. This relationship can be applied to modulus values and to viscosity as well. Applying the relationship to viscosity we get Equation 16C.5:

$$\log\left(\eta_{T1}/T_1\right) - \log\left(\eta_{T_g}/T_g\right) = \log\left(a_{T1}\right) \qquad (16C.5)$$

where
   η_{T1} = viscosity at absolute temperature T_1
   η_{Tg} = viscosity at glass transition temperature
   a_T = shift factor from data or from estimate

Obviously, determining viscosity at T_g would be difficult or nearly impossible. However, if we are just interested in comparing the viscosities at two different process temperatures, T_2 and T_1, we can simplify and use relative comparisons:

$$\log\left(\eta_{T2}/T_2\right) - \log\left(\eta_{T1}/T_1\right) = \log\left(a_{T2}\right) - \log\left(a_{T1}\right)$$

Where the values of a_{Tx} can be estimated using the following:

$$\text{Log}(a_{T_x}) = [\{-17.44\,(T_x - T_g)\}/\{51.6 + (T_x - T_g)\}]$$

This is shown in Figure 16C.4 for the theoretical polymer from Figure 16C.1 using $-20°C$ as the glass transition temperature. For a material with a very low glass transition temperature such as silicone, the reduction in viscosity with increasing temperature is less dramatic. As the materials get further away from glass transition, the viscosity reduction with temperature becomes less.

16.3.2 HEAT BUILD UP DURING FLOW

One of the consequences of rubber viscoelastic behavior is heat build up during high shear flow. During relatively low shear rate compression molding processes, this is probably not a significant factor. However, during higher shear molding such as transfer and injection molding, this heat build up can be the cause of premature scorch and quality problems. Here again, exact relationships have not been

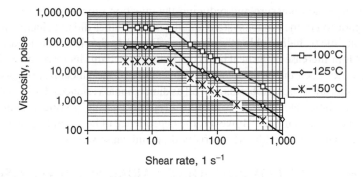

FIGURE 16C.4 Effects of temperature on viscosity.

established for describing the heat build up during high shear flow, but Equation 16C.6 can be used to establish correlations and predict the effects of viscosity increases and shear rate increases.

$$6\Delta T = \int_0^t (\eta/\rho c)\gamma^2 dt \qquad (16C.6)$$

where
  ΔT = temperature change per second in °C
  η = viscosity in Pa-s
  γ = shear rate in 1/s

  Using Equation 16C.5, we can establish the following proportionalities to estimate the effects of viscosity and shear rate on heat build up:

$$\Delta T \propto \eta \quad \text{and} \quad \Delta T \propto \gamma^2 \qquad (16C.7)$$

Using the proportionalities in Equation 16C.7, we can graph the effects of changing viscosity and shear rate on the heat build up and this is shown in Figure 16C.5. Using the example from the previous section, we decreased an injection cycle time by

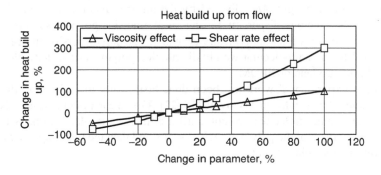

FIGURE 16C.5 Relative effects of viscosity and shear rate on heat build-up.

increasing injection pressure and reducing the injection time from 20 to 10 s. Since the configuration of the runner system and the mold cavities did not change, reduction of injection time by half doubled the shear rate. Doubling the shear rate, in turn, caused a factor of four increase in the heat build up during flow. If this increase in heat build up is sufficient to reach the critical scorch temperature of the vulcanization system, we can see premature scorch problems as well as dimensional variances. Obviously, as we increase the shear rate, the viscosity will decrease causing a corresponding linear decrease in the heat build up. However, the dependence of heat build up on the square of the shear rate can override the viscosity drop and cause scorch issues.

16.4 EFFECTS OF COMPOUNDING VARIABLES ON VISCOSITY, ELASTICITY, AND FLOW

16.4.1 EFFECTS OF POLYMER MOLECULAR WEIGHT

This is the starting point for practical rubber compounds and will be, obviously, the first item discussed. Commercial polymers including rubber show a distribution of molecular weights. Generally, the molecular weight distribution of commercial polymers is skewed toward the higher molecular weights. Various molecular weight averages are used to describe these distributions and the width of the distribution. The general formula for molecular weight averages is given in Equation 16C.8:

$$M_j = \left(\sum N_i M_i^j\right) \Big/ \left(\sum N_i M_i^{j-1}\right) \qquad (16C.8)$$

where

N_i = number of molecules with molecular weight M_i

If $j = 1$, then this is the number average or simple arithmetic mean, M_n

If $j = 2$, then this is the weight average molecular weight, M_w, which is sensitive to the higher molecular weight fractions

If $j = 3$, this is called the z average molecular weight, M_z

If $j = 4$, then this is called the $z + 1$ average molecular weight, M_{z+1}

A typical distribution of molecular weights with the various averages is shown in Figure 16C.6. The breadth of the distribution is often expressed as the weight average molecular weight, M_w, divided by the number average molecular weight, M_n, and is called the polydispersity. For a typical emulsion polymer such as SBR or NBR, polydispersity values are typically around 2. The polydispersity number can be larger than 2 for broad molecular weight distribution polymers such as Ziegler–Natta catalyzed polyethylene or EPDM.

A lot of experimental work has been done historically to understand the influence of molecular weight on viscosity [14–22]. For linear polymers, the viscosity varies linearly with molecular weight until a critical molecular weight is reached where entanglement of the polymer chains occurs and then the viscosity varies by the 3.4 power of molecular weight.

$$\eta \propto M \text{ where } M < M_c \quad \text{and} \quad \eta \propto M^{3.4} \text{ where } M > M_c \qquad (16C.9)$$

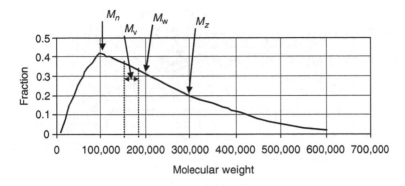

FIGURE 16C.6 Typical molecular weight distribution of commercial polymers.

where
   η = viscosity
   M = molecular weight
   M_c = critical molecular weight for entanglement

The value of critical molecular weight for entanglement varies with the type of polymer and shows dependence on the bulk of the pendant groups in vinyl (carbon–carbon backbone) polymers. The value of M_c for polyethylene (carbon–carbon backbone with no bulky pendant groups) is approximately 4,000 while the M_c for polystyrene (carbon–carbon backbone with a bulky, phenyl pendant group) is approximately 38,000. In general, rubber polymers have molecular weights well above M_c so viscosity shows a 3.4 power dependence on molecular weight.

The question then arises, since rubber polymers have a distribution of molecular weights, which average molecular weight should be used to establish the dependence of viscosity. Fox and coworkers found that the weight average molecular weight, M_w, best described the dependence of viscosity on molecular weight:

$$\eta_0 = kM_w^{3.4} \tag{16C.10}$$

where
   η_0 = zero shear rate viscosity
   k = constant for a particular polymer system

An interesting analysis done on polyethylene with very broad molecular weight distributions [23] indicated that viscosity average molecular weight gave a better correlation to experimental data. This may be relevant to EPDM polymers made with similar polymerization catalysts giving very broad molecular weights. A convenient way to estimate viscosity and molecular weight is with the traditional Mooney viscometer. Since this operates in the approximate range of 2 s^{-1} shear rate, it is very close to the zero shear rate viscosity and should show proportionality to $M_w^{3.4}$. Thus, the data reported by polymer suppliers as Mooney viscosity can be used to estimate the relative weight and average molecular weight. For a first approximation, we can assume that two polymers varying by Mooney viscosity will have the same

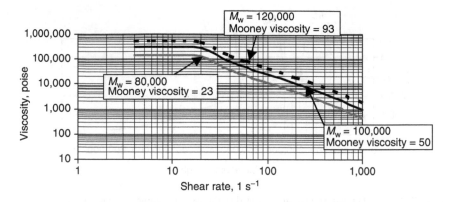

FIGURE 16C.7 Effects of weight average molecular weight on viscosity.

degree of branching or linearity and the viscosity shear rate curves will be parallel. Therefore, if we know the viscosity difference between the two polymers at low shear rates, we can estimate the viscosity at higher shear rates if we know the viscosity shear rate curve for one of the polymers. For example, if two polymers from the same supplier and same polymerization process have Mooney viscosities of 35 and 50 and we know that the relative viscosities of the two polymers at high shear rates should follow the ratio of Mooney viscosities. In other words, the 50 Mooney polymer has roughly 43% higher viscosity at low shear rate and should have roughly 43% higher viscosity at higher shear rates as well since the curves will parallel each other. This is shown as an example in Figure 16C.7.

16.4.2 EFFECTS OF FILLERS

Unfortunately, this is one area that is not really well understood. The basic theory of particle reinforcement of liquids was developed as refinements of the Einstein Equation (24,25):

$$\eta_1 = \eta_0^*(1 + A\phi + B\phi^2) \tag{16C.11}$$

where
η_0 = unfilled viscosity
η_1 = viscosity of filled material
ϕ = volume fraction filler in the mixture
A = constant which is 2.5 for spherical particles
B = constant which is in the range of 10–14 for spherical particles

$$G_1 = G_0^*(1 + A\phi + B\phi^2) \tag{16C.11a}$$

where
G_0 = unfilled shear modulus
G_1 = shear modulus of filled material
ϕ = volume fraction filler in the mixture

As you can see from Equation 16C.11, the type of filler should not have any impact on the viscosity nor should it have any impact on Modulus of elasticity. In the practical world that we live in, we know that N110 carbon black (SAF) at the same volume or phr loading gives much higher viscosities and much higher Modulus values compared with N990 carbon black (MT). Thus, Equation 16C.11 does not do a very good job of predicting the effects of fillers. From here, the situation becomes unclear since comprehensive studies of various fillers and their effects on viscosity and Modulus are not extensive.

White and Crowder [26] studied the effects on viscosity of various low-structure carbon blacks and found that the smaller particle size carbon blacks gave increases in viscosity and shifted the viscosity versus shear rate curves. Use of low-structure carbon blacks kept the analyses from being confounded by the structure effects (agglomeration of particles). Unfortunately, many of the useful commercial Elastomer compounds involve high-structure carbon blacks such as N550 (FEF).

In unpublished work, until now, the author has studied the combined effects of structure and particle size of carbon black on the properties of an NBR compound. Obviously, particle size, by itself could not characterize the results seen. Fortunately, the DBP number, per ASTM D2414, combines the effects of particle size and structure. The DBP number is a measure of the amount of dibutyl phthalate that the carbon black will absorb. Essentially, the DBP number is a titration to a mechanically sensed end point—when the carbon black surface can no longer absorb the liquid DBP, a paste is created in the mixing chamber and the conversion from free-flowing powder to paste is sensed by torque. The basic experimental design program used is summarized in Table 16C.1. The results for compression modulus and for low shear rate viscosity are shown in Figures 16C.8 and 16C.9, respectively. The correlations between experimental values and the plotted response surfaces was very good (greater than 0.900) and indicates that the DBP number is a good composite value for predicting the reinforcing nature of various carbon blacks. This work is far from extensive and indicates that this relationship should be developed for various polymer families and for the range of carbon blacks and other fillers. Before formulating compounds for compression, transfer, or injection molding, the compounder should consider a similar set of experiments to optimize the best fillers for a combination of properties such as compression set, strength of vulcanizate, and mold flow. In addition, in this study, elasticity was seen to increase with increasing filler loadings and increasing carbon black DBP values. This is not surprising and has been observed previously [27,28]. Loss of elasticity was measured as the hysteresis in dynamic compression testing and is shown in Figure 16C.10. Of course, this makes sense with many years of experience in the rubber industry where high structure, small particle size carbon blacks have been observed to smooth extrusions.

The situation can become even more complicated with high aspect ratio fillers and platy fillers. High-structure carbon blacks are not, overall, spherical, but some of the short fiber mineral fillers could have very unusual effects on viscosity and flow. Unfortunately, this is an area that does not have a wealth of data accumulated. One would expect some directional nature to properties after flow with short fiber or high aspect ratio fillers. In other words, the properties in the direction of flow should be

TABLE 16C.1

Experimental Design Formulations, NBR Compound. Triplicate Data Points Were Used for Each Formula

Ingredient	phr
Basic Recipe	
33% Acrylonitrile, 35 ML(1 + 4) at 100°C Polymer	100.00
Stearic acid	2.00
AgeRite Resin D	1.00
Vanox ZMTI	1.00
Carbon black (see below)	30.00 to 50.00
Zinc oxide	2.00
Varox DCP-40C	3.00

Compound	phr Carbon Black	DBP Number of Carbon Black
1	30.00	43 (N-990)
2	50.00	43
3	70.00	43
4	30.00	65 (N-762)
5	50.00	65
6	70.00	65
7	30.00	90 (N-660)
8	50.00	90
9	70.00	90
10	30.00	121 (N-550)
11	50.00	121
12	70.00	121

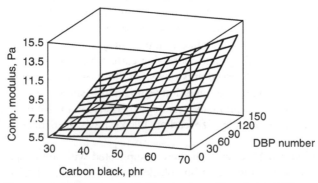

Response surface for compression modulus

R-squared correlation adjusted for degrees of freedom = 91%

FIGURE 16C.8 Effects of loading and DBP number of carbon black on NBR compression modulus.

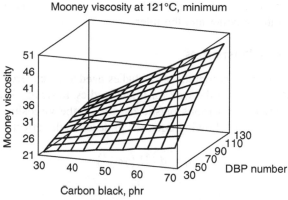

FIGURE 16C.9 Effects of loading and DBP number of carbon black on NBR Mooney viscosity.

significantly different than the properties perpendicular to the direction of flow with high aspect ratio fillers. Hopefully, the information given here will provide some guidance for further testing and evaluations.

Another complicating factor could be any bonding or strong interaction between the polymer and the filler. Coupling agents act by inducing some covalent bonding between the polymer and the filler. This, in turn, would increase the effective volume of rubber bound and immobilized by the filler particles. By current theory, the amount of interaction and the effective volume of polymer immobilized by the filler is believed to explain the deviation from Equations 16C.11 and 16C.11a. Particularly interesting is the interaction between high surface area fumed silica and silicone elastomers. Without coupling agents or any other polymer-filler bonding

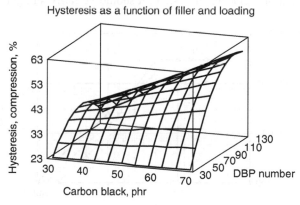

FIGURE 16C.10 Effects of loading and DBP number of carbon black on NBR hysteresis.

chemicals, solvent swell, and bound rubber testing would indicate some covalent bonding between the polymer and the filler.

16.4.3 Effects of Plasticizers

This is also an area where comprehensive studies need to be undertaken. Some good work has been done to point the way for future studies, however. Kraus and Gruver [29] studied the effects of various plasticizer oils on the viscosity of rubber compounds and proposed the following equation to describe the effects on viscosity:

$$\eta = \phi_2^{3.4} F\left(\gamma \phi_2^{1.4}\right) \qquad (16C.12)$$

where
$\phi_2 =$ volume fraction polymer in compound (i.e., ϕ_2 would be 0.80 for 80% polymer with 20% plasticizer which would be 100 phr polymer and 25 phr plasticizer)
$F \ =$ constant for the polymer and plasticizer systems used
$\gamma \ =$ shear rate

From Equation 16C.12, we can establish proportionality between plasticized and unplasticized materials as follows:

$$\eta_1/\eta_0 = \left[\phi_1^{3.4} F\left(\gamma \phi_1^{1.4}\right)\right] / \left[\phi_0^{3.4} F\left(\gamma \phi_0^{1.4}\right)\right] = \phi_1^{4.8}/\phi_0^{4.8}$$

where
$\phi_1 =$ volume fraction polymer of plasticized material
$\phi_0 =$ volume fraction polymer of unplasticized material

As more plasticizer is added, the viscosity at a given shear rate drops. This is plotted in Figure 16C.11 for our example material from Figure 16C.1. For a particular polymer and plasticizer system, it should be fairly straightforward to run

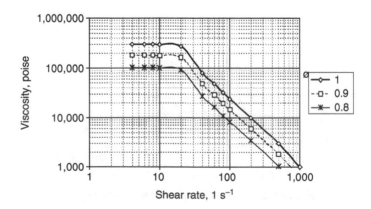

FIGURE 16C.11 Plasticizer effects on viscosity.

the experiments and establish the parameters of Equation 16C.12 for further compound design.

Some plasticizers have limited solubility in the rubber polymer and have combined functions as plasticizers and as process aids. Particularly at high shear rates, some plasticizers can separate out of the matrix giving surface lubrication and a process aid type effect. In the following section on process aids, techniques are discussed for studying and quantifying these types of effects. If there are questions or concerns about the solubility of the plasticizer in the polymer, this can be determined by conventional volume change measurements per ASTM D471. For example, if a new plasticizer is being evaluated in a particular NBR compound, volume change measurements of an unfilled, unplasticized version of the compound at room temperature or at molding temperatures will define the limits of how much plasticizer can be used. In our example, we simply make gum vulcanizate samples of the polymer and its cure system and perform volume change measurements at room temperature and molding temperature to define the upper limits of plasticizer phr that can be used in the polymer.

16.4.4 EFFECTS OF PROCESS AIDS

For the purposes of mold flow and mold release, most process aids function by having only low or partial solubility in the polymer matrix. For example, microcrystalline waxes and polyethylene waxes make good process aids for highly polar polymers such as FKMs since they deposit on the surface of the shear flow front and on the molded part providing lubrication and release. Likewise, special FKM polymers are manufactured as process aids and slip agents for use in polyethylene manufacture. In general, very effective process aids for mold flow and mold release also need to be used at very low and tightly controlled levels in the polymer. Using our example of FKM materials using microcrystalline wax as a process aid, in transfer or injection molding 0.500 phr is very effective, 1.00 phr is effective but may lead to mold deposits during long-term operations over several hundred cycles, and 1.50 phr or higher will give knit lines and parts that are wet with excess wax when they first emerge from the mold.

There are techniques to study the effects of process aids on mold flow. Typically, a capillary rheometer is very useful for studying the effects of process aids on flow. The pressure drop through the die of a capillary rheometer is the total pressure drop at the ends and the pressure drop within the die:

$$P_T = \Delta p_e + \Delta p \qquad (16C.13)$$

where

P_T = total pressure drop through the die
Δp_e = pressure drop at the ends of the die
Δp = pressure drop within the die

The shear stress, at the wall of the capillary relates to the total pressure drop and the length and diameter of the die:

$$P_T = \Delta p_e + 4(\sigma_{12})_w L/D \qquad (16C.14)$$

where

$(\sigma_{12})_w$ = shear stress at the wall
L = die length
D = die diameter

By running experiments with a constant D, different L/D ratios, with total volumetric flow rate constant, a plot of P_T versus L/D can be used to get the shear stress at the wall from the slope of the plot. From this, the effectiveness of process aids in improving flow can be measured as the reduction in shear stress at the wall due to the lubrication effects.

Measuring the effectiveness of a process aid in terms of mold cleanliness and release is a very challenging situation. In many cases, the only accurate measure may be a carefully monitored, extended production run comparing long-term release and mold cleanliness with and without the process aid or comparing different process aids.

16.5 USING PREVIOUSLY DEVELOPED PRINCIPLES IN PROBLEM SOLVING

The goal of this section is to use the relationships described above in example molding problems. Many are situations that I have encountered over the years and many will seem familiar to those in the molding industry. This is intended to emphasize the knowledge base already accumulated to lead to disciplined problem solving of molding difficulties. The sequence suggested is not the only sequence by any means and is intended only as an example. Disciplined problem-solving techniques will speed resolution, improve responsiveness to plant and customer concerns, and contribute quickly to the success of the organization.

16.5.1 PROBLEM—SUDDEN LACK OF FLOW AND SCORCH PROBLEMS IN COMPRESSION, TRANSFER, OR INJECTION MOLDING

Step 1. In solving this problem is to review the rubber compound and the trend in data to determine if the material has become more scorchy in recent history. A good example of how a recipe can be well under control, yet give periodic problems occurs with CR and FKM compounds. The metal oxide + thiourea cure system for CR and the metal oxide activated bisphenol cure system for FKM compounds are very sensitive to ambient moisture and humidity. Particularly in systems where the rubber is mixed in one location and shipped long distances or stored for extended time periods, this ambient summer/winter effect can be pronounced. Examination of records may indicate that the recipe and ingredient levels are well controlled, but scorch problems arise when warm and humid weather occurs in May or June. Perhaps the best fix to this problem is to track a moving average scorch time from Rheometer or Mooney viscometer data and adjust the accelerator level when the scorch time moving average moves outside a range of acceptable values. Thus, when cooler, drier weather occurs in winter, the accelerator level is adjusted so that scorch

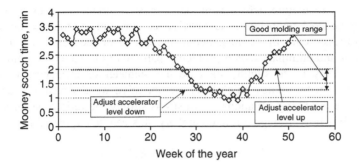

FIGURE 16C.12 Five Point moving average scorch time, CR compound.

times are dropped to keep cure times from extending and, when warmer, wetter weather occurs in summer, the accelerator level is adjusted so that scorch times are extended to prevent premature scorch in molding. An example of such a response system is shown in Figure 16C.12 for a metal oxide/thiourea cured CR.

Step 2. If no seasonal or special cause can be found in compound, scorch conditions may be to look at mold closure rates. From Equation 16C.7 above and by examination of Figure 16C.4, we can see that the heat build-up during shear flow is proportional to the square of the shear rate. Check to see if final mold closure rates have been accelerated, which would cause a major increase in heat build-up during flow. If this is the case, return to the older, slower closure rates if possible.

Step 3. If Steps 1 and 2 do not resolve the issues may be to consider preheating the rubber performs. From Equation 16C.5 and Figure 16C.6, we can see that we can shift viscosity down if we increase the temperature of the rubber material. Reducing the viscosity will linearly reduce the temperature build-up in the rubber material during flow as given in Figure 16C.4. The trick then becomes developing a method to preheat the rubber, especially in compression or transfer molding. Many companies have evaluated microwave preheating, which can be quite effective. Basically, the preforms are placed in a microwave chamber for a controlled time period immediately before loading in the press. Unfortunately, if the rubber material has a tendency to absorb moisture and ambient conditions are high humidity, this can cause excessive heating in the preforms during warm, humid weather. Perhaps a simpler and more effective measure is to introduce a delay of closure on the press mechanism. By this method, the operator loads the preforms, presses the cycle start mechanism, and the press delays closure for a set period to preheat the rubber in the hot mold. This will also alleviate parting line rupture problems in very lightly filled compounds caused by excessive thermal expansion and contraction during molding and demolding. In injection molding, getting higher temperatures in the raw rubber is fairly easy to accomplish just by increasing the temperature of the extrusion fill chamber. I have noticed extreme reluctance to do this because of concerns that increasing the temperature of the rubber before injection may increase the incidence of premature scorch. Figure 16C.4 shows a typical reduction of viscosity by increasing the stock temperature and Figure 16C.5 shows the effects of reducing viscosity on heat build up during flow. Typically, it is the rapid heat build-up, not the initial stock temperature, which contributes most to premature scorch. Moreover, increasing the stock temperature can help reduce injection and transfer times and can also help reduce cure times. In general, this is one of the best ways to improve productivity and eliminate scorch problems, particularly in injection molding.

Step 4. May be to slightly increase the diameter or clearances of the flow channels or transfer sprues. Slightly is the operative word which will be demonstrated in the following equations. Viscosity is simply the ratio of shear stress and shear rate. For flow through a round capillary, this can be expressed as follows:

$$\text{Shear Stress} = (\Delta P * R)/(2 * L) \qquad (16C.15)$$

where
$\Delta P =$ pressure drop
$R \ \ =$ radius of capillary
$L \ \ =$ length of capillary

$$\text{Shear Rate} = (4Q)/(\pi R^3) \qquad (16C.16)$$

where
$Q =$ volumetric flow rate
$R =$ radius of capillary

$$\eta_{app} = (\text{Shear Stress})/(\text{Shear Rate}) = \text{apparent viscosity}$$
$$= (\Delta P * R^4 * \pi)/(4 * Q * 2 * L)$$
$$Q = (\Delta P * R^4 * \pi)/(4 * \eta_{app} * 2 * L) \qquad (16C.17)$$

Thus, by a small increase in the radius of the flow channel, the volumetric flow rate will increase by the fourth power of this radius. This is shown in Figure 16C.13. Decreasing the apparent viscosity of the compound, η_{app}, will not have nearly as profound of an effect on increasing flow rate as small changes in the radius or gap of the flow channel.

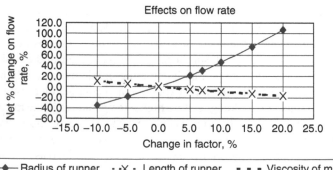

FIGURE 16C.13 Relative effects of various parameters on volumetric flow through a capillary.

Step 5. May be to investigate process aids in the formula to reduce shear stress at the walls of the flow channels. However, if a method of testing outside of production molding is not readily available, this can be a difficult and time-consuming process. If you are currently using the recommended or industry standard process aids for the polymer, then a thorough examination of the preceding factors may be more productive in solving the problems.

16.5.2 PROBLEM—BLISTERING, FLAT SPOTS, AND EXCESSIVE FLASH IN COMPRESSION, TRANSFER, OR INJECTION MOLDING

Contrary to the problem above, this is usually due to too much flow and not enough backpressure in the mold flow. Therefore, it may be necessary to work through the steps in the previous section in the opposite direction.

Step 1. May be to examine the scorch time of the compound for seasonal or special cause trends. In rubber molding, we are trying to flow the material and vulcanize the material in one process. The onset of vulcanization, scorch, may be providing some of the viscosity and pressure increases we need in the product cavity to eliminate blisters and minimize flash. We may need to establish a range of acceptable scorch rates and adjust the accelerator level in the formula to keep the scorch rate in the good molding window. This adjustment may need to be a one-time affair, or it may need to be seasonal as shown in Figure 16C.12.

Step 2. May be to reduce the heat build-up or starting temperature of the stock to increase viscosity. Slower mold or transfer pot closure in compression molding or transfer molding will decrease shear rate and reduce heat build-up in the stock. Unfortunately, this may also result in longer cure times since some of the heat added when the cavity is filled will then be required to bring the stock up to vulcanization temperature. In injection molding, this may involve reducing the temperature of the stock in the feed extruder to increase viscosity. In compression or transfer molding, slower final closure with a delayed bump cycle may solve the problem. Many people add 4–5 bumps after closure and assume that the number of bumps has eliminated trapped air. However, more often than not, only the last bump is the solution. By the time the last bump occurs, the vulcanization reaction has started and the viscosity of the stock has increased dramatically by Equation 16C.10 so that the last bump occurs on a much stiffer stock and finishes the proper formation of the finished part. Thus, delaying one bump may accomplish as much or more than adding 4–5 bumps after closure and will give less wear and tear on the press.

Step 3. May be to examine the type and level of process aids used. By their nature, many process aids have limited solubility in the rubber matrix and separate on the surface during high shear flow. The type and level of process aids may be reducing wall shear stress during flow to too low a level and the material flows in and out of the cavity before vulcanization.

Step 4. May be to decrease the diameter or clearances of the flow channels into the mold cavity, or especially in the flash area. As mentioned above, the volumetric flow through a capillary varies by the fourth power of the capillary diameter so that reducing the sprue or runner diameter by a small amount will dramatically reduce the flow rate. In this case, our material flows too quickly in and out of the cavity and we do not build sufficient back-pressure to avoid flat spots, non-fills, or blisters on the finished part. Again, Figure 16C.13 shows the relative effect of various parameters on changing the volumetric flow. The sprue or runner diameter or the clearance of the flow channels is the most sensitive "knob" to turn to impact volumetric flow.

If none of the above work, Step 5 may be to reformulate the compound with higher molecular weight polymers (higher Mooney viscosity) or with higher overall structure fillers. If, for example, the primary filler is N660, low-structure carbon black, an equivalent amount of N550 carbon black will increase viscosity significantly (see Figure 16C.7). Remembering Equation 16C.10 above, viscosity is proportional to the 3.4 power of weight average molecular weight. We may want to consider blending some high Mooney viscosity polymer with the base polymer already used to move the weight average molecular weight of the base polymer significantly. For example, if we are using a 33 mol% acrylonitrile, 50 Mooney viscosity polymer as the base, we may want to consider an 80/20 or 60/40 blend of 33 mol% acrylonitrile polymers with Mooney viscosities of 50 and 80. Adding a small amount of high-molecular weight polymer will have a big influence on weight average molecular weight and on viscosity.

16.5.3 PROBLEM—SUDDEN DIMENSIONAL SHIFT IN FINISHED PART

This problem faces many molding operations in production, especially in parts with linear or diametric dimensions approximately above 75 mm. Step 1 would be to look at the mold or platen temperatures. By far, the biggest factor in dimensions of the finished part is the actual temperature of the mold. Certainly, the production personnel and the tooling source will claim that something has happened to the compound causing the change, but temperature is the predominant factor in dimensional results for the finished part. For example, the typical coefficient of linear expansion for an NBR compound is 0.00015 mm/mm/°C. If room temperature is 25°C and molding temperature is 180°C, the total shrinkage after molding would be 0.023 mm/mm. If the mold were cut allowing for 0.023 mm/mm shrinkage at a molding temperature of 180°C on a 100.0 mm part and the molding temperature had drifted to 190°C, the shrinkage would increase to 0.025 mm/mm. While this does not seem like a major change, on the 100.0 mm part, this change would move the finished dimension from 100.0 to 99.8 mm. On a 125.0 mm target dimension, the finished part would move from 125.0 to 124.8 mm. This effect is amplified by lightly filled elastomers with high coefficients of linear expansion such as FKM elastomers. A 75 Shore A, carbon black filled FKM compound has a coefficient of linear expansion of 0.00023 mm/mm/°C. If the mold were designed for a vulcanization temperature of 180°C and the temperature was actually 190°C, this would shift a 75.0 mm nominal dimension to 74.8 mm.

Step 2 may be to examine the flow conditions and any shifts in molding. As mentioned earlier and shown in Figure 16C.3, shear rate has a major impact on the elastic recovery of the material. This elastic recovery can be pronounced after demolding where the part acts like the stretched chewing gum and retracts in the direction of flow. In addition, from Figure 16C.5, the effects of increasing shear rate on heat build-up in the stock are pronounced. Essentially, decreasing injection time from 20 to 10 s in injection molding would be a 100% increase in shear rate, which would cause a 300% increase in heat build-up in the compound during flow as shown in Figure 16C.5, ignoring the reduction in viscosity at higher shear rate. In effect, this actually increases the molding temperature of the material without changing the mold or platen temperatures and molding temperature is the biggest factor in dimensional results for the finished part.

One of the most surprising ways to affect the dimensions of an injection-molded part is to vary the injection pressure hold time. Generally, the injection pressure is applied and then released after a specified time to allow the injector to recharge for the next shot. Releasing the injection pressure early versus late can have a profound effect on the dimensions of the finished part. This has been observed many times but has not yet been thoroughly studied or quantified. However, on large diameter parts such as O-Rings, this can be the difference between dimensionally compliant and dimensionally rejected parts. In the initial setup of the injection process, injection pressure hold time should be a variable studied for its effects on the finished parts. If the relationship of injection hold time is established in the initial production approval process, it is more likely to be monitored and controlled in routine production.

16.5.4 Problem—One Runner or Sprue Has Insufficient Flow While Another Has Too Much Flow

This is a fairly common problem when multiple runners are used in injection molding or multiple sprues in transfer molding. Typically, despite adjusting flow times, injection pressures, or mold closure, one runner or sprue consistently gets too much flow and excess flash while another is consistently starved for rubber and the parts never fill properly. While many will try to make this a complicated problem, this is one of the simplest to solve. Once again, refer to Equation 16C.17 regarding the dependence of volumetric flow on the radius of the capillary or the clearance of the flow restriction. To further emphasize this, the relative effects of various parameters on the volumetric flow, Q, are plotted in Figure 16C.13.

From Figure 16C.13, based on Equation 16C.17, we can see that the flow rate varies by the inverse of runner length and viscosity of material. However, the volumetric flow varies by the fourth power of capillary radius. From Figure 16C.13, we can see that increasing the capillary radius by 5% will increase the volumetric flow by 20%. Now, we could spend lots of time and money modifying the compound for better flow, but it would be a fairly low return investment compared to slight increases in the runner or sprue radius.

16.6 CONCLUSION

By discussion of the theories of rubber viscoelasticity and flow, the reader should be better able to solve the common problems in compression, transfer, and injection molding of specialty elastomers. Some examples are given of common problems that are typically encountered and some guidelines are given for analyzing and solving the problems. It is also important to use statistical experimental design in using these principles. Statistical experimental design will not only point out the significant factors involved in the particular situation, but it will also give indications or quantification of factor interactions. For example, we can assume that we have a new part going into injection molding and we need to define the critical process control factors for success. Many will recognize this as a process failure mode and effects analysis (PFMEA) and it is a truly useful exercise. In our example, we have developed a compound that appears to flow well and will yield the short cure times dictated by price pressures on the finished part. Now, we need to document and involve the operations personnel in the PFMEA process. Typical variables that would be important would be:

1. Injection pressure (injection time)
2. Injection pressure hold time
3. Injector barrel temperature
4. Mold or platen temperature
5. Vulcanization time

We could perform a 5 factor, 2 level design or a fractional factorial design to study these variables on a variety of final part properties such as:

1. Physical properties, if practical, on the finished part such as modulus in tension or compression set
2. Dimensional variability and conformance to specification limits
3. Appearance criteria if they can be quantified

If we choose to do a fractional factorial, we can narrow the critical variables to, hopefully, 2–3 highly significant variables, which can then be studied further in a full factorial or response surface type experimental design. Training in experimental design is not the intention of this chapter and many good sessions are available. Computer software also expedites much of the time-consuming design and calculation duties. To research the computer software and training courses available, the reader is encouraged to use Internet search engines with the following type searches: Statistical Experimental Design, Statistical Experimental Design Training, Statistical Experimental Design Software, or Experimental Design. A good text to use as a starting point is "Statistics for Experimenters: An Introduction to Design, Data Analysis, and Model Building," edited by Box et al. [30]. This book has been widely recognized as an excellent beginner's text and a good reference book for experienced users as well. If you are not yet familiar with Statistical Experimental Design techniques, when you gain this knowledge, you will realize very quickly how appropriate these techniques are for multicomponent systems and mixtures such as elastomer compounds. Using the principles discussed early in this chapter with statistical experimental design will provide thorough and final resolution of many of these issues as well as up-front PFMEA.

REFERENCES

1. E.A. Collins and J.T. Oetzel, *Rubber Age* (NY), 102, 64 (1970).
2. E.A. Collins and J.T. Oetzel, *Rubber Age* (NY), 103, Feb. (1971).
3. S. Einhorn and S.B. Turetzky, *J. Appl. Polymer Sci.*, 8, 1257 (1964).
4. J.R. Hopper, *Rubber Chem. Technol.*, 40, 462 (1967).
5. M. Mooney, *Physics* (NY), 7, 413 (1936).
6. M. Mooney, in *Rheology: Theory and Applications* (F.R. Eirich, ed.), Vol. 2, Academic Press, New York, 1958.
7. N. Nakajima and E.A. Collins, *Polymer Eng. Sci.*, 14, 137 (1974).
8. F.C. Weissert and B.L. Johnson, *Rubber Chem. Technol.*, 40, 590 (1967).
9. J.L. White and J.W. Crowder, *J. Appl. Polymer Sci.*, 18, 1013 (1974).
10. J.L. White and N. Tokita, *J. Appl. Polymer Sci.*, 9, 1929 (1974).
11. W.E. Wolstenholme, *Rubber Chem. Technol.*, 38, 769 (1965).
12. R.I. Tanner, *J. Polymer Sci.*, Part A-2, 8, 2067 (1970).
13. M.L. Williams, R.F. Landel, and J.D. Ferry, *J. Am. Chem. Soc.*, 77, 3701 (1955).
14. P.J. Flory, *J. Am. Chem. Soc.*, 62, 1057 (1940).
15. J.R. Shaefgen and P.J. Flory, *J. Am. Chem. Soc.*, 70, 3709 (1948).
16. T.G. Fox and P.J. Flory, *J. Am. Chem. Soc.*, 70, 2384 (1949).
17. T.G. Fox and V.R. Allen, *J. Chem. Phys.*, 41, 344 (1964).
18. T.G. Fox and S. Loschaek, *J. Applied. Phys.*, 26, 1082 (1955).
19. W.F. Busse and R. Longworth, *Trans Soc. Rheol.*, 6, 179 (1962).
20. T. Masuda, K. Kitagawa, T. Inoue, and S. Onogi, *Macromolecules*, 3, 116 (1970).
21. R.S. Porter and J.F. Johnson, *Proceedings of the Fourth International Rheology Congress*, Vol. 2, 467 (1965).
22. R.S. Porter and J.F. Johnson, *Chem. Rev.*, 66, 1 (1966).
23. W.F. Busse and R. Longworth, *Trans. Soc. Rheol.*, 6, 179 (1962).
24. E. Guth and R. Simha, *Kolloid-Z.*, 74, 266 (1936).
25. D. Barthes-Biesel and A. Acrivos, *Int. J. Multiphase Flow*, 1, 1 (1973).
26. J.L. White and J.W. Crowder, *J. Appl. Polymer Sci.*, 18, 1013 (1974).
27. C.D. Han, *J. Appl. Polymer Sci.*, 18, 821 (1974).
28. N. Minagawa and J.L. White, *J. Appl. Polymer Sci.*, 20, 501 (1975).
29. G. Kraus and J.T. Gruver, *Trans. Soc. Rheol.*, 9(2), 17 (1965).
30. *Statistics for Experimenters: An Introduction to Design, Data Analysis, and Model Building*, Edited by G.E.P. Box, W.G. Hunter, and J.S. Hunter, Wiley, New York, 1978.

Index

A

Abrasion resistance, of HNBR materials, 97
Accelerators, types of, 15–17
Acetone–diphenylamine (ADPA), 472
Acetylene process, 3–4
Acetyl methyl ricinoleate, 390, 395
Acetyl tributyl citrate, 390, 393, 395
Acid acceptor, for epichlorhydrin
    compounds, 259–262
ACN, 481
Acrylic elastomers-ACM, 404, 453–454
  air oven aging of
    formulations, 458–459
      soap and sulfur-cured compounds, 460
      TCY-cured compounds, 460
  compounds extracted in IRM 901 oil
    before air oven aging, 461–464
  formulations and properties, 456–457
  processing aids incorporation
    in, 484, 486–488
Acrylic elastomers (ACMs and AEMs), 98
Acrylonitrile-butadiene-isoprene
    (ACN/BR/IR) elastomer,
    terpolymer of, 51
Acrylonitrile-butadiene rubber (Buna N
    and Perbunan), *see* NBR
Adkcizer RS-735, 393
AEM, 404, 473
  processing aids incorporation in, 487
Aerosil® P–972, 144–145
Aflas® 100 series, 140
AGERITE MA, 413
AGERITE RESIN D, 413
AgeRite Stalite S, 172
Aging stabilizers, for HNBR, *see*
    Hydrogenated Nitrile-Butadiene
    Rubber (HNBR)
Alkylated chlorosulfonated polyethylene,
    335–337
Alkylphenol disulfide, 416
ALTAX (MBTS), 420
Aluminum trihydrate (ATH), 362

Aminox, 445
Ammonium benzoate cured compounds, 178
Anti-drain back valve, 189
Antioxidants
  and comparative study, 18
  in Gechron 2000 ECO, 262
  solubility in heptane, 446
  volatility of, 445
Antiozonants, 17–19
  in Gechron 2000/100, 263
API TR 6JI Elastomer Life Estimation Test
    Procedures, 515
Aranox, 445–446
Armeen 18D, 212, 479–480, 484, 487
Aromatic process oils, 390, 393, 398
Arrhenius plot, of EVM insulation
    compound, 359
Arrhenius' theory of the thermal
    acceleration, 515
Ascium, 302
ASTM D2000 compounds, 383
  classification system, 96, 135
  elastomers comparison, 97
ASTM D1418 designations, 246
ASTM 901 Oil, 445, 449
Avoid this mistake (ATM), 389, 394, 397,
    402–404, 406
Axel Int 216, 172

B

Bayer, 105
Benzoates, 394
4,4'-Bis (a-dimethylbenzyl)diphenylamine,
    440
Bisphenol A, 420
A,α'-Bis(t-butylperoxy)-diisopropyl
    benzene, 410, 472
1,1-Bis(t-butylperoxy)-3,3,
    5-trimethylcyclohexane, 410
Black thiokol FA printing rollers, 380
Blanc Fixe, 144